Introduction to Analytical Electron Microscopy

Introduction to Analytical Electron Microscopy

Edited by
John J. Hren
University of Florida
Gainesville, Florida

Joseph I. Goldstein
Lehigh University
Bethlehem, Pennsylvania

and

David C. Joy
Bell Laboratories
Murray Hill, New Jersey

PLENUM PRESS • NEW YORK AND LONDON

SPONSORS:

Electron Microscopy
Society of America

Microbeam Analysis
Society

Library of Congress Cataloging in Publication Data

Main entry under title:

Introduction to analytical electron microscopy.

Includes index.
1. Electron microscopy. I. Hren, John J. II. Goldstein,   Joseph, 1939-
III. Joy, David, 1943-
TA417.23.I57   502'.8   79-17009
ISBN 0-306-40280-7

This volume contains the proceedings of a workshop on Analytical Electron
Microscopy, held in San Antonio, Texas, August 13-14, 1979, as part of the
joint meeting of the Electron Microscopy Society of America and the Microbeam
Analysis Society.

©1979 Plenum Press, New York
A Division of Plenum Publishing Corporation
227 West 17th Street, New York, N.Y. 10011

Printed in the United States of America

PREFACE

The birth of analytical electron microscopy (AEM) is somewhat obscure. Was it the recognition of the power and the development of STEM that signaled its birth? Was AEM born with the attachment of a crystal spectrometer to an otherwise conventional TEM? Or was it born earlier with the first analysis of electron loss spectra? It's not likely that any of these developments alone would have been sufficient and there have been many others (microdiffraction, EDS, microbeam fabrication, etc.) that could equally lay claim to being critical to the establishment of true AEM. It is probably more accurate to simply ascribe the present rapid development to the obvious: a combination of ideas whose time has come.

Perhaps it is difficult to trace the birth of AEM simply because it remains a point of contention to even define its true scope. For example, the topics in this book, even though very broad, are still far from a complete description of what many call AEM. When electron beams interact with a solid it is well-known that a bewildering number of possible interactions follow. Analytical electron microscopy attempts to take full qualitative and quantitative advantage of as many of these interactions as possible while still preserving the capability of high resolution imaging.

Although we restrict ourselves here to electron transparent films, much of what is described applies to thick specimens as well. Not surprisingly, signals from all possible interactions cannot yet (and probably never will) be attained simultaneously under optimum conditions. As a consequence there is a variety of analytical electron microscopes (home-built and commercial) with a wide array of possible accessories. Since most current AEM's are converted TEM's, this book allots considerable space to the special problems inherent to the transformation of a TEM into an AEM. It is likely that most of these problems will be absent from future generations of commercial instruments. User feedback (i.e., your complaints) has been an important feature of instrument development and will continue to remain so. This process will never end, of course, since we will never have the ideal microscope that does everything perfectly for everyone at all times!

A few words about the development of this book and its purposes are in order. The need for such a volume became apparent to the societies and to the editors through the widespread response of attendees to the annual meetings of EMSA and MAS at sessions organized under the various guises of AEM. In effect, the membership voted for greater coverage with their feet. Having established the need, it remained to define a suitable mechanism for satisfying it. It was clear that no single author, or even a reasonable number of co-authors, could write a suitable textbook within the forseeable future. A combined multi-authored and edited book accompanying a tutorial was selected as the most efficient means to achieve the proper level and scope of coverage. The selection of the subjects and their coverage is the responsibility of the editors. Like the instrument manufacturers, we look to the readers for comments (that is, "constructive criticism") so that future volumes of a similar nature may be improved.

Because of the press of publication deadlines for the annual meeting, there will undoubtedly remain minor errors, both technical and typographical in this volume. With 22 co-authors and the variety of material presented, the depth of coverage also varies somewhat. We judged these to be tolerable imperfections under the circumstances, since the alternatives were not satisfactory.

Analytical electron microscopy is ripe for application today. The proven methods described in this volume can and should be widely utilized by the research community as soon as possible. If this occurs, as seems likely, AEM promises to open up still more research doors down to the atomic level. We eagerly await this response.

A book like this (and the associated workshop) could not take place without the generous cooperation and assistance of many people. We first deeply thank our co-authors without whom nothing could have been done. Their efforts will be remembered and welcomed. Secondly, we are truly heartened by the fine cooperation of both societies who were willing to risk such a venture and then to back it wholeheartedly. We hope their confidence has been well-placed. Thirdly, we thank Ellis Rosenberg and Steve Dyer of Plenum Press for their flexible and understanding attitudes and for their cooperation throughout this enterprise. Finally, we thank the multitude of editorial assistance at the University of Florida. In particular, we are deeply indebted to Jerry Lehman, his mother, Irene Lehman, and JoAnne Upham, who labored heroically over mounds of paper with the word processor, and to Pam Fugate, Linda King, and Vicki Turner for stellar secretarial assistance.

John Hren

David Joy

Joe Goldstein

TABLE OF CONTENTS

CHAPTER 1

PRINCIPLES OF IMAGE FORMATION

J. M. COWLEY

DEPARTMENT OF PHYSICS, ARIZONA STATE UNIVERSITY

TEMPE, ARIZONA 85281

1.1 INTRODUCTION

Whatever their backgrounds, electron microscopists tend to bring to the topic of image interpretation a set of preconceptions based on their particular education and experience. The difficulty in presenting a summary of the principles of image interpretation for a volume such as the present one arises from the wide variety of backgrounds of the readers.

In the early days of electron microscopy the background of optical microscopy was familiar to most microscopists and provided useful although incomplete analogies. Electron microscopists of the newer generation often have little experience of optical microscopy and tend to base their assumptions on working experience with a particular area of electron microscopy and the convenient concepts of that field. Biologists, for the most part, can get along with the useful assumption that the image intensity depends on the amount of scattering matter present, with the darker regions in positive prints corresponding to thicker regions or heavier atoms. The limitations due to radiation damage, which prevent them from using very high resolution, shield them from too much exposure to the complications of phase-contrast imaging. The essentially non-crystalline nature of their specimens shields them from the complications of diffraction and the dominance of the daunting dynamical scattering effects.

Materials science microscopists, on the other hand, are brought up in a world of dynamical diffraction contrast, where thickness fringes, bend contours and strain contrast predominate. A standardized set of simplifying assumptions provide a relatively simple and very effective guide through the labyrinth of three-dimensional diffraction processes which are dominant for the useful specimen thicknesses and the usual ranges of resolution.

Yet for all electron microscope users the challenge of the many potential advantages of using the high resolution capabilities of modern microscopes, or of applying some of the recently developed techniques of STEM, microdiffraction and microanalysis, requires that a more fundamental understanding be sought of the basic principles of image formation. In this review, therefore, we will summarize the essentials of electron scattering and imaging, referring the reader, where necessary, to appropriate texts for the more detailed expositions and for a wider range of illustrative examples. For the sake of those who prefer words to symbols, each topic will be first discussed in a loose, verbal fashion. The mathematical statements which follow in each case are included for the sake of those who prefer more exact formulations and may safely be ignored without loss of continuity or concept.

For the basic ideas of geometric and physical optics, standard textbooks of optics such as those by DITCHBURN (1963) and STONE (1963) may be consulted while for those with more background in physics, LIPSON and LIPSON (1969) is suitable. Introductory treatments of electron optics are found in such books as BOWEN and HALL (1975) and HIRSCH et al. (1965). The mathematical sections of this review follow the treatment of COWLEY (1975) which should be consulted for more complete developments of most of the topics. More specific references to these and other sources will be given in relation to particular subject areas, when appropriate, but we do not attempt to provide an exhaustive bibliography.

Electron Scattering and Diffraction

The terms "scattering" and "diffraction" are often used interchangeably. However, a common convention is that when "scattering" is not used in a general, all inclusive sense, it refers to the case of a beam of incident radiation striking a small particle (an atom or cluster of atoms acting as a unit) and giving rise to an angular distribution of emergent radiation which depends on the nature of the individual particles and not on their relative positions. This can be called the incoherent scattering case. The intensity distribution is the sum of the intensities given by the individual particles acting independently.

The term "diffraction" is used when interference effects between waves scattered by many atoms give rise to modulations of the intensity distribution which may be measured to give information on the relative positions of atoms. Diffraction may be considered as "coherent scattering" when wave amplitudes, rather than intensities from individual atoms, are added. However, it must be emphasized that it is the correlation of atom positions, not the coherence of the incident radiation, which is different for diffraction and scattering. The interaction of electrons with atoms is the basis of both and there is a common mathematical formulation. It is only the set of approximations made as a convenience in practice which distinguishes the two aspects.

Electrons are scattered much more strongly by matter than x-rays or visible light. A single atom can scatter enough electrons to allow it to be detected in an electron microscope. Monomolecular layers can give strong diffraction effects with electrons but thicknesses of a micrometer or more are needed for comparable relative diffraction intensities for x-rays.

For even very large assemblies of atoms, the amplitude of scattered x-rays is very much less than the amplitude of the incident beam. This provides the basis for the common, very useful approximation, called the "kinematical" or "single-scattering" approximation, justifiable because if the amplitude of single scattering is very small, the amplitude, and even more so the intensity, of doubly scattered radiation will be negligible. Then the amplitude of the scattered or diffracted x-rays is given as a function of the scattering angle by a simple mathematical operation, the Fourier transform, applied to the electron density distribution in the sample. When this angular distribution is observed very far from a scattering atom (or group of atoms scattering independently because there is no significant correlation in their positions), it is described in terms of an atomic scattering factor, characteristic of the type of atom and listed conveniently in tables. When diffraction by an assembly of atoms is considered, one merely adds together the scattered amplitudes for all atoms with phase factors depending on their relative positions. For crystals, as is well-known, the regularity of the atom arrangement leads to a reinforcement of the scattered amplitude in a regularly spaced set of strong diffracted beams. The process of working back from the observed intensities of these beams, through the Fourier transform relationship, to the relative atomic positions is the foundation of x-ray crystal structure analysis.

This kinematical scattering approximation is so elegant and so relatively simple, as compared with multiple-scattering theory, that it is used for electrons whenever it seems even half-way reasonable (and often when it does not). But for electrons the scattering by atoms is very strong. There can be appreciable multiple scattering within a single moderately heavy atom and for very heavy atoms double and triple scattering can rarely be neglected.

An important factor in this respect is that because electrons have short wavelengths (0.037 Å for 100 keV electrons) relative to atomic dimensions the scattering angles are small (around 10 milliradians). When electrons enter a crystal in the direction of one of its principal axes, the electrons travel along rows of atoms and electrons scattered by one atom can be rescattered by any of the subsequent atoms in the row. For crystals then, multiple scattering is very important. The diffraction effects are strongly affected so that for electron diffraction by crystals one must usually look beyond the kinematical approximation and deal with the complications of the dynamical (or coherent multiple scattering) theory.

There are favorable circumstances, involving the averaging over crystal orientations and thicknesses, especially for light-atom materials, which can make the kinematical theory a good working approximation for crystal structure analysis from electron diffraction patterns (see VAINSHTEIN, 1964). For electron microscopy a similar sort of averaging can simplify things in the case of very thick specimens with very small crystal sizes, but for the more usual thin specimens there is no such convenient means of escape since the intensity of each point in the image of a crystal depends on the full dynamical diffraction effects in the immediate neighborhood of the point.

It is only because the scattering of electrons by atoms is very strong that fine detail, at an atomic level, can be seen in electron microscopes, but because it is strong, severe complications can arise for all but the ideal cases of very thin specimens containing only light atoms.

The Physical Optics Analogy

It was BOERSCH (1946) who first stated clearly the alternative view that, instead of considering electrons as particles being bounced off atoms, one could consider an electron wave being transmitted through the potential field of the charged particles in a sample and the main effect of the potential field is to change the phase of the electron wave. Atoms or assemblies of atoms then constitute phase objects for electron waves, as thin pieces of glass or unstained biological sections do for light waves. In the usual idealized picture, a plane wave enters the sample. The wave at its exit surface has a distribution of relative phase values depending on the variations of the potential field it has traversed, plus relatively small changes of amplitude corresponding to the loss of electrons by inelastic scattering processes. The angular distribution of electron wave intensity (the number density of electrons detected) at a large distance from the object is then given by squaring the magnitude of the wave amplitude in the Fraunhofer diffraction pattern, given by a Fourier transform of the exit complex wave amplitude distribution. This diffraction pattern intensity distribution will be the same as is given by the single-scattering kinematical theory only if the phase changes of the wave in the specimen are small. Large phase changes correspond to the presence of strong multiple scattering.

On this basis we can build up a consistent theory of electron diffraction and of electron microscope image contrast using direct analogies with the concepts of elementary physical optics, including Fraunhofer diffraction, Fresnel diffraction, phase contrast imaging and the wave optical formulation of the theory of lens action, aberrations, image contrast and resolution.

Diffraction Patterns

If a plane parallel beam of radiation strikes a specimen the angular distribution of emergent radiation, as seen from a distance large compared with the specimen dimensions, is the Fraunhofer diffraction pattern. It is convenient to consider the intensity distribution as a function not of the scattering angle ϕ, but of the parameter $u = 2\lambda^{-1} \sin(\phi/2)$.

For electrons the wavelength λ is small, the angles ϕ are small and in the diffraction pattern on any plane of observation the distances between features are closely proportional to differences of u values [Fig. 1.1(a)]. In the two dimensions of the plane of observation, the coordinates x,y are proportional to parameters u,v derived from the components of the diffraction angles ϕ_x, ϕ_y.

For single atoms, or for random arrays of many atoms, the diffraction pattern intensities are proportional to the square of the atomic scattering factor, f(u), and fall off smoothly with scattering angle [Fig. 1.1(b)]. Any systematic correlation between atom positions is reflected in a modulation of scattered electron intensity distribution. For most biological and nonbiological materials considered to be "amorphous," the only correlation of atom positions is that due to interatomic bonds of closely prescribed lengths. This gives a modulation of the diffraction intensities with a periodicity roughly proportional to the reciprocal of the bond lengths, but since there is no preferred direction for the bonds, these modulations are smeared out into diffuse circular halos and the percentage modulation of the intensities is usually quite low [Fig. 1.1(c)].

The extreme case of correlated atom positions is the strictly periodic arrangement of atoms in a perfect crystal. To the periodicity, a, of the repetition in the crystal there corresponds the set of regularly spaced sharp diffraction spots with separations proportional to 1/a [Fig. 1.1(d)]. For a very thin crystal lying nearly perpendicular to the incident beam and having periodicities a and b in two perpendicular directions, the diffraction pattern will be a regular cross-grating of sharp spots with separations proportional to 1/a and 1/b in the two directions [see Fig. 1.10(b)].

The perfectly periodic two dimensional case is an idealization which is seldom realistic. If the lateral dimensions of the crystal are small, the diffraction spots will be smeared out by an amount inversely proportional to the crystal dimensions. If the crystal is distorted or bent, the spots will be spread into arcs or the intensities will be changed. If there are crystal defects, if the atoms are disordered on the lattice sites or if the atoms have thermal vibrations, there will be diffuse scattering in the background between the spots. Usually, of course, the crystal is three-dimensional and the dimension in the beam direction can have a strong effect on the presence or absence of diffraction spots, can make the diffraction pattern intensities highly sensitive to small crystal tilts and can give rise to a multitude of complicated dynamical diffraction effects.

Fig. 1.1. (a) The formation of a Fraunhofer diffraction pattern and the intensity distribution along one diameter for (b) a single atom or a random array of atoms; (c) an amorphous material, and (d) a single crystal.

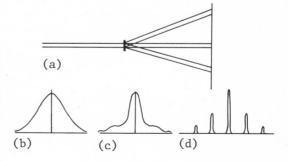

Mathematical Formulation

The kinematical approximation is equivalent to the approximation of excluding all but the first term in the Born series. Starting from the integral formulation of the Schrödinger equation for a plane incident wave of wave vector $\underline{k}_0$ the wave function for the scattered wave is

$$\psi(\underline{r}) = \psi^o(r) + \frac{\mu}{4\pi}\int\frac{\exp\{-ik(r-r')\}}{(r-r')} \qquad \phi(r')\psi(r')dr' \qquad (1.1)$$

If we assume that in the integral we can replace $\psi(\underline{r}')$ by the incident plane wave $\exp\{-i\underline{k}_0 \cdot \underline{r}'\}$ and consider the solution at a large distance r = R, this can be written

$$\psi(r) = \exp\{-i\underline{k}_0\underline{R}\} + (\mu/R)\exp\{-i\underline{k}_0\underline{R}\}\cdot g(u) \qquad (1.2)$$

where $g(u) = \int\phi(r)\exp\{-i(\underline{k}-\underline{k}_0)\cdot\underline{r}\}dr$

$$= \int\phi(r)\exp\{2\pi i\underline{u}\cdot\underline{r}\}dr \ldots \qquad (1.3)$$

and $|\underline{u}| = 2\lambda^{-1}\sin\phi/2$

i.e. the scattering amplitude g(u) is given by the Fourier transform of the specimen potential distribution. This applies for any distribution of scattering matter, as well as for the case of a single atom when the scattered amplitude is the atomic scattering amplitude f(u).

For any assembly of atoms at positions $\underline{r}_i$, one can write

$$\phi(r) = \Sigma_i\phi_i(\underline{r}) * \delta(\underline{r}-\underline{r}_i), \qquad (1.4)$$

where the * sign denotes a convolution, so that

$$g(u) = \Sigma_i f_i \, e^{2\pi i\underline{u}\cdot\underline{r}_i}, \qquad (1.5)$$

and the intensity distribution is proportional to

$$I(\underline{u}) = |g(u)|^2 = \Sigma_i\Sigma_j f_i f_j^* \exp\{2\pi i\underline{u}(\underline{r}_i-\underline{r}_j)\}. \qquad (1.6)$$

For the special case of a periodic object

$$\phi(\underline{r}) = \Sigma_h F_h \exp\{2\pi i\underline{h}\cdot\underline{r}\}, \qquad (1.7)$$

where $\underline{h}$ represents the vector to a reciprocal lattice point denoted by the set of indices h,k,l, and the diffraction pattern amplitudes are given by the intersection of the Ewald sphere ($\underline{u} = 2\pi(\underline{k}-\underline{k}_0)$) with the distribution

$$g(\underline{u}) = \Sigma_h F_h \delta(\underline{u}-\underline{h}) \qquad (1.8)$$

which is the set of weighted reciprocal lattice points.

In the physical optics formulation, for an incident wave ψ_o(xy) and an object transmission function q(xy), the wave at the exit face of the specimen ψ_e(xy) is ψ_0(xy)·q(xy) and the Fraunhofer diffraction pattern is given by the two-dimensional Fourier transform

$$\Psi(uv) = \iint \psi_e(xy)\exp\{2\pi i(ux+vy)\}dx \; dy \tag{1.9}$$

For a thin object, the transmission function represents the change of phase of the electron wave on traversing the potential field of the specimen;

$$q(xy) = \exp\{-i\sigma\phi(xy)\}, \tag{1.10}$$

where $\phi(xy) = \int\phi(\underline{r})dz$ and $\sigma = \pi/\lambda E$ where E is the accelerating voltage. The assumption that $\sigma\phi(xy) \ll 1$ then gives the equivalent of the single-scattering approximation.

The wave at any finite distance R from the specimen may be calculated by using the Fresnel diffraction formula

$$\psi(xy) = \psi_e(xy) * p(xy), \tag{1.11}$$

where p(xy) is the propagation function, given in the usual small angle approximation by

$$p(xy) \cong (i/R\lambda)\exp\{-ik(x^2+y^2)/2R\}. \tag{1.12}$$

For thicker specimens, when Eqn. 1.10 does not apply, one can calculate diffraction effects by dividing the specimen into thin slices and applying alternately the phase change for each slice, Eqn. 1.10, and the propagation, Eqn. 1.12, to the next slice. This is the basis for the dynamical diffraction formulation of Cowley and Moodie (see COWLEY, 1975). Alternatively one can go back to the Schrödinger equation in the differential form and find solutions for the wave in a crystal subject to suitable boundary conditions as in Bethe's original dynamical theory (see HIRSCH et al., 1965).

1.2 THE ABBE THEORY OF IMAGING

One of the most important properties of a lens is that it forms a Fraunhofer diffraction pattern of an object at a finite distance. In Fig. 1.2 we suggest (using for convenience a ray diagram to indicate a wave-optical process) that an ideal lens brings parallel radiation to a point focus at the back-focal plane. If a specimen is placed close to the lens, radiation scattered through the same angle ϕ by different parts of the specimen will be brought to a focus at another point in the back-focal plane, separated by a distance proportional to ϕ (in the small-angle approximation). The intensity distribution on the back-focal plane is thus the Fraunhofer diffraction pattern, suitably scaled.

The more usual function of a lens is to form the image of a specimen. Then radiation scattered from each point of the object is brought together at one point of the image, as in Fig. 1.3. At the same time, however, the Fraunhofer diffraction pattern is formed at the back-focal plane, as in Fig. 1.2. The imaging process may then be described as the formation of a diffraction pattern of the object in the back-focal plane (a Fourier transform operation) plus the recombination of the diffracted beams to form the image (a second Fourier transform operation).

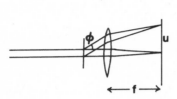 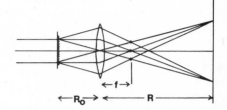

Fig. 1.2.   The use of a lens to produce a focussed Fraunhofer diffraction pattern in the back-focal plane.

Fig. 1.3.   A ray diagram used to
illustrate the wave-optical Abbe
theory of imaging.

For a perfect, ideal lens, the reconstruction of the object trans-mitted wave in the image plane would be exact as all radiation leaving the object is brought back with exactly the right phase relationships to form the image. For a real lens some of the diffracted radiation is stopped when it falls outside the lens aperture. The phase relation-ships are upset by lens aberrations. These changes can be considered to take place on the back-focal plane where the wave function is multiplied by a "transfer function" which changes its amplitude and phase. Corres-pondingly, on the image plane the complex amplitude distribution of the wave function can be considered as smeared out or convoluted by a smearing function which limits the resolution and affects the contrast.

It would be convenient if we could make an equivalent statement for image intensities, i.e. that the effect of lens apertures and aberra-tions is to smear out the intensity distribution by means of a smearing function which can be related (by a Fourier transform operation) to a Contrast Transfer Function characterizing the lens. This is the usual practice for light optics but it does not apply in general for electron microscopy. The difference is that the usual imaging with light is in-coherent; each point of the object emits or scatters light independently with no phase relationship to the light from neighboring points. In electron microscopy the conditions usually approach those for coherent imaging, with a definite phase relationship between the waves trans-mitted through neighboring parts of the specimen. It is only under very special circumstances (which fortunately occur quite often) that the relatively simple ideas developed for light optics can be used as reason-able approximations for the electron case.

For the coherent, electron optical case, the transfer function which modifies the amplitudes of the diffraction pattern in the back-focal plane can be written

$$T(uv) = A(uv)\exp\{i\chi(uv)\}$$ (1.13)

where the aperture function $A(uv)$ is zero outside the aperture and unity within it. The phase factor $\chi(uv)$ is usually written as including only the effects of defocus Δf and the spherical aberration constant C_s since it is assumed that astigmatism has been corrected and other aberrations have negligible effect. Then

$$\chi(uv) = \pi\Delta f\cdot\lambda(u^2+v^2) + \tfrac{1}{2}\pi C_s\lambda^3(u^2+v^2)^2$$ (1.14)

The corresponding smearing functions, given by Fourier transform of Eqn. 1.13, are in general complicated and have complicated effects on the image intensities except in the special cases we will discuss later.

Incident Beam Convergence

So far we have considered only the ideal case that the specimen is illuminated by a plane parallel electron beam. In practice the incident beam has a small convergence and this may have important effects on the diffraction pattern and on high resolution images.

Two extreme cases can be considered. First, it may be appropriate to assume that the electrons come from a finite incoherent source, with each point of the source emitting electrons independently [see Fig. 1.4(a)]. This is a useful assumption for the usual hot-filament electron gun.

Then for each point in the source the center point of the diffraction pattern, and the whole diffraction pattern intensity distribution, will be shifted laterally (and also modified in the case of relatively thick specimens). The main effect on the image will result from the different effects of the transfer function, Eqn. 1.13, when applied to the differently shifted diffraction patterns. In the extreme case that the incident beam convergence angle is much greater than the objective aperture angle, we approach the incoherent imaging situation common in light optics, and all interference effects, including the production of out-of-focus phase contrast, disappear.

The other extreme case of convergent illumination is that occurring when the incidence convergent wave is coherent, e.g. when it is formed by focussing a point source on the specimen [see Fig. 1.4(b)]. This situation is approached when a field emission gun is used. Then it is the amplitude distribution in the diffraction pattern which is smeared out and there will be complicated additional interference effects within the pattern (SPENCE and COWLEY, 1978). Provided that the convergence angle is not too great, the effects on the image will be small except in the case of images taken far out of focus.

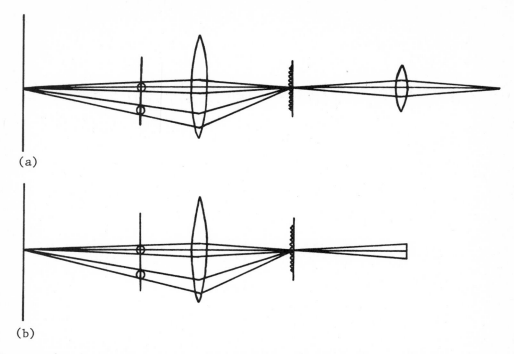

(a)

(b)

Fig. 1.4. Illumination of the specimen in a microscope with a convergent beam for the cases of (a) a finite incoherent source and (b) a point source and a lens used to give coherent radiation with the same convergence.

Chromatic Aberration

The focal lengths of electron lenses depend on the electron energies and also on the currents in the lens windings. A variation in the accelerating voltage (ΔE), or a variation in the lens current (ΔI), during the recording time will have the effect of smearing the intensity distribution in the image. The intensities for different electron energies are added incoherently. The loss of resolution is given by

$$\Delta = C_c \alpha \left(\frac{\Delta E}{E} + \frac{2\Delta I}{I} \right) \tag{1.15}$$

where C_c is the chromatic aberration constant and α is the objective aperture angle.

In addition a change of electron energy gives a change of electron wavelength $\lambda \propto E^{-1/2}$ and this in turn affects the phase factor, Eqn. 1.14.

There are important chromatic aberration effects associated with losses of energy of electrons which suffer inelastic scattering. These will be considered in the subsequent sections.

Mathematical Formulation

For an incident wave amplitude $\psi_0(xy)$ and an object transmission function $q(xy)$ the diffraction pattern amplitude on the back-focal plane of an ideal lens is

$$\Psi_e(uv) = Q(uv) * \Psi_o(uv) \tag{1.16}$$

where capital letters are used for the Fourier transforms of the corresponding real space functions. With the transfer function

$$T(uv) = A(uv) \exp\{i\chi(uv)\} \tag{1.17}$$

the image amplitude is given by a second Fourier transform as

$$\psi(xy) = \psi_e(xy) * t(xy) \tag{1.18}$$

where the spread function $t(xy)$ is given by

$$t(xy) \equiv c(xy) + is(xy)$$

$$= FA(uv)\cos(\chi(uv)) + i\ FA(uv)\sin(\chi(uv)) \tag{1.19}$$

and the magnification factor $(-R/R_0)$ has been ignored. For an isolated point source the intensity of the image is $|t(xy)|^2$ which is the spread function for incoherent imaging. The corresponding contrast transfer function for incoherent imaging is then given by Fourier transform as $T(uv)* T^*(-u,v)$.

For a plane incident wave from a direction (u_1,v_1) the diffraction plane amplitude, Eqn. 1.16, becomes

$$[\Psi_e(uv) * \delta(u-u_1, v-v_1)]\ A(uv)\exp\{i\chi(uv)\} \tag{1.20}$$

and the image, Eqn. 1.18, becomes

$$\Psi_e(xy) \exp\{2\pi i(u_1x+v_1y)\} * t(xy)$$

For a coherent source the image intensity from a finite source is thus

$$\left|\iint S(u_1v_1)[\psi_e(xy)\exp\{2\pi i(u_1x+v_1y)\} * t(xy)]\ du_1dv_1\right|^2$$

and for an incoherent source it is

$$\iint |S(u_1v_1)|^2 \cdot |\psi_e(xy)\exp\{2\pi i(u_1x+v_1y)\} * t(xy)|^2du_1dv_1 \tag{1.21}$$

Here $S(u_1v_1)$ is the source function.

The effect of beam convergence can thus be expressed as in Eqn. 1.20 by a convolution in the diffraction plane in the case of a thin object for which the effect on the incident wave can be represented by a two-dimensional transmission function. For thicker objects the formulation is much more complicated.

The effect of chromatic aberration cannot be represented by a convolution in the diffraction plane. It is given by summing the image intensity distributions for all wavelengths and focal length variations.

1.3 INELASTIC SCATTERING

Other chapters in this volume (e.g. Chapter 10 by Silcox) will deal with the inelastic scattering processes in detail. Here we briefly summarize the effects on the image formation.

Electrons may be scattered by phonons, i.e. by the waves of thermal vibration of atoms in the specimen. The energy losses suffered by the incident electrons are so small (about 10^{-2} eV) that no appreciable chromatic aberration effect is introduced. For an ideal lens of infinite aperture, all thermally scattered electrons would be imaged along with the elastically scattered electrons and the effect on the image would be negligible. For a lens with a finite aperture the effect of thermal scattering is not significant for amorphous specimens. For crystalline specimens, however, the elastic scattering is often concentrated into a few sharp diffraction spots while the thermal scattering is diffusely spread over the background. The objective lens aperture is used to select the incident beam spot or one of the diffracted beam spots. This may cut off much of the thermal diffuse scattering and the loss of these electrons can be attributed to an absorption function, included in the calculations of image contrast by adding an out-of-phase term or by adding a small imaginary part to make the effective crystal potential into a complex function.

The other important means by which incident electrons can loose energy is by excitation of the bound or nearly-free electrons of the sample into higher energy states. This may involve collective excitations, as when a plasmon oscillation is generated, or else single-electron excitations. The excitations most relevant for imaging involve energy losses ranging from a few eV to about 30 eV and scattering angles which are much smaller than the usual elastic scattering angles.

For light-atom materials the number of electrons inelastically scattered in this way may exceed the number elastically scattered.

The electrons which have lost energy will be imaged by the microscope lenses in exactly the same way as the elastically scattered electrons except that, because of the change of energy, the image will be defocussed. The consequent smearing of the image will usually be about 10 Å for thin specimens and so will not be important for low resolution or medium resolution electron microscopy. For very high resolution imaging the inelastically scattered electrons will produce a slowly-varying background to the image detail (on a scale of 5 Å or less) produced by elastically scattered electrons.

Even if the electrons inelastically scattered from a thin specimen are separated out by use of an energy filter and brought to the best focus, the image which they produce does not have good resolution. This follows because the inelastic scattering process is not localized.

The interaction with the incident electrons is through long-range Coulomb forces so that the position of the scattering event cannot be determined precisely and the image resolution cannot be better than about 10 Å.

For thicker specimens the inelastically scattered electrons can be scattered again, elastically and give a good high resolution image if properly focussed. The situation is almost the same as for an incident beam, of slightly reduced energy and slightly greater beam divergence, undergoing only elastic scattering.

1.4 STEM and CTEM

It is easy to yield to the temptation of thinking about scanning transmission electron microscopy (STEM) as an incoherent imaging process since one can visualize in terms of geometric optics and simple scattering theory how a narrow beam is scanned across the specimen and some fraction of the transmitted or scattered intensity is detected to form the image signal. This viewpoint can lead to serious errors. The imaging process depends just as much on coherent, interference phenomena in STEM as in conventional transmission electron microscopy (CTEM) given equivalent geometries. For some of the special geometries used for STEM in practice it is possible to approach incoherent imaging conditions but it is unwise to assume this to be the case without a thorough analysis.

The relationship between STEM and CTEM imaging is demonstrated readily by use of the reciprocity principle (COWLEY, 1969). Since this principle is often misapplied, leading to statements that "reciprocity fails," it is necessary to state it carefully, as follows. The wave amplitude at a point B due to a point source at A in a system is the same as the amplitude at A due to a point source at B. This holds provided that transmission through the system involves only scalar fields (no magnetic fields for electrons) and only elastic scattering processes. It holds to a very good approximation for electrons in electron microscope instruments if the image rotations and associated distortions due to the magnetic fields of the lenses can be ignored and if inelastic scattering effects are negligible. It holds in relation to intensities, not amplitudes, in the presence of inelastic scattering if the change of energy of the electrons is not detectable under the experimental conditions (DOYLE and TURNER, 1968). Within these restrictions we see that the image intensity must be the same for the equivalent points in Figs. 1.5(a) and (b) representing idealized STEM and CTEM systems. We have drawn lines to indicate beam geometries in these figures only as a matter of convenience: the wave-optical reciprocity theorem does not imply the geometric-optics notion of the reversibility of ray paths.

The reciprocal relationship is extended to finite sources of detectors by the assumption that a detector adds incoherently the intensities at all points of the detector aperture and the finite source is ideally incoherent so that all points of it emit independently to give intensities which are added incoherently at all points of the detector. It is clear that STEM with a finite detector aperture as in Fig. 1.5(a) is equivalent to CTEM with a finite incoherent source as in Fig. 1.4(a) but is not equivalent to CTEM with a coherent source as in Fig. 1.4(b).

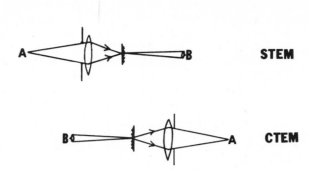

Fig. 1.5. The reciprocal relationship between the essential elements for STEM and CTEM instruments.

As a consequence of the reciprocity relationship it can be confidently predicted that for identical electron optical components and equivalent sources and detectors, the contrast of STEM images will be the same as for CTEM. In fact the whole range of phase-contrast and amplitude-contrast effects, Fraunhofer and Fresnel diffraction phenomena and bright-field and dark-field imaging behavior, familiar in CTEM, have been reproduced in STEM.

The main differences in practice between the two forms of microscopy arise from the differences in the techniques used for the detection and recording of image intensities. In CTEM a static two-dimensional image is formed and the intensity distribution is recorded by the simultaneous integration of the dose at all points in a two-dimensional detector, usually a photographic emulsion. The STEM image information is detected for one image point at a time. The detector, usually a phosphor-photomultiplier combination, provides an electrical image signal in serial form for display on a cathode ray tube or for recording in any analog or digital form. A very important practical limitation for STEM is that, in order to get the good signal-to-noise ratio needed for high quality imaging, the number of electrons scattered from each picture element within a short period must be large (10^4 or more). For high resolution imaging this number of electrons must be concentrated within a very small probe size. The requirement is therefore for a very high intensity electron source, such as can only be approached by use of a field-emission gun.

STEM also suffers from the practical inconvenience that the picture size is relatively small. The usual field of view of 1000 x 1000 picture elements or less is much smaller than is provided by a photographic plate.

On the other hand the fact that the STEM image appears as a serial, electrical signal allows enormous flexibility for on-line image evaluation and image processing and for recording of the image in an analog or digital form for subsequent processing.

Other practical aspects of STEM will be treated in detail in Chapter 11 by Humphreys in this volume.

STEM Imaging Modes

Where applicable, it is often convenient to discuss STEM image contrast in terms of the equivalent CTEM imaging configuration. For many of the more important STEM modes, however, there is no convenient CTEM analogue. Then it is preferable to discuss image formation and calculate intensities on the basis of the STEM optics without reference to CTEM.

Figure 1.6 illustrates the STEM system. The beam incident on the specimen has a convergence defined by the objective aperture. For every position of the incident beam a diffraction pattern is produced on the detector plane. This is a convergent beam diffraction (CBED) pattern in which the central beam and each crystal diffraction spot will form a circular disc, which will be of uniform intensity for a very thin crystal but may contain complicated intensity modulations for thicker crystals (see Fig. 1.7).

The intensity distribution in the CBED pattern contains a wealth of information concerning the atomic arrangements within the small regions illuminated by the beam. The interpretation and use of these "microdiffraction" patterns obtained when the incident beam is held stationary on the specimen is the subject of Chapter 14 by Warren (see also COWLEY, 1978a). For the coherent convergent incident beam produced with a field emission gun, CBED intensities may be strongly modified by interference effects (see SPENCE and COWLEY, 1978).

If the detector aperture is small and is placed in the middle of the central spot of the CBED pattern (Fig. 1.6) the imaging conditions will be those for the usual CTEM bright-field imaging mode. As the detector diameter is increased the signal strength is increased. The bright-field phase-contrast term for a thin specimen passes through a maximum value and then decreases to zero as the collector aperture size approaches the objective aperture size. For strongly scattering, thin objects the second order bright-field "amplitude contrast" term increases continually with collector aperture size. The collector aperture size and defocus giving the best contrast and resolution can be determined for any type of specimen by use of appropriate calculations (see COWLEY and AU, 1978).

The most common STEM mode is the dark-field mode introduced by Crewe and colleagues (see CREWE and WALL, 1970) in which an annular detector collects nearly all the electrons scattered outside of the central beam. This mode of dark-field imaging is considerably more efficient that the usual dark-field CTEM modes (i.e. there is a larger image signal for a given incident beam intensity) and so is valuable for the study of radiation sensitive specimens. Images may be interpreted under many circumstances as being given by incoherent imaging with an image intensity proportional to the amount of scattering matter present. This convenient approximation breaks down for crystalline specimens when the amount of scattering depends stongly on crystal structure, orientation and thickness. Also it fails near the high resolution limits when interference effects can modify the amount of scattered radiation which is undetected because it falls within the central beam (COWLEY, 1976).

The annular-detector dark-field mode has been used with great effect for the detection of individual heavy atoms and their movements (ISAACSON et al., 1976) and also for the quantitative measurement of the mass of biological macromolecules (LAMVIK and LANGMORE, 1977).

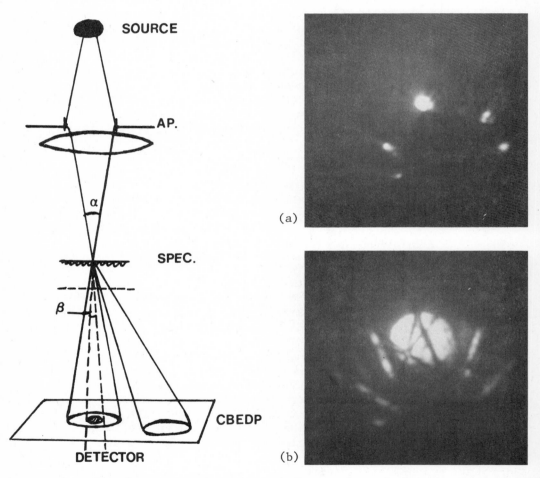

Fig. 1.6.   The formation of a convergent beam diffraction pattern in the detector plane of a STEM instrument showing the position of the bright-field detector aperture.

Fig. 1.7.   Convergent beam electron diffraction patterns from a small region (less than 50Å diameter) of a MgO crystal for objective aperture sizes (a) 20μm and (b) 100μm.

When the annular detector is used to collect all scattered radiation, however, the information contained with the diffraction pattern is lost. In principle, the use of a two-dimensional detector array to record the complete CBED pattern for each image point could add considerably to the information obtained concerning the specimen. Structural details smaller than the resolution limit could be deduced or recognized by means of structure analyses or pattern recognition techniques (COWLEY and JAP, 1976). More immediate and practical means for using some of this CBED information include the use of special detectors and masks to separate out particular features of the pattern (diffraction spots, diffuse background, Kikuchi lines) for form image signals. Equipment designed to allow the exploitation of these possibilities is now in its initial stages of operation (COWLEY and AU, 1978).

A further means for dark-field imaging in STEM is provided by the convenient addition of an electron energy filter to separate the elastically and inelastically scattered electrons or to pick those electrons from any part of the CBED pattern which have lost any specified amount of energy. The resolution of such images may be restricted by the limited localization of the inelastic scattering process but the method has important possibilities in providing means for locating particular types of atom or particular types of interatomic bonds within specimens (see Chapters 7, 8, and 9 on ELS). The versatility of the STEM detection system allows the inelastic dark-field image signals to be combined with the signals from other dark-field or bright-field detectors in order to emphasize particular features of the specimen structure.

Mathematical Description

For a point source, the wave incident of the STEM specimen is given by Fourier transform of the objective lens transfer function.

$$\psi_o(\underline{r}) = F[A(\underline{u})\exp\{i\chi(\underline{u})\}] = c(\underline{r}) + is(\underline{r}) \qquad (1.22)$$

where $\underline{r}$ and $\underline{u}$ are two-dimensional vectors. For a thin specimen with transmission function $q(\underline{r}-\underline{R})$, where $\underline{R}$ is the translation of the specimen relative to the incident beam (or vice versa), the wave amplitude on the detector plane is

$$\Psi_R(\underline{u}) = [Q(\underline{u})\exp\{2\pi i\underline{u}.\underline{R}\}] * A(\underline{u})\exp\{i\chi(\underline{u})\} \qquad (1.23)$$

and the intensity distribution is $I_R(\underline{u}) = |\Psi_R(\underline{u})|^2$. The image signal corresponding to the beam position $\underline{R}$ is then

$$J(\underline{R}) = \int I_R(\underline{u}) \, D(\underline{u}) \, d\underline{u} \qquad (1.24)$$

where $D(\underline{u})$ is the function representing the detector aperture.

In the limiting case of a very small axial detector, $D(\underline{u})$ is replaced by $\delta(u)$ and Eqn. 1.24 becomes

$$J(\underline{R}) = |q(\underline{R}) * t(\underline{R})|^2 \qquad (1.25)$$

which is equivalent to Eqn. 1.18 which applies for plane-wave illumination in CTEM.

1.5 THIN, WEAKLY SCATTERING SPECIMENS

For a very thin object we may assume (neglecting inelastic scattering) that the only effect on an incident plane wave is to change its phase by an amount proportional to the projection of the potential distribution in the beam direction. Writing this in mathematical shorthand, if the potential distribution in the specimen is $\phi(xyz)$, the phase change of the electron wave is porportional to

$$\phi(xy) = \int\phi(xyz)dz. \tag{1.26}$$

If the value of the projected potential $\phi(xy)$ is sufficiently small it can readily be shown that the image intensity can be written as

$$I(xy) = 1 - 2\sigma\phi(xy) * s(xy), \tag{1.27}$$

where σ is the interaction constant. The * sign represents a convolution or smearing operation.

This means that image contrast, given by the deviation of the intensity from unity, is described directly as the projected potential smeared out by the spread function or smearing function $s(xy)$ which determine the resolution. Provided that the spread function is a clean, sharp peak, the image will show a well-shaped circular spot for each maximum in the projected potential, i.e. for each atom or group of atoms. The smearing function depends on the defocus of the objective lens and on its aberrations. For the Scherzer optimum defocus (SCHERZER, 1949), which depends on the electron energy and the spherical aberration constant, the smearing function is a sharp, narrow negative peak [Fig. 1.8(b)] so that the image shows a small dark spot for each atom and this spot is the clearest, sharpest peak attainable with a given objective lens. For 100 keV electrons and C_s = 2 mm, for example, the optimum defocus is about 950 Å underfocus and the corresponding width of the spread function limits the resolution to about 3.5 Å.

The imaging conditions are usually discussed in terms of the modification of the wave amplitude on the back-focal plane of the objective lens. The equivalent of Eqn. 1.27 is that the diffraction pattern amplitude is multiplied by $A(uv)\sin(\chi(uv))$ where $A(uv)$ is the aperture function and $\chi(uv)$ is the phase factor given in Eqn. 1.14. In order to give a good representation of the object in the image, the function $\sin\chi$ should be as close to unity over as large a region of the diffraction pattern as possible so that the diffracted beams will have the correct relative amplitudes when they recombine to form the image. For the optimum defocus it is seen in Fig. 1.8(a) that the $\sin\chi$ function is small for low-angle scattering but is near unity for a wide range of scattering angles (corresponding to spacings in the object in the range of roughly 20 to 3.5 Å) before it goes into wild oscillations at high angles. This suggests that in the image slow variations of potential

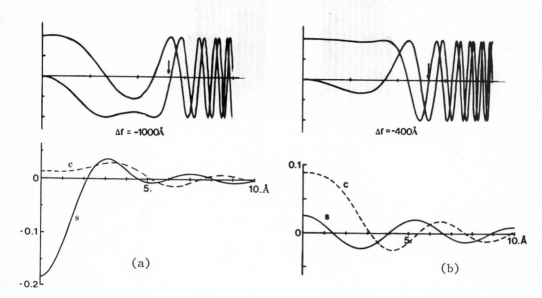

Fig. 1.8. The real and imaginary parts of the transfer function (cosχ (u) and sinχ(u)) and the spread functions c(r) and s(r) for defocus (a) -1000Å and (b) -400Å. 100keV, C_s = 2mm, aperture radius u = 0.265Å^{-1} (arrows).

will not be seen, detail in the range of 3.5 to about 20 Å will be well-represented and finer detail will be represented in a very confused manner.

This mode of imaging is thus of very little use for biologists who wish to see detail on the scale of 10 Å or greater with good contrast. Their needs are discussed in the following section.

It is the common practice to insert an objective aperture to eliminate all the high angle region past the first broad maximum of the sinχ function so that all scattered waves contributing to the image will do so in the correct phase. Then the image will be interpretable in the simple intuitive manner in that, to a good approximation, it can be assumed that dark regions of the picture correspond to concentrations of atoms, with a near-linear relationship between image intensity and the value of the smeared projected potential.

The restrictions to very thin weakly scattering objects [σφ(xy)<<1] is often not so severe as might be assumed from initial calculations, especially for light-atom materials. It is always possible to subtract a constant or slowly-varying contribution to the projected potential, since a constant phase change has no effect on the imaging (except to make a very small change in the amount of defocus).

For an amorphous specimen (for example a layer of evaporated carbon, 100 Å thick) the phase change due to the average potential, ϕ_0, may be 1 or 2 radians but since there are many atoms in this thickness, overlapping at random, the projected potential may not differ greatly from ϕ_0 and the relative phase changes may be only a fraction of a radian [Fig. 1.9(a)]. Then the weak phase object approximation, Eqn. 1.27, holds quite well and this simple approximation is more likely to fail because of the neglect of three-dimensional scattering effects than because of large phase changes.

On the other hand a single very heavy atom [Fig. 1.9(b)] may have a maximum of projected potential no greater than for the amorphous light-atom film [Fig. 1.9(a)] but the deviation from the average projected potential is much greater and the error in the use of the weak phase object approximation may be large. Similarly large deviations from the average occur in the projections of the potential for crystals when the incident beam is parallel to the rows or planes of atoms. If the same atoms as in the amorphous film [Fig. 1.9(a)] were rebuilt into a crystal, the projected potential could be as in Fig. 1.9(c) and the weak-scattering approximation could fail badly.

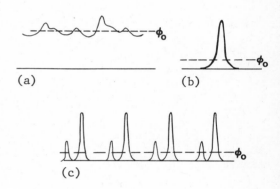

Fig. 1.9. Illustrating the subtraction of the average potential value ϕ_0 from the projected potential distribution for the cases of (a) a thin amorphous film of light atoms, (b) a single heavy atom and (c) a thin crystal.

Beam Convergence and Chromatic Aberration

For the case of weakly scattering objects the consideration of these complicating factors is greatly simplified. It has been shown by FRANK (1973) and by ANSTIS and O'KEEFE (1976) for the particular case of crystals (including cases when the weak scattering approximation fails) that the effects of beam divergence and chromatic aberration can be included by applying an "envelope function" which multiplies the transfer function and reduces the contribution of the outer, high angle, parts of the diffraction pattern. For most high resolution imaging it is one or other of these envelope functions, rather than the ideal transfer function which effectively limits the resolution.

For current 100 keV electron microscopes used with the incident beam focussed on the specimen to give the high intensity needed for high-magnification imaging, the incident beam convergence may limit the resolution to 4 Å or more. For the 1 MeV microscopes which have recently been used for high resolution imaging with apparent point-to-point resolution approaching 2 Å (HORIUCHI et al., 1976), the limiting factor seems to be the chromatic aberration effect due to lens current and high voltage instabilities.

The results of calculations by O'Keefe, illustrating the forms for these envelope functions and the consequent limitations on resolution, are shown in Fig. 1.11.

The use of an envelope function to take account of convergence is simple and convenient but is strictly applicable only for small angles of illumination in CTEM or small collector angles in STEM. Calculations for STEM (COWLEY and AU, 1978) have shown that if the objective aperture size is correctly chosen, an increase in collector aperture angle can give a slight improvement in resolution as the signal strength increases to a maximum. Then for relatively large collector angles, approaching the objective aperture angle, there may be conditions of reversed contrast with an appreciably better resolution. The influence of CTEM illumination angle or of STEM collector size is, however, a rather complicated function of the defocus, the objective aperture size, the spherical aberration constraint and the strength of the scattering by the specimen.

With care the resolution limitations due to beam convergence, chromatic aberration and such other factors as mechanical vibrations or stray electrical or magnetic fields may be reduced. It is possible, for example, to achieve a very small beam convergence with adequate intensity for high magnification images by use of a very bright source such as a field emission gun.

Then the resolution limitations represented by the envelope functions will be less important. If a sufficiently large or no objective aperture is used, diffracted beams from the outer parts of the diffraction pattern, where the transfer function is oscillating rapidly, will contribute to the image. This will produce detail in the image on a very fine scale, but since the diffracted beams are recombined with widely different phase changes, this fine detail will not be directly interpretable in terms of specimen structure. Obviously some information about the specimen structure is contained in this detail but in order to extract the information it will be necessary to resort to indirect image processing methods or else to seek agreements between observed image intensities and intensities calculated on the basis of postulated models.

Most of the image processing techniques which have been proposed depend on the idea that the difficulties due to the ambiguities of phase and loss of information around the zero points of the transfer function can be overcome by the use of two or more images obtained with different amounts of defocus (see SAXTON, 1978). While some progress has been made along these lines, few clear indications of improved resolution have been obtained. The alternative of calculating images from models of the specimen structure is realistic only for crystals.

Mathematical Formulation

For a weak phase object, we assume $\sigma\phi(xy) \ll 1$ and Eqn. 1.10 becomes

$$q(xy) = \exp\{-i\sigma\phi(xy)\} = 1-i\sigma\phi(xy)$$

For parallel incident radiation, the wave function in the back-focal plane of a lens modified by the transfer function is

$$\Psi(uv) = [\delta(uv) - i\sigma\phi(uv)] \cdot A(uv)\exp\{i\chi(uv)\} \qquad (1.28)$$

and on the image plane the wave function is

$$\psi(xy) = [1-i\sigma\phi(xy)] * [c(xy) + i\, s(xy)] \qquad (1.29)$$

where $c(xy)$ and $s(xy)$ are the Fourier transforms of $A(uv)\cos\chi(uv)$ and $A(uv)\sin\chi(uv)$. Neglecting terms of second order in $\sigma\phi$, the image intensity becomes

$$I(xy) = 1 + 2\sigma\phi(xy) * s(xy) \qquad (1.30)$$

so that we may consider the image to be produced by a system with contrast transfer function $A\sin\chi$.

For dark-field imaging in CTEM with a central beam stop which removes only the forward scattered beam, the δ function and $\phi(0,0)$, from Eqn. 1.28, we are left with only second order terms and obtain

$$I_{DF}(xy) = \{\sigma\phi'(xy) * c(xy)\}^2 + \{\sigma\phi'(xy) * s(xy)\}^2 \qquad (1.31)$$

where $\phi'(xy) \equiv \phi(xy) - \bar{\phi}$ and the average potential $\bar{\phi}$ is equal to $\phi(0,0)$.

For the case of STEM with a small axial detector, Eqn. 1.25 becomes

$$J(\underline{R}) = |(1-i\sigma\phi(\underline{R}) * \{c(\underline{R}) + i\, s(\underline{R})\}|^2$$

$$= 1+2\sigma\phi(\underline{R}) * s(\underline{R}) \qquad (1.32)$$

which is identical with Eqn. 1.30.

If the detector angle is not very much smaller than the objective aperture angle, the bright-field image is given by replacing the spread function $s(\underline{R})$ by $t_1(\underline{R})$ where

$$t_1(\underline{R}) = s(\underline{R})[d(\underline{R}) * c(\underline{R})] - c(\underline{R})[d(\underline{R}) * s(\underline{R})] \qquad (1.33)$$

where $d(\underline{R})$ is the Fourier transform of the collector aperture function $D(\underline{u})$ of Eqn. 1.24. The form of $t_1(\underline{R})$ and its dependence on defocus and collector aperture size has been explored by COWLEY and AU (1978). For the CTEM case, Eqn. 1.21 reduces to the same form, representing the effect of beam convergence from a finite source. For a small collector angle in STEM or a small finite convergence angle in CTEM, the function $d(\underline{R})$ will be a broad, slowly varying peak and $d(\underline{R})*s(\underline{R}) = 0$ since $\int s(\underline{R})$

d$\underline{R}$ = 0. Then

$$t_1(\underline{R}) \cong s(\underline{R})[d(\underline{R}) * c(\underline{R})]$$

or the effective transfer function is

$$A(\underline{u}) \sin(\chi(u)) * [D(u) \cdot A \cos(\chi(u))] \qquad (1.34)$$

Convolution with this narrow function will have the effect of smearing out the transfer function and reducing the oscillations by a greater amount as the period of the oscillations becomes smaller, i.e. the effect will be to multiply transfer functions such as those of Fig. 1.8(a) by a rapidly decreasing envelope function.

If it is assumed that a dark-field STEM image is obtained by use of an annular detector which collects all the scattered radiation (neglecting the loss of scattered radiation which falls within the central beam disc) the expression of Eqn. 1.23 with $Q(uv) = i\sigma\phi(uv)$ gives

$$J_{DF}(\underline{R}) = \sigma^2\phi^2(\underline{R}) * \{c^2(\underline{R}) + s^2(\underline{R})\} \qquad (1.35)$$

which is equivalent to the incoherent imaging of a self-luminous object having an intensity distribution $\sigma^2\phi^2(R)$. The result of Eqn. 1.35 is, of course, different from Eqn. 1.31. The two dark-field imaging modes are not equivalent.

The approximation of incoherent imaging fails significantly for imaging of detail of dimensions near the resolution limit, i.e. when the oscillations of $\phi(\underline{u})$ are of size comparable to that of the hole in the annular detector.

1.6 THIN, STRONGLY SCATTERING SPECIMENS

As suggested above, the simplifying approximation of Eqn. 1.27 can fail even for a single heavy atom. For thin crystals, and for amorphous materials containing other than light atoms, we must use a better approximation. For sufficiently thin specimens and for sufficiently high voltages (e.g. thicknesses of less than 50 to 100 Å for 100 keV or 200 to 500 Å for 1 MeV, depending on the resolution being considered and the required accuracy) we can use the approximation that the transmission function of the specimen involves a phase change proportional to the projected potential. The phase changes, even referred to an average due to the average projected potential, may be several radians. We must then consider not only the first-order term in $\sigma\phi(xy)$ as in Eqn. 1.27 but also higher-order terms. For the case of parallel incident radiation, Eqn. 1.27 is replaced by

$$I(xy) = 1 + 2\sin\sigma\phi(xy) * s(xy) - 2(1-\cos\sigma\phi(xy)) * c(xy) \qquad (1.36)$$

or taking only second-order terms into account

$$I(xy) = 1+2\sigma\phi(xy) * s(xy) - \sigma^2\phi^2(xy) * c(xy) \qquad (1.37)$$

This means that there are two components to the image intensity. The first, as before, is given by the projected potential smeared out by the smearing function s(xy), which is equivalent to multiplying the diffraction pattern amplitudes by $A(uv)\sin(\chi(uv))$. The new term is given by the square of the projected potential (or, more accurately, by the square of the positive or negative deviation from the average projected potential) smeared out by the smearing function c(xy). Use of this smearing function is equivalent to a multiplication of a component of the diffraction pattern by the function $A(uv)\cos(\chi(uv))$ illustrated in Fig. 1.8(a).

It is seen from Fig. 1.8(b) that for the optimum defocus, while the function s(r) has a sharp negative peak, c(r) is rather broad and featureless. This means that for high resolution detail the $\sin\sigma\phi$ term in Eqn. 1.36 will give good contrast but the $\cos\sigma\phi$ term will be smeared out into the background.

The situation for medium or low resolution imaging is best seen from the curves of Fig. 1.8(a) or (b). The part of the diffraction pattern corresponding to $\sin\sigma\phi$ is multiplied by $\sin\chi$ which is close to zero for the small scattering angles which correspond to slow variations in projected potential. The portion corresponding to $(1-\cos\sigma\phi)$ is multiplied by $\cos\chi$ which is close to unity for small scattering angles and hence this will dominate the image contrast.

This situation is familiar in biological electron microscopy. In order to get good contrast in the useful range of resolution (details on a scale greater than 10 to 20 Å) it is advisable to use a heavy atom stain which will give a strong $(1-\cos\sigma\phi)$ signal which is then strongly imaged because of the relatively large values of the $\cos\chi$ function. A small objective aperture size is used in order to remove the high-angle scattered radiation which adds only a confused background to the image and reduces the image contrast. Detailed calculations on models of stained biological objects by BRIDGES and COWLEY (to be published) have confirmed this interpretation.

The case of thin crystals is distinctive in that it involves more quantitative consideration of more specialized diffraction patterns and so will be treated separately.

Mathematical Formulation

Provided that we may use the phase-object approximation, the transmission function of the object is written

$$q(\underline{r}) = \exp\{-i\sigma\phi(\underline{r})\} \qquad (1.38)$$

and the image intensity for a parallel beam incident is

$$I(\underline{r}) = \left|\exp\{-i\sigma\phi(\underline{r})\} * \{c(\underline{r}) + i\,s(\underline{r})\}\right|^2 \qquad (1.39)$$

Extending this to the case of a finite incoherent source in CTEM or a finite collector aperture, with aperture function $D(\underline{u})$, in STEM, we find that, for bright-field imaging when we neglect the amount of "scattered" radiation included in the collector aperture, the image intensity can be written

$$I_{BF}(\underline{r}) = D_o + 2\sin\sigma\phi(\underline{r}) * t_1(\underline{r})$$
$$- 2(1-\cos\sigma\phi(\underline{r})) * t_2(\underline{r}) \qquad (1.40)$$

where D_o is the integral over $D(\underline{u})$, $t_1(\underline{r})$ is given by Eqn. 1.34 and

$$t_2(\underline{r}) = s(\underline{r})\{d(\underline{r}) * s(\underline{r})\} + c(\underline{r})\{d(\underline{r}) * c(\underline{r})\} \qquad (1.41)$$

The form of the functions $t_1(\underline{r})$ and $t_2(\underline{r})$ under various conditions of defocus and aperture size has been calculated by COWLEY and AU (1978).

For a very small collector aperture size, Eqn. 1.40 reduces to Eqn. 1.37. If the collector aperture is increased until it is equal to the objective aperture size,

$$t_1(\underline{r}) = 0,$$

$$t_2(\underline{r}) = c^2(\underline{r}) + s^2(\underline{r}).$$

Then the bright-field signal has the form

$$I_{BF}(\underline{r}) \cong D_o|1-\sigma^2\phi^2(\underline{r}) * \{c^2(\underline{r}) + s^2(\underline{r})\}|$$
$$= D_o|1-I_{DF}(\underline{r})|. \qquad (1.42)$$

where $I_{DF}(\underline{r})$ is, in this case, the dark-field signal calculated on the assumption that all scattered radiation, including that contained in the central beam spot, is detected to form the signal.

1.7 THIN, PERIODIC OBJECTS: CRYSTALS

For very thin specimens, there is no difference between the phase contrast imaging of periodic and nonperiodic objects, provided that we limit ourselves to the Scherzer optimum defocus and use an objective aperture to remove the higher angle parts of the diffraction pattern beyond the flat part of the transfer function. The projected potential, or the sine and cosine of the scaled projected potential, will be imaged according to Eqn. 1.27 or 1.36. Thus for thin crystals viewed down one of the unit cell axes, the image is periodic with the periodicity of the unit cell projection and shows the distribution of the atoms within the unit cell in projection, within the limitations of the point-to-point resolution of the microscope. With present day electron microscopes it is possible to make a direct structure analysis of a crystal by direct visualization of atoms in this way for a wide range of materials for which the heavier atoms are separated in projection by distances of the order of 3 Å or more (COWLEY and IIJIMA, 1977). Since the imaging does not depend on the periodicity, crystal defects may be imaged with the same clarity as the perfect crystal structure, provided that the defects do not have a three-dimensional structure which gives an unduly complicated two-dimensional projection [see Fig. 1.10]. This provides a unique opportunity for the study of the configurations of atoms in individual imperfections of the structure and is rapidly broadening our understanding of the nature of the defects in a number of types of material [COWLEY, 1978(b)].

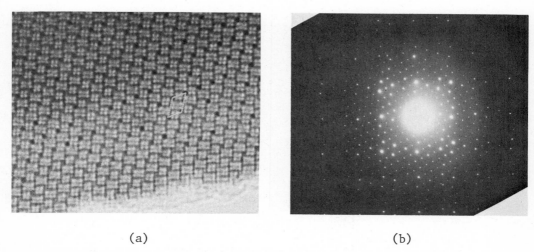

(a)   (b)

Fig. 1.10. The image, (a), and the diffraction pattern, (b), of a crystal of $Nb_{22}O_{54}$. The image shows varying contrast of some parts of the unit cell due to atomic disorder. The image contrast changes with increasing thickness. 100keV, $C_s \cong 2.8mm$, unit cell dimensions (indicated) 21.2Å by 15.6Å (courtesy Sumio Iijima).

The imaging of crystals, on the other hand, makes it possible for the first time to make quantitative correlations of image intensities with known specimen structures. Such a possibility can provide an enormous expansion of the power and range of applications of electron microscopy. To achieve this, however, it is necessary to refine both the experimental techniques and the interpretive methods.

Crystals must be aligned with an accuracy of a small fraction of a degree in two directions by use of a tilting stage. The crystal thickness must be determined with reasonable precision. The amount of defocus, the aberration constants and the aperture size of the objective lens must be accurately known.

On the other hand, it is rarely sufficient to use a simple approximation such as Eqn. 1.36 to calculate image intensities. This phase-object approximation has been shown to give significant errors for

crystal thicknesses of about 20 Å for 100 keV electrons. Reliable calculations for image interpretation must involve the use of three-
dimensional, many-beam dynamical diffraction theory using either the
matrix formulation of Bethe's original dynamical theory of electron diffraction or, more usually, the multislice formulation of dynamical
theory due to Cowley and Moodie (see COWLEY, 1975). In the latter, the
crystal is subdivided into a number of very thin slices perpendicular
to the beam. Each slice acts as a thin phase object. Between slices
the electron wave propagates according to the usual laws of Fresnel diffraction. Standard computer programs for these operations are now available.

The degree to which agreement can be obtained between observed image intensities and intensities calculated by these methods, taking into
account practical experimental parameters such as incident beam convergence and chromatic aberration, is demonstrated in Fig. 1.11. The computer programs can be extended to deal with cases of defects in crystals
by use of the assumption of periodic continuation, i.e. it is assumed
that the image of a single defect will be exactly the same as for a defect in a periodic array of well separated defects which forms a superlattice having a large unit cell.

This method for the study of crystal structures and crystal defects
has recently been extended to a wide variety of inorganic compounds and
minerals. Improved resolution, approaching 2 Å has been achieved by use
of the high resolution, high voltage microscopes now in operation (e.g.
HORIUCHI et al., 1976; KOBAYASHI et al., 1974).

Special Imaging Conditions

It is well known that images of crystals can show details on a much
finer scale than the point-to-point or Scherzer resolution limits.
Fringes with spacings well below 1 Å have been observed. HASHIMOTO
et al. (1977) have shown pictures of gold crystals having details on a
scale approaching 0.5 Å within the intensity maxima at the positions of
the rows of gold atoms. Pictures of silicon by IZUI et al. (1977) show
clearly separated spots 1.36 Å apart at the projected positions of
silicon atoms.

These pictures are taken without the objective aperture limitation
considered above. They correspond to situations in which the diffraction pattern consists of only a few sharply defined diffraction spots.
The requirement for clear imaging of the crystal periodicities is then
that these fine diffraction spots should be recombined with maximum amplitude and well-defined relative phases. This is quite different from
the requirement for a flat transfer function needed for imaging of general non-periodic objects or periodic objects with large unit cells. In
order to maintain large amplitudes for the relatively high angle diffraction spots the effects of beam convergence and chromatic aberration must
be minimized, not only by careful control of the instrumental parameters
but also by choosing the values of the defocus which make the transfer
function less sensitive to these effects.

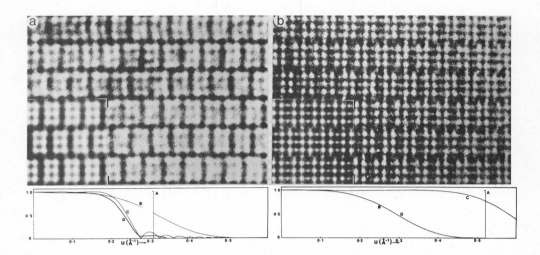

Fig. 1.11. Structure images of $Nb_{12}O_{29}$ taken at (a) 100kV (courtesy of S. Iijima) and (b) 1 MeV (courtesy of S. Horiuchi). The inserts at lower left of each micrograph are calculated images for 38Å thick crystal of $Nb_{12}O_{29}$. The aperture functions below show the resolution conditions under which each calculation was carried out. At 100 kV the physical aperture (A) at u = 0.308/Å limits resolution of 3.2Å; the aperture function due to a defocus-depth halfwidth of 100Å (B) limits resolution to 2.4Å; while the aperture function due to an incident beam convergence of 1.4 milliradian (C) restricts it to 3.8Å. The combined effect of these functions (D) results in an image of 3.8Å resolution. At 1 MV the physical aperture (A) and convergence aperture function (C) limit resolutions to 1.9Å and 1.5Å, respectively. For the calculated image to match the experimental result required a defocus-depth halfwidth of 500 Å resulting in the B curve shown. The combined effect (D) is virtually identical to and yields an image of 2.5Å resolution (courtesy of M.A. O'Keefe).

It is possible to achieve these special imaging conditions only for crystals of simple structure in particular orientations. As HASHIMOTO et al. (1978) have pointed out, the image intensities may then be sensitive indications of the details of potential distributions in the crystals even though they are in no way to be regarded as providing direct pictures of the structure. The special imaging conditions, however, become rapidly more difficult to achieve as the size of the unit cell increases and are not relevant for the imaging of defects of the crystal structure.

On the other hand, an improved appreciation of the special conditions for imaging of periodic structures has lead to the realization that some compromises are possible between the extreme situations. For crystals having defects which disrupt the periodicity to a limited extent (relatively small changes of lattice constant) one can image the structure with its defects to see detail which is finer than for a non-periodic object although not as fine as for a strictly periodic object. This is demonstrated, for example, by the pictures of defects in silicon due to SPENCE and KOLAR (1979).

Mathematical Formulation

If a crystal is divided into very thin slices perpendicular to the incident beam, the transmission function of the nth slice is

$$q_n(xy) \cong \exp\{-i\sigma\phi_n(xy)\}$$

where

$$\phi_n(xy) = \int_{z_n}^{z_{n+1}} \phi(\underline{R})dz \tag{1.43}$$

Propagation through a distance $\Delta = z_{n+1} - z_n$ to the $n+1$ th slice is given by convolution with the propagation function [Eqn. (1.12)]. Then we have the recursion relationship

$$\psi_{n+1}(\underline{r}) = \{\psi_n(\underline{r}) * p_\Delta(xy)\}q_{n+1}(\underline{r}) \tag{1.44}$$

or, in terms of Fourier transform,

$$\Psi_{n+1}(\underline{u}) = \{\Psi_n(\underline{u}) \cdot P_\Delta(\underline{u})\} * Q_{n+1}(\underline{u})$$

which, for a periodic object can be written

$$\Psi_{n+1}(h,k) = \Sigma_{h,k}\Psi_n(h_1k_1^1)P_\Delta(h_1k_1)Q_{n+1}(h-h,k-k) \tag{1.45}$$

and this is an operation readily programmed for a computer with

$$P_\Delta(h,k) \cong \exp\{i\pi\Delta\lambda\left(\frac{h^2}{a^2} + \frac{k^2}{b^2}\right)\} \tag{1.46}$$

for a unit cell with dimension a,b.

The image intensity is then calculated from the exit wave function or its Fourier transform $\Psi_N(\underline{u})$ as

$$I(\underline{r}) = |F[\Psi_N(\underline{u}) \cdot A(u)\exp\{i\chi(\underline{u})\}]|^2 \tag{1.47}$$

The condition that the image wave amplitude should be identical with the wave at the exit face of a crystal is that $\exp\{i\chi(\underline{u})\} = 1$ for all diffracted beams $u = h/a$, $v = k/b$ i.e. that

$$\pi\Delta f\lambda\left(\frac{h^2}{a^2} + \frac{k^2}{b^2}\right) + \tfrac{1}{2}\pi C_s\lambda^3\left(\frac{h^2}{a^2} + \frac{k^2}{b^2}\right)^2 = N\pi \tag{1.48}$$

The case for N odd is included because it represents the case of an identical wave function shifted by half the periodicity.

It is not possible to satisfy this condition unless $\underline{a}^{2}$ and $\underline{b}^{2}$ are in the ratio of integers. For the special case that $\underline{a} = \underline{b}$, (Eqn. 1.48) is satisfied for $\Delta f = na^{2}/\lambda$ and $C_{s} = 2ma^{4}/\lambda^{3}$ for integral n,m (see KUWABARA, 1978). The extent to which these conditions may be relaxed is a measure of the limitations on the degree of crystal perfection or on the finite crystal dimensions for which the image may represent the ideal in-focus image of the crystal (see COWLEY, 1978c).

1.8 THICKER CRYSTALS

When the crystal thickness exceeds 50-100 Å for 100 keV electrons or 200-500 Å for 1 MeV electrons there is in general no relationship visible between the image intensities and the projected atom configurations, (although for some particular cases the same thin-crystal image is repeated for greater thicknesses). It is suggested that the image intensities will be increasingly sensitive to details of crystal structure, such as the bond lengths, ionization or bonding of atoms and thermal vibration parameters but since the images are also more sensitive to experimental parameters such as the crystal alignment and the lens aberrations, the refinement of crystal structures by use of high resolution thick crystal images remains an interesting but unexploited possibility.

Most of the important work done on thicker crystals has been on the study of the extended defects of crystals having relatively simple structures with medium resolution (5 Å or more) and no resolution of crystal structure periodicities. The extensive work done on the form and behavior of dislocations and stacking faults, in metals, semiconductors and an increasingly wide range of inorganic materials, together with the theoretical basis for image interpretation in these cases are very well described in such books as HIRSCH et al. (1965) and BOWEN and HALL (1975). Because dynamical diffraction effects are of overwhelming importance for these studies it is essential to simplify the diffraction conditions as much as possible to make it relatively easy to calculate and to appreciate the nature of the diffraction contrast. For thin crystals with large unit cells, viewed in principal orientations it may be necessary to take hundreds or even thousands of interacting diffracted beams into account [see Fig. 1.10(a)]. For the thicker crystals of relatively simple structure it is often possible to choose orientations for which the two-beam approximation is reasonably good; namely, when the incident beam gives rise to only one diffracted beam of appreciable amplitude and these two beams interact coherently. The image intensity can vary with crystal thickness because interference between the two beams gives both of them a sinusoidal variation with thickness. Crystal defects show up with strong contrast because changes in the relative phase of the two beams result in different interference effects and so different intensities. The phase changes can be sudden as when the crystal suffers a shear displacement at a stacking fault; or the phase change can be more gradual as when the lattice is strained around a dislocation line or other defect and the deviation from the Bragg angle varies as the incident and diffracted beams travel through the strain field.

The standard dynamical diffraction theory, as originated by Bethe and developed by many other authors (see HIRSCH et al., 1965) is the theory of the interaction of electron waves with a perfectly periodic potential distribution bounded by plane faces. To adapt this to the study of crystal defects it is usual to make a simplifying "column approximation." For very thin crystals we have made the assumption that the electron wave at a point on the exit faces of the crystal is influenced only by the potential along a line through the crystal to that point in the beam direction. For thicker crystals we may assume that the electron wave at a point on the exit face of the crystal is affected by the diffraction, not along a line, but within a thin column extending through the crystal in the beam direction.

The width of this column may be estimated in various ways which all agree that it may be surprisingly narrow. For 100 keV electrons, for example, and crystals several hundred Å thick, the column width may be taken as small as 5-10 Å with errors which are not serious for most purposes. For a microscope resolution limit of 10 Å or more the column approximation serves very well.

To calculate the image of a defect it is necessary to calculate the amplitudes of the incident and diffracted beams for each column of crystal passing near the defect. The amplitudes for a column containing a particular sequence of lattice strains and disruptions can be calculated on the assumption that all surrounding columns are identical, i.e. that the crystal is perfectly periodic in directions perpendicular to the beam direction. The calculations are usually made by the difference equation method of Howie and Whelan (see HIRSCH et al., 1965) in which the progressive changes of the wave amplitudes are followed as the waves progress through the crystal. Convenient computer techniques developed by Head and colleagues (HEAD et al., 1973) have provided systematic techniques for the identification of defects of many types. In many cases, especially for thick crystals, it is necessary to introduce the effects of inelastic scattering (mostly thermal diffuse scattering) on the elastically scattered waves. This is done usually by the simple expedient of adding a small imaginary part to make the structure amplitudes of the crystal complex. It leads to a variety of easily observable effects.

Recent refinements of these dynamical diffraction studies of crystal defects include the "weak beam method" which gives much finer details of defect structure at the expense of very low image intensities (COCKAYNE, RAY and WHELAN, 1969). This method relies on the fact that in a situation where both strong and weak diffracted beams, or only weak diffracted beams, are present, the weak beam intensities vary much more rapidly with crystal thickness or with change of incident beam orientation than do the strong beams. Thus the details of dislocation structure in an image formed by allowing only a weak beam through the objective aperture may be on the scale of 10-15 Å whereas for dark-field images formed with strong reflections the intensity variations may be stronger but show no detail finer than about 50 Å.

For the interpretation of detail on a very fine scale, the column approximation may not be sufficient. A review of the more exact

treatments which avoid this approximation has been given recently by
ANSTIS and COCKAYNE (1979).

Because the image intensity modulations for crystal defects are
strongly dependent on the angle of incidence of the electron beam in relation to the Bragg angle for the operative reflections, the visibility
of defects may be reduced and the characteristic features of their images may be lost if the range of angles of incidence is too large in
CTEM or if the collector aperture size is too large in STEM. For CTEM
this usually does not constitute a serious restriction but for STEM, especially if a high-brightness source is not used, it may produce an undesirable low intensity of the useful images. One means for overcoming
this limitation of the STEM method has been suggested by COWLEY (1977).
If a slit detector is used in place of a circular detector aperture a
relatively large signal can be obtained for incident beam directions
having only a narrow range of angles of incidence on the operative reflecting planes. This is one of the situations for which the flexibility in collector aperture configuration, inherent in the STEM mode,
offers considerable advantage.

Lattice Fringes

Within the limits set by incident beam divergence and the mechanical and electrical stabilities of the electron microscope, it is possible to produce interference fringes in the image with the periodicity of
the diffracting lattice planes for any crystal thickness provided that
the objective aperture allows two or more diffracted beams to contribute
to the image. Under the usual operating conditions, with no careful control of the experimental parameters of the electron microscope or of the
specimen material, the information content of such images is very
limited. The position of the dark or light fringes relative to the atomic planes is indeterminate since this is strongly dependent on the crystal orientation and thickness, the objective lens defocus and the cen-
tering of the incident or diffracted beams with respect to the objective
lens axis. The spacing of the fringes is usually close to that of the
relevant lattice planes but may vary appreciably if the crystal varies
in thickness or is bent or, in particular, if other strong reflections
are excited locally.

If care is taken to avoid complications from all of these factors,
however, some very useful data may be obtained by observations of lattice fringes. Variations of lattice plane spacings corresponding to variations in the composition or degree of ordering in alloys have been observed (WU, SINCLAIR and THOMAS, 1978). Also the presence of defects
may be detected even though the perturbations of the fringe spacings or
contrast can usually give no direct evidence on the defect structure.

Lattice fringes may, of course, be observed in STEM, as in CTEM.
The accessibility of the convergent beam diffraction pattern in a STEM
instrument allows a rather clearer picture to be obtained of the conditions under which lattice fringes are formed (SPENSE and COWLEY, 1978).
If the disc-shaped diffraction spots corresponding to the individual
reflections do not overlap, no interference is possible between incident

beam directions with sufficient angular separation to produce interference effects with the lattice plane periodicity. The region where diffraction spot discs overlap is the region where such interference effects can take place, so that if the detector aperture includes the region of overlap, the image can show the lattice plane spacing.

It is not difficult to specify the detector aperture size and shape
which will give maximum lattice fringe visibility and image intensity
for any particular conditions of beam incidence, objective aperture size
and lens aberration.

Mathematical Considerations

In order to calculate the wave function at the exit face of a crystal, using the column approximation, when there is a distortion of the
crystal which is a continuous function of distance in the beam direction, the multi-slice formulation of Cowley and Moodie, described above,
may be used. The solution of the wave equation, following the Bethe
formulation, in each section of the crystal which can be considered as
periodic is rarely feasible. Most commonly the difference equation form
due to Howie and Whelan is used. The changes in incident and diffracted
wave complex amplitudes due to diffraction from other waves in the crystal, absorption effects and excitation errors, can be written

$$\frac{d}{dz}\,\underline{\psi} = 2\pi i(\underline{\underline{A}}+\underline{\underline{\beta}})\underline{\psi} \tag{1.49}$$

where $\underline{\Psi}$ is the column vector whose elements Ψ_h are the amplitudes of the
diffracted waves, $\underline{\underline{\beta}}$ is a diagonal matrix whose elements are $\beta_h =
d(\underline{h}\cdot\underline{R}(z))/dz$ and $\underline{R}(\underline{z})$ is the vector giving the displacements of the lat-
tice points. The matrix $\underline{\underline{A}}$ has diagonal and off-diagonal elements:

$$A_{hh} = \zeta_h + i\sigma\phi_o'/4\pi$$

$$A_{hg} = \sigma(\phi_{h-g} + i\phi_{h-g}')/4\pi \tag{1.50}$$

where ζ_h is the excitation error for the h reflection and ϕ_h' is the imaginary part added to the structure amplitude to represent the effect of
absorption.

For the two-beam case this simplifies to a simple pair of coupled
equations which may be integrated through successive slices of crystal.

For other than very thin crystals it is not possible to represent
the effect of the crystal on the incident wave function by multiplication by a scalar transmission function, as was assumed above. Instead
we may consider the action of a crystal to be represented by the action
of a scattering matrix on an incident wave vector, $\Psi_0(\underline{h})$, representing
the Fourier coefficients of the incident wave. Thus

$$\underline{\Psi} = \underline{\underline{S}}\underline{\Psi}_0 \tag{1.51}$$

and the matrix $\underline{\underline{S}} = \exp\{iz\underline{\underline{M}}(\underline{h})/2k\}$ where $\underline{\underline{M}}(\underline{h})$ is similar to the matrix $\underline{\underline{A}}$
of Eqn. 1.49. This equation can also be iterated through successive

slices. For n similar slices $\Psi_n = \underline{S} \Psi_0(\underline{h})$. This formulation follows the concepts developed by STURKEY (1962) and others and forms the basis for a number of sophisticated and powerful treatments of diffraction and imaging problems.

1.9 VERY THICK SPECIMENS

As the thickness of a specimen increases, the distribution of intensity in the diffraction pattern is dominated increasingly by multiple scattering effects. Diffracted beams traversing further regions of the specimen may be diffracted again and again, both elastically and inelastically. For crystals the spot patterns given by thin crystals are gradually submerged under the diffuse background scattering. Diffraction of the diffusely scattered electrons by the crystal lattice gives complicated Kikuchi line configurations which in turn gradually lose contrast and are lost in an overall broad background of scattering. For non-crystalline specimens the initial rather featureless scattering distribution is successively broadened by multiple elastic scattering. Also because the electrons lose energy through successive inelastic scatterings, the distribution of electron energies becomes broader and broader and the number of electrons which have not lost any energy becomes very small.

For CTEM the image resolution becomes poorer as a result of the increasing angular spread of the electrons because the position at which scattering occurs becomes less and less well-defined as the electron beam spreads in the specimen. Also as the energy spread of the transmitted electrons increases, the resolution suffers as a result of the chromatic aberration of the lenses. Usually, in order to improve the contrast, the objective aperture size is made small but this has the effect of reducing the image intensity rather drastically because, as the angular range of the scattering is increased by multiple elastic scattering, the fraction of the transmitted radiation remaining near the central spot is rapidly reduced (Fig. 1.12).

For STEM there is the same loss of resolution due to the angular spreading of the beam in the specimen. However, loss of energy of the electrons by multiple inelastic scattering in the specimen does not affect the resolution, since there are no imaging lenses after the specimen. Hence, the effect of specimen thickness on the resolution will be less severe for STEM than for CTEM. For 100 keV electrons and specimen thicknesses, of the order of a few micrometers, estimates suggest that STEM will have an advantage over CTEM by a factor of about 3 (SELLAR and COWLEY, 1973) although for special instrumental configurations this factor has been estimated to be as high as 10 (GROVES, 1975). With increasing accelerating voltage this factor will decrease, being about 2 for 1 MeV electrons.

It has been shown that for STEM the best contrast is obtained for very thick specimens if a very large detector aperture is used (or the order of 10^{-1} radians or more) (see Fig. 1.12). This has the added advantage that the image signal intensity can be as high as half the inci-

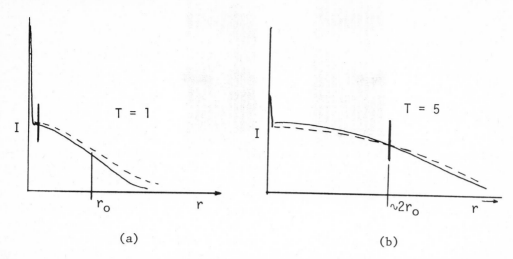

(a)   (b)

Fig. 1.12. Diagrams suggesting the change in the angular distribution
of scattered electrons and the reduction of the central peak of unscattered electrons, for amorphous specimens of thickness T equal to (a) the
mean free path for elastic scattering and (b) five times this thickness.
The optimum objective aperture size, giving the greatest transfer of intensity from inside to outside the aperture with a small change of thickness, is indicated in each case relative to the mean scattering angle
for single scattering.

dent beam intensity so that there is no problem of decreasing image
intensity (SMITH and COWLEY, 1975). Also for STEM it is possible to get
dark field image contrast by use of an energy filter to separate electrons which have lost less energy than the average from those which have
lost more. The desirable energy cut-off for maximum contrast obtained
in this way may be as high as several hundred volts energy loss for
thick specimens (PEARCE-PERCY and COWLEY, 1976). Again it appears that
STEM has a potential advantage in that the flexibility inherent in the
STEM detection system and the associated signal processing possibilities, allows the optimum imaging conditions to be achieved in a relatively straightforward manner.

In calculating image contrast for very thick specimens it is usual
to assume that the multiple scattering will mix up the relative phases
of diffracted beams to the extent that interference effects will be
washed out and a simple incoherent imaging theory can be used. The intensities, rather than the amplitudes of multiply scattered waves are
added together. This assumption is so very convenient and the alternative of a proper coherent scattering theory is so forbidding, that it is
easy to ignore the fact that coherent scattering effects may be significant for any quantitative image evaluation for even quite large thicknesses (SELLAR, 1977).

Mathematical Descriptions

In the incoherent scattering approximation it is assumed that the intensity distribution from the first slice of crystal is spread further by scattering in a second slice and so on. For a single slice of thickness Δz the intensity distribution may be written

$$I_1(u) = \exp\{-\mu\Delta z\}[\delta(u) + \Delta z \cdot f^2(u)]$$

where $f^2(u)$ is the square of the scattering amplitude, and the absorption coefficient μ is given by $\int f^2(u)du$. Fourier transforming this in terms of some arbitrary variable w,

$$G_1(w) = \exp\{-\mu\Delta z\}[1 + \Delta z P(w)]$$

The effect of transmission through n layers to give a total thickness T $= n\Delta z$ is to convolute $I_1(u)$ by itself n times or correspondingly to raise $G_1(w)$ to the nth power

$$G_1{}^n(w) = \exp\{-\mu T\}[1 + \Delta z P(w)]^n$$

$$\rightarrow \exp\{-\mu T + TP(w)\}, \qquad (1.52)$$

so that the angular distribution of scattered intensity becomes

$$I_T(u) = F^{-1}[\exp\{T(P(w)-\mu)\}]. \qquad (1.53)$$

The optimum detector aperture size for a STEM instrument with a very thick specimen is found by finding the value of $|u|$ for which the differential of the intensity $I_T(u)$ with respect to the thickness T changes sign.

The distribution of the number of electrons with energy loss is found in the same way in terms of the mean free path for inelastic scattering and the optimum energy cut-off for an energy filter is found by finding the energy loss value for which the differential of this number with respect to T changes sign.

1.10 CONCLUSIONS

One conclusion to be drawn from our discussions of the high resolution imaging of thin specimens is that no simple definition of resolution is possible unless the concept of resolution is severely restricted. One can ask, for example, what is the smallest distance between two distinct maxima or minima of intensity in an image. This provides an operational definition of resolution of one kind, useful as a convenient criterion to be used by instrument designers but it makes no reference to the main function of an electron microscope which is to provide information regarding the structure of the specimen. In practice this type of definition requires qualification in that, as is well-known, point-to-point imaging is different from lattice fringe imaging. In the latter, strong diffracted beams occur at large diffraction angles so that some contrast may be seen provided that the transfer function of the lens is not zero. For non-periodic objects the diffracted wave amplitudes fall off rather uniformly so that outer non-zero parts of the transfer function will multiply weak scattering amplitudes and the resultant contributions of the outer parts of the diffraction pattern to the image intensity will be so small that the corresponding fine detail of the image will be of too little contrast to be detected. Thus, in order to provide a reliable resolution criterion of this sort it is necessary to specify the nature of the ordering and the degree of crystallinity in the specimen, but no independent evidence on these questions, other than from electron microscopy, is available.

It is, of course, possible to consider only the extreme case of near perfect periodicity and take the minimum observable lattice fringe spacing as a measure of the "resolution limit." This is, in fact, a good test of some instrumental parameters such as the mechanical stability of the column, the specimen drift, interference from stray electrical or magnetic fields and incident beam convergence. It is not a sensitive measure of chromatic aberration and is insensitive to spherical aberration.

The use of a test object of completely random structure would provide a different basis for testing. But most "amorphous" materials, including amorphous carbon films, are known to contain small regions which are relatively well ordered to the extent that the diffraction patterns given by individual picture elements (of diameter comparable with the resolution limit) may contain quite strong maxima at high angles. Therefore, the assumption of a scattered amplitude falling off fairly uniformly with scattering angles is invalid.

From a different point of view we may choose to measure resolution in terms of the ability of the electron microscope to provide a recognizable image of the known structure of a test object. For very thin, weakly-scattering objects the usual Scherzer criterion applies. The resolution is assumed to be given by the reciprocal of the value of $u = 2\lambda^{-1}\sin(\phi/2)$ at the outside limit of the flat part of transfer function, $\sin(\chi(u))$, for the optimum defocus. The objective aperture is chosen to eliminate all the radiation scattered at higher angles, for which the transfer function is oscillatory. As we have seen, this criterion for measuring resolution cannot be used for strongly scattering and thicker samples. Nor can it be used in general for dark-field imaging since for detail near the resolution limit the intensity distribution of dark field images often has no direct relationship to the atomic arrangement in the specimen.

The situation is rather more favorable when the images of weakly scattering objects can be reasonably well interpreted by incoherent imaging theory as in the case of STEM with an annular dark-field detector, STEM bright field imaging with a large detector aperture (see Eqn. 1.42) or bright field CTEM with a large angle of illumination (NAGATA et al., 1976). It is a well-known result of light optics (BORN and WOLF, 1964) that for incoherent illumination the resolution can be better than for coherent illumination by a factor of about 1.5 (actually $2^{1/2}$ for a Gaussian spread function as is evident from Eqn. 1.35).

Probably the best way to characterize the performance of an electron microscope is to determine the transfer function for the objective lens for a thin phase object. This is consistent, although not identical, with the current practice in light optics. If the transfer function can be determined experimentally the resolution and contrast of images produced by the instrument for any specimen can be evaluated according to any of the criteria which seem useful or by detailed calculation.

The most convenient method for achieving this is by use of an optical diffractometer which provides the Fourier transform (or, more accurately, the squared amplitude of the Fourier transform) of the image intensity distribution. Provided that the specimen is a thin, weakly-scattering object, the optical diffraction pattern intensity will give, to a good approximation, the square of the transfer function, $\sin(\chi(u))$, multiplied by the intensity distribution in the diffraction pattern of the object. Usually the specimen used for this purpose is a thin "amorphous" carbon film. Caution is necessary in practice to ensure that none of the restrictions we have mentioned on the use of this method are violated.

It has been found in practice that the transfer function measured in this way may depend very strongly on a number of experimental factors and so will show variations even for the same lens used on the same specimen at the same voltage and defocus. One important requirement for the development of more quantitative high resolution electron microscopy is therefore a means for obtaining a rapid measurement of the transfer function, preferably as a continuous "on-line" monitoring of the microscope performance. Some limited success in this respect has been achieved for example by BONHOMME and BEORCHIA (1978) but further work in this direction is desirable.

The quest continues for even better and better resolution. One approach is to find ways of interpreting the image detail coming from the outer parts of the diffraction pattern, beyond the Scherzer limit, by use of image processing techniques. The other approach is to extend the Scherzer limit. This involves the attainment of smaller spherical aberration or smaller wavelengths without the sacrifice of other important factors, since the directly interpretable resolution for the Scherzer optimum defocus is given by

$$\Delta \cong 0.6 \ c_s^{1/4} \ \lambda^{3/4} \tag{1.54}$$

The factor 0.6 is approximate and may be replaced by various factors depending on the assumptions made. The improvement of C_s is difficult and has a limited effect because of the 1/4 power. The decrease of the wavelength by increasing of the accelerating voltage therefore appears to be the most promising approach provided that the resolution is not limited by the stability of the accelerating voltage or lens currents. These latter factors do seem to be effective in limiting the resolution of the current high voltage, high resolution instruments to about 2 Å whereas the theoretical resolution given by Eqn. 1.54 is more like 1.5 Å or less.

If the resolution is to be improved by image processing techniques it is desirable, of course, to start from a situation where the directly interpretable resolution is as good as possible. The extent to which image processing methods may succeed depends then on the extent to which the fineness of the image detail exceeds the limit of directly interpretable detail. This in turn depends on the extent to which the image detail is limited by chromatic aberration, beam convergence, and similar effects, in relationship to the limitations on interpretability set by spherical aberration.

The next stage in the improvement of electron microscope resolution will clearly be a very significant one from the point of view of applications in materials science. For most inorganic materials the interatomic distances, seen in projection along favorable crystallographic directions, are mostly in the range 1.5 to 2.0 Å. Improved resolution means more contrast for the imaging of atoms. Thus the next factor of 2 in resolution from the present commonly attainable limit will enormously enhance the power of the electron microscope to give detailed information on crystal structures and the all-important perturbations of crystal structures due to the various types of defects.

An equally important direction of development is the addition of microanalytical methods to the imaging capabilities. The addition of information on crystal structure from microdiffraction patterns and on chemical composition from the microanalysis techniques, based on the detection of characteristic x-rays or characteristic electron energy losses, forms the subject matter of most of the other contributions to this volume. The recent commercial production of instruments to combine all of these capabilities, with a limited sacrifice of performance in any one respect, provides the basis for a major reorganization of research, especially in the materials sciences, aimed at a much more complete understanding of physical and chemical properties in terms of the atom configurations and atom interactions.

CLASSICAL and GENERAL REFERENCES

Anstis, G. R. and Cockayne, D. J. H. (1979), Acta Cryst., in press.
An excellent analysis of the various theoretical approaches to the description of dynamical diffraction by crystals with defects.

Boersch, H. (1947), Z. Naturforsch 2a, 615.
One of the early works of this often neglected scientist who contributed many important ideas to the subject of electron microscopy, including much discussion of the possibility of observing individual atoms.

Born, M. and Wolf, E. (1964), "Principles of Optics", Pergamon Press, London.
The standard reference work for many years on the contemporary approach to optics.

Bowen, D. K. and Hall, C. R. (1975), "Microscopy of Materials", John
    Wiley and Sons, New York.
    A more modern treatment than Hirsch et al. (1965) but on a more in-
    troductory level, keeping as closely as possible to a nonmathemat-
    ical description style: a little shakey on some points of funda-
    mental physics.

Cowley, J. M. (1975), "Diffraction Physics", North Holland Publishing
    Company.
    A rather formidable book for the non-theorist which attempts to cor-
    relate x-ray and electron diffraction with electron microscopy by
    use of a common theoretical basis. Needs to be read in conjunction
    with more detailed accounts, or prior knowledge, of the experi-
    mental situations.

Doyle, P. A. and Turner, P. S. (1968), Acta Cryst. A24, 390.
    The earliest and, for many purposes, definitive discussion of the
    application of the reciprocity principle in electron diffraction.

Head, A. K., Humble, P., Clarebrough, L. M., Morton, A. J., and Forward,
    C. J. (1973), "Computer Electron Micrographs and Defect Identifica-
    tion", North Holland, Amsterdam.
    A clear, systematic account of the methods and typical results for
    calculating images of dislocations, stacking faults, etc., with
    computer programs.

Hirsch, P. B., Howie, A., Nicholson, R. B., Pashley, D. W., and Whelan,
    N. J. (1965), "Electron Microscopy of Thin Crystals", Butter-
    worth's, London.
    This is the "Yellow Bible" of materials science electron micro-
    scopists. Produced as the result of a summer school, it contains
    an excellent account of the explanation of the contrast effects for
    crystal defects in crystals. A second edition published in 1977
    has a chapter added to summarize the progress since 1965.

Lipson, S. G. and Lipson, H. (1969), "Optical Physics", Cambridge Uni-
    versity Press, Cambridge.
    A book biased by the fact that both authors are physicists and one
    is an outstanding crystallographer who contributed greatly to the
    use of optical diffraction analogues for x-ray diffraction pro-
    cesses.

Saxton, W. O. (1978), "Computer Techniques for Image Processing in Elec-
    tron Microscopy", Academic Press, New York.
    A definitive but difficult book on the basic concepts and methods
    of image processing and a detailed description of the contributions
    of the author and his colleagues to this subject, with computer
    programs for the main operations.

Scherzer, O. (1949), J. Appl. Phys. 20, 20.
    The clear, original statement on how to produce the optimum phase
    contrast imaging for weakly-scattering objects.

Vainshtein, B. K. (1964). "Structure Analysis by Electron Diffraction", Pergamon Press, Oxford.
> An account of the work of the Soviet Union school which developed and applied electron diffraction methods of crystal structure analysis: contains excellent accounts of the geometry of electron diffraction patterns and the simpler approximations for electron diffraction intensities.

OTHER REFERENCES

Anstis, G. R. and O'Keefe, M. A. (1976), in Proc. 34th Annual Meeting Electron Micros. Soc. Amer., Ed. G. W. Bailey, Claitor's Publ. Div., Baton Rouge, p. 480 and in press.

Bonhomme, P. and Beorchia, A. (1978), in Electron Microscope 1978, Ed. J. M. Sturgess, Microscopical Society of Canada, Vol. 1, p. 86.

Cockayne, D. J. H., Ray, I. L. F., and Whelan, M. J. (1969), Phil. Mag. 20, 1265.

Cowley, J. M. (1969), Appl. Phys. Letters 15, 58.

Cowley, J. M. (1976), Ultramicroscopy 2, 3.

Cowley, J. M. (1977), in High Voltage Electron Microscopy 1977, Eds. T. Imura and H. Hashimoto, Japanese Soc. Electron Micros., Tokyo, p. 9.

Cowley, J. M. (1978a), Advances in Electronics and Electron Physics, Ed. L. Marton, Academic Press, New York, 46, 1-53.

Cowley, J. M. (1978b), Annual Reviews of Physical Chemistry, Ed. B. S. Rabinovich, Annual Review, Inc., Palo Alto, 29, 251-283.

Cowley, J. M. (1978c), in Electron Microscopy 1978, Vol. III, Ed. J. M. Sturgess, Microscopical Society of Canada, Toronto, p. 207.

Cowley, J. M. and Au, A. Y. (1978), in Scanning Electron Microscopy/1978, Vol. 1, Om Johari, Ed., SEM Inc., AMF O'Hare, Illinois, p. 53.

Cowley, J. M. and Iijima, S. (1977), Physics Today 30, No. 3, 32.

Cowley, J. M. and Jap, B. K. (1976), in Scanning Electron Microscopy/1976, Vol. 1, Om Johari, Ed., IITRI, Chicago, p. 377.

Crewe, A. V. and Wall, J. (1970), J. Mol. Biol. 48, 375.

Ditchborn, R. W. (1963), Light (2nd Edition), Blackie & Sons, London.

Frank, J. (1973), Optik, 38, 519.

Groves, T. (1975), Ultramicroscopy $\underline{1}$, 15.

Hashimoto, H., Endoh, H., Tanji, T., Ono, A., Watanabe, E. (1977), J. Phys. Soc. Japan $\underline{42}$, 1073.

Hashimoto, H., Kumao, A., and Endoh, H. (1978), in Electron Microscopy 1978, Vol. III, Ed. J. M. Sturgess, Microscopical Society of Canada, Toronto, p. 244.

Horiuchi, S., Matsui, Y., Bando, Y. (1976), Jpn. J. Appl. Phys. $\underline{15}$, 2483.

Isaacson, M. S., Langmore, J., Parker, W. W., Kopf, D., and Utlaut, M. (1977), Ultramicroscopy $\underline{1}$, 359.

Izui, K., Furono, S., Otsu, H. (1977), J. Electron Microsc. $\underline{26}$, 129.

Kobayashi, K., Suito, E., Uyeda, N., Watanabe, M., Yanaka, T., Etoh, T., Watanabe, H., Moriguchi, M. (1974), in Electron Microscopy 1974, Eds. J. V. Sanders, D. J. Goodchild, Aust. Acad. Sci., Canberra, Vol. 1, p. 30.

Kuwabara, S. (1978), J. Electron Microsc. $\underline{27}$, 161.

Lamvik, M. K. and Langmore, J. P. (1977), in Scanning Electron Microscopy/1977, Ed. Om Johari, IIT Res. Inst., Chicago, Vol. 1, 401.

Nagata, F., Matsuda, T., Komoda, T., and Hama, K. (1976), J. Electron Microsc. $\underline{25}$, 237.

Pearce-Percy, H. T. and Cowley, J. M. (1976), Optik $\underline{44}$, 273.

Sellar, J. R. (1977), in High Voltage Electron Microscopy, 1977, Eds. T. Imura and H. Hashimoto, Japanese Soc. Electron Microsc., Tokyo, p. 199.

Sellar, J. R. and Cowley, J. M. (1973), in Scanning Electron Microscopy/1973, Ed. Om Johari, IIT Res. Inst., Chicago, p. 143.

Smith, D. J. and Cowley, J. M. (1975), Ultramicroscopy $\underline{1}$, 127.

Spence, J. C. H. and Cowley, J. M. (1978), Optik $\underline{50}$, 129.

Spence, J. C. H. and Kolar, H. (1979), in press.

Stone, J. M. (1963), Radiation and Optics, McGraw-Hill Book Co., New York.

Wu, C. K., Sinclair, R., and Thomas, G. (1978), Metal Trans. $\underline{9A}$, 381.

CHAPTER 2

INTRODUCTORY ELECTRON OPTICS

R.H. GEISS

IBM RESEARCH LABORATORY

SAN JOSE, CALIFORNIA

2.1 INTRODUCTION

The purpose of this chapter is to present an introductory, nonmathematical background in electron optics. The level is geared to the user of an electron microscope who is interested in understanding the qualitative features of the electron optical column but does not want to design a microscope. Because of this, mathematical statements are made and not derived in general and, in fact, mathematics is kept to a minimum with little more than high school background required. Although the discussion will center on magnetic lenses, as they are the most popular, an introduction to the terminology, laws and techniques will be given using light optical principles. Electrostatic lenses will be discussed briefly, especially as the electron gun is an electrostatic lens. Although aberrations are discussed in some detail, it is important to remember that a good approximate picture of the image formation process may be easily derived ignoring aberrations and adhering only to Gaussian optics. Subsequent chapters in this book will describe applications to image formation. For those interested in more complete and/or mathematical treatments, a list of texts on electron optics is included at the end.

2.2 GEOMETRICAL OPTICS

Refraction

The fundamental law of geometrical optics as it pertains to light optics is Snell's law. This law describes the refraction of a light wave at the interface between two media with differing indices of refraction, n_i. It is written

$$\frac{\sin\theta_1}{\sin\theta_2} = \frac{n_2}{n_1}$$

where θ_1 and θ_2 are the angles of incidence and refraction with respect to the interface normal. The physical construction is diagrammed in Fig. 2.1.

Fig. 2.1 Geometrical construction showing Snell's law.

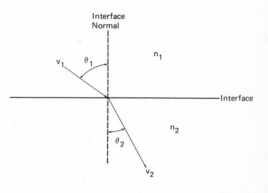

The refractive index, n , of a substance is simply the ratio of the velocity (speed) of light in a vacuum to the velocity of light in the substance, $n = c/v$. In vacuum the speed of light, $c \cong 3 \times 10^8$ m/sec, while in liquid or solid media such as oil or glass, light travels more slowly. Hence, the refractive index of a substance is always greater than one, and may be as large as 2.5. More commonly, n lies in the range of 1.3 to 1.7. The index of refraction of some common materials is given in Table 2.1.

Table 2.1. Index of refraction of some common materials for light with $\lambda = 5893$ Å (yellow light from sodium flame)

Diamond	2.42
Crown Glass	1.52
Fused Quartz	1.46
Water (at 20°C)	1.333
CO_2	1.00045
Air	1.00029
Vacuum	1.00000

From Snell's law it then follows that a light wave passing from air into a glass lens will be bent closer to the interface normal on entering and vice versa on exiting. This effect is illustrated for three lens configurations in Fig. 2.2.

The convergent or divergent properties of a lens are described with respect to incident light rays parallel to an axis passing through the center of the lens about which the lens is rotationally symmetric. It follows that lenses with greater curvature or larger refractive index will deflect light more. Such properties are associated with the

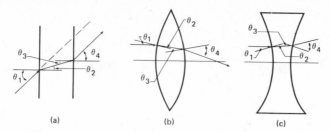

(a)   (b)   (c)

Fig. 2.2. (a) A plane parallel sided lens simply shifts the incident light waves; $\theta_1 = \theta_4 > \theta_2 = \theta_3$, (b) a convex, or positive, lens acts to converge the light waves; $\theta_1 > \theta_2$, $\theta_4 > \theta_3$, and (c) a concave, or negative, lens acts to diverge the incident light waves; $\theta_1 > \theta_2$, $\theta_4 > \theta_3$.

"strength" of a lens. As will be discussed later, the concept of refractive index may be applied to electron optics in the case of both electrostatic and magnetic lenses.

Cardinal Elements

Consider a bundle of light rays parallel to the axis of rotational symmetry incident on a converging lens as shown in Fig. 2.3.

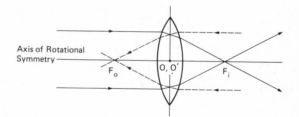

Fig. 2.3. Geometrical construction used to determine image cardinal elements.

The rays are converged by the action of the lens and pass through the point, F_i, lying on the axis. This point is called the <u>image focal point</u> of the lens and is one of the cardinal elements. All rays parallel to the lens axis will pass through this point.

The intersection of the incident parallel rays with the extended exit rays passing through F_i forms a plane perpendicular to the lens axis, intersecting this axis at O. This plane is called the image principal plane and the point at O is the image principal point. The <u>image principal point</u> is another cardinal element of the lens.

A thi<u>rd</u>, and most familiar, cardinal element is the length described by $\overline{OF_i}$. This is the <u>image focal length</u>, f_i, and is the distance from the image principal point to the image focal point.

A similar construction for a bundle of parallel rays incident from the opposite side of the lens would define the <u>object focal point</u>, F_0; the <u>object principal point</u>, O^1, and the <u>object focal length</u>, $O^1F_0 = f_0$.

In the case drawn in Fig. 2.3, the image principal point coincides with the object principal point, $O = O^1$. When this occurs we have what is called a thin lens. On the other hand, a more common construction especially in electron lenses is the thick lens, where the image and object principal planes do not coincide. The geometrical construction for a thick lens is shown in Fig. 2.4.

Fig. 2.4. Cardinal elements for a thick lens showing image and object focal points, F_i and F_0, principal points O and O^1 and focal lengths f and f_0 respectively. Note the principal plane is determined by the intersection of the extended incident and existing rays.

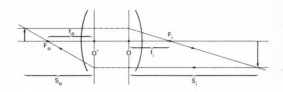

Real and Virtual Images

If an object is placed in front of a converging lens in a plane intersecting the axis at a point, A, beyond the object focal point, F_0, a real image will be formed after the lens in a plane intersecting the axis at point B beyond the image focal point F_0. This is illustrated in Fig. 2.5.

Fig. 2.5. Geometrical construction showing the formation of a real image by a thin lens. Ray 1 leaves from Q parallel to the axis and is refracted at the image principal plane through the image focal point, F_i. Ray 2 passes through the object focal point, F_0, and is refracted parallel to the axis at the object principal plane. Ray 3 passes undeviated through the center of the lens.

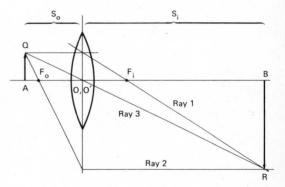

By convention, rays are always considered to travel from left to right or top to bottom on any diagram. The object is always in the space to the left or above the lens and is described as being "in front of" or "before" the lens. The image space is described as "after" or "behind" the lens. The coordinate origin is the center of the lens with "plus" to the right and above the horizontal axis describing the lens symmetry as in Figs. 2.3, 2.4 and 2.5. When using the graphical approach, the size of the lens doesn't matter as far as locating the image is concerned. It is only necessary to define the principal planes and allow them to extend as far as necessary to intersect the appropriate rays. This is demonstrated in Fig. 2.5 by the path taken by Ray 2. Obviously, even though the lens may not be large enough to let the diagrammed rays go through, in practice other rays will pass through and form the image. A third ray, the ray which passes through the center of

the lens is often convenient to use in diagramming the object-image relationship, especially when determining virtual images.

If the object is placed between the object focal point, F_0, and
principal point, O^1, as shown in Fig. 2.6, a different kind of image
will be formed. That is, if a screen were placed in the image plane at
B, no real image would be observed, but an apparent, or virtual, image
is formed. Virtual image formation is actually used in some electron
microscopes by the intermediate lens in the formation of low magnification images.

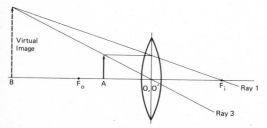

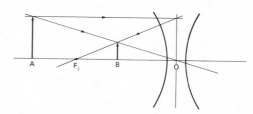

Fig. 2.6. Formation of a virtual
image by a converging lens. It is
required that the object, A, be located between F_0 and O^1 (=0). The
image is formed in the plane derived by the intersection of Ray 1
and Ray 3 (previously described in
Fig. 2.5, there is no Ray 2 possible here).

Fig. 2.7. Formation of a virtual
image at B from an object located at A. The image plane at B
is determined by the intersection
of the exit rays.

Virtual images are formed by diverging lenses since only diverging
rays appear to come from them. The formation of a virtual image by a
diverging lens is shown in Fig. 2.7. Since all electron lenses we deal
with are converging lenses, the diverging lens will not be discussed
further.

Lens Equations

There are a few important relationships for lenses which can be expressed in simple mathematical terms. These relationships are true for
light and electron optical lenses. Consider the refractive indices of
the media in front of and behind a lens to be n_0 and n_i, respectively,
then

$$\frac{n_o}{n_i} = \frac{f_o}{f_i}$$

where f_0 and f_i correspond to the object and image focal lengths. It
can also be shown that

$$\frac{n_o}{s_o} + \frac{n_i}{s_i} = \frac{n_o}{f_o} = \frac{n_i}{f_i}$$

where s_0 and s_i are given in Figs. 2.4 and 2.5. Here, s_0 is the distance O^1A between the object principal point and the object; and s_i is the distance OB between the image principal point and the image. Since in most instances $n_0 = n_i$, we have $f_0 = f_i = f$ and the lens equation becomes

$$\frac{1}{s_o} + \frac{1}{s_i} = \frac{1}{f}$$

The magnification, M, of the object is given by

$$M = \frac{Y_i}{Y_o} = -\frac{s_i}{s_o}$$

where Y_i and Y_0 are the height of the image and object, respectively. Sign convention here allows that for real image formation the image is inverted. Thus if Y_0 is positive, Y_i will be negative; while both s_i and s_0 are positive numbers. In the more general case where $f_0 = f_i$, the expression for transverse magnification becomes

$$M = \frac{Y_i}{Y_o} = \frac{f_o}{s_o - f_o} = \frac{s_i - f_i}{f_i} = -\frac{s_i}{s_o} \frac{f_o}{f_i}$$

Paraxial Rays

Thus far all the discussion of image formation has assumed an idealized case, that is a lens with no abberations which maps the object in a point-to-point correspondence on to the image. More specifically, we have ignored second and higher orders of inclination of the illumination with respect to the optical axis. Mathematically, if the angle of inclination is β then we may replace the tangent β by its sine or its arc, i.e., in the series expansion of $\sin \beta$

$$\sin\beta = \beta - \frac{\beta^3}{3!} + \frac{\beta^5}{5!} - \cdots$$

the higher order terms may be neglected and the approximation $\sin \beta \cong \beta$ used. (Note, β is in radians, but $\sin \beta = \sin \theta$ when θ is in degrees.) Rays intercepting an object point through a range of β satisfying this approximation are called paraxial rays and the image formed by these rays is called the Gaussian image. The study of optics neglecting the second and higher order terms is often referred to as: the ideal case, paraxial ray case, stigmatic imaging, first order theory or Gaussian imaging. Image formation by a bundle of incident paraxial rays is shown in Fig. 2.8. Here all the paraxial rays emerging from any point in the object at A will again pass through a single point in the image at B. That is, each point in the image is conjugate to a point in the object. For example, P is conjugate to P_0, A is conjugate to B and the image plane is said to be conjugate to the object plane.

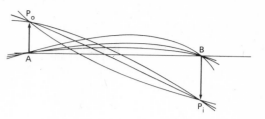

Fig. 2.8. Gaussian image formation by paraxial rays emerging from P_0 all arriving at the single point P_i. The image at B is conjugate to the object at A.

2.3 ELECTROSTATIC LENSES

Refraction

The foregoing discussion applied in particular to light optics and glass lenses, but for the most part can be applied equally as well to electron optics. The concepts of image formation and cardinal elements are identical in electron and light optics. And the basic laws of refraction (Snell's law) and rectilinear propagation under constant refractive index are the same. The only problem is to translate these basic concepts from light optics into electron optics.

An electron of charge -e experiences a force $\underline{F} = -e\underline{E}$ when exposed to an electrostatic field $\underline{E}$. (The designation $\underline{E}$ means the field strength is a vector, that is, has magnitude and direction.) In more convenient terms, the field strength is described in terms of the change of scalar potential, ϕ, over incremental distance s, i.e., $\underline{E} = -\delta\phi/\delta s$, where δs is in the direction of $\underline{E}$ and may be written in cartesian coordinates as $\delta s^2 = \delta x^2 + \delta y^2 + \delta z^2$. Thus the work done by the force when an electron moves through the electrostatic field from point A to B is given by

$$W = \sum_{A}^{B} \underline{F}\delta s = e\sum_{A}^{B} \frac{\delta\phi}{\delta s} \cdot \delta s = e\phi_B - e\phi_A$$

and the result is independent of the path but depends only on the potential at the end point. Conservation of energy also requires that this change in potential energy of the electron in going from point A to B is equal to the change in kinetic energy, i.e.,

$$e\phi_B - e\phi_A = \tfrac{1}{2} mv_B^2 - \tfrac{1}{2} mv_A^2$$

where m is the mass of the electron and v the velocity. If point A is at zero potential with the electron at rest ($v_A = 0$), then

$$\tfrac{1}{2} mv^2 = e\phi$$

thus

$$v = 2\sqrt{\frac{e}{m} \phi} = k\sqrt{\phi}$$

assuming we do not require a relativistic correction for the electron mass. In electron (particle) physics the index of refraction, n, is directly proportional to the velocity. (The inverse relationship was shown for light waves.) Thus,

$$n\alpha \quad v\alpha \quad k\sqrt{\phi}$$

and the electron equivalent of Snell's law, known as Bethe's law of refraction, becomes

$$\frac{\sin\theta_1}{\sin\theta_2} = \frac{n_2}{n_1} = \frac{v_2}{v_1} = \sqrt{\frac{\phi_2}{\phi_1}}$$

where v is the potential expressed in the more familiar units of practical volts. Bethe's law of refraction tells us that as the refractive index is increased (i.e., the potential increased), the electrons will be accelerated toward the higher potential, which is exactly what we expect.

Action of Electrostatic Lenses

As we have just shown, an electron moving from one region of electrostatic potential to another experiences refraction in a manner similar to a light wave encountering a glass lens. However, the change in potential, or refractive index, in an electrostatic lens is continuous in space compared to the abrupt change that occurs at the interface of a glass lens. It thus becomes necessary to determine the potential distribution within the electrostatic lens in order to trace the trajectories of the electrons. Unfortunately, the mathematics are very complex and complete analytical solutions are found for only the most simple electrode configurations. Instead, experimental methods are usually employed to determine the equipotentials. Further discussion of this topic is beyond the scope of this discussion. We can, however, approximate the path of an electron through a simple potential distribution by imposing a discontinuous change in refractive index as we encounter each equipotential line. This would be similar to having a group of glass lenses back to back with continuously increasing or decreasing curvature. Consider the case of two coaxial cylinders of the same diameter at different potentials as shown in Fig. 2.9. The equipotential lines are indicated by the dotted curves and the proportionate change in potential across the lens is indicated at a few increments. The electrostatic field would be normal to these equipotential lines.

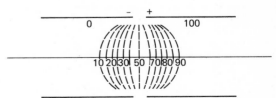

Fig. 2.9. Schematic of coaxial electrostatic lens cylinders showing equipotential lines.

An electron entering the lens from the left, at the lower potential, will be refracted toward the axis at each convex equipotential and away from the axis at each concave equipotential as it proceeds past the lens center to the right. The opposite will be true if the electron were to enter the lens from the right at the higher potential and proceed toward the lower potential. Here the convex equipotential lines diverge the electron and convex lines converge the electron. The net effect for any electrostatic lens, however, will be to act as a converging lens since an electron is accelerated by the longitudinal (parallel to the axis) component of the potential as it proceeds toward a higher potential; or decelerated if it is proceeding toward a lower potential. Consequently, the electron always spends more time in regions where there is a convergent effect than in regions where there is a divergent effect.

The cardinal elements of an electrostatic lens are defined as previously discussed for a glass lens. This is shown in Fig. 2.10 for a cylinder lens and Fig. 2.11 for a unipotential lens.

Notice from the figure that the lens action is different depending on the direction of the incident electron with respect to the direction of the potential gradient. As can be seen, the image and object focal

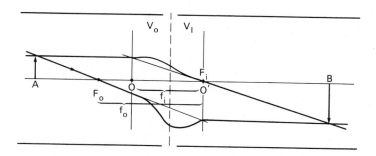

Fig. 2.10. Schematic of an electrostatic cylinder lens with $V_0 > V_I$ showing cardinal elements of the lens as previously defined.

lengths are different. The relevant lens equations from which one can obtain the focal lengths and image magnification are

$$\frac{f_o}{s_o} + \frac{f_i}{s_i} = 1$$

$$\frac{Y_i}{Y_o} = - \frac{f_o}{f_i} \frac{s_i}{s_o}$$

and

$$\frac{f_i}{f_o} = - \sqrt{\frac{\phi_i}{\phi_o}} = - \sqrt{\frac{v_i}{v_o}}$$

This last expression is the Lagrange-Helmholtz relation and is derivable from the relation between refractive index and potential shown previously.

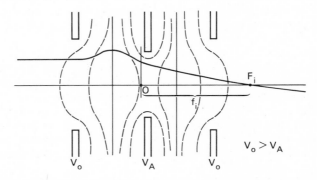

Fig. 2.11. Unipotential or Einsel lens showing typical electron trajectory and image cardinal elements. This lens has a symmetrical potential gradient, thus the electron trajectory through the lens is independent of the direction of incidence.

Types of Electrostatic Lenses

(a) Single Aperture Lens

Probably the simplest type of electrostatic lens is the single aperture lens which consists of a single aperture separating two regions of different but uniform fields.

(b) Cylinder Lens

Another straight-forward lens is the lens shown in Figs. 2.9 and 2.10 consisting of two coaxial cylinders of equal diameter. In these lenses the potentials in object and image space are constant and unequal. Cylinder lenses are used in cathode-ray tubes of various kinds.

(c) Unipotential or Einsel Lens

This lens is the most important electrostatic lens used in electron microscopy aside from the electron gun, which is an electrostatic accelerating lens. Electron guns will be discussed later. An Einsel lens generally consists of three circular apertures with the potential on the outer two electrodes the same, usually at the same potential as the anode of the gun. The central electrode is insulated from the outer electrodes and is held at a different potential, often the same as the filament which is at a high negative voltage. An example of an einsel lens is shown in Fig. 2.11.

Electrostatic lenses are not used very frequently in electron microscopes because of the hazards associated with the use of the very high voltages and the fact that even the best electrostatic lenses are inferior to the best magnetic lenses. Electrostatic lenses have been used for electron diffraction cameras though, since the electrons do not spiral about the optical axis as they do in magnetic lenses and therefore the diffraction pattern does not rotate with respect to the image.

2.4 MAGNETIC LENSES

Action of a Homogeneous Field

The action of a homogeneous magnetic field on an electron is quite different than previously discussed for either the electrostatic field or the light optical lenses. In the latter two cases the velocity of the electron or light wave was changed in the direction of propagation by the action of the refractive medium. When an electron encounters a magnetic field, however, the magnitude of its velocity is unchanged but its direction is changed to be normal to both the field direction, $\underline{B}$, and the original electron velocity, $\underline{v}$, In mathematical terms the electron experiences a force given by

$$\underline{F} = - e \ (\underline{v} \times \underline{B})$$

where the x indicates the vector or cross product between $\underline{v}$ and $\underline{B}$. Compare with the force from an electrostatic field $\underline{E}$ given previously as $\underline{F} = -e\underline{E}$ which acts in the direction of the field. The magnitude of the magnetic force is

$$F = B \ e \ v \ \sin\theta$$

where θ is the angle between $\underline{v}$ and $\underline{B}$. Physically, the action can be described by the "right hand rule" where the thumb, first and second fingers of the right hand are held at right angles to each other approximating a rectangular coordinate system as shown in Fig. 2.12. The direction of the Field is along the direction of the Fore-finger, the electron velocity, or Speed, in the direction of the Second finger and the resultant force on the electron, or Thrust, in the direction of the Thumb.

Fig. 2.12. Right hand coordinate system using thumb, first and second fingers on the right hand.

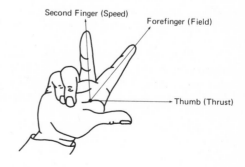

Second Finger (Speed)

Forefinger (Field)

Thumb (Thrust)

If the field were uniform and infinite the electron would follow a circular path with radius

$$R = {}^{mv}/_{eB} = \frac{1}{B} \left(\frac{2m_o V_r}{e} \right)^{\frac{1}{2}}$$

where $v = 2e/m^0$ V as shown before. V is the relativistically corrected accelerating voltage given by

$$v_r \approx V_o (1 - 10^{-6} V_o)$$

and m_0 is the rest mass of the electron $= 9.1 \times 10^{-31}$ kg.

If the electron was injected into this homogeneous field at some angle, Θ, to the axis of the magnetic field, the velocity could be broken into two components; one a longitudinal component, $\underline{v}_1$, along the field axis and the other, $\underline{v}$, a transverse component perpendicular to the field. As $\underline{v}_1$ is parallel to $\underline{B}$, $\underline{v}_1$ x $\underline{B}$ = 0 and the longitudinal component will be unchanged. The transverse component will describe a circle of radius $r = mv /eB$ perpendicular to the field direction. The resultant path of the electron will therefore be a helix of fixed radius proceeding in the direction of the field.

Action of an Inhomogeneous Field

In all lens configurations encountered in an electron microscope,
the magnetic field distribution is
inhomogeneous, that is, varying in
space. Since almost all magnetic
lenses are rotationally symmetrical
it is convenient to describe the
field in a cylindrical coordinate
system, r, Θ and z as shown in Fig.
2.13, referenced to a rectangular
coordinate system.

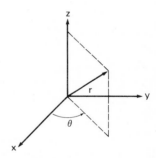

Fig. 2.13. Cylindrical coordinates r, Θ and z.

Footnote: A reminder on magnetic units and relationships:
  i) $\underline{B}$ is the magnetic induction or flux density and in the MKS system has units of Webers/meter^2 or teslas.
  ii) $\underline{H}$ is the magnetic field intensity and has units amperes/meter
in the MKS system.
  iii) In all systems $\underline{B} = \mu\underline{H}$, where μ is the magnetic permeability.
In vacuum $\mu = \mu_0 = 4\pi \times 10^{-7}$ Wb/amp·meter in the MKS units.
  iv) For the more familiar CGS system, B is in gauss, H is in oer-
steds and $\mu = 1$ in vacuum, thus $\underline{B} = \underline{H}$ and the two are used interchangeably.
  v)   The conversion units from MKS to CGS is 1 Wb/m^2 (= 1 Tesla) =
10^4gauss.

The simplest form of a magnetic lens is a single turn of wire forming a circular conductor of radius R, centered on z_0 and carrying a current I as shown in Fig. 2.14(a). The magnetic field at any point, z, on the axis is given by

$$B(z) = \frac{\mu_0 I}{2[1 + (z/R)^2]^{\frac{3}{2}}}$$

which reduces to

$$B(z_0) = \frac{\mu_0 I}{2R}$$

at the center of the wire.

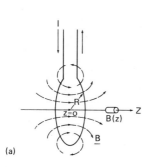

(a)

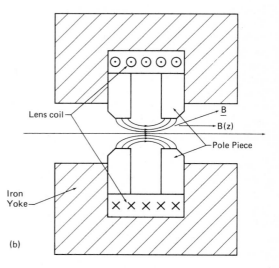

(b)

Fig. 2.14. Magnetic field distribution in (a) a single turn of wire and (b) a magnetic lens with pole pieces.

To a first approximation, if there were N closely wound turns of radius R, instead of just a single turn, the field at z_0 would be

$$B(z_0) = \frac{\mu_0 NI}{2R} \quad (Wb / m^2)$$

(MKS units are used throughout, thus $\mu_0 = 4\pi \times 10^{-7}$ Wb/amp·m, B(z) is Wb/m^2, I is amperes, and all dimensions are in meters.)

A complete magnetic lens is made if the coil is surrounded by a yoke and pole pieces made of ferromagnetic materials, such as iron, as shown in Fig. 2.14(b). The magnetic field produced in the gap of this lens is the sum of two components, (i) the field produced by the coil, B_c, and (ii) the field produced by the ferromagnetic pole pieces, B_{pp}. Usually, $B_{pp} \gg B_c$, so that the maximum field in the gap is limited by the saturation magnetization of the pole pieces material to values on the order of 2.0 to 2.5 Wb/m^2 (20 to 25 k gauss).

The field distribution in the gap may be represented by a longitudinal component B(z) and a radial component B(r). (Note: B(z) denotes the magnetic field distribution along the z axis only, e.g., B(z) = B(r=0,z).) The variation in these components along the lens axis, z, is depicted in Fig. 2.15 for a lens with pole pieces having bore = D and gap = S.

The action of a magnetic lens on an electron may be easily described in terms of these two components of the magnetic field. If an electron with velocity parallel to the z-axis enters the field, it initially encounters only the radial component, B(r). Applying the right hand rule we find the force on the electron will cause it to assume a trajectory in the θ direction. This follows from $\underline{F} = -e\ \underline{v}_2 \times \underline{B}_r = \underline{F}_\theta$. Note that if there were no radial component of the magnetic field in the lens, the field would be homogeneous and the electron with velocity in the z-direction would be completely unaffected by the magnetic lens. The radial component of the field causes the electron to change direction such that its velocity is now normal to $\underline{B}_2$ and it now experiences a force $\underline{F} = -e\ \underline{v}_\theta \times \underline{B}_2 = \underline{F}_r$, which is radially directed toward the z-axis. The combined components of the field thus cause the electron to spiral toward the z-axis until it crosses it at some point and then spiral back out again as depicted in Fig. 2.16. In other words, a bundle of parallel electrons will be focussed to a point. This focussing action is caused by the B(z) component of the field, thus B(z) is the component most frequently discussed. It is also the component that is most easily measured.

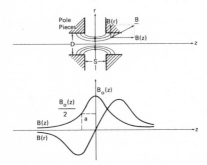

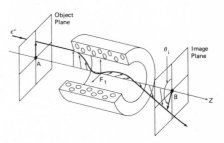

Fig. 15. Field distribution B(z) and B(r) in a symmetrical magnetic lens with bore = D and gap = S.

Fig. 16. Image formation with a magnetic lens depicting electron rotation about the z - axis.

Paraxial Ray Equations

The radial force on the electrons, $\underline{F}_r$, can be calculated from the right-hand rule to be

$$\underline{F}_r = -\ e\ \underline{v}_\theta \times \underline{B}\ (z)$$

Footnote. A word on semantics, as all magnetic lenses considered are axially symmetric, the description of a symmetrical, or asymmetrical, magnetic lens refers to whether or not the bores of the two pole pieces are equal. Also, since a lens is axially symmetric, B(θ) = 0.

and after some mathematical substitution and rearranging becomes

$$\underline{F}_r = - \left(\frac{e^2}{4m}\right) B^2(z) \; \underline{r}$$

This equation says that the radial force drives the electrons toward the z-axis (because of the negative sign in front of the expression) and is directly proportional to the distance, r, of the electron from the axis; which is after all, the principle of a focussing lens. The equation also shows that all magnetic lenses are converging lenses because the equation is independent of the sign of B(z) since the field is contained in $B^2(z)$, e.g. $(-B)^2 = (+B)^2 = B^2$.

Since force is equal to mass times acceleration, F = ma, and the acceleration is defined as the second derivative of the spatial coordinate, r, with respect to time, e.g., $a = d^2r/dt^2$, we have

$$\underline{F}_r = ma_r = m \frac{d^2r}{dt^2} = - \left(\frac{e^2}{4m}\right) B^2 \; r$$

or, rearranging

$$\frac{d^2r}{dt^2} + \frac{e^2}{4m^2} B^2 \; r = 0$$

gives the equation of motion of an electron in a plane normal to the z-axis. It can also be shown that

$$\frac{dz}{dt} = \left(\frac{2eV}{m}\right)^{\frac{1}{2}}$$

along the z-axis. Combining these two equations under the assumption of paraxial rays and that r is not too far from the z-axis gives the "paraxial ray equation" which describes the motion of an electron in a rotating meridional plane

$$\frac{d^2r}{dz^2} + \frac{e}{8m_o V_r} \; B^2(z) \; r = 0$$

A meridional plane is determined by the original location of the electron in field free space and the z-axis. The function d^2r/dz^2 may be thought of as the change in slope of the electron path with respect to the z-axis. Rewriting the equation as

$$\frac{d^2r}{dz^2} = - \frac{e}{8m_o} \cdot \frac{B^2(z)}{V_r} \; r$$

one can see that the greater $B^2(z)/V_r$, the faster the electron is bent toward the axis, and vice versa.

The rotation of the electron about the z-axis as it proceeds in the z-direction may be written as

$$\frac{d\theta}{dz} + \left(\frac{e}{8m_o V_r}\right)^{\frac{1}{2}} B(z) = 0$$

This equation may be integrated to give the total rotation $\Delta\Theta$ of the electron as it moves from point z_0 to z_1:

$$\Delta\Theta = - \int_{Z_0}^{Z_1} \left(\frac{e}{8m_0V_r} \right)^{\frac{1}{2}} B(z) \ dz$$

The notation $\int_{Z0}^{Z1} B(z)dz$ means the sum of all the increments $B(z)dz$ from $z = z_0$ to $z = z_1$.

The important point is that $\Delta\Theta$ is a function of $B(z)$, so that if $B(z)$ is of the opposite sign, $\Delta\Theta$ will be in the opposite direction. This property is often used in the design of electron microscopes by incorporating sequential lenses having fields of opposite sign to nullify the image rotation.

Bell Shaped Field

An analytical solution to the paraxial ray equation is obtained only if the form of the axial field distribution $B(z)$ is known. Following the form of the equation describing the field distribution around a single turn of wire shown previously, the field distribution

$$B(z) = \frac{B_o}{1 + \left(\dfrac{z}{a}\right)^2}$$

known as Glaser's bell-shaped field is found to be a good approximation to the field distribution in the gap of a symmetrical magnetic lens. In the equation, a is half the FWHM of the field distribution $B(z)$ at $B_0(z)/2$, and $B_0(z)$ is the maximum field strength along the axis at $z = 0$. Substituting this bell-shaped field distribution into the paraxial ray equation yields

$$\frac{d^2r}{dz^2} + \frac{eB_o^2}{8m_oV_r} \frac{r}{\left[1 + \left(\dfrac{z}{a}\right)^2\right]^2} = 0$$

The conventional method for finding solutions to such equations is to make substitutions for the variables, r and z, which puts the equation into a form which may be solved analytically. For example, if we make the substitutions $x = z/a = \cot \phi$ and $y = r/a$ the equation becomes

$$\frac{d^2y}{d\phi^2} + 2 \cot \phi \ \frac{dy}{d\phi} + k^2y = 0$$

where $k^2 = (e B_o^2/8m_0V_o)a^2$, is a frequently quoted lens parameter. One solution to this equation is

$$y = \frac{1}{\omega} (\sin \omega\phi/\sin \phi)$$

where $\omega^2 = k^2 + 1$. You can see that y is simply a reduced parameter relating to the radial distance, r, of the electron from the z-axis, that is, $y = r/a$; $\cot \phi = z/a$ is a function converting linear distance into angular measure as shown in Fig. 2.17. "a" is a straight line parallel to the z-axis.

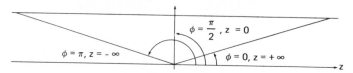

Fig. 2.17. Shows relationship between ϕ and z.

The slope of y at any point is given by

$$y' = \frac{dy}{dx} = \frac{dy}{d\phi}\frac{d\phi}{dz} = \frac{1}{\omega a} \ (\sin \omega\phi \cos \phi - \omega \sin \phi \cos \omega\phi)$$

Lens Excitation Parameters ω and k^2

It is instructive to determine under what conditions an incident beam of electrons parallel to the axis exits the field also parallel to the z-axis, e.g., calculate under what conditions $y' = 0$ for $\phi = 0$ and π.

Calculation of y' for $\phi = 0$ ($z = +\infty$) shows $y' = 0$ identically. Thus we will consider this the incident beam. It remains to calculate under what conditions $y' = 0$ for $\phi = \pi$ ($z = -\infty$). It is easily seen this yields $y' = 1/\omega a \sin \omega\pi = 0$ which requires $\sin \omega\pi = 0$. From trigonometry this condition is satisfied when ω = integer, e.g., 0, 1, 2, 3, ...

For a beam to enter and exit the lens parallel to the axis requires that it either be unaffected by the field or cross the axis. Actually, the beam may cross the axis a number of times as shown below. Remembering $\omega^2 = k^2 + 1$ where $k^2 = (eB_0^2/8m_0V_r) \ a^2$ we can tabulate the number of axial crossings of the electron trajectory for various values of ω and thus k^2. This is done in Table 2.2.

Obviously, the cases $\omega = 0$, 1 are uninteresting. The case $\omega = 2$, $k^2 = 3$, is the condition under which the beam crosses the z-axis one

Table 2.2. Relationship between ω and k^2 and the number of beam cross-overs of the z-axis

ω	ω^2	k^2	comment
0	0	-1	meaningless
1	1	0	no field, beam is undeviated
2	4	3	beam crosses z-axis one time
3	9	8	beam crosses z-axis two times
4	16	15	beam crosses z-axis three times

time which, by symmetry arguments, must occur in the center of the lens. This operating condition is called the telefocus condition. It may be observed in the image mode of a TEM if the objective lens has sufficient "excitation," Cl strong and C2 off. Almost all S(TEM) microscopes have a sufficiently strong objective lens for this operation, but only a few, if any, TEM microscopes do. And most conventional objective lenses operate under the condition $k^2 < 3$.

The case $\omega = 3$, $k^2 = 8$ provides for two axial cross-overs of the electron beam, each at the same distance from the lens center. Some of the present day S(TEM) microscopes are using lenses with this excitation and locating the specimen near the first cross-over.

The case $\omega = 4$, $k^2 = 15$, provides three cross-overs of the z-axis. To achieve this would require such a high magnetic field, assuming reasonable focal properties are maintained, that such lenses have not been realized for commercial use. The table could be continued for $\omega = 5$, 6, 7, ..., etc., with an increase in the number of cross-overs, but such a discussion would be purely academic.

For non-integer values of ω the beam crosses the z-axis but does not exit parallel to it. In the conventional lens $k^2 < 3$, so the beam crosses the optical axis one time. For $3 < k^2 < 8$ the beam crosses the axis two times and either may be used as a focal point for image formation, etc. As almost all TEM and SEM objective lenses operate at lens excitations such that $k^2 < 3$, we will restrict our discussion to lenses of this type. Some typical electron trajectories are drawn in Fig. 2.18 for various values of k^2.

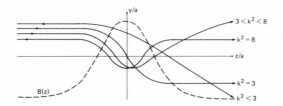

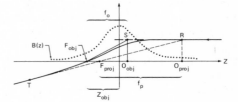

Fig. 2.18. Plot of electron trajectories for various lens excitation parameters

$$k^2 = \left(\frac{e\,B_o^{\,2}}{8m_o V_r}\right) a^2 \ .$$

Fig. 2.19. Typical electron trajectory in a bell-shaped magnetic field with excitation parameter $k^2 < 3$ depicting cardinal elements for projector and objective lens operation.

Cardinal Elements of Magnetic Lenses

From our discussion of lens cross-overs for various values of lens excitation, it is easily seen that the cross-over for all but the weakest lens will occur while the electron is still under the influence of the magnetic field. This is different than previously discussed for electrostatic lenses where the focal point is usually outside the

potential field of the lens. A typical electron trajectory in a magnetic lens operated such that $k^2 < 3$ (but not too weak) is shown in Fig.
2.19.

As one can see in the figure, there are two focal lengths, focal
points and principal planes defined for a single trajectory through the
lens. Because the trajectory intersects the z-axis while still in the
magnetic field, the path will continue to curve after passing the axis
such that the asymptote, RT, to the trajectory at a large distance from
the lens will not intersect the z-axis at the cross-over point. The intersection of this asymptote and the extended incident beam defines the
principal plate at O_{proj} and the distance $F_{proj} - O_{proj}$ the projector
focal length, f_p. These cardinal elements are those used to define the
lens properties for lenses used as magnification lenses, condenser
lenses, etc., and are defined as the general class of projector lenses.
The object for this lens configuration is outside or just inside the magnetic field.

If the object is placed well into the magnetic field, as is the
case in many modern electron microscope objective lenses, a different
set of cardinal elements is required to describe the lens behavior.
These cardinal elements are determined by the intersection of the beam
trajectory with the z-axis defining the focal point, F_{obj}. The intersection, s, of a tangent to F_{obj} and the extension of the incident beam
direction defines the principal plane at O_{obj} and the distance $O_{obj} F_{obj}$
the objective focal length, f_0.

Quantitatively, the cardinal elements can be described in terms of
the solution to the paraxial ray equation. The projector focal length
is given by

$$f_p = \frac{1}{y'}, = +\frac{a\omega}{\sin\omega\pi} = a\omega \csc\omega\pi$$

where y' is evaluated at $z = -\infty$ (or $\phi = \pi$). Obviously, a projector
focal length is undefined when the exit beam is parallel to the axis,
e.g., when $\omega = 2, 3, 4$, since $\sin\omega\pi = 0$ here.

The objective focal point is determined by the zero of the function
$y = (1 \omega a)(\sin \omega\phi/\sin\phi)$ e.g., where the electron beam crosses the axis.
This occurs when $\sin\omega\phi = 0$ or $\omega\phi = n\pi$, with n any integer. The location
of this focal point with respect to the lens center is obtained from
evaluating $z/a = \cot\phi$ at $\phi = n\pi/\omega$. Addressing the case n = 1 only, we
obtain $Z_{obj} = a \cot\pi/\omega$ as the distance of the objective focal point from
the center of the lens and $Z_{im} = -a \cot\pi/\omega$ as the corresponding distance
of the image focal point from the center of the lens. z_{obj} and z_{im} are
often called the midfocal points.

In an electron microscope objective lens, therefore, the diffraction pattern appears in the back focal plane (BFP) of the lens located
at z_{im} and the objective aperture is positioned there. The specimen is
usually situated in front of the objective midfocal point, z_{obj}.

The objective focal length f_0 as determined by the length $\overline{F_{obj} \, O_{obj}}$
is

$$\frac{1}{f_0} = - \frac{1}{y} \frac{dy}{dz} = - \frac{1}{a} \sin \frac{\pi}{\omega} \qquad \text{for the case n = 1}$$

or

$$f_0 = - a \csc \frac{\pi}{\omega}$$

Similarly, the image focal length is given by $f_i = a \csc \pi/\omega$. The minimum value of the focal length is obtained when $\sin \pi/\omega = 1$ or $\omega = 2$, corresponding to the telefocal lens excitation $k^2 = 3$. At $k^2 = 3$, $f_0 = - a$ and $f_i = a$ with $z_{obj} = -z_{im} = 0$, the midfocal points are located at the lens center $z = 0$. For this unique operating condition, the specimen is placed at the center of the lens, $z = 0$, and the BFP is located at $z = a$.

The location of the principal points with respect to the lens center is given by

$$z_{pi} = - a \cot \frac{\pi}{2\omega}$$

and

$$z_{po} = a \cot \frac{\pi}{2\omega}$$

for the image and object mid-principal points, respectively.

In Table 2.3 normalized values of these cardinal elements have been calculated for various lens excitation parameters, k^2, using the relationships derived above.

Table 2.3. Values of the cardinal elements with respect to lens center and the focal length (normalized to the half width of the magnetic field, a) as functions of k^2. The image cardinal elements are determined assuming the incident beam is comimg from $z = -\infty$.

k^2	Midprincipal Point $z_{po}/a = -z_{pi}/a$	Midfocal Point $z_{im}/a = -z_{obj}/a$	Focal Length f/a
1	0.496	0.761	1.257
1.5	0.651	0.442	1.093
2	0.782	0.248	1.020
2.5	0.897	0.109	1.006
3	1.000	0.	1.000
4	1.181	-0.167	1.014
5	1.340	-0.297	1.043
6	1.482	-0.403	1.079

Defining $l_i = s_i - f_i$ and $l_0 = s_0 - f_0$ where s_i and s_0 were defined in Figs. 2.4 and 2.5 we can show that

$$l_0 l_i = f_0 f_i$$

which is the electron optical equivalent to Newton's formulas in light optics. l_i represents the distance between the image and the image focal point, l_0 the corresponding distance for the object. In the usual case where $f_0 = -f_i$ we have $l_0 l_i = -f^2$. Newton's formula demonstrates that there will be an image formed in real space independent of whether the object is immersed in the lens field.

It can also be shown that the magnification is given by

$$M = -\frac{f_0}{l_0} = -\frac{l_i}{f_i}$$

The rotation of the image with respect to the object is found to be

$$\Delta\theta = \frac{n\ k\ \pi}{(k^2 + 1)^{\frac{1}{2}}} = \frac{n\ k\ \pi}{\omega}$$

Since $n \leq (k^2 + 1)^{1/2}$, the most frequently encountered case will be with $n = 1$, therefore $\Delta\theta = k\pi/\omega$. For example, if $k^2 = 1$, $\omega^2 = 2$ and $\Delta\theta \cong \pi/\sqrt{2} \cong 127°$ rotation.

Objective Lenses

The excitation of objective lenses found in most of the commercial electron microscopes is $k^2 < 3$. Thus the specimen is positioned above the lens and the focal point is below lens center. The electron ray path for such a lens was shown in Fig. 2.19. Since the specimen lies in front of lens center, the prefield of the lens is not strong enough to image a plane conjugate to the specimen in real space before the lens. Hence the plane conjugate to the specimen is virtual and lies behind the lens and, as a result, the illumination conditions at the specimen are defined by the second condenser lens of the microscope. This means that to obtain the nearly parallel illumination needed for highest resolution C2 must be strongly over focussed resulting in a large probe. Small probes may be obtained by forming the crossover of the C2 lens at the specimen plane, but only under conditions which result in a very convergent beam.

Spherical aberration, which shall be discussed shortly, essentially determines the resolving power of a lens. In the "normal" lens ($k^2 < 3$) discussed here, it has been found that the spherical aberration coefficient is somewhat greater than the focal length of the lens and therefore limits the ultimate resolution for a fixed geometry of lens pole pieces. Resolution can be improved, however, by immersing the specimen further into the field. In particular two types of lenses with specimens deeply immersed have been studied.

If the specimen is positioned at the exact center of the lens, the excitation is given by $k^2 = 3$. This corresponds to the telefocal condition previously discussed with parallel incident and exit rays. A lens in this configuration is known as the single field <u>condenser objective lens</u>. In this lens, developed by Riecke, the first half of the magnetic field lies in front of the specimen and acts like a short focal length condenser while the second half of the field lies behind the specimen and acts like an objective lens, thus the terminology condenser-objective. The prefield of this lens is sufficiently strong to image the plane conjugate to the specimen in real space before the lens, usually at the C2 aperture position. By using a special alignment procedure developed by Riecke, it is thus possible to form a demagnified image of the C2 aperture at the specimen plane. This provides a small probe with nearly parallel illumination.

By immersing the specimen even further into the lens field where $3 < k^2 < 8$, the electron trajectory will now have two crossovers. Such a lens is called a second zone lens. The specimen may be placed near the second crossover, but its position is not as critical as in the condenser objective lense of Riecke. Again the prefield of this lens is sufficiently strong to image the plane conjugate to the object in real space in front of the lens; however, the location of this conjugate plane is within the gap of the lens so it is not readily accessible and the procedure to obtain parallel illumination at the specimen closely resembles that used in a "normal" objective lens although smaller probes may be used. Second zone lenses have not been used frequently in microscopes because of the large excitation required to produce the magnetic field. They do provide a distinct advantage in microscopes combining STEM with TEM when the specimen is placed at the first crossover of the lens. Operating the lens under strong excitation provides a small convergent probe for STEM and a reduction in the excitation allows normal TEM operation.

Frequently, the lens excitation, k_1^2 is given as the ratio of the square of the number of ampere-turns, $(NI)^2$, to the relativistically accelerating voltage, V_r, that is

$$k_1^2 = \beta (NI)^2/V_r$$

The proportionality constant β is related to the lens geometry and field, and is given by

$$\beta = \frac{e\mu_o^2}{32m} \left(\frac{B_o}{B_p}\right)^2 \left(\frac{D}{S}\right)^2 = 0.0087 \left(\frac{B_o}{B_p}\right)^2 \left(\frac{D}{S}\right)^2$$

where B_0 is the maximum field in the gap, $B = \mu_0 NI/S$ is the magnetic flux density that would exist in the gap if $S \gg D$ and D and S are the pole piece bore and gap dimensions, respectively. Hence

$$k_1^2 = 0.0087 \left(\frac{B_o}{B_p}\right)^2 \left(\frac{D}{S}\right)^2 \frac{(NI)^2}{V_r}$$

The lens excitation, k_1^2 used here is that described by Liebmann et al. in their extensive discussion of lens properties and is related to the lens excitation k^2 used throughout this lecture by $k_1^2 = (R/a)^2 k^2$; k^2 is the parameter used by Glaser and most modern-day authors; R is the bore radius, $R = D/2$. Some computed values of β, B_0/B_p and R/a are given for a few typical values of S/D in Table 2.4.

Table 2.4. Values of β, B_o/B_p and R/a as $f(S/D)$.

S/D	β	B_o/B_p	R/a
1	0.0064	0.860	0.9099
2	0.0021	0.987	0.4975
3	0.0010	0.999	0.3332

The optical parameters of objective and projector lenses have been calculated under a variety of geometrical conditions, S/D, and excitation conditions, k^2 or k_1^2. These data are available in many of the references given and the serious reader is advised to look there.

Lens Aberrations and Defects

i)   Spherical Aberration

Thus far we have assumed paraxial illumination and Gaussian imaging, that is a conjugate point-to-point correspondence between the object and the image. Such is not the case in real optical systems and performance is usually determined by the minimization of aberrations. Figure 2.20 illustrates one of the aberrations.

Fig. 2.20. Paraxial ray P A Q and non-paraxial ray P B Q_1 illustrating spherical aberration in the Gaussian image plane.

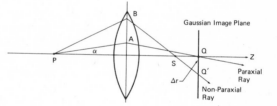

Fig. 2.21. Distortion in a magnetic lens (a) object, (b) pincushion distortion, (c) barrel distortion, (d) spiral distortion.

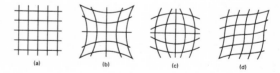

The image of object point P will be formed at point Q in the Gaussian image plane for all paraxial rays P A Q. However, if the rays are not paraxial, that is if $\sin \alpha \neq \alpha$, they will be bent more at the periphery of the lens. The result is that the image point Q will be displaced a distance Δr to Q', as shown in the figure for the nonparaxial ray P B Q'. Point P will thus have an apparent radius $\Delta r/M$, where M is the image magnification. This effect is called spherical aberration and is one of the principle factors limiting the resolution of an electron microscope.

The addition of a third order term, $\alpha^3/3!$ is usually sufficient to allow the approximation $\sin\alpha \cong \alpha - \alpha^3/3!$ for magnetic lenses. The image distortion, Δr, is related to α^3 through $\Delta r = C \alpha^3$ where C_S is defined as the coefficient of spherical aberration. Compared to the objective focal length, $C_S > f$ for k^2 small and $C_S < f$ for $k^2 \geqq 3$ approaching a constant value for large k^2.

ii) Pincushion, Barrel and Spiral Distortion

When spherical aberration is present in a lens, e.g., $\Delta r = C_S \alpha^3$, the image magnification will vary in proportion to the cube of the distance of the image point from the axis. The result is a distorted image. There are two kinds of distortion, pincushion and barrel, which are rather common in electron microscopes, especially at low magnification; and a third kind, spiral, which is not frequently seen. Pincushion distortion occurs when the magnification of the image increases with the distance of the image point from the center as shown in Fig. 2.21(b). Barrel distortion is the opposite of pincushion and occurs when the magnification decreases with distance from the center as shown in Fig. 2.21(c). Spiral distortion occurs when the angular rotation of an image point depends on the distance of the point from the axis. The images are sigmoidal shaped as shown in Fig. 2.21(d).

iii) Astigmatism

Even with the high technology of today it has not been possible to produce a magnetic lens with perfect axial symmetry. The result is that the image plane for objects lying in one direction will be different than the image plane for objects lying in another direction. This means, for example, that the " — " in an H will be focussed in one plane and the " | | " in another. Hence, there is no sharp image plane, only a plane of least confusion between the two sharply focussed images. Lenses with this effect are said to be stigmatic; however, devices called stigmators used in all modern electron microscopes can completely compensate for this lens imperfection. Stigmators are usually used in both the objective lens and the illumination system and sometimes in the magnification system of lenses.

iv) Chromatic Aberration

As a consequence of refraction by the electric or magnetic fields of a lens, electrons leaving a point, P, with different velocities will be brought to a focus at different points, Q and Q', in accordance with their velocity as shown in Fig. 2.22, e.g., the greater the velocity the longer the focal length.

Fig. 2.22. Electrons leaving point P with different velocities, dut to potentials V and V + ΔV, will be brought to a focus at Q and Q', respectively. This effect is called chromatic aberration.

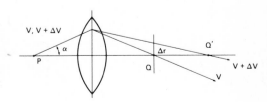

The result is, as in the case with spherical aberration, that the point P will be imaged as a disc of radius $\Delta r/M$. By geometrical arguments it can be shown that

$$\Delta r = C_c \alpha \frac{\Delta V}{V}$$

where C_c is defined as the coefficient of chromatic aberration. Its numerical values are similar to C_S and are usually on the order of f (a few mm.). Differences in electron velocity or electron potential, V, may come from many sources. Among the most prevalent are fluctuations of the accelerating voltage and the inelastic scattering within the specimen. Chromatic aberration usually limits the resolution in thick specimens.

v)   Boersch Effect

Another, although minor, source of the spread in electron velocities in the beam comes from the Boersch effect. Laws of physics provide that particles having the same charge repel, thus the electrons may interact with each other creating transverse velocity components and thereby changing the longitudinal components. This effect is only noticeable when the electrons are at a very high current density crossover, such as at the gun. For a heated tungsten filament the Boersch effect gives rise to an overall energy spread of the beam of 1 to 2 eV, while for a cold field emitter the energy spread is only a few tenths of an eV.

Special Magnetic Lenses

There are so many different designs of magnetic lenses that it is impossible to discuss them all here. However, a brief description of a few of the most interesting will be given.

i)   Quadrapole and Octapole Lenses

Quadrapole lenses are formed by arranging electrodes of alternate potential or magnetic poles of opposite polarity as shown in Fig. 2.23 in a plane normal to the electron trajectory.

Since the field lines are normal to the electron velocity, much stronger focussing action may be achieved with must less power expenditure. Since the action of the lens is to form a line image from a point

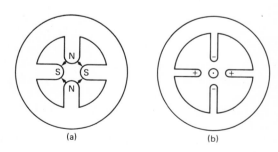

Fig. 23. Schematic of (a) magnetic and (b) electrostatic quadrapole lenses.

(a)   (b)

source, however, a single quadrapole is not useful as an imaging lens. Crossing quadrapoles of opposite polarity will yield point focus, but other properties are unattractive.

Octapoles are similar to quadrapoles but have eight electrodes or poles of opposite sign. Combining octapoles and quadrapoles it is possible to design a lens system free of spherical aberration, but mechanical alignment factors prohibit their practical use.

ii)   Pancake and Snorkel Lenses

In the quest to attain greater resolution, the requirements of low aberration lenses have dictated pole pieces with extremely small bore and gap. The problem with this approach is that the gap becomes increasingly small, difficulties in the specimen manipulation become significant. Thus practical limits are established on S and D depending on the degree of specimen manipulation required. To overcome this problem, Mulvey has proposed two lenses of unconventional design that may offer new hope for overcoming these problems.

The first is the "pancake" lens, Fig. 2.24, which consists of a partly shielded flat or helical coil producing a very high magnetic field in front of the lens. Calculations incidate that the axial field distribution, B(z), can have an appreciably lower C than that of a comparable standard pole piece lens. Because of the physical geometry of the lens there is ample space above the lens at the field maxima for specimen manipulation.

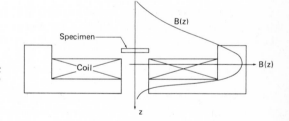

Fig. 24. The geometry and magnetic field distribution of a "pancake" lens.

The second is the "snorkel" lens, which consists of a central magnetic core and outer shroud with an enclosed coil that can be excited to a high current density. The lens, as shown in Fig. 2.25, may be thought of as a modified pancake lens, modified by the addition of the central iron core in the general shape of a conical snout with a hole in the center.

The field distribution in this lens is such that the electrons may be focussed in front of the lens and thus the specimen located there allowing easy access for study. Munro has calculated the field distribution in the snorkel lens and found that, to a good approximation, snorkel lenses may be regarded as "half-pancake" lenses. He assumed a solid iron core as might be used for an SEM rather than one with a hole in it for transmission microscopy as pictured here.

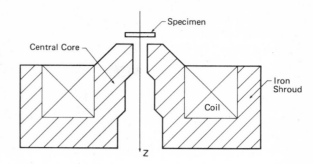

Fig. 25. Snorkel lens.

2.5 PRISM OPTICS

Magnetic Sectors

As previously discussed, the magnetic force on a moving electron is described by the expression

$$\underline{F}_e = e \, \underline{v} \times \underline{B}$$

and the electron path will be normal to both $\underline{v}$ and $\underline{B}$. In a uniform magnetic field, $\underline{B}$, the electron will describe a circular path of radius, r, which depends on the velocity or energy of the incident electron.

$$r = \frac{m_o v}{eB} = \frac{1}{B} \left(\frac{2m_o}{e} \cdot V_r \right)^{\frac{1}{2}}$$

If the energy of the electron is changed by an amount ΔV, the radius of the electron trajectory will be correspondingly changed to

$$r \pm \Delta r = \frac{1}{B} \left(\frac{2m_o}{e} \right)^{\frac{1}{2}} (V_r \pm \Delta V_r)^{\frac{1}{2}} = \frac{m_o}{eB} (v \pm \Delta v)$$

Consider a magnet consisting of two identical poles in the form of a circular sector of angle ϕ, with mean radius r_0 and having a homogeneous magnetic field $B_0(R)$ normal to the sector along each circle R of radius r as shown in Fig. 2.26. This is the ideal case and is sufficient for our purposes.

In this simple case the sector may be considered as equivalent to a thin lens with coincident object and image principle points at O, focal points at F_0 and F , respectively, and corresponding focal lengths $f_0 = 0 \, F_0$ and $f = 0 \, F$. An object at P will thus be imaged at Q. It can be shown that the points P, C and Q are collinear and that

$$\phi_P + \phi_Q + \phi = \pi$$

Once P has been selected and the sector geometry known, we only have to determine PC to find Q. The linear magnification is given by $-CQ/CP$. In the case where $CQ = CP$ the linear magnification is -1 and we have "sigmatic-operation." Also

$$p = q = r_o \cot (\phi/2)$$

When a beam of electrons passes through a specimen, it is expected that some, or all, of the electrons will lose a portion of their incident energy through one or more of the many mechanisms of inelastic scattering that can take place in the specimen. Thus a monoenergetic beam incident on a specimen will probably exit with a spectrum of energy. (The details of this topic, electron energy loss in solids, will be discussed in Chapters 7, 8 and 9). Suffice it to say it is often important to study the energy loss spectrum and magnetic sectors as described here as being used widely for this purpose today.

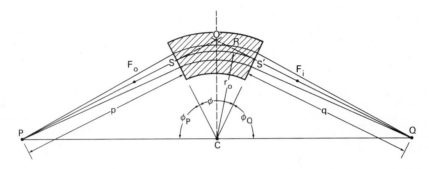

Fig. 2.26. Diagram of magnetic sector of angle ϕ showing object and image relationships, P and Q, respectively.

Recalling that the radius of the electron trajectory depends on the electron energy, it is easy to see that a beam of electrons containing a spectrum of energies (velocities) at point P in Fig. 2.26 will be dispersed along a line through point Q. Generally, it is preferred to place a small slit at Q and record only those electrons passing through the slit with an electron sensitive detector. Since the radius r is a function of v/B, the ratio of the electron velocity (energy) to the applied magnetic field, (if we keep v/B constant) will remain the same and by changing the magnetic field electrons having different energy will selectively be passed through the slit at Q.

Electrostatic Sectors

In a manner analogous to the magnetic field, an electron entering a uniform electric field E_0 will be deflected into a circle of radius r_0, where

$$r_o = \frac{m \, v^2}{e \, E_o} = \frac{2 \, V_o}{E_o}$$

All the quantities have been described before. The major difference between electrostatic and magnetic fields is that the force on the electron by an electrostatic field is in the direction of the field (in the negative sense in the case of an electron) that is, $\underline{F}_e = -e\,\underline{E}$, rather than perpendicular to the field as is the case for a magnetic field.

One can qualitatively describe electrostatic sectors in much the same way as magnetic sectors. The optics are a bit more complicated but can still be discussed in terms of cardinal elements. Because of this and the fact that electrostatic sectors are not often used in electron microscopes, they will not be discussed here. One major problem in their use is worth mentioning, though. For an electric field to have any measurable effect on deflecting an electron with an accelerating voltage, V_0, it is essential that the potential of the field be approximately V_0. For 100 keV electrons this then requires that the potential across the sector would have to be about 100 keV, so that one encounters problems of high voltage supply and breakdowns, etc.

Wein Filter

Despite the problem associated with high voltage, there is one popular use for the electrostatic field in electron velocity analyses, that is, by crossing magnetic and electric fields it is possible to create a situation in which electrons with particular values of energy will be unaffected by the fields and electrons at all other energies will.

For example, if a homogeneous magnetic field, B, acts over a distance ℓ, electrons entering the field will be deflected through an angle α at right angles to the electron velocity $\underline{v}$ and field $\underline{B}$, given by

$$\sin\alpha = \frac{\ell}{R} = \frac{\ell\,B\,e}{m\,v}$$

as shown in Fig. 2.27(a). Similarly, an electron incident on a region of length ℓ, containing an electric field E, will be deflected through an angle β given by

$$\sin\beta = \frac{\ell}{R} = \frac{\ell\,e\,E}{m\,v^2}$$

as seen in Fig. 2.27(b).

If the magnetic and electric fields are applied at right angles to each other (crossed) as shown in Fig. 2.27(c), the magnetic field will deflect the electron to the left and the electric field will deflect it to the right. If the fields are properly adjusted the respective deflections will cancel and certain electrons will pass through undeflected. This requires $\sin\alpha = \sin\beta$, that is

$$\frac{\ell\,B\,e}{m\,v_o} = \frac{\ell\,e\,E}{m\,v_o^{\,2}}$$

which reduces to $v_0 = E/B$. Thus, for a particular set of field values $\underline{B}$ and $\underline{E}$, only electrons with velocities $v_0 = E/B$ will pass through the crossed fields undeflected. By appropriately placing entrance and exit slits, a device so constructed acts as an electron velocity filter. Such a device is called a Wien filter. Various models have been built, some obtaining very high resolution, i.e., on the order of a few meV.

Fig. 2.27. (a) Deflection by a magnetic field, B, (b) deflection by an electric field, E and (c) crossed fields yielding no deflection for $v_0 = E/B$.

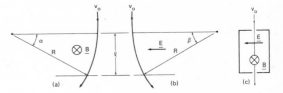

2.6 OPTICS OF THE ELECTRON MICROSCOPE

Introduction

The optical components of an electron microscope may be conveniently broken into three components according to function. The illumination system, composed of the electron gun and two condenser lenses, provides the electrons to "illuminate the specimen." The objective lens, which we've discussed in detail, is the heart of any electron microscope and forms the image of the specimen usually with a magnification of 50X to 100X. The magnification system consists of three or four projector lenses, often identified as diffraction, intermediate and projector lenses, and provides the final magnification of the image. The total magification in some transmission electron microscopes may exceed one million times, with each lens contributing about 20X in a three lens system.

In SEM and dedicated STEM microscopes there is no need for the magnification system and such microscopes will have only one or two condenser lenses and an objective lens. Magnification in scanning microscopes is inversely related to the extent of the area scanned, e.g., large area scan gives low magnification and small area scan gives high magnification.

Before discussing the operation of the complete electron microscope, it is necessary to discuss the illumination system in some detail.

Electron Gun

The biased electron gun is equivalent to a triode electrostatic lens and consists of a filament at the negative accelerating potential, a surrounding electrode called the Wehnelt at a slightly more negative bias (a few hundred volts) than the filament, and an anode at ground potential. As a result of the negative bias on the Wehnelt, the electrons emitted from a small region of the filament are focussed to a crossover in front of the anode as shown in Fig. 2.28.

The diameter of the beam crossover, d , and the angle of the beam divergence, α_i, are crucial in determining the brightness of the gun and hence the resolution and visibility of the image. Brightness at the gun is defined as the current density per unit solid angle and is given by

$$\beta = \frac{J_o \, e \, V_o}{\pi \, k \, T} = \frac{4 \, I_B}{\pi^2 d_c^2 \alpha_c^2} \quad A/m^2 \text{ steradian}$$

J_0 is the current density at the filament, V_0 is the accelerating voltage, k is Boltzmann's constant (= 8.61×10^{-5} eV/°K), and T the temperature of the filament; I_B is the beam current, d_c and α_c are defined in Fig. 2.28.

Typical operating values at 100 keV for a heated tungsten filament are $J_0 = 2 - 3 \times 10^4$ A/m², d = 30 μm, $I_B = 100$ μA and $\alpha_c = 0.01$ rad giving $\beta = 4 \times 10^8$ A/m²·sr. Comparable values for a heated LaB$_6$ gun are $J_0 = 25 \times 10^4$ A/m², $d_c = 5$ μm, $I_B = 200$ μA giving $\beta = 5 \times 10^{10}$ A/m²·sr and for a cold field emission gun $J_0 = 10^7$ A/m², $d_c = 0.5$ μm, $I_B = 10^{-8}$ A, $\alpha_c = 10^{-4}$ rad and $\beta = 2.5 \times 10^{12}$ A/m²·sr at 100 keV. Filament lifetimes and required operating vacuum are approximately 30 hours at 10^{-5} Torr, 500 hours at 10^{-6} torr and 10^3 hours at 10^{-10} torr for the thermionic tungsten, LaB$_6$ and cold field emission guns, respectively.

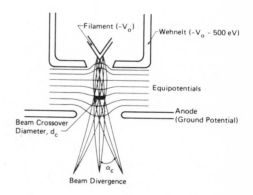

Fig. 2.28. Electron gun showing how the electrons emitted from the filament are brought together to a crossover by the electrostatic lens action of the gun elements.

Condenser Lens System

The purpose of the condenser lens system is to deliver electrons from the gun crossover to the specimen under the multitude of conditions needed by the microscopist. For conventional transmission microscopy this requires focussing a high intensity, reduced image of the gun crossover at the specimen plane at one extreme, to focussing the beam at infinity at the other extreme, providing nearly parallel illumination of considerably reduced intensity at the specimen. For scanning electron microscopy, the singular requirement is for a small diameter, high intensity probe at the specimen.

In principle, a single condenser lens situated about half way between the gun and specimen would be adequate for TEM, but not for STEM unless a FEG (field emission gun) was used. With a single condenser lens the smallest spot at the specimen would be approximately the same size as the gun crossover since the magnification of the lens would be approximately unity. Modern microscopes universally incorporate two condenser lenses with one, the second condenser, C2, placed approximately halfway between the specimen and the gun, and the other, C1, halfway between the gun and C2. The demagnifying factor of the combined lenses is usually sufficient to produce submicron probes at the specimen. Since Cl is located closer to the gun than a single condenser lens would be, a larger number of electrons are collected, thus a higher current may be focussed into the probe.

With STEM microscopes the specimen is immersed in the magnetic field of the objective lens to the extent that this prefield acts as another condenser lens, C3. For STEM operation Cl and C2 are adjusted to provide a small beam of almost parallel electrons to the prefield of the objective lens which, in turn, focusses the beam to probes as small as 15 Å in diameter on the specimen. (We are not considering FEG here.) Ray diagrams illustrating these different situations are drawn in Fig. 2.29.

As shown in Fig. 2.29, the angular aperture of the illumination is defined by the angle α_i. When the gun crossover image is not focussed at the specimen plane, α_i is defined as shown in Fig. 2.29(c). The angular aperture of the illumination is a measure of the parallelism of the incident probe. The effective angle of the source is defined by α in Fig. 2.29(c). This is the angle of the undeviated ray from the crossover with respect to the optical axis and is useful for calculating the crossover image at the crossover of Cl. Subsequent angles can be defined at each lens crossover as the electrons proceed down the column; of particular significance is the angle α_i, previously defined.

Fig. 2.29. Electron ray paths for (a) single condenser transfer of filament crossover to specimen plane, (b) double condenser formation of reduced image of crossover at specimen, (c) double condenser formation of nearly parallel illumination, and (d) double condenser plus objective prefield formation of very small probe for STEM.

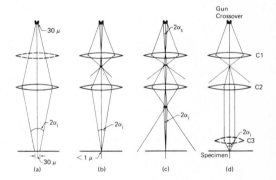

The angular aperture of illumination, α_i, is determined by the combined strengths of Cl and C2. That is, to have a small α_i it is essential to operate Cl with a short focal length (strong). C2 is then adjusted to have its crossover above, at, or below the specimen plane depending on the miminum illumination required to view the specimen. For a fixed Cl, the relationship between C2 excitation and angular aperture follows the general trend shown in Fig. 2.30.

The intensity of illumination is proportional to the square of the angular aperture of illumination, α_i^2. Consequently, the illumination at the specimen drops off very rapidly as the second condenser excitation is changed from the crossover condition (point A in Fig. 2.30). From the figure it can be seen that the smallest α_i are obtainable with C2 over-focussed, e.g., with the crossover above the specimen.

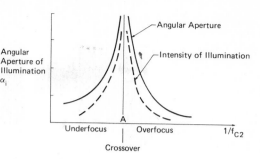

Fig. 2.30. Relationship between the strength of C2 (1/f) and α_i as indicated by the axial position of C2 crossover with respect to the specimen plane.

Coherence

The term coherence refers to the range of phase differences in the illuminating beam as it approaches the specimen. If the electrons come from a single point source, then all the waves in the incident beam are in phase with one another and the illumination is said to be coherent. On the other hand, if the source of electrons is so large that there is no phase relationship between the incident waves, the illumination is incoherent. The situation at the filament in an electron microscope is somewhere in between these two extremes and the incident illumination is defined as partially coherent.

Consider two waves coming from a source of width d, passing through separate slits, S_1 and S_2, in an otherwise opaque screen. If d = 0, the waves are coherent and the intensity at some plane after the slits is given by

$$I = I_1 + I_2 + I_{12}{}^*$$

I_1 and I_2 represent the intensity distribution in the individual waves and $I_{12}{}^*$ is the intensity resulting from interference between the two waves. As d gets larger $I_{12}{}^*$ becomes smaller until at some point d is so large that $I_{12}{}^* = 0$ and the two waves are said to be incoherent.

In electron microscopy, coherence is related to the angular aperture of the illumination, α_i, and is related to the dimensions of the region on the specimen over which the illumination appears coherent. This is an extremely important concept when imaging objects which give rise to interference effects. The length, a, over which the illumination is said to be coherent is given by

$$a = \frac{\lambda}{2\alpha_i}$$

Typical illumination conditions for crossover at 100 keV have α_i = 10^{-3} rad, thus a = $\lambda/2\alpha_i$ = 0.037 x $10^{-10}/2$ x 10^{-3} = 18 x 10^{-10} m (= 18 Å) and the illumination is not very coherent. With C2 crossover considerably overfocussed, it is reasonable to have α_i = 2.5 x 10^{-5} rad giving the coherence length a = 740 Å at the specimen. This illumination condition would allow interference effects such as Fresnel fringes, magnetic domains, or lattice images to be obtained.

Magnification Lens System

The properties of the individual lenses that make up the magnification lens system have been previously described. All the lenses are designed as projector lenses and their most important role is to magnify the image of the specimen formed by the objective lens. Usually, the collective lens action in the magnification process has been programmed into the microscope, so that the operator has little control over the individual lenses. The total magnification of the final image is the product of the individual lens magnifications, e.g., in a four lens system (including the objective lens)

$$M_{total} = M_{obj} \times M_{diff} \times M_{int} \times M_{proj}$$

In normal operation the excitation and therefore the magnification of the objective lens is almost constant since it is used for providing the primary image of an object located in a prescribed plane. For the same reason, the excitation of the projector lens is usually fixed since the location of the final image is fixed at the viewing screen. Consequently, almost the whole range of magnification is obtained by varying only the diffraction and intermediate lenses. At very low magnification, say 50X, a slightly different optical situation is required. Here, because of the intrinsic high magnification of the objective lens it is either turned off or operated at a reduced excitation with the diffraction lens now effectively acting as a long focal length objective lens. Ray diagrams showing low and high magnification ranges and standard diffraction pattern formation are shown in Fig. 2.31.

In obtaining diffraction patterns the object plane for the diffraction lens is adjusted to be the back focal plane of the objective lens. Recall from previous discussion that the diffraction pattern of the specimen is at the back focal plane of the objective lens. This plane inter-

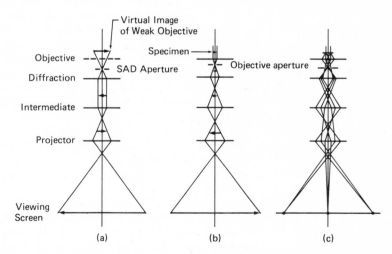

Fig. 2.31. Ray diagrams for (a) low magnification, (b) high magnification and (c) diffraction.

sects the axis at the objective focal point and is the location of the objective aperture. The selected area aperture is positioned at the image plane of the objective lens which is the object plane of the diffraction lens in Fig. 2.31(b), normal magnification operation. In terms of paraxial imaging, we may say that the final viewing screen is conjugate to the objective image plane under normal magnification operation and conjugate to the back focal plane (sometimes called the image focal plane) of the objective lens in diffraction imaging. By suitable alteration of the diffraction and intermediate lens excitation it is possible to have the final viewing screen conjugate to any plane in the optical column.

2.7 COMPARISON OF CTEM AND STEM OPTICS

The optical systems of the CTEM have been described in detail here. Utilizing Figs. 2.29 and 2.31, one should be able to follow the electron path down the column for the standard modes of microscope operation. For the sake of completeness, an electron ray diagram for a complete microscope in the high magnification mode is drawn in Fig. 2.32.

When using the CTEM the best image quality will be obtained by operating C1 as strong as possible while still providing the necessary illumination at the object plane with C2 defocussed.

All previous discussion pertains to CTEM with an objective lens excitation $k^2 \neq 3$ and usually with $k^2 < 3$. For the unique excitation $k^2 = 3$ the operation of the condenser lenses to control illumination conditions is very restricted. Again, C1 is operated as strongly as possible to form the smallest effective source, but C2 is now fixed to the value of excitation that focusses its crossover to the front focal plane of the objective lens. This is the plane before the objective lens which intersects the optical axis at F_0 and is conjugate to the objective back focal plane. Since it is conjugate to the BFP, the front focal plane may be viewed on the final screen by observing the BFP in diffraction. With properly adjusted illuminating conditions, the probe of diameter, d, at the specimen will be defined by the demagnified image of the C2 aperture of diameter D_{C2} (which is conjugate to the specimen). The demagnification factor is obtained from $M_d = \ell_0/f_0$, where ℓ_0 is the distance from the C2 aperture to F_0 and f_0 is the focal length of the objective lens, thus $d = \ell/M \, D_{C2}$. The electron optical path from the gun to the objective lens is shown in Fig. 2.33 for this case when $k^2 = 3$. Riecke has described an alignment procedure for this optical configuration and has used it to obtain small area, 100 Å in diameter, diffraction patterns having an angular resolution of approximately 10^{-5} rad.

The illumination requirements of the STEM microscope require as small a probe as possible, usually 5-20 Å, with sufficient beam current to yield a usable image. As discussed previously, this small probe is obtained by immersing the specimen in the field of a highly excited objective lens to be extent that the prefield of this lens will form the small probe from the nearly parallel incident illumination. If the specimen is placed in the center of a properly energized symmetrical lens, $k^2 = 3$, the lens operates in the telefocal condition described before. That is, all bundles of parallel rays passing through F_0 will be focussed on the specimen, regardless of their inclination to the optical axis. Thus a set of scan coils adjusted such that the pivot point of the coils is at F_0 will provide the necessary incident probe conditions. To obtain the parallel illumination, it is convenient to turn off C2 and use a small C2 aperture to define the angular aperture of the probe.

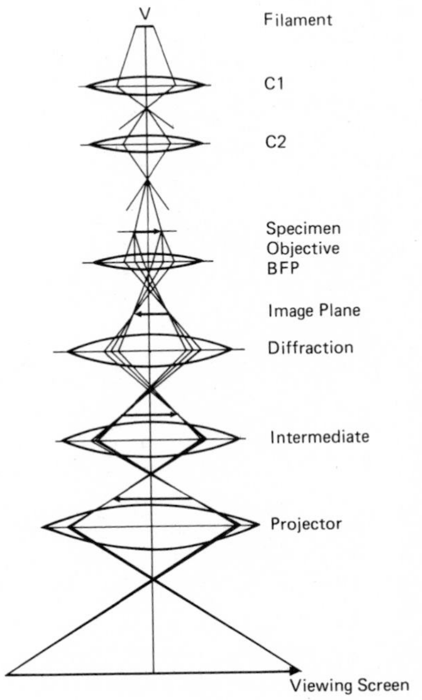

Fig. 2.32. Ray diagram for a complete CTEM operating in the high magnification mode. CA, OA and SAA are the second condenser, objective. and selected area aperture, respectively.

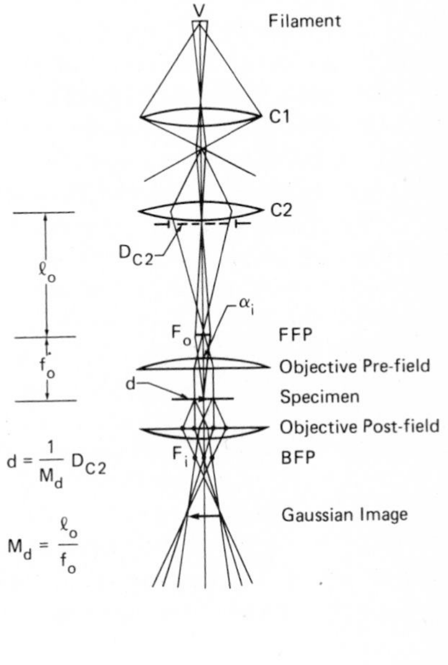

Fig. 2.33. Electron ray paths for the case $k^2 = 3$ and C2 focussed to F_0.

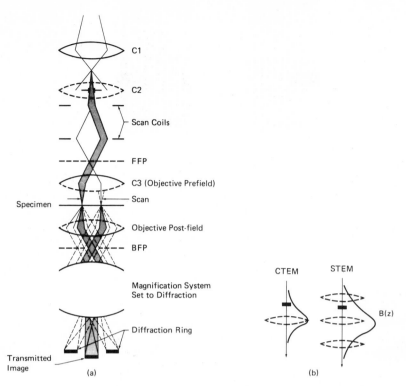

Fig. 2.34. (a) Ray paths to form STEM image on CRT using objective lens with $k^2 = 3$, (b) objective lens only for $3 < k^2 < 8$ showing field strengths in CTEM and STEM mode. STEM mode is approximated by three equivalent lenses.

Since F_0 is conjugate to F_i the incident beam after passing through the specimen will pivot about F_i, which defines the BFP of the objective lens. A detector placed at F_i (as is the case in a dedicated STEM) or anywhere in the optical path after the specimen where F_i is imaged [the case for S(TEM)] will gather the required signal to form the image. In S(TEM) microscopes this is achieved by placing a detector near the final viewing screen and adjusting the magnification lens system to focus the diffraction pattern on the viewing screen. A schematic electron ray path for the optical condition is drawn in Fig. 2.34(a). The optical requirements are very similar when using a very strongly excited objective lens, $3 < k^2 < 8$, thus only the objective lens is drawn in Fig. 2.34(b).

The angular aperture of the incident probe, α_i, assuming nearly parallel illumination from the condenser lens system with a C2 aperture of diameter D_{c2} is given as

$$\alpha_i = D_{c2}/2f_0$$

For most STEM microscopes this angle is on the order of 5×10^{-3} radians compared to normal TEM imaging with $\alpha_i < 10^{-3}$ rad.

CONCLUSION

The basic principles of CTEM and STEM microscope operation have been outlined. Using the simple concepts of geometrical optics it is, in theory, possible to approximate the path of any electron in the microscope column. Drawing a few ray diagrams for the standard operational conditions will undoubtedly help the operator visualize the predicament of the electron anywhere in the column.

REFERENCES

The following is a list of books on electron optics and related topics listed in the author's order of perference with some general comments on each.

Hawkes, P.W., Electron Optics and Electron Microscopy, Taylor and Francis Ltd., London, 1972.
Excellent book describing elements of electron optics and application to microscopes. Scanning microscopy and image interpretation included.

Grivet, P., Electron Optics, Pergamon Press, Oxford, 1972, 2nd Edition.
An excellent monograph dealing with the principles of electrostatic and magnetic lenses and the applications of electron optics to various electron applications of electron optics to various electron instruments.
Available as a paperback in two volumes.

ElKareh, A.B. and ElKareh, J.C.J., Electron Beams, Lenses, and Optics, 2-Vols., Academic Press, New York, 1970.
Extensive treatment with many tables for calculation of lens parameters including excellent treatment of aberrations. Recommended for designers. Kempler, O. and Barnett, M.E., Electron Optics, Cambridge Univ. Press, London, 1971.
In same vein as Grivet, recommended for further reading by those with good background in physics and mathematics.

Hall, C.E., Introduction to Electron Microscopy, McGrawHill, New York, 1966.
Fundamentals of light and electron optics and limited applications. Good beginning book.

Septier, A., ed., Focusing of Charged Particles, 2 vols., Academic Press, New York, 1967.
Volumes cover the whole field of particle optics, including electrons and ions. High intensity beams, particle accelerators and prism optics are well covered.

Oatley, C.W., The Scanning Electron Microscope, Cambridge Univ. Press, London, 1972.
Good introduction to electron optics of SEM. Especially good discussion of probe formation.

Bauer , R., and Cosslett, V.E., eds. Advances in Optical and Electron
    Microscopy, Academic Press, London.
    Series of books with research and review articles by esteemed
    authors covering a variety of topics relating to electron optics and
    microscopy.

Marton, L., Advances in Electronics and Electron Physics, Academic
    Press, London.
    Series of books with review articles on a wide range of topics many
    relating to electron optics.
    Below are listed in no particular order some rather complete
    articles relating to electron optics.

Agar, A.W., Alderson, R.H. and Chescoe, D., "Princ. and Prac. of E.M.
    Operation", Prac. Meth. in Elec. Mic., ed. Glauert, North-Holland,
    Amsterdam, 1974.
    Good introduction to the electron microscope with discussions per-
    taining to general principles of use and operation.

Mulvey, T. and Wallington, M.J., "Electron Lenses", Reports Prog. Phys.,
    36, 347421, 1973.
    Extensive article concerning electron lenses and their design. Many
    curves and tables. Extensive references.

Glaser, W.,"Elektronen und Tonenoptik", Handbuch der Physik, 33, 1956,
    123 395.
    Thorough discussion of theoretical electron optics. Discussion of
    bellshaped field. In German.

Siegel, B., SEM / 1975 (part III), Proc. of SEM, 1975, IITRI, Chicago,
    ed. O. Johari, 647 659.
    Tutorial on electron optics with special reference to the SEM.

Listed alphabetically are technical references either referred to
in this chapter or generally recommended to those with advanced
backgrounds.

Boersch, H., Z. Phys., 139, 1954, 115 46.

Francken, J.C. and Heeres, A., Optik, 37, 1973, 483 500.

Kammenga, W., Verster, J.L. and Frenchen, J.C., Optik 28, 1968, 442.

Koike, H., JEOL News, 16E, 1978, 26.

Liebmann, G. Proc. Phys. Soc. B, 62, 1949, 75372.
        ibid 64,   1951, 9727.
        ibid 65,   1952, 18892.
        ibid 66,   1953, 44858.
        ibid 68,   1955, 73745.
        ibid 68,   1955, 68285.
        Br. J. Appl. Phys. 1,   1950, 92103.
        Phil. Mag. 41,   1950, 114351.

Liebmann, G. and Grad, E.M., Proc. Phys. Soc. B, 64, 1951, 95671.

Mulvey, T., SEM/1974 (part 1), 11TRI, Chicago, ed O. Johari, 1974, 4350.

Mulvey, T. and Newman, C.D., Scanning E.M.: Systems and Applications 1973, Inst. of Phys., London, 1973, 1621.

Munro, E. and Wells, D.C., SEM/1976 (part I), Proc. 9th SEM Symposium, Chicago, IITRI, Chicago, Ed. O. Johari, 1976, 2736.

Munro, E., SEM/1974 (part I), IITRI, Chicago, ed. O. Johari, 1974, 3549.

Riecke, W.D., Optik, 19, 1962, 81116.
    ibid 169207.
    Proc. Fifth European Congress on E.M., Manchester, Institute of Phys., London, 1972, 7677.
    ibid 98103.

Riecke, W.D. and Ruska, E., Proc. 6th Int. Conf. Electorn Microscopy, Kyota, ed. R. Uyeda, Maruzen Press, Tokyo, 1966, 1, 1920.

Ruska, E., Optik, 22, 1965, 319348.

Suzuki, S., Akashi, K. and Tochigi, H., Proc. 26th EMSA Meeting, Claitar's Pub. Div., Baton Rouge, La., 1968, 320321.

CHAPTER 3

PRINCIPLES OF THIN FILM X-RAY MICROANALYSIS

J.I. GOLDSTEIN

DEPARTMENT OF METALLURGY AND MATERIALS ENGINEERING
LEHIGH UNIVERSITY
BETHLEHEM, PENNSYLVANIA 18015

3.1 INTRODUCTION

The scanning electron microscope-electron probe microanalyzer (SEM-EPMA) and transmission electron microscope (TEM) are two established tools for microscopy. The first instrument enables one to obtain high magnification pictures as well as microchemical information from micron sized areas in solid samples. The second instrument enables one to obtain high magnification pictures and diffraction data from electron transmission thin specimens. Over 10 years ago DUNCUMB (1968) mounted a wavelength dispersive x-ray spectrometer on a TEM in order to obtain chemical and structural as well as diffraction data from the same area of a thin specimen. This idea of a combination instrument has developed rapidly in the last few years into the scanning transmission electron microscope--analytical electron microscope (STEM-AEM) instrument of today. In the modern version of this instrument a 60 to 200 kV electron beam is focused to < 100 and often to < 10 nm diameter at the specimen surface. Scanning coils move the focused beam over the specimen to obtain a STEM image. The emitted x-rays are measured with an energy dispersive x-ray spectrometer (EDS). Quantitative electron probe microanalysis can be accomplished when the focused beam is positioned at selected points on a specimen. In addition, particle identification and x-ray scanning can be performed with this instrument.

There are several other advantages of the STEM instrument. Because of the high accelerating voltage and the thin specimens used, only minimal spreading of the focused electron beam occurs. Hence a very small x-ray source region is obtained. A schematic drawing of the distribution of electrons and the activated volume in a thick and thin specimen is shown in Fig. 3.1 (LORIMER, 1976). The x-ray source size in STEM microanalysis can almost equal that of the beam diameter < 10 - 100 nm for very thin specimens. Another advantage of using a thin foil speci-

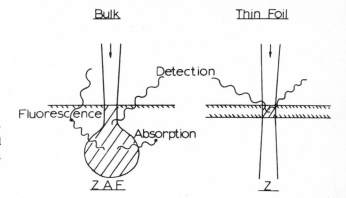

Fig. 3.1. Schematic distribution of electrons and activated volume in a thick and thin specimen. [LORIMER, (1976).]

men is that in most cases no absorption or fluorescence correction need be applied to the x-ray data. This allows for the application of relatively simple procedures for obtaining quantitative analyses of samples.

There are however several problems which still make STEM microanalysis a challenge. Among these is the presence of spurious x-rays generated in the electron column, specimen holder, and x-ray collimator. These x-rays often make their way in one form or another into the EDS detector. In addition the small currents in the electron beam combined with the small excitation volume in the thin foil, lead to low x-ray count rates. These low count rates limit the detection of small amounts of elements in a specimen as well as the accuracy and precision of an analysis. Lastly the method of specimen preparation may be crucial to an analysis as contamination layers may build up on the surfaces of thin films.

In this chapter the basic theory of x-ray production will be reviewed and methods to obtain quantitative analyses from thin foil samples will be described. In addition the spatial resolution and minimum mass sensitivity will be discussed as these are limiting factors in any x-ray microanalysis measurement. Instrumental factors which may limit the quality of the measured x-ray data are reviewed by Zaluzec in Chapter 4 of this book. Several comprehensive review papers on STEM thin film x-ray microanalysis have been written to which the serious reader is referred (LORIMER, 1977; GOLDSTEIN and WILLIAMS, 1977; and BEAMAN, 1978).

3.2 QUANTITATIVE X-RAY ANALYSIS

Primary Emitted X-ray Intensities

The average number of ionizations n per primary beam electron with energy E_0 incident on a sample containing element A is (CASTAING, 1951):

$$n = \frac{N\ C_A\rho}{A_A} \int\limits_{E_c}^{E_o} \frac{Q_A}{dE/dX}\ dE \qquad\qquad (3.1)$$

where dE/dX is the mean energy change of an electron in traveling a distance X, N is Avogadro's number, ρ is the density of the material, A_A is the atomic weight of A, C_A is the concentration of element A, E_K is the critical excitation energy for K, L, or M characteristic x-rays from element A, and Q_A is the ionization cross section (the probability per unit path length of an electron of a given energy causing ionization of a particular shell (K, L, or M) of an atom A in the specimen). The ionization cross section is a function of the energy E along the path length X of the electron. In solid samples some fraction of electrons, $1 - R$, is backscattered from the target and does not generate ionizations. Multiplying n by the fluorescence yield ω_A for element A characteristic K or L lines, the fraction a of the total K or L line intensity that is measured as K_α or L_α radiation and the backscatter factor R gives the characteristic x-ray intensity in photons per incident electron generated in the sample:

$$I_A = \frac{const}{A_A}\ C_A R\ \omega_A a_A \int\limits_{E_c}^{E_o} \frac{Q_A}{dE/dX}\ dE \qquad\qquad (3.2)$$

The measured intensity of the characteristic x-ray line must also be corrected for absorption of x-rays within the specimen and fluorescence effects from x-rays generated from other elements in the target. Classical microprobe analysis considers all of the so called ZAF factors in order to obtain accurate local chemical analyses. Methods of quantitative x-ray analysis have been discussed in several recent books (GOLDSTEIN and YAKOWITZ, 1975; REED, 1975).

In electron microscope thin films, few electrons are backscattered and the electrons lose only a small fraction of their energy in the film. Therefore Q_A can be assumed constant in the film. The trajectory of the electron can also be assumed to be essentially the same as the thickness of the thin film t. Therefore the characteristic x-ray intensity can be given by a much more simplified formula namely:

$$I_A = const\ C_A\ \omega_A\ Q_A\ a_A\ t/A_A \qquad\qquad (3.3)$$

If one assumes that the analyzed film is "infinitely" thin, the effects of x-ray absorption and fluorescence can be neglected, and the generated x-ray intensity and the x-ray intensity leaving the film are identical. This assumption is known as the thin film criterion.

A number of techniques have been developed for quantitative thin film analysis using the thin film criterion. These techniques are outlined in the following sections.

Quantitative X-ray Analysis Using the Ratio Technique and Thin Film Criterion

From Eqn. 3.3 it appears that the composition of an analyzed region can be obtained simply by measuring the emitted intensity, I_A, and by calculating the constant and other terms. Unfortunately this cannot be done easily since many of the geometric factors and constants cannot be evaluated exactly. In addition, the film thickness varies from one point in the specimen to another, making it inconvenient to measure film thickness t continuously. Several investigators (DUNCUMB, 1968; PHILIBERT and TIXIER, 1968; TIXIER and PHILIBERT, 1969; and CLIFF and LORIMER, 1975) have proposed analysis techniques in which the x-ray intensity ratio I_A/I_B of two elements A and B in a foil is measured simultaneously. This intensity ratio is related directly to the mass concentration ratio C_A/C_B. Using Eqn. 3.3 for elements A and B results in an equation of the form:

$$\frac{C_A}{C_B} = \frac{A_A(Q\omega a)_B}{A_B(Q\omega a)_A}\frac{I_A}{I_B}$$ (3.4a)

$$\frac{C_A}{C_B} = k_{AB}\frac{I_A}{I_B}$$ (3.4b)

The term k_{AB} varies with operating voltage but is independent of sample thickness and composition if the two intensities are measured simultaneously and if the thin film criterion is satisfied. The measurement of x-ray intensity ratios I_A/I_B, is called the ratio technique. The technique has been adopted by many analysts and has also become known as the Cliff-Lorimer method. The advantages of the ratio technique are that thickness variations can be ignored and fluctuations in electron beam intensity do not alter the results. HALL (1968) proposed an alternative approach in which the intensity of characteristic radiation I_A is proportional to the atomic density of the element A of interest. This approach is used for biological thin section analyses and is discussed by Hall in Chapter 5 of this book. The ratio technique (Eqn. 3.4) is commonly used for unsupported thin films of ceramics or metals. One can use either bulk standards, thin standards, or standardless techniques to obtain the true concentration ratios. These various approaches to quantitation are described in the following sections.

(1). Analysis using bulk standards--DUNCUMB (1968), PHILIBERT and TIXIER (1968), and NASIR (1972) have shown that it is possible to refer to bulk standards for quantitative thin foil analysis. The method takes advantage of x-ray microanalysis techniques which are well developed. However the method requires the preparation of bulk standards. In the bulk standards approach, thin specimen intensities are computed by using Eqn. 3.3 for I_A and I_B. The bulk standard intensities are corrected by employing the usual ZAF corrections.

In the bulk standards technique, the mass concentration ratio C_A/C_B can be related to the x-ray intensity ratio I_A/I_B which is measured simultaneously in a thin foil by:

$$\frac{C_A}{C_B} = [\frac{I(B)}{I(A)}] \cdot [\frac{\log U_o^B}{E_c^B} \cdot \frac{E_c^A}{\log U_o^A}] \cdot [\frac{R_B/S_B'}{R_A/S_A'}] \cdot \frac{f(\chi)_A}{f(\chi)_B}] \cdot [\frac{I_A}{I_B}] \qquad (3.5)$$

where $I(A)$ and $I(B)$ are the measured x-ray intensities generated by the bulk standards (measured at the same probe current and beam potential E as the sample), using $U_o^A = E_c/E_c^A$ is the over-voltage ratio. The $\log U_o$ and E_c terms enclosed in the parenthesis give the ratio Q_A/Q_B using a form of the Bethe ionization cross section expression for calculation of thin film emitted x-ray intensities, Eqn. 3.3. Other ionization cross sections might be used in Eqn. 3.5 but these have not been employed. The term in brackets corresponds to the usual bulk standards correction as described for example in REED (1975) or GOLDSTEIN and YAKOWITZ (1975). It is also assumed that the foil is thin enough so that the thin film criterion is observed and the effects of x-ray absorption and electron backscattering can be neglected.

DUNCUMB (1968) suggested measuring the standard intensities $I(A)$ and $I(B)$ at lower beam potentials. This procedure is used in order that ZAF correction procedures which have been developed for low operating potentials can be employed. Duncumb used the bulk standards method with good results at operating potentials of 15 to 35 kV while JACOBS and BABOROVSKA (1972) and JACOBS (1973) used the method successfully on several metal and oxide samples with the EMMA-3 instrument operating at 100 or 40 kV.

(2). Analysis using thin film standards--The advantage of using thin film standards is that measurements can be made for the same operating conditions as for the specimen. In one approach the mass thickness $(\rho t)_{std}$ of the standard must be known as accurately as possible. Assuming that the thin film criterion is applicable, it has been shown by TIXIER (1979) that by using Eqn. 3.3 the concentration ratio for elements A and B is:

$$\frac{C_A}{C_B} = \left[\frac{I(B)}{I(A)} \cdot \frac{(\rho t)_{std.A}}{(\rho t)_{std.B}}\right] \cdot \frac{I_A}{I_B} \qquad (3.6)$$

The terms $(\rho t)_{std\ A}$, $(\rho t)_{std\ B}$ denote the mass thickness of pure element standards A and B. $I(A)$ and $I(B)$ are the intensities of elements A and B in the standards. PHILIBERT et al. (1970) have used the above technique with pure element standards obtained by thermal evaporation. Since the correction factor for thin specimens does not depend on concentration, Eqn. 3.4, a thin standard of known mass thickness which contains both elements of interest can be used (TIXIER, 1979). The major disadvantages of this thin film standard technique are the necessity for preparing the standards and for accurately measuring the mass thickness. These disadvantages are so serious that this technique has seen little practical use.

CLIFF and LORIMER (1972) have shown that if the composition of a thin film is known, the thin film specimen can be used as its own standard. If the characteristic x-ray intensities of two elements in the standard, I_A and I_B, are measured simultaneously from the same point, and the composition C_A and C_B in the standard is known, k_{AB} in Eqn. 3.4(b) can be determined experimentally. The constant k_{AB} varies with operating voltage but is independent of the sample thickness and composition as long as the thin film criterion is satisfied. If k_{AB} scaling factors are available for various combinations of elements then weight fraction ratios of various elements in thin films can be obtained simply by multiplying the measured intensity ratio I_A/I_B by the appropriate k_{AB} factor. Once k_{AB} values are obtained for various combinations of elements, thin film standards are no longer needed for an analysis.

Experimental values of k_{AB} using an EMMA-4 instrument at 100 kV have been determined by CLIFF and LORIMER (1975) for a series of elements A, relative to Si. The element Si is considered as element B in Eqn. 3.4 and the constant k_{ASi} is commonly called k. Using Eqn. 3.4, k is given by:

$$k = k_{ASi} = [\frac{C_A}{C_{Si}}] \cdot [\frac{I_{Si}}{I_A}] \tag{3.7}$$

Experimental values of k for K_α lines at 100 kV (CLIFF and LORIMER, 1975) are plotted in Fig. 3.2 as a function of the energy of the measured characteristic line of element A.

Fig. 3.2. Experimental k values for 100 kV plotted as a function of x-ray energy (keV) for K_α lines, [CLIFF and LORIMER, (1975).] Experimental error limits are also given in the Figure.

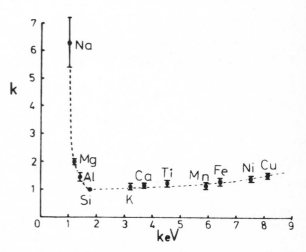

Experimental k values for L_α lines at 100 kV using an EMMA-4 instrument are reported by GOLDSTEIN et al. (1977). These experimental data for K_α

TABLE 3.1

Calculated and Measured k Values, 100 kV

Z	K_α line - experimental k(CLIFF and LORIMER, 1975)	K_α line - calculated k(GOLDSTEIN et al., 1977)
11 Na	5.77	1.66
12 Mg	2.07 ±0.1	1.25
13 Al	1.42 ±0.1	1.12
14 Si	1.0	1.0
15 P	---	0.99
20 Ca	1.0 ±0.07	1.02
22 Ti	1.08 ±0.07	1.16
23 V	1.13 ±0.07	---
24 Cr	1.17 ±0.07	---
25 Mn	1.22 ±0.07	1.31
26 Fe	1.27 ±0.07	1.33
28 Ni	1.47 ±0.07	1.44
29 Cu	1.58 ±0.07	1.59
0 Zn	1.68 ±0.07	---
32 Ge	1.92	2.01
42 Mo	4.3	4.65
47 Ag	8.49	7.91
50 Sn	10.6	12.1

Z	L_α lines - experimental k(GOLDSTEIN et al., 1977)
47 Ag	2.32
62 Sn	3.07
74 W	3.11
79 Au	4.19
82 Pb	5.3

and L_α lines are summarized in Table 3.1 along with appropriate error limits where available. Numerical values of k were obtained directly from the authors. In addition experimental k curves for total L spectra ($L_\alpha + L_\beta$ ...) have been reported for a Philips EM300 instrument by SPRYS and SHORT (1976).

Combinations of experimental k values can be used along with Eqn. 3.7 to factor out the presence of Si and to obtain k_{AB} values for various combinations of elements AB. However, the k_{AB} values may change from instrument to instrument since the EDS detectors and specimen environs are not the same on all instruments. It is surprising, therefore, that there is such a paucity of k values considering the present popularity of the Cliff-Lorimer technique and the experimental errors associated with the measured values.

(3). Standardless analysis--Values of k_{AB} and k can be calculated directly using Eqns. 3.4(a),(b) (RUSS, 1973; GOLDSTEIN et al., 1977). However, the characteristics of the EDS detector must be considered. All x-ray photons less than 15 keV energy entering the active region of the Si(Li) detector are detected. However, some x-ray photons entering the detector may be absorbed in the Be window, gold surface layer or silicon dead layer. Of these absorbers, the Be window is the most important. Window absorption is particularly significant for the light elements Na, Mg and Al. Therefore, values of k can be calculated after correction for Be window absorption in Eqn. 3.4. From Eqns. 3.4 and 3.7;

$$k = \frac{A_A (Q\omega a)_{Si} \cdot e^{-\mu/\rho \frac{Si}{Be}} \rho_{Be} y}{A_{Si} (Q\omega a)_A \cdot e^{-\mu/\rho) \frac{A}{Be}} \rho_{Be} y} \tag{3.8}$$

where $\mu/\rho)_{Be}^{Si}$ and $\mu/\rho)_{Be}^{A}$ are the mass absorption coefficients of Si and element A in the Be foil, ρ_{Be} is the density of Be and y is the Be foil thickness.

Equation 3.8 has been used to calculate values of k by Goldstein et al. (1977). The following paragraphs briefly summarize the choice of values of ω, a, μ/ρ and Q used by GOLDSTEIN et al. (1977) for the calculation of K shell ionizations.

The fluorescence yield for K shell fluorescence is given to a sufficient accuracy by the expression first proposed by WENTZEL (1927) as

$$\omega_K = Z^4/(b + Z^4) \tag{3.9}$$

where b is a constant equal to 10^6. The values of a for atomic numbers 22 and above are taken from SLIVINSKY and EBERT (1972) and for atomic numbers 15 and 20 from HEINRICH et al. (1976). The a values for Na, Mg, Al and Si (0.99, 0.98, 0.98, 0.98) were extrapolated using data for higher atomic numbers. The mass absorption coefficients $\mu/\rho)_{Be}$ were taken from the tables of HEINRICH (1966). The Be window thickness of most EDS units including that of the EMMA-4 is 7.5 μm (0.3 mils).

The formula for K shell cross section Q can be written in the following general form (POWELL, 1976)

$$Q = \text{const} (1/E_K^2 U_K) \ln (c_K U_K) \tag{3.10}$$

In this equation the overvoltage U_k, is the ratio of operating energy E_o to E_k the binding energy of electrons in the K shell. The parameter c_k is assumed to be constant for the K shell.

Several K shell ionization cross-section formulas have been proposed each with a different value of c_k. POWELL (1976) evaluated various ionization cross-section formulas where U is greater than 4. The formulas for Q include those of MOTT and MASSEY (1949) Q_{MM}, GREEN and COSSLETT (1961) Q_{GC} and a linearized form of the Bethe equation which best fits the inner-shell ionization cross-section data called the Bethe-Powell cross section Q_{BP} (POWELL, 1976). The values of c_k in Eqn. 3.10 are 2.42, 1.0 and 0.65 for Q_{MM}, Q_{GC} and Q_{BP} respectively. The calculated and measured k values (GOLDSTEIN et at., 1977) are plotted versus the energy of the K_α lines in Figs. 3.3 and 3.4.

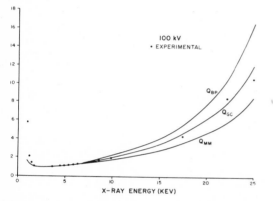

Fig. 3.3. Calculated and measured k values at 100 kV as a function of the energy of the K_α lines. [GOLDSTEIN et al., (1977)].

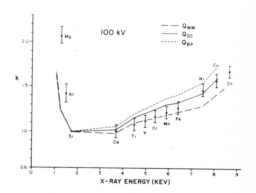

Fig. 3.4. Calculated and measured k values at 100 kV as a function of the energy of the K_α lines. Various Q values are used [GOLDSTEIN et al., (1977)].

In Figure 3.4 the error bars for the experimental data given by Cliff and Lorimer are included. The spread in experimental k values is reported to depend mainly on the quality of the chemical analyses of the standards rather than on counting statistics.

The calculated values using the GREEN-COSSLETT (1961) Q appear to fit the data best above x-ray energies of 1.7 keV. The calculated k values of GOLDSTEIN et al. (1977) using Q_{GC} are listed in Table 3.1 for the K_α lines at 100 kV. These calculated values can be directly compared with the experimental values of CLIFF and LORIMER (1975). The Na, Mg and Al data are affected by absorption in the Be window of the EDS detector, which decreases the measured intensities and hence increases the calculated k values. The calculated values of GOLDSTEIN et al. (1977) have taken this absorption problem into account, Eqn. 3.8. Nevertheless, the calculated k values for Na, Mg and Al fall well below the measured values of CLIFF and LORIMER (1975). No satisfactory explanation has been given in the literature for these discrepancies.

However, the effect of contamination could have an effect on the intensity ratios measured for these 3 elements.

In the EMMA-4 instrument used by CLIFF and LORIMER (1975) and in
all first generation TEM/STEM's the carbon contamination build-up on the
surfaces is significant. The mass absorption coefficient of carbon for
Si, Al, Mg and Na K_α radiation is large and could reduce the intensity of
the x-ray intensity of these elements significantly over that generated
in the thin film. ZALUZEC and FRASER (1979) have observed measurable
absorption effects by monitoring the characteristic intensity ratio
Ni$_{K_\alpha}$/Al$_{K_\alpha}$ in a NiAl thin film as a function of time (and thus contamination). The intensity of the Ni$_{K_\alpha}$ radiation is little affected while
the Al$_{K_\alpha}$ radiation is increasingly absorbed by the accumulation of carbon. ZALUZEC and FRASER (1979) observed an increasing Ni$_{K_\alpha}$/Al$_{K_\alpha}$ intensity ratio with time, clearly demonstrating the absorption of Al$_{K_\alpha}$ x-ray
intensity by carbon contamination. The contamination problem is discussed in some detail by Hren in Chapter 18.

To calculate the effect of C contamination, Eqn. 3.8, which is used
to calculate k, should be multiplied by the term,

$$\exp \left[- \frac{\mu}{\rho})_C^{Si} \rho_C X\right] \bigg/ \exp \left[- \frac{\mu}{\rho})_C^A \rho_C X\right] \tag{3.11}$$

where ρ_C is the density of the carbon contamination film, X is the x-ray
path length through the carbon film and $\mu/\rho)_C^A$ is the mass absorption coefficient of Al, Mg or Na in carbon. The mass absorption coefficients
are 357, 557, 904 and 1534 cm^2/g for Si, Al, Mg and Na in carbon
(HEINRICH, 1966). Multiplying Eqn. 3.8 by Eqn. 3.11, yields k values of
5.71, 2.23 and 1.38 for Na, Mg and Al if the x-ray path length through
the carbon contamination is 5 μm. These calculated values of k are in
excellent agreement with the measured k values of CLIFF and LORIMER
(1975), 5.77, 2.07 and 1.42, Table 3.1. In a later publication, LORIMER
et al. (1977) show newly measured k values of Na, Mg and Al of ~3.2, 1.6
and 1.2 respectively. The new values of LORIMER et al. (1977) can be
reconciled by the presence of a 2.5 μm thick carbon contamination layer.
Calculations, using Eqns. 3.8 and 3.11 and X equal to 2.5 μm, yield k
values of 3.08, 1.67 and 1.24 for Na, Mg and Al respectively. Contamination layers 2.5 to 5.0 μm in thickness, have not been observed to spread
uniformly over the surfaces of thin films analyzed in the TEM-STEM.
However contamination spikes > 1 μm in height have been observed when a
static beam has been placed for a long period of time on the specimen.
Since these spikes only minimally intercept the x-ray path between specimen and detector, they are probably not completely responsible for the
loss of Na, Mg and Al intensity relative to Si. Since k values for Na,
Mg and Al vary with the amount of carbon contamination or some other yet
unknown factor, the numbers listed in Table 1 can hardly be used for
quantitation of X-ray data. Values of k for Na, Mg and Al should be measured as a function of contamination rate for a specific instrument if
quantitative measurements are desired. This procedure should be adopted
until such time that serious contamination can be removed from AEM instruments.

The calculation of k values for $L\alpha$ radiation is more complex and is not discussed in this chapter. The calculation procedures are given in detail by GOLDSTEIN et al. (1977). These authors found that by using either the Mott-Massey or Bethe-Powell Q values for $L\alpha$ lines, calculated k values are in reasonably close agreement with experimental values at 100 kV. Figure 3.5 shows the calculated and measured k values for 100 kV as a function of the energy of the L lines. The agreement between calculated and measured data is quite good considering the uncertainties in the constants for the x-ray crosssections and the quality of the data available for the $L\alpha$ intensity ratio. It appears that one can use the calculation scheme for k values of $L\alpha$ lines as a first approximation although much work remains to be done.

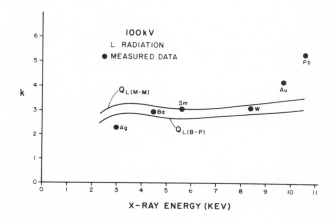

Fig. 3.5. Calculated and measured k values at 100 kV as a function of the energy of the L_α lines [GOLDSTEIN et al., (1977)].

(4). Practical analysis techniques--The use of calculated or measured k values in the ratio technique, Eqn. 3.4 appears to be straightforward. The relative simplicity of using k or k_{AB} values in Eqn. 3.4(b) rather than using bulk or thin film standards has made the ratio or Cliff-Lorimer method quite popular for quantitative analysis. In practice, most analyses require the determination of more than 2 elements in a given thin film. Equation 3.4(b) can be used to obtain a quantitative analysis if only two elements are present in a thin film since $C_A + C_B = 1$. For 3 elements or more one must measure intensity ratios for the other elements and the element concentrations must sum to 100%. For example in a 3 element analysis

$$\frac{C_A}{C_B} = k_{AB} \frac{I_A}{I_B} \tag{3.12a}$$

$$\frac{C_B}{C_C} = k_{BC} \frac{I_B}{I_C} \tag{3.12b}$$

$$C_A + C_B + C_C = 1 \tag{3.12c}$$

All element intensities I_A, I_B, I_C .... must either be measured or the weight fraction of a particular element in the sample must be known in advance of the analysis.

In many analyses, the elements of interest will not include Si. To obtain the correct k_{AB} factor for any two elements, the measured or cal-culated k value for element A relative to Si is divided by the k value for element B relative to Si. Using Eqn. 3.7, one obtains

$$k_{AB} = \frac{k_{ASi}}{k_{BSi}} = \frac{C_A}{C_{Si}} \cdot \frac{I_{Si}}{I_A} \frac{C_B}{C_{Si}} \cdot \frac{I_{Si}}{I_B} = \frac{I_B}{I_A} \cdot \frac{C_A}{C_B} \tag{3.13}$$

The k_{AB} values obtained by using k values are the same as the k_{AB} values defined by Eqn. 3.4(b). The analyst may either obtain k_{AB} values using Eqn. 3.13 and measured or calculated values of k (Table 3.1) or calcu-late k_{AB} values directly as in the following equation:

$$k_{AB} = \frac{A_A \ (Q\omega a)_B \cdot e^{-\mu/\rho)_{Be}^B} \ \rho_{Be}^y}{A_B \ (Q\omega a)_A \cdot e^{-\mu/\rho)_{Be}^A} \ \rho_{Be}^y} \tag{3.14}$$

In practice, various analysts have measured k_{AB} values directly from Eqn. 3.4(b) by using thin films of known composition C_A, C_B and mea-suring intensity ratios (I_A/I_B) directly on the sample. PANDE et al. (1977), RAO and LIFSHIN (1977) and LYMAN et al. (1978) have all deter-mined k_{AB} values for stainless steel containing Cr, Ni and Fe in this manner; but they have not reported their values. In our laboratory, ROMIG (1979) has done extensive STEM x-ray analysis work in the FeNi sys-tem. The measured k_{AB} values were obtained by using a Philips 300 TEM-STEM operating at 100 kV and equipped with a rear entry NSI-EDS detec-tor. To determine k_{NiFe} a homogeneous alloy containing 14.5 wt% Ni-85.5 wt%, Fe was used. The analyzed thin foil was 500 to 2000 Å thick. Nine separate k_{NiFe} measurements were made using a 200 Å diameter probe and a counting time of 60 sec. The measured k_{NiFe} value is 1.13 ± 0.06 at the 95% confidence level.

Any error in the measurement of the Ni content of the homogenous alloy must be added to the measured uncertainty. The uncertainty in the measurement was due to the relatively low NiK_α count rate from the thin sample. The k_{NiFe} value calculated from CLIFF and LORIMER (1975) k values, Table 3.1, is 1.16 while the calculated k_{NiFe} value from GOLDSTEIN et al. (1977) k values, Table 3.1, is 1.08. Both these k_{NiFe} values fall within the uncertainty in the experimental measurements of ROMIG (1979).

The wide range of acceptable k_{NiFe} values for the FeNi system will clearly lead to relatively large errors in a quantitative analysis using Eqn. 3.4(b). Only in a minority of cases can one claim, at this time, a level of quantitation similar to that achieved by electron probe micro-analysis (EPMA), 1-2% of the amount of the element present. For the 14.5 wt% Ni alloy, one might be able to measure the composition to ± 0.7 wt% by STEM-AEM techniques. On the other hand one can obtain the composition to ± 0.2 wt% by EPMA techniques.

In summary the preferred method of STEM x-ray microanalysis is to measure k_{AB} values using the Cliff-Lorimer technique, Eqn. 3.4(b), on suitable standards of the material of interest. Often homogenous thin films of known composition can be prepared from well characterized homogeneous bulk specimens. Analyses of other specimens at the same operating conditions can be obtained accurately using the measured k_{AB} values and the Cliff-Lorimer technique. If standards are not available k_{AB} can be obtained from measured k values, Eqn. 3.13, or from calculated k values, Eqn. 3.14.

Limitations of the Thin Film Criterion

The thin film criterion states that no absorption or fluorescence correction need be applied if the thin film is "infinitely" thin. Many analysts interpret this criterion to mean that if the film is transparent to electrons at the operating potential being employed, the film satisfies the thin film criterion. TIXIER and PHILIBERT (1969) have noted that absorption effects can occur even when the foil is transparent to the impinging electron beam if the combination of mass absorption coefficient and film thickness is large enough. Experiments have shown that for the case where the mass absorption coefficient for one of the x-ray lines I_A or I_B in Eqn. 3.4 is large, for example, $Au_{M\alpha}$ in Cu-75.5 wt% Au (JACOBS and BABOROVSKA, 1972), $Al_{K\alpha}$ in $CuAl_2$ (LORIMER et al., 1976), $Al_{K\alpha}$ in NiAl (ZALUZEC and FRASER, 1976), and $P_{K\alpha}$ in $(FeNi)_3P$ (GOLDSTEIN et al., 1976), absorption effects were observed. In these examples, the ratio C_A/C_B, Eqn. 3.4, will vary with foil thickness since one or both measured intensities I_A, I_B vary with thickness. Therefore, for some analyses, the ratio or Cliff-Lorimer technique will be inaccurate without a correction for the effect of absorption.

Both TIXIER and PHILIBERT (1969) and GOLDSTEIN et al. (1977) have given criteria whereby the analyst can judge if the thin film approximation has broken down and an absorption correction must be made. Both thin film criteria are based on a maximum allowable limit; that absorption will not alter the observed characteristic x-ray intensity ratio I_A/I_B by more than 5 to 10% from the ratio obtained from an infinitely thin film. The thin film criterion of TIXIER and PHILIBERT (1969) is that for each element measure in the thin film,

$$\chi_A \: \rho t < 0.1 \qquad\qquad (3.15)$$

or an aborption correction is necessary. The term χ_A is equal to $\mu/\rho)_{spec}^A$ · CSC α where $\mu/\rho)_{spec}^A$ is the mass absorption coefficient for the characteristic x-ray of element A in the specimen composed of elements

A, B, C ... and α is the take-off angle, the angle between the specimen and the x-ray detector. The density ρ is calculated by the relation

$$1/\rho = \Sigma \, \frac{C_i}{\rho_i} \qquad\qquad (3.16)$$

where C_i are the mass concentrations and ρ the densities for the elements in the thin film. The thin film criterion of GOLDSTEIN et al. (1977) is that for any set of two elements A and B considered in the ratio method, the absolute value of

$$(\chi_B - \chi_A) \cdot \rho t/2 < 0.1 \qquad\qquad (3.17)$$

or an absorption correction is necessary.

To see if absorption is a problem in thin foil analysis several alloy systems which have been studied by STEM (JACOBS and BABOROVSKA, 1972; GOLDSTEIN et al., 1976; LORIMER et al., 1976; and ZALUZEC and FRASER, 1976) will be discussed. Table 3.2 summarizes the values of $\mu/\rho)^A_{spec}$ and $\mu/\rho)^A_{spec}$ for each alloy system. Except for the FeNi system the mass absorption coefficients for element A in the specimen are large. It might be expected that the intensity of element A, I , would be decreased over that expected if the thin film criterion applied.

TABLE 3.2

Absorption Data for Several Alloy Systems

Alloy	A	μ/ρ^A_{SPEC} (cm^2/g)	B	μ/ρ^B_{SPEC} (cm^2/g)	ρ (g/cc)
Cu-75.5 wt% Au	Au$_{M\alpha}$	1352	Cu$_{K\alpha}$	170.6	15.0
CuAl$_2$	Al$_{K\alpha}$	3056	Cu$_{K\alpha}$	51.8	4.31
NiAl	Al$_{K\alpha}$	3435	Ni$_{K\alpha}$	59.5	5.80
FeNi	Ni$_{K\alpha}$	218	Fe$_{K\alpha}$	80.8	7.85
(FeNi)$_3$P	P$_{K\alpha}$	1645	Ni$_{K\alpha}$	199.4	7.0

Table 3.3 gives the values of the thin film criteria of TIXIER and PHILIBERT (1969) for element A and GOLDSTEIN et at. (1977) considering both elements A and B for various values of film thickness. The take-off angle α is set at 45°. The TIXIER and PHILIBERT (1969) criterion is

more conservative than that of GOLDSTEIN et al. (1977). The GOLDSTEIN et al. (1977) criterion emphasizes that the difference in x-ray absorption coefficients is the more important parameter. For example, if two characteristic x-ray lines have similar values of μ/ρ the two lines will be absorbed equally and their intensity ratio will be unaffected.

TABLE 3.3

Thin Film Criteria for Five Alloy System

Alloy Composition	(Tixier & Philibert 1969) Film Thickness (Å)			(Goldstein et al. 1977) Film Thickness (Å)		
	1000Å	3000Å	5000Å	1000Å	3000Å	5000Å
Cu-75.5 wt% Au	0.287*	0.860*	1.43*	0.125*	0.376*	0.625*
$CuAl_2$	0.186*	0.559*	0.931*	0.091	0.275*	0.458*
NiAl	0.282*	0.845*	1.41*	0.138*	0.415*	0.692*
FeNi	0.024	0.072	0.121*	0.0076	0.023	0.038
$(FeNi)_3P$	0.163*	0.488*	0.814*	0.072	0.215*	0.358*

*exceeds thin film criterion

Thin films with a density of 6 to 8 g/cc are transparent to 100 kV electrons up to ~ 3000Å, while films with a density of less than 5g/cc and more than 12g/cc are transparent to 100 kV electrons up to ~ 5000Å and 1000Å respectively. In the five examples cited in Tables 2 and 3 all but FeNi will require an absorption correction even when a good image of the specimen is apparent in the TEM-STEM.

In practice it is necessary for the analyst to test whether the thin film criterion applies for the alloy system under investigation. A table of mass absorption data similar to that of Table 3.2 should be drawn up. The mass absorption coefficients should be listed for each characteristic x-ray measured in the alloy matrix. In applying either of the two thin film criteria cited above element A should be the element in the alloy with the highest mass absorption coefficient. The GOLDSTEIN et al. (1977) criterion will have the largest value when the element in the alloy with the smallest mass absorption coefficient is chosen as B. The thin film criterion should be applied for the maximum film thickness at which it is still possible to obtain an image of the specimen. If the analyst desires a more conservative thin film criterion, the limit of < 0.1 in Eqns. 3.15 and 3.17 can be decreased. A limit of < 0.01 allows for no more than 0.5 to 1% absorption of emitted x-rays in the thin film.

Absorption Correction

    If the thin film criterion has broken down, an absorption correction for thin film analysis must be applied. Using the Cliff-Lorimer or ratio technique, Eqn. 3.4, the following type of correction can be made:

$$\frac{C_A}{C_B} = k_{AB} \frac{I_A}{I_B} \frac{f(\chi)_B}{f(\chi)_A} \qquad (3.18)$$

where $f(\chi)$ and $f(\chi)$ are the absorption corrections for elements A and B in the thin film. Both TIXIER and PHILIBERT (1969) and GOLDSTEIN et al. (1977) have derived the absorption correction by independent means. Both groups obtain a correction of the following form:

$$\frac{C_A}{C_B} = k_{AB} \frac{I_A}{I_B} \cdot \left(\frac{(\mu/\rho)^A_{SPEC}}{(\mu/\rho)^B_{SPEC}} \right) \cdot \left(\frac{1 - e^{-(\mu/\rho)^B_{SPEC} \csc \alpha (\rho t)}}{1 - e^{-(\mu/\rho)^A_{SPEC} \csc \alpha (\rho t)}} \right) \qquad (3.19)$$

in which the ratio $f(\chi)_B/f(\chi)_A$ is combined. As is often the case, absorption is only significant for one of the elements and various simplified forms of Eqn. 3.19 have been proposed. It does not appear necessary to use a simplified form of Eqn. 3.19 since the solution of the equation is not complex.

    The absorption correction given in Eqn. 3.19 is a function of C_A, C_B, C_C etc. since the absorption coefficients and the density ρ are a function of composition. To solve Eqn. 3.19 for the concentration ratio C_A/C_B, the ratio I_A/I_B is measured, k_{AB} is either obtained from measured k values Eqn. 3.13 or calculated from Eqn. 3.14 and the absorption correction is applied using assumed values of C_A, C_B, etc. Several iterations of Eqn. 3.19 may be necessary in order that assumed and calculated concentrations converge to the same value.

    Complicating the calculation of the thin film absorption correction is the fact that it is necessary to measure the specimen thickness at each analysis point. Several techniques for measuring specimen thickness have been developed. LORIMER et al. (1976) have a technique of tilting the specimen after analysis to measure the separation between pairs of contamination spots on top and bottom of the film and to calculate the corresponding thickness. JOY and MAHER (1975) have suggested a technique for measuring crystal thickness using a calibration curve developed from the relative transmission of the primary electron beam through the specimen. Other techniques include the trace method, counting extinction contours in a wedge shaped crystal or stereomicroscopy. These techniques are seldom accurate to within ± 10%. KELLY et al. (1975) have described a method which makes use of measurements of spacing of intensity oscillations in convergent beam diffraction patterns obtained with the STEM. Foil thicknesses to ± 2% can be determined as shown by RAO (1976). LOVE et al. (1977) have reviewed some of these techniques and have pointed out that the contamination technique determines the total foil thickness including any surface film while the convergent beam diffraction method established only the thickness of foil which diffracts. They also point out the use of characteristic and continuum x-rays to calibrate foil thickness. Conversion to absolute thicknesses necessitates, however, calibrating the x-ray intensity measurement as a function of foil thickness by the above techniques. Depending on the specimen distortion, spot analysis position and specimen tilt, it should be possible to use one of these techniques for thickness measurement. In practice, the measurement of foil thickness may require more work than obtaining the intensity ratio data itself.

When element A can be excited by continuum x-ray radiation or by characteristic fluorescence due to some lines emitted by element B in the foil, a fluorescence correction might be needed. PHILIBERT and TIXIER (1975) have found that continuum fluorescence is negligible and that characteristic fluorescence will be negligible if $(\mu/\rho)^B_{spec} \cdot \rho t \ll 1$. TIXIER (1979) has stated that the characteristic fluorescence correction is small if $(\mu/\rho)^B_{spec} \cdot \rho t < 0.1$.

A binary alloy system in which characteristic fluorescence is significant in solid samples is FeCr. In this system, $Cr_{K\alpha}$ is fluoresced by $Fe_{K\alpha}$ radiation. For a very small amount of Cr in FeCr, the value of $(\mu/\rho)^{Fe}_{spec}$ is 373 cm^2 g. For 2000Å thick films the value of $(\mu/\rho)^{Fe}_{spec} \cdot \rho t$ equals 0.057. Such a value indicates that a fluorescence correction might be necessary for these binaries.

Figure 3.6 shows the apparent increase in wt% Cr for thick and thin foil specimens for a series of Fe-Cr alloys (LORIMER et al., 1977). The thin foil specimens were estimated to be approximately 2000Å thick. The bulk analyses have been corrected for absorption in the specimen. Although the fluorescence effects in the thin foil are less than those in the bulk sample, they are still significant.

A correction for characteristic fluorescence of element A by element B in thin films has been developed by PHILIBERT and TIXIER (1975) and is given as:

$$\frac{I^A_f}{I_A} = 2\ \omega_B\ C_B\ (\frac{r_A - 1}{r_A}) \cdot \frac{A_A}{A_B} \cdot (\mu/\rho)^B_A \cdot (\mu/\rho)^B_{SPEC} \cdot \frac{E_{c_A}}{E_{c_B}}\ (\rho t)^2 \qquad (3.20)$$

where I^A_F is the fluorescence intensity which is added to the primary intensity I_A, ω_B is the fluorescence yield for element B, r_A is the absorption jump ratio for element A, $(\mu/\rho)^B_A$ is the mass absorption coefficient of element B radiation in element A, A_A and A_B are the atomic weights of

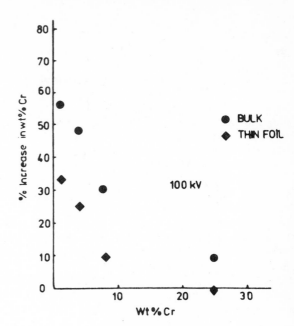

Fig. 3.6. Variation in % increase in Cr concentration as
a function of Cr concentration
for bulk and thin specimens at
100 kV [LORIMER et al.,
(1977)].

elements A and B and E_{c_A} and E_{c_B} are the critical excitation energies for
the characteristic lines of elements A and B.

If the thin film criterion has broken down (see Limitations above)
for a particular analysis, it may be necessary to apply both absorption
and fluorescence corrections. The absorption and fluorescence correction require the measurement of the thickness of the film at each point
of analysis and both corrections are a function of the unknown concentrations. In using both corrections, Eqns. 3.18-3.20, the following
formula is developed:

$$\frac{C_A}{C_B} = k_{AB} \frac{I_A}{I_B} \left(\frac{f(\chi)_B}{f(\chi)_A} \right) \cdot \frac{1}{\left(1 + \frac{I_f^A}{I_A} \right)} \tag{3.21}$$

To minimize the errors in such corrections, it is very important to analyze as thin a portion of the film as possible.

Using STEM x-ray microanalysis one can measure the composition of
very small particles, establish the presence of elements segregated at
or very close to grain boundaries and determine the interface compositions of various adjoining phases. Because the electron beam diameter
can be reduced to $\leq 50\overset{\circ}{A}$ in size, one might expect to obtain compositional analyses at this level of spatial resolution. Unfortunately,
even in thin films electrons are elastically scattered and the regions
from which x-rays are produced are often much larger than the beam dia-

meter. In this section, we will review calculations of the x-ray spatial resolution and show that a simple equation will give first order
values of spatial resolution for thin films. In addition, experimental
methods and measurements of x-ray spatial resolution will be described.

Analytical and Computer Models

The spatial resolution for chemical analysis in a thin foil is a
function of atomic number, specimen thickness and the accelerating voltage. GOLDSTEIN et al. (1977) have estimated the effective electron beam
spreading or broadening by assuming that electron scattering takes place
at the center of the thin film, that the dominant process causing electron beam spreading is elastic scattering by atomic nuclei and that the
electron beam is a point source. Figure 3.7 shows the model used to calculate beam broadening (b) in a thin film of thickness t due to single
scattering through an angle ϕ.

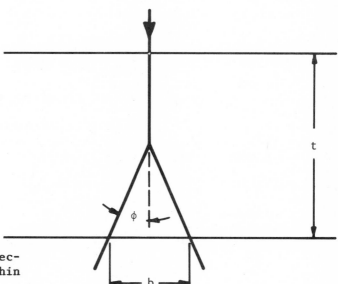

Fig. 3.7. Model for electron beam broading in a thin
film of thickness t.

The equation which relates beam broadening as a function of E_0 and t is

$$b = 625 \; \frac{Z}{E_o} \; \frac{\rho^{\frac{1}{2}}}{A} \, t^{3/2} \qquad (3.22)$$

where b is in cm, Z is the atomic number, A is the atomic weight, E_0 is
in kV, ρ is the density of the film in g/cm^3 and t is in cm. The beam
broadening varies inversely as E_0 and increases with film thickness.
The beam broadening equation is derived by assuming that the source size
can be defined as that within which 90% of the electron trajectories
lie.

GOLDSTEIN et al. (1977) calculated values of beam broadening b for
a zero diameter electron beam using Eqn. 3.22. Table 3.4 gives these
values as a function of various film thickness for pure element targets
at E_0 = 100 kV. According to GOLDSTEIN et al. (1977), the single scattering model breaks down at $\phi \geq 20°$ (Fig. 3.7) on the basis that at
larger angles the path length, and hence the probability of scattering
becomes significantly increased. Therefore, the model cannot be applied
at large scattering angles. For typical electron microscope thin film
specimens observed at 100 kV, (500Å of Au, 1000Å of Cu and 3000Å of Al)
the predicted beam spreading is 173Å, 214Å and 422Å respectively. To
obtain the x-ray spatial resolution or x-ray source size the electron
beam diameter d must be added to the amount of beam broadening b predicted by Eqn. 3.22. With the amount of beam spreading calculated for
typical thin films of Al, Cu and Au, there is little advantage in using
electron beams less than 50Å in diameter to improve x-ray spatial resolution. The loss of x-ray intensity at beam diameters < 50Å is too
great in terms of the marginal improvement in x-ray spatial resolution
obtained. Only when very thin films, < 100Å for Au, < 200Å for Cu and <
500Å for Al, are used, can a smaller electron beam size be justified.

TABLE 3.4

Electron Beam Spreading in Thin Films

E_o = 100 kV, Broadening b(Å)

Film Thickness

Element	100Å	500Å	1000Å	3000Å	5000Å
Carbon	1.6	18	51.3	267	574
Aluminum	2.6	29	81.2	422	909
Copper	6.8	76	214	1117	*
Gold	15.5	173	*	*	*

* $\phi \geq 20°$

In summary, the broadening equation, Eqn. 3.22, indicates that maximum
x-ray spatial resolution will be obtained when film thickness and electron beam size is minimized and the accelerating voltage is maximized.
According to Eqn. 3.22, the spatial resolution can be improved by a
factor of 2 by employing a 200 kV rather than a 100 kV beam.

Monte Carlo calculations of beam broadening and x-ray source size
in thin film x-ray microanalysis have been performed by FAULKNER et al.
(1977), KYSER and GEISS (1977), GEISS and KYSER (1979) and NEWBURY
(1979). The paths of individual electrons are calculated in a step-wise
manner through the solid as the electron undergoes elastic and inelastic

scattering. NEWBURY (1979) and GEISS and KYSER (1979) have calculated
the diameter of a cylinder containing 90 percent of the x-ray generation
events for various thin foils at beam excitations of 100 kV and for a
zero diameter electron beam. The results of their calculation are shown
in Table 3.5 and can be compared directly with the results of the single
scattering model, Table 3.4. The Monte Carlo values are slightly higher
than the analytical model calculations for 100Å foils. For foils
greater than 1000Å thick, the Monte Carlo values are less than the
values predicted by the analytical model. The deviations are no greater
than 40% for any of the calculations. The Monte Carlo method provides
the best calculation of broadening (b) and x-ray source size because the
model includes such effects as electron backscattering, multiple scattering, and non-normal electron beams and can account for the actual
width of the electron probe. An interesting point, however, is that
above a certain mass thickness, $\rho t \cong 185$ $\mu g/cm^2$, which is equivalent to
1000Å of Au or 2000Å of Cu, the beam spreading becomes independent of incident electron probe size (for small probes) (GEISS and KYSER, 1979).

The fact that the results of the analytical model agree moderately
well with the Monte Carlo model (within 40%), allows the analyst to make
a first order estimate of the x-ray source size using Eqn. 3.22. Although the Monte Carlo calculations yield more accurate results of x-ray
source sizes for thicker films, they are more tedious and the necessary
computer programs may not be readily available. Because of the importance of the Monte Carlo method and its relative sophistication, a description of the model and calculation scheme is given by Kyser in Chapter 6 of this book. In addition, various applications of the method are
given by Kyser including some calculations of x-ray spatial resolution.

TABLE 3.5

Electron Beam Spreading in Thin Films Calculated

with Monte Carlo Method

$$E_o = 100 \text{ kV, Broadening b (Å)}$$

Element	\multicolumn{6}{c}{Film Thickness}
	100Å	500Å	1000Å		3000Å	5000Å
Carbon	2.2*	19*	41*	50[+]	160*	330*
Aluminum	4.1*	30*	76*	90[+]	300*	664*
Copper	7.8*	58*	175*	160[+]	970*	2440*
Gold	17.1*	150*	522*	380[+]	5990*	17250*

* NEWBURY (1979)

[+] GEISS and KYSER (1979)

Measurements of Spatial Resolution

Various determinations of x-ray spatial resolution have been made
using 3 general experimental approaches: 1) Measurements of small precipitates in matrices of differing composition, 2) Determinations of element segregation at grain boundaries, and 3) Measurements of composition
profiles across phase interfaces. Each of these experimental approaches
will be described and the advantages and disadvantages of each method
will be pointed out.

(1). Small precipitates - In many practical problems, an analysis
of a precipitate particle embedded in the matrix of a suitable TEM/STEM
foil is desired. Naively one would expect that if one places a focused
electron beam ($\leq$ 200$\overset{\circ}{A}$) on a particle > 200$\overset{\circ}{A}$ in apparent diameter, that
one would be able to perform a quantitative x-ray analysis since none of
the x-ray lines from the matrix will be measured. Figure 3.8 shows the
EDS spectrum from a 150 - 200$\overset{\circ}{A}$ diameter FeNi metal particle embedded in
the glass matrix of Apollo 16 sample 60095 (MEHTA and GOLDSTEIN, 1979).
The CuKα peaks are contributed by specimen continuum excitation of the
Cu supporting grid bars. The SiKα, CaKα and AlKα peaks in the EDS spectrum are generated from the surrounding glass. Although these metal particles are approximately the same size as the focused electron beam,
they are far smaller than the nominal 1500$\overset{\circ}{A}$ thin foil. Therefore, it is
not surprising that x-rays are generated from the surrounding glass matrix. In this case, it is very difficult to obtain even a qualitative
analysis of the metal particle because the FeKα peak is contributed by
both particle and matrix.

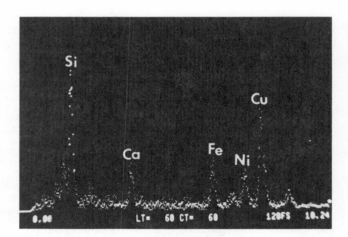

Fig. 3.8. EDS spectrum from a 150-200A diameter metal particle in the
glass matrix of Apollo 16 sample 60095.

Figure 3.9 shows a TEM image of two 2 phase (αFeNi, FeS) metal
spheres $\sim$ 3500$\overset{\circ}{A}$ in diameter embedded in a 1500$\overset{\circ}{A}$ film of a silicate
glass, Apollo 15 sample 15286 (MEHTA et al., 1979). Figure 3.9   also

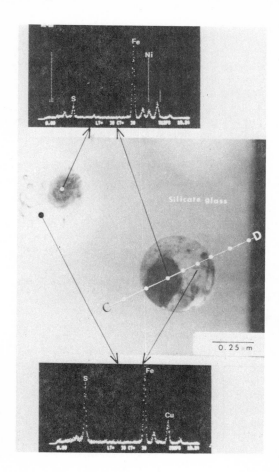

Fig. 3.9. TEM photomicrograph of ∿3500A metal particles in Apollo 15 glass 15286. The EDS spectra are obtained by STEM from each of the two metal phases. Line CD indicates the positions where a composition profile was obtained across the particle.

shows EDS spectra obtained by STEM at E_0 = 100 kV from a FeNi rich (dark) area and a sulfide rich (light) area in the metal particle. Each spectrum was obtained from a point analysis ∿ 1000Å from the edge of metal particle. The $Cu_{K\alpha}$ peaks are also caused by specimen continuum excitation of the supporting Cu grid bars. Using an estimated value of beam spreading b from Eqn. 3.22 of 300Å and an electron beam size d of 200Å the x-ray source size, b + d, is ∿ 500Å. The x-ray spatial resolution is much smaller than the particle size and no x-ray signal should be measured from the matrix in this case. A close look at the x-ray spectra in Fig. 3.9 shows a small $Al_{K\alpha}$ peak to the left of the Si peak and perhaps some indication of a $Ca_{K\alpha}$ peak. Although it appears from this measurement that the calculated x-ray spatial resolution is under-estimated, it is more probable that the metal particle does not extend through the whole thickness of the foil. On the other hand, GEISS and KYSER (1979) have pointed out that a spike of contamination build-up will have an effect on spatial resolution. Even before the focused beam hits the specimen, it will be spread by elastic interactions with the relatively thick contamination spike. Therefore, for several reasons, there are some real limits on the size of particles which can be ana-lyzed if they are embedded in various matrices.

FAULKNER et al. (1977) have taken advantage of the matrix contribution to an EDS spectrum of a particle and have developed a technique for obtaining x-ray spatial resolution. The method consists of measuring the x-ray output from a foil of known thickness containing particles of known diameter. The model used by FAULKNER et al. (1977) is shown in Fig. 3.10. The electron beam is normal to a film of thickness t, and a particle of radius r_1 is embedded in the film. The x-ray emission volume source size is assumed to be cylindrical with a diameter of $2r_2$.

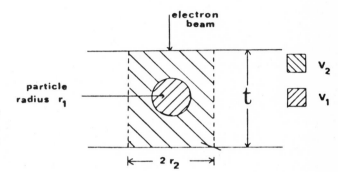

Fig. 3.10. Method for determining x-ray spatial resolution (FAULKNER et al., (1977)1.

From this figure the x-ray spatial resolution, $2r_2$ is given by

$$2\ r_2 = \left[\frac{4r_1^{\ 3}}{3t} \cdot \left(1 + \frac{V_2}{V_1}\right)\right]^{\frac{1}{2}} \cdot 2r_1 \qquad (3.23)$$

where V_1 and V_2 are the particle volume and x-ray emission volume respectively. To obtain V_1 and V_2, measurements are made of the intensities of one element uniquely present in the precipitate and one element uniquely present in the matrix. These intensities are proportional to the weight fractions of these elements in the irradiated phase and, hence, the volume of the phase. This model also assumes that no fluorescence or absorption corrections are necessary and that the Cliff-Lorimer k values of the two measured elements are similar. FAULKNER et al. (1977) used Eqn. 3.23 to measure the x-ray source diameter for TiC particles, 450-700Å in diameter, in an austenitic alloy. The values obtained were ∿ 500Å at E_0 = 200 kV and ∿ 1000Å at E_0 = 100 kV for the 1500 ± 200Å films.

The measured x-ray source sizes are more than twice those calculated using either Eqn. 3.22 or the Monte Carlo model, Table 3.5, assuming the diameter of the focused beam d to be ∿ 100Å. It is not clear why calculated and measured results differ by so much but several assumptions in the FAULKNER et al (1977) model should be examined, namely that the x-ray source size is cylindrical in shape and that the Cliff-Lorimer k values are similar. The x-ray source size actually increases continually as the electron beam passes through the sample. Therefore, not only is the calculation of V_2 more complex; but the distance of the particle from the top of the foil is important and enters into the calculation. In addition, the extent of electron beam tailing is significant in such a calculation. However, the use of small precipitates of known composition to measure spatial resolution appears to be worth pursuing. In its present stage of development, the method probably yields results within a factor of 2 of the correct spatial resolution.

(2). Segregation at grain boundaries - Local compositional changes at or near the grain boundaries may exist in equilibrium with the matrix at elevated temperatures where diffusion is rapid or in nonequilibrium with the matrix where preferential phase transformations or preferential diffusion of solute atoms along the boundary occurs. It is generally thought that the segregation is localized, within less than 100Å of the boundary. Although it may be possible to focus the electron beam of the STEM instrument to $\leq$ 50Å, the x-ray source size will inevitably be larger than the localized region of segregation unless the foil is exceedingly thin.

DOIG and FLEWITT (1977) have demonstrated the presence of grain boundary films $\leq$ 100Å thick by STEM techniques. They studied grain boundary segregation of Sn and P in Fe-3 wt% Ni alloys containing 0.09 wt% Sn and 0.06 wt% P. A model was developed (DOIG and FLEWITT, 1977) to obtain film compositions in these alloys. This model should be applicable generally to problems of grain boundary segregation. Figure 3.11 shows a schematic diagram from which the model was developed.

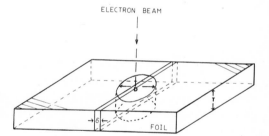

ELECTRON BEAM

Fig. 3.11. Schematic diagram showing the microanalysis of a grain boundary region using STEM.

The x-ray source is assumed to be cylindrical with a spatial resolution D the grain boundary film thickness is δ and the foil thickness is t. A mass balance for any element can be constructed for the measured concentration C within the excited x-ray volume:

$$C_m = [C_{gb}(D\ \delta\ t) + C_{matrix} \left[\frac{\pi D^2 t}{4} - D\ \delta\ t \right]] \Big/ \frac{\pi D^2 t}{4}$$ (3.24)

For a film at a grain boundary, the concentration of the segregated species is C_{GB}. The concentration within the matrix is C_{matrix}. The term C_m is measured using the quantitative analysis techniques discussed earlier in this chapter. DOIG and FLEWITT (1977) measured C_m for the two embrittled alloys and obtained C_m = 1.4 At% P and C = 0.22 At% Zn at the grain boundary. If δ, D and C_{matrix} are known, the value of C_{gb} can be obtained from Eqn. 3.24. Since none of these values are known precisely, the C_{gb} values obtained by these authors are imprecise. In the P containing alloy, Doig and Flewitt argued that the film is composed of

the phase $(FeNi)_3P$. If one assumes that $C_{matrix} \sim 0$ and that the film
is a $(FeNi)_3P$ phase, the value of $C_{meas} = 1.4$ At% yields a ratio of δ/D
of ~ 0.05 using Eqn. 3.24. DOIG and FLEWITT (1977) used a 150Å focused
beam at $E_0 = 80$ kV and analyzed foils $\sim 2000 \pm 500$Å in thickness. Using
Eqn. 3.22(b) is 650Å and after accounting for the electron beam diameter
d the x-ray spatial resolution, b + d, in Fe3Ni is estimated to be $\sim$
800Å. Therefore, the film thickness detected is ~ 35Å wide. If C_{matrix}
is not $\cong 0$, the value of δ is smaller.

Although the x-ray spatial resolution of the STEM instrument is not
sufficient to resolve and quantitatively measure compositions in ≤ 100Å
grain boundary films, it is capable of detecting these films, even if
they are ≤ 35Å in width. Equation 3.24 is useful for determining the C_{gb}
composition but the equation requires too much detailed information such
as the value of C_{matrix} and δ to yield a quantitative value. The fact
that the x-ray source size is not cylindrical further complicates the
analysis. Future calculations using Monte Carlo techniques should allow
more accurate formulations of the type given in Fig. 3.11 and Eqn. 3.24.
It appears, however, that this approach may not be the most ideal method
to obtain the x-ray spatial resolution.

(3). Composition profiles across phase interfaces - The most successful method used in electron probe microanalysis to obtain the x-ray
resolution is to move the electron beam point by point across an interface between two phases which show a concentration discontinuity
(GOLDSTEIN and YAKOWITZ, 1975). The finite volume from which x-rays are
excited in a sample causes the true concentration profile to appear
"smeared." The width of the x-ray excitation region can be determined
by drawing a tangent to the measured concentration profile at the midpoint of the discontinuity and by measuring the intercept on the distance axis at the composition of each phase. Such a measurement is
shown in Fig. 3.12 where the Ni concentration profile was measured by
STEM techniques in 500Å steps across a two phase interface between α and
γ in a 14.7 wt% Ni-Fe alloy heat treated at 500°C for 127 days (ROMIG,
1979). Contamination prevented smaller steps from being used. The operating conditions are $E_0 = 100$ kV, and a 200Å diameter electron beam was
employed. The foil is ~ 1500Å in thickness. Measurements were made
within a Philips 300 TEM/STEM and the interface was approximately parallel to the electron beam. The total width of the x-ray excitation region is ~ 500Å.

Several other experiments have been reported for ferrous materials,
for example, LYMAN et al., 1978; RAO and LIFSHIN, 1977; WILLIAMS and
GOLDSTEIN, 1978; and PANDE et al., 1977, in which the focused electron
beam was moved across a two phase interface in ~ 500Å steps and the Ni,
Cr and/or Fe x-ray intensity measured. Figures 3.13-3.15 show results
from LYMAN et al. (1978), RAO and LIFSHIN (1977) and LIN et al. (1979),
respectively. Table 3.6 summarizes the measured spatial resolutions obtained by the method and the calculated x-ray spatial resolutions using
the single scattering model, Eqn. 3.22. Considering that the film thickness is estimated in several cases and the measurement of spatial resolution is not too precise in part due to contamination, the agreement between measured and calculated x-ray resolution is quite good. One instrumental difficulty, that of specimen drift, has been pointed out by

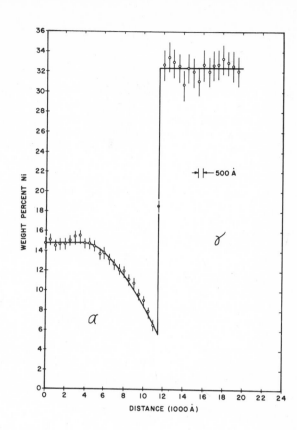

Fig. 3.12. Ni profile across α/γ interface in a 14.7 wt% NiFe alloy.

RITTER et al. (1979). The authors measured drift of the specimen due to movement of the tilt holder of 100 - 200Å/minute in their JEOL instrument. This effect will surely enlarge the effective x-ray spatial resolution and will severely limit the ability to do high resolution measurements across interfaces on a routine basis. Nevertheless, the method of measuring composition profiles across phase interfaces appears to be the most successful procedure to date for obtaining x-ray spatial resolution. The results so far tend to confirm both the approach of GOLDSTEIN et al. (1977) and the Monte Carlo calculations of KYSER and GEISS (1977) and NEWBURY (1978).

3.4 SENSITIVITY LIMITS

The sensitivity limits for STEM x-ray analysis are defined by the minimum detectable mass (MDM) and the minimum mass fraction (MMF). These limits are determined by several factors such as the spurious x-ray count rate, the efficiency and collection geometry of the EDS detector and the cross sections for the production of characteristic and continuum x-radiation. The spurious x-ray effects as discussed by Zaluzec, Chapter 4 must be minimized and the spurious x-ray count rate must be decreased below the continuum count rate from the sample of interest if minimum sensitivity limits are to be achieved. JOY and MAHER in a recent paper (1977) have discussed the optimization of sensitivity limits and some of their material is summarized here.

The MDM is the measure of sensitivity in analyzing individual particles on extraction replicas, sputtered thin films, stained biological sections, etc. The material of interest is either free standing or in a weakly scattering matrix. The background from the STEM system is, therefore, the limiting factor in the analysis.

The MDM that can be measured can be lowered by increasing the count rate of the element of interest in the specimen. Using the appropriate equations developed by JOY and MAHER (1977)

$$MDM \sim \frac{1}{P_A \cdot \tau \cdot J} \qquad (3.25)$$

where τ is the counting time, J is the electron current density impinging on the sample and

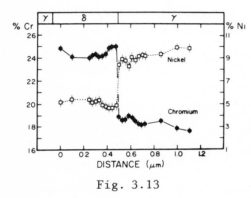

Fig. 3.13 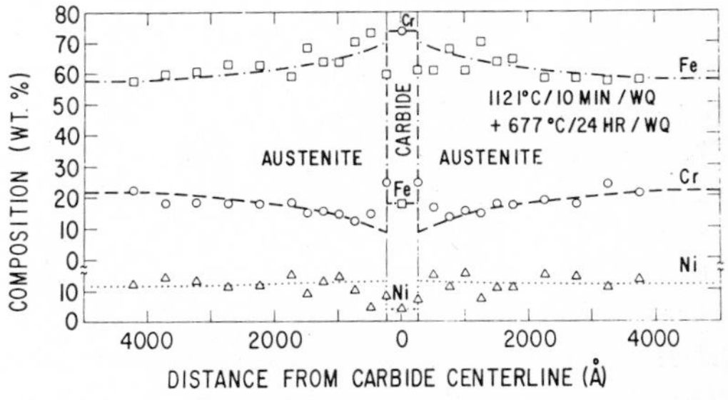   Fig. 3.14

Fig. 3.13. Cr and Ni profile across w~q interface in 304 stainless steel.

Fig. 3.14. Cr, Fe and Ni profile across carbide/matrix interface in a heavily sensitized steel.

Fig. 3.15. Ni profile across α/γ interface in the Edmonton iron meteorite.

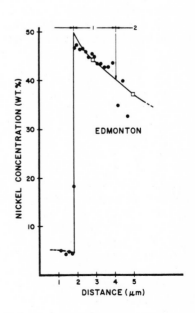

Fig. 3.15

$$P_A = Q_A \, \omega_A \, a_A \, T \qquad\qquad (3.26)$$

The term T is the EDS detector efficiency (including solid angle and energy effects).

To improve the MDM, only τ and J can be increased since P is constant for a given element and detector configuration. In practice, one can increase J by increasing emission of the filament or by using a brighter electron source (LaB_6 vs. W for example). The counting time can be increased but this is limited by contamination, specimen and electronics drift. According to JOY and MAHER (1977), if the material to be analyzed is sufficiently well dispersed, then normal TEM operation may be preferable to STEM operation. The current density is higher in TEM and often the entire mass of the object to be analyzed can be illuminated by the electron beam. In STEM only a small portion of the sample will be analyzed at one time.

JOY and MAHER (1977) have calculated MDM for various elements using measured ionization cross sections and Poisson statistics to assess the validity of the presence of a characteristic peak. Figure 3.16 shows the calculated MDM plotted versus atomic number Z for τ = 100 seconds. The effect of E_0 and J are also considered in the figure. These calculations show that the MDM of an element (10<Z<40) is of the order 5 x 10^{-20} g for 100 sec counting time, an incident current density from a thermionic emitter of $20 A/cm^2$ and accelerating voltages in the range 60 - 100 kV. SHUMAN et al. (1976) have also developed a method to optimize the MDM and have shown the MDM, in a 100 sec. collection time using a thermionic gun, to be $\sim 10^{-19}$ g in agreement with previous analyses of the iron core of single ferritin molecules (SHUMAN and SOMLYO, 1976). The use of a field emission gun will decrease the MDM because of the higher electron current density J (Eqn. 3.25). OPPOLZER and KNAUER (1979) have measured a current density of $10^4 A/cm^2$ for a 100Å diameter beam using a field emission gun on a dedicated STEM. With their instrument it should be possible to measure a MDM of $\sim 2 \times 10^{-21}$ g.

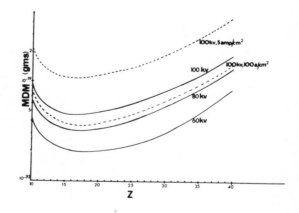

Fig. 3.16. Calculated minimum detectable mass (MDM) versus Z for 100 sec counting time. The solid lines are for J = 20 amg/cm^2 [JOY and MAHER, (1977)].

The minimum mass fraction MMF is a measure of the sensitivity of detecting one element in the presence of another. It is an important

TABLE 3.6

Calculated and Measured X-ray Resolution for Ferrous Materials

Reference	Foil Thickness (Å)	E (kV)	Beam Size d(Å)	Spatial Resolution b(Å), Eqn. 3.22	Total X-ray Source Size (Å) calculated (d+b)	Total X-ray Source Size (Å) measured
Lyman et al. (1978	1500	100	10-50	350	360-400	≈ 300
Rao and Lifshin (1977)	1000-2000	200	50	100-270	150-320	≈ 500
Lin et al. (1979)	1500	100	200	350	550	≈ 500
Romig (1979)	≈ 1500	100	200	350	550	≈ 500
Pande et al. (1977)	1000-1500	100	100	190-350	290-450	≈ 500

parameter in analyzing precipitates or compositional gradients in multi-phase alloys, silicates, ceramics or other materials. In determining the MMF, the important and limiting factor is the continuum radiation obtained from the exited x-ray volume.

ZIEBOLD (1967) has shown that trace element sensitivity or the minimum mass fraction limit can be expressed as

$$MMF \sim \frac{1}{(P/B \cdot P \cdot \tau)^{\frac{1}{2}}} \qquad (3.27)$$

where P is the pure element counting rate, P/B is the peak/background ratio of the pure element (the ratio of the counting rate of the pure element to the background counting rate of the pure element) and τ is the counting time.

To improve MDM, all three terms in Eqn. 3.27 should be maximized. As discussed previously, τ is limited by contamination and electronic-mechanical stability of the instrument. The peak intensity P can be increased by increasing the electron current density J and by optimizing the x-ray detector configuration (see Chapter 4 by Zaluzec).

Several authors have developed expressions for P/B ratios in thin film analysis (JOY and MAHER, 1977; ZALUZEC, 1978). These expressions are quite complex and the interested reader is directed to the original papers for specific details. Figure 3.17 is taken from ZALUZEC (1978) and shows the predicted variation in P/B as a function of atomic number (Z) for various values of E_0 with the x-ray detector observation angle at 90° relative to the forward scattering direction (the geometry found in most present TEM/STEM instruments). These calculations show a direct increase in P/B for pure elements with E_0 for all elements. The sharp drop at Z = 45 corresponds to a change from K to L lines. Clearly then P/B ratios can be increased and MMF improved at higher operating voltages. Calculations by ZALUZEC (1978) also show that P/B increases with x-ray detector observation angle, and levels out after $\sim$ 125° at E_0 = 100 kV. Therefore, P/B ratio will be increased and MDM improved if

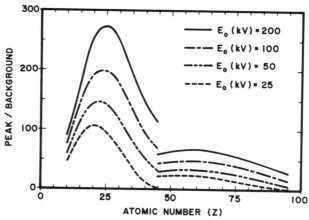

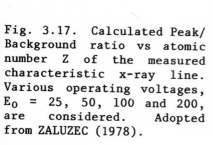

Fig. 3.17. Calculated Peak/Background ratio vs atomic number Z of the measured characteristic x-ray line. Various operating voltages, E_0 = 25, 50, 100 and 200, are considered. Adopted from ZALUZEC (1978).

the x-ray detector is placed at $\geq$ 125° relative to the forward scattering direction. In this configuration, the detector looks down at the specimen and the thin film can be placed at zero degree tilt. Several AEM manufacturers are adopting their instruments to incorporate these high take-off angle x-ray detectors.

The P/B ratio is a function of the average atomic number of the film and will decrease with increasing atomic number since the intensity of the continuum background radiation B is a direct function of average atomic number Z of the matrix. Therefore, one can optimize MMF by making measurements of elements $20 \leq Z \leq 45$ in low Z matrices. A choice of such optimum conditions is often not possible since the problem itself dictates both the elements to be analyzed and the matrix to be considered.

JOY and MAHER (1977) have developed theoretical expressions for the minimum mass fraction in STEM microanalysis. Figure 3.18 shows the calculated MMF versus the atomic number Z of the characteristic x-ray line analyzed in a 1000Å thick crystal of silicon at E_o = 100 kV. Two sets of probe diameters d and current densities J are considered (d = 100Å, J = 20 A/cm^2 in STEM and d = 1000Å, J = 100 A/cm^2 in TEM). For the smaller probe diameter of 100Å the MMF (15 < Z < 40) is only $\sim$ 3 wt% at 100 kV. If the region of interest is larger, > 1000Å, or if the compositional gradient occurs over distances of a micron or more, it is advantageous to use a larger spot size, $\sim$ 1000Å, and higher current densities. In this case, the MMF (15 < Z < 40) is reduced to 0.2 wt% at 100 kV. RITTER et al. (1979) have shown several experimental measurements in which increasing the beam current improves the x-ray intensities and counting statistics. As a comparison to the thin film analysis, MMF values of $\sim$ 0.01 wt% at 30 kV are typical for a SEM-EPMA using a wavelength dispersive detector (GOLDSTEIN and YAKOWITZ, 1975) where x-ray source size is $\sim$ 1 μm.

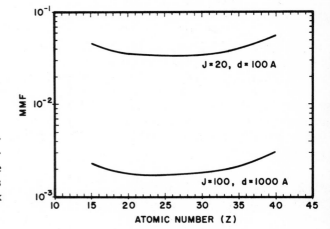

Fig. 3.18. Calculated minimum mass fraction (MMF) versus atomic number Z of the element whose x-ray line is measured in a silicon matrix [JOY and MAHER, (1977)].

The analyst can determine MMF directly for a sample of interest by making several relatively simple measurements on the sample thin film. The following discussion develops the necessary equations. As the elemental composition C approaches low values the number of x-ray counts of element A, I_A, from the sample is no longer much larger than the number of x-ray counts from the background continuum radiation I_b^A for element A. The analysis requirement is to detect significant differences between the sample and continuum background generated from the sample. To say that I_A is "significantly" larger than I_b^A the value of I_A must exceed I_b^A by $3(2I_b^A)^{1/2}$. To convert the number of x-ray counts to MMF in wt%, we use the ratio technique, Eqn. 3.4 where

$$\frac{I_A - I_b^A}{I_B - I_b^B} \geq \frac{3(2I_b^A)^{\frac{1}{2}}}{I_B - I_b^B} = k_{AB} \frac{MMF}{C_B} \qquad (3.28)$$

The term I_b^B is the background continuum radiation for element B. The x-ray intensity for element B in the above equation is expressed in terms of total accumulated x-ray counts above background, $I_B - I_b^B$, (not counts/sec).

To use Eqn. 3.28, it is assumed that the concentration of element B, C , is accurately known. Since I and I vary with foil thickness there are some difficulties in establishing MMF experimentally. However if both I , I and I are measured simultaneously and the ratio technique, Eqn. 3.28 is employed, a value of MMF can be established at each film thickness.

Another approach to determine MMF is to employ a thin film where element A is present and its concentration in the sample, C , is known. For this type of analysis,

$$\frac{3(2I_b^A)^{\frac{1}{2}}}{(I_A - I_b^A)} \cdot C_A = MMF \qquad (3.29)$$

FAULKNER et al. (1977) have made some measurements of background intensity for Fe in a pure Fe foil 1500Å thick. A ∿ 50Å diameter electron beam was used to determine the detection limit of Fe. At 100 kV after $\tau = 10^3$ sec the value of $I_b^{Fe} = 350$ counts. The minimum value of I_A that can be measured to demonstrate that the element Fe is present is $3(2I_b^{Fe})^{1/2} = 79$ counts. For a 1500Å thick foil $(I_A - I_b^{Fe})$ for pure Fe was 10,000 counts. Using Eqn. 3.29 MMF = (79/10,000) x (100%) = 0.79 wt%.

In our laboratory, ROMIG (1979) has determined the MMF of Ni in an FeNi matrix. An Fe - 5.13 wt% Ni alloy was used and a Philips 300 STEM was employed for the analysis. The instrument was operated at 100 kV, with a 200Å probe spot and a 60 sec counting time. For a 200 - 300Å thick foil, $I_b^{Ni} = 257$ counts, $I_{Fe} - I_b^{Fe} = 3519$ counts, $C_{Fe} = 94.87$ wt% and $k_{NiFe} = 1.13$. The MMF using Eqn. 3.28 was 1.62 wt%. For a thicker foil, 1500 -2000Å, $I_b^{Ni} = 1174$ counts, $I_{Fe} - I_o^{Fe} = 25,393$ and the MMF using Eqn. 3.28 was 0.48 wt%. The 1500 - 2000Å film thickness is more typical of the specimen as prepared by ion thinning. Therefore for thermal emission guns, a MMF of less than 1 wt% and perhaps 0.5 wt% can be expected under practical analysis conditions.

BENGTSSON and EASTERLING (1978) have shown experimentally that MMF can be improved by up to a factor 2 by increasing the operating voltage from 100 to 200 kV and hence the P/B ratio. OPPOLZER and KNAUER (1979) have used a dedicated STEM with a field emission gun to measure the minimum mass fraction of a foil of an amorphous metallic glass. The instrument was operated at 100 kV with a 100Å probe spot and a 100 sec counting time. The foil was 500Å thick. The measured data is shown in Table 3.7. The MMF using Eqn. 3.29 is $\sim$ 0.15 wt% for the major elements (P, Cr, Fe, Ni) which comprise the glass. The higher count rates obtained at small beam diameters with the field emission gun allow for the improved MMF.

TABLE 3.7

STEM X-ray Analysis of Metglas^R 2628A

(Oppolzer and Knauer, 1979)

Element	C_A(wt%)	I_A(counts)	I_b^A(counts)	$I_A - I_b^A$(counts)	MMF(wt%)
P	7.3	11556	2702	8854	0.18
Cr	14.4	15916	1323	14593	0.14
Fe	35.3	39048	977	29071	0.12
Ni	41.7	41485	904	41581	0.13

Although MMF for thin film x-ray microanalysis, with a $\leq$ 200Å spot size is generally < 1 wt%, several experimental problems may occur. Assuming that the spurious x-ray signal is minimized, if the EDS spectrum is relatively clean with few characteristic peaks, then background subtraction and the determination of I_b^A is relatively straightforward. In some cases peak overlaps, continuum background subtraction and EDS spectrometer artifacts must be considered. If peak overlaps occur or backgrounds cannot be easily determined so that they can be subtracted from peak intensities, the actual MMF will be much larger than the values given by Eqn. 3.28. Methods for treating continuum background subtraction, peak overlaps, and artifacts in EDS detectors have been reviewed by various authors (GEISS and HUANG, 1975; SHUMAN et al., 1976; STATHAM, 1976; FIORI and NEWBURY, 1978) and will not be discussed here. In practice, if the x-ray EDS spectrum contains a large number of peaks or the x-ray peak of interest is at low energies where the continuum is a maximum, a more sophisticated treatment of the data is necessary.

3.5 SUMMARY

The methods for obtaining quantitative x-ray analysis from thin film specimens in the STEM-AEM have been presented in this chapter. The Cliff-Lorimer ratio method appears to be the most useful of the techniques available. Such x-ray analyses are relatively straightforward even if absorption and fluorescence corrections are found to be necessary. The potential difficulties in obtaining characteristic x-ray intensities, which are not affected by specimen and/or instrument related effects, are in many cases more worrisome than the analysis techniques discussed in this chapter. Methods for estimating and measuring the x-ray spatial resolution as a function of Z, E_O, ρ and film thickness t and for determining the minimum detectable mass and mass fraction in thin foil samples have also been described in this chapter. Given information on how to obtain a quantitative analysis and how to obtain the optimum spatial resolution and mass sensitivity, the analyst is now ready to tackle the many problems which can be solved using the STEM-AEM. The time has now come for the use of thin film x-ray microanalysis technique for a wide variety of applications. Several selected examples of applications will be given in Chapter 4 by Zaluzec.

ACKNOWLEDGMENTS

The author wishes to thank D. B. Williams, S. Mehta and A. D. Romig of Lehigh for helpful reviews of the manuscript and for contributions of their own data and results. Financial support was provided by NASA grants NGR 39-007-043 and NGR 39-007-056 and NSF grant EAR 74-22518-A01.

REFERENCES

Beaman, D. R., 1978, Environmental Pollutants, ed. T. Y. Toribara, J. R. Coleman, B. E. Dahneke and I. Feldman, Plenum Pub. Co., 255.

Bengtsson, B. and Easterling, K. E., 1978, SEM/1978, ed. O. Johari, SEM, Inc. 655.

Castaing, R., 1951, Thesis, University of Paris, ONERA Publ. #55.

Cliff, G. and Lorimer, G. W., 1972, Proc. 5th European Congress on Electron Microscopy, Institute of Physics, Bristol, 140.

Cliff, G. and Lorimer, G. W., 1975, J. Microscopy, 103, 203.

Doig, P. and Flewitt, P. E. J., 1977, J. Microscopy, 110, 107.

Duncumb, P., 1968, J. de Microscopie, 7, 581.

Faulkner, R. G., Hopkins, T. C. and Norrgård, K., 1977, X-ray Spectrometry, 6, 73.

Fiori, C. E. and Newbury, D. E., 1978, SEM/1978, 1, ed. O. Johari, SEM, Inc., 401.

Geiss, R. H. and Huang, T. C., 1975, X-ray Spectrometry, 4, 196.

Goldstein, J. I., Costley, J. L., Lorimer, G. W. and Reed, S. J. B., 1977, SEM/1977, 1, ed. O. Johari, IITRI, Chicago, Ill., 315.

Goldstein, J. I., Lorimer, G. W. and Cliff, G., 1976, Proc. Sixth Europ. Congr. on EM, TAL Intern., Ramat Gan, Israel, 56.

Goldstein, J. I. and Williams, D. B., 1977, SEM/1977, ed. O. Johari, IITRI, Chicago, Ill., 651.

Goldstein, J. I. and Yakowitz, H., 1975, Practical Scanning Electron Microscopy, Plenum, New York.

Green, M. and Cosslett, V. E., 1961, Proc. Phys. Soc., 78, 1206.

Hall, T. A., 1968, Quantitative Electron Probe Analysis, ed. K. F. J. Heinrich, NBS Spec. Publ. 298, 269.

Heinrich, K. F. J., 1966, The Electron Microprobe, T. D. McKinley, K. F. J. Heinrich and D. B. Wittry, eds., Wiley, New York, 296.

Heinrich, K. F. J., Fiori, C. E. and Myklebust, R. L., 1976, Proc. 11th Annual Conf. MAS, 29.

Jacobs, M. H., 1973, J. Microscopy, 99, 165.

Jacobs, M. H. and Baborovska, J., 1972, Proc. 5th European Cong. on E. M., The Institute of Physics, Bristol, 136.

Joy, D. C. and Maher, D. M., 1975, Proc. 33rd Annual EMSA Meeting, 242.

Joy, D. C. and Maher, D. M., 1977, SEM/1977, ed. O. Johari, IITRI, Chicago, Ill., 325.

Kelly, P. M., Jostsons, A., Blake, R. G. and Napier, J. G., 1975, Phys. Stat. Sol., 31, 771.

Kyser, D. F. and Geiss, R. H., 1977, Proc. 12th Annual Conf. MAS, 110.

Lin, J., Williams, D. B. and Goldstein, J. I., 1979, Geochim. Cosmochim Acta., in press.

Lorimer, G. W., 1976, Analytical Electron Microscopy, Proceedings of a Workshop, Cornell University.

Lorimer, G. W., 1977, Proc. 12th Annual Conf. MAS, 108.

Lorimer, G. W., Al-Salman, S. A., and Cliff, G., 1977, Dev. in EM and Analysis, ed. D. L. Misell, Inst. Phys. Conf. Ser. No. 36, The Institute of Physics, 369.

Lorimer, G. W., Cliff, G. and Clark, J. N., 1976, Developments in Electron Microscopy and Analysis, EMAG 75, J. A. Venables, ed., Academic Press Inc., London, 153.

Love, G., Cox, M. G. C. and Scott, V. D., 1977, Dev. in EM and Analysis, ed. D. L. Misell, Inst. Phys. Conf. Ser. No. 36, The Institute of Physics, 347.

Lyman, C. E., Manning, P. E., Duquette, D. J. and Hall, E., 1978, SEM/1978, 1, ed. O. Johari, SEM Inc., Chicago, Ill., 213.

Mehta, S., and Goldstein, J. I., (1979), Analytical electron microscopy study of submicroscopic metal particles in glassy constituents of 15015 and 60095 lunar breccias, Lunar and Planetary Science X, in press.

Mehta, S., Goldstein, J. I. and Friel, J. J., 1979, Submicron sized metal particles in glass coatings of lunar breccia 15286, Lunar and Planetary Science X, in press.

Mott, N. F. and Massey, H. S. W., 1949, The Theory of Atomic Collisions, Oxford Univ. Press, London, 2nd ed., 243.

Nasir, M. J., 1972, Proc. 5th European Cong. on E. M., Manchester, The Institute of Physics, Bristol, 142.

Newbury, D. and Myklebust, R. L., 1979, Ultramicroscopy, 3, 391.

Oppolzer, H. and Knauer, U., 1979, SEM/1979, 1, ed. O. Johari, SEM, Inc., Chicago, Ill., in press.

Pande, C. S., Suenaga, M., Vyas, B., Isaacs, H. S. and Harling, D. F., 1977, Scripta Met., Scripta Met., 11, 681.

Philibert, J., Rivory, J., Bryckaert, D. and Tixier, R., 1970, J. Phys. D., 3, 70.

Philibert, J. and Tixier, R., 1968, Brit. J. Appl. Phys., 1, 685.

Philibert, J. and Tixier, R., 1975, Physical Aspects of Electron Microscopy and Microbeam Analysis, ed. B. M. Siegel and D. R. Beaman, J. Wiley, New York, 333.

Powell, C. J., 1976, NBS Special Publication 460, ed. K. F. J. Heinrich, D. E. Newbury and H. Yakowitz, 97.

Rao, P., 1976, Proc. 34th Annual EMSA Meeting, 546,

Rao, P. and Lifshin, E., 1977, Proc. 12th Annual Conf. MAS, 118.

Reed, S. J. B., 1975, Electron Microprobe Analysis, Cambridge Univ. Press, Cambridge.

Ritter, A. M., Morris, W. G. and Henry, M. F., 1979, SEM/1979, 1, ed. O. Johari, SEM, Inc., Chicago, Ill., in press.

Romig, A., 1979, Low temperature phase equilibria in Fe-Ni alloys, Ph.D. Dissertation, Lehigh, in progress.

Russ, J. C., 1973, Proc. 8th National Conf. on Electron Probe Analysis, 30.

Shuman, H. and Somlyo, A. P., 1976, Proc. Nat. Acad. Sci., 1193.

Shuman, H., Somlyo, A. V. and Somlyo, A. P., 1976, Ultramicroscopy, 1, 317.

Slivinsky, V. W. and Ebert, P. J., 1972, Phys. Rev. A., 5, 1581.

Sprys, J. W. and Short, M. A., 1976, Proc. 11th Annual Conf. MAS, 9.

Statham, P. J., 1976, X-ray Spectrometery, 5, 16.

Tixier, R., 1979, "Electron Probe Microanalysis of Thin Samples" to be published in "Microbeam Analysis in Biology" ed. C. Lechene and R. Warner, Academic Press.

Tixier, R. and Philibert, J. 1969, Proc. 5th Int. Cong. on X-ray Optics and Microanalysis, eds. G. Mollenstedt and K. H. Gaukler, Springer-Verlag, Berlin, 180.

Wentzel, G., 1927, Zeit. Phys., 43, 524.

Williams, D. B. and Goldstein, J. I., 1978, Ninth Int. Cong. on EM, 1, 416.

Zaluzec, N. J., 1978, Ninth Int. Cong. on EM, 1, 548.

Zaluzec, N. J. and Fraser, H. L., 1976, Proc. 34th EMSA Meeting, ed. G. W. Bailey, Claitor's Publishing Div., Baton Rouge, 420.

Zaluzec, N. J. and Fraser, H. L., 1979, Proc. 8th Int. Cong. on X-ray Optics and Microanalysis, Boston, 1977, ed. D. Beaman, R. E. Ogilvie and D. Wittry, in press.

Ziebold, T. O., 1967, Anal. Chem., 39, 858.

CHAPTER 4
QUANTITATIVE X-RAY MICROANALYSIS:
INSTRUMENTAL CONSIDERATIONS AND APPLICATIONS TO MATERIALS SCIENCE

NESTOR J. ZALUZEC
METALS AND CERAMICS DIVISION, OAK RIDGE NATIONAL LABORATORY
RADIATION EFFECTS AND MICROSTRUCTURAL ANALYSIS GROUP
OAK RIDGE, TENNESSEE 37830

4.1 INTRODUCTION

X-ray microanalysis using a modern transmission or scanning transmission electron microscope (TEM/STEM) is an extremely powerful analytical tool. The ability to observe and characterize the morphology,
crystallography, and elemental composition of regions of a specimen as
small as 20 nm in diameter is a major breakthrough for materials
science. In this chapter the practical aspects of the application of
x-ray microanalysis to nonbiological systems will be considered (the use
of this technique in biological research is the topic of Chapter 5).
This chapter will first deal with optimizing the instrumental factors
which influence microanalysis, namely: the specimen/detector geometry;
the choice of accelerating voltage, electron gun, imaging mode, and most
importantly the elimination of instrumental artifacts. The application
of the principles of thin film analysis to simple single phase samples
will then be considered, followed by a discussion of multiphase systems
where analysis and interpretation becomes more complex. These examples
will be drawn primarily from the author's own experience, because of the
difficulties inherent to reconstructing complete experimental details
from the literature. The published applications of several other
authors are cited in the bibliography.

4.2 INSTURMENTAL LIMITATIONS IN AEM BASED X-RAY MICROANALYSIS

The general principles of thin film x-ray microanalysis in an analytical electron microscope (AEM) have been outlined in Chapter 3. Before applying the techniques described there to quantitative analysis,
it is important to consider the instrumental factors which influence the
information recorded by the x-ray detector system. Thus, this section
will first consider the problems associated with the interfacing of a
solid state energy dispersive x-ray spectrometer (EDS) to the column of
an electron microscope, and then concentrate on optimizing the AEM

system to minimize spectral artifacts. For the most part, the limitations in TEM/STEM based microanalysis are purely instrumental and can be corrected through judicious modification of the electron-optical column. Complications arise because the present generation of analytical electron microscopes have evolved from the addition of analytical attachments to TEM or STEM instruments rather than being designed from the outset as true AEMs.

4.3 INSTRUMENTAL ARTIFACTS: SYSTEMS BACKGROUND

Ideally, the measured x-ray spectrum should be due solely to the interaction of the incident electron probe with the microvolume of material it irradiates. Unfortunately, in the majority of instruments in use today this is not the case, although it is now being corrected by the manufacturers in the new instruments. The phenomenon is most easily observed by allowing the incident probe to pass through a small hole in a suitable specimen. In such a situation one would expect all characteristic emission to cease. This is usually the exception rather than the rule. Remotely produced uncollimated radiation (electrons and/or x-rays) from the electron-optical column continuously bombards the sample and its immediate surroundings independent of the electron probe position. This radiation can induce fluorescence of the specimen and its environment thereby producing additional x-ray emission. The nature, strength and source of this uncollimated radiation varies from instrument to instrument; however, the result is the same - degradation of the high spatial resolution and the accuracy of the obtained quantitative information. Clearly, the reduction or elimination of this radiation in an AEM is an important aspect of x-ray microanalysis. It has been the subject of much study during the last few years: (GEISS and HUANG, 1975; ZALUZEC and FRASER, 1976, 1977, 1978; JOY and MAHER, 1976; SHUMAN et al., 1976; HREN et al., 1976; GOLDSTEIN and WILLIAMS, 1977; MORRIS et al., 1977; NICHOLSON et al., 1977; ZALUZEC et al., 1978; HEADLEY and HREN, 1978) as well as a recent comprehensive tutorial (BENTLEY et al., 1979) from which this discussion draws heavily. In addition to fluorescence of the specimen, x-rays characteristic of the support grids, goniometer stage, anti-contamination devices, pole pieces, and any other materials in the immediate vicinity of the sample are frequently detected. These x-rays are a result of fluorescence by the remotely produced uncollimated radiation as well as by locally scattered electrons and/or x-rays produced by the interaction of the focussed electron probe with the specimen. All of this nonlocalized fluorescence of the specimen and its environment can be classified as "systems background" (JOY and MAHER, 1976).* The discussion which follows will be loosely divided into two parts (with some overlap): (1) background resulting from remote sources of uncollimated fluorescing radiation, and (2) that resulting from sources local to the specimen.

---

*Systems background is used in its most general sense to include all measured characteristic or continuum x-rays which do not result directly from the excitation of the specimen by the focussed electron probe. The term spurious x-rays used in Chapter 3 is roughly equivalent.

Fluorescence by Uncollimated Radiation: Remote Sources

In order to produce the fine electron probes used in an AEM it is usually necessary to intercept a relatively large electron flux by a series of beam defining apertures. These apertures can act as excellent sources of uncollimated radiation: electrons (not confined to the incident probe) and x-rays (both characteristic and continuum).

Depending on the specific design of the condensor aperture system, it is possible for electrons to scatter around the periphery of the fixed and variable aperture support mechanisms. These electrons, together with any that are scattered from the bore of contaminated or thick apertures, can result in the formation of significant electron tails (i.e., electrons not confined to the immediate diameter of the focussed probe) and thus can produce characteristic x-rays from regions of the specimen outside the primary electron probe. A second type of uncollimated radiation can also be traced directly to the various beam defining apertures. Typical condensor apertures are relatively thin, high atomic number materials of sufficient mass thickness to completely stop the incident electrons but substantially transparent to hard x-rays. Conditions such as this are nearly perfect for these apertures to act as thin target x-ray sources (both characteristic and Bremmstrahlung). The situation is further aggreveted by the fact that the emission of Bremmstrahlung is highly anistotropic. At the electron energies typical of AEM investigations ($\geq$ 100 keV) this emission is peaked in the forward scattering direction. Thus, in an AEM one can find the conditions and geometry nearly ideal for x-ray fluorescence of the specimen and its environment. Fig. 4.1a schematically illustrates these remote sources.

Figure 4.1. (a) Schematic illustration of remote sources of uncollimated fluorescing radiation from fixed (A) and variable(B) beam defining apertures. (b) Minimization techniques using thick fixed apertures (A) above the variable beam defining aperture (B), thick spray apertures (C) below the C_2 aperture, and non-beam defining apertures (D) just above the objective lens pole piece.

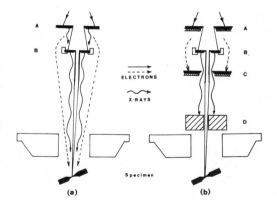

The choice of the test specimen used to assess the prescence and nature of the uncollimated radiation is critical. One of the best is of the self-supporting, thinned disc variety. The periphery of the specimen is effectively a thick ($\geq$ 0.1 mm) rim of material representing a bulk

sample while the thin electron transparent region is confined to the
edge of a small hole.   The composition of the test specimen is also critical.   The most pronounced effects are observed in specimens of medium-
to-high atomic number elements (or homogeneous alloys) having both low
energy (<3 keV) and medium energy (7 - 15 keV) characteristic x-ray emission lines.   The latter since above 20 keV the detector efficiency decreases substantially.   Suitable test specimens may be pure or alloyed
molybdenum, zirconium, silver, or some of the homogeneous aluminum alloys such as β NiAl.   These materials have the additional advantage that
their emission lines are well separated from any systems peaks characteristic of the specimen holder material (usually copper).

In the remainder of this text we define the term "in-hole" spectrum
as any x-ray spectrum detected when the incident electron probe is not
impinging on the specimen.   The origins of the "in-hole" spectrum are
from the entire specimen and its environment excited by the uncollimated
radiation.   Even though this radiation may be weaker than the primary
electron beam, it can produce measureable x-ray emission due to the
large mass of material it fluoresces (diameters of several millimeters).
It can be seen at once that uniformly thin films or extraction replicas
minimize the amount of material which can undergo fluorescence.

There are essentially four steps in the elimination of remotely produced uncollimated radiation: (1) identification of the fluorescing
species; (2) consultation with the instrument manufacturer concerning
their solutions to the problem; (3) modifications to the electron optical column; and (4) corrections to measured spectra for quantitative
analysis.

The identification of the fluorescing source is probably the most
difficult task.   The following sequence of tests is suggested (BENTLEY
et al., 1979; GOLDSTEIN, 1978).   First, examine the probe forming system
to ascertain if it is possible for electrons to reach the specimen by
scattering around the periphery of various apertures and holders.
Second, with the variable condensor aperture holder in position and a
thick solid disc in place of the normal aperture and attempt to detect
the presence of any electron current at the specimen using a suitable
Faraday cup and sensitive electrometer (note the currents may be $\leq$ 10
A).   Third, acquire a representative in-hole spectrum with a suitable
test specimen.   If x-rays are the principle fluorescing source, the intensity of the low energy characteristic line should be extremely small
relative to the medium energy line.   Cross sections for fluorescence by
x-ray photons favors excitation of the higher energy line.   For example,
the cross section ratio of M K /L for fluorescence by 60 keV x-ray photons is on the order to 200 while the ratio for N K /L is nearly 600.
The same ratios for fluorescence by 100 keV electrons are on the order
of 0.02.   In addition, characteristic x-ray production will be due primarily to fluorescence deep within the sample and the lower energy line
will undergo siginficant absorption relative to the higher energy line.
In comparison, electron excitation favors the lower energy lines and
x-ray generation will occur nearer the sample surface.   Thus the ratio
of the intensities of the high/low lines will be greater.   A final test
is to irradiate the bulk regions of the specimen with the probe.   The

overall characteristics of this electron excited spectrum can then be compared qualitatively with the "in-hole" spectrum and these combined results will usually be sufficient to identify the source or sources.

The various manufacturers now offer a variety of modifications that reduce the magnitude of uncollimated radiation. These include: spray apertures above and below the variable (C2) drive assembly to reduce electron tails (SHUMAN et al., 1976), thick apertures in the probe forming system to reduce the emitted hard x-ray flux (JOY and MAHER, 1976) and thick (∿ mm) non-beam defining collimators between the condensor aperture system and the top of the objective lens pole piece (ZALUZEC and FRASER, 1976). Figure 4.1(b) illustrates these solutions. The prospective user is referred to a recent tutorial devoted to the subject for further details (BENTLEY et al., 1979).

Fluorescence effects by uncollimated radiation have been reported in a very wide variety of electron microscopes which were interfaced with an EDS system. However, the situation is largely correctable. The published data includes the following: Philips EM400 FEG (CARPENTER and BENTLEY, 1979); Philips EM400 (CLARKE, 1978); Philips EM301 (GEISS and HUANG, 1975; HREN et al., 1976); Philips EM300 (SHUMAN et al., 1976; GOLDSTEIN and WILLIAMS, 1977); JEOL 100B (JOY and MAHER, 1976); JEOL 100C (BENTLEY and KENIK, 1977); JEOL 100CX (HEADLEY and HREN, 1978); JEOL JSEM200 (ZALUZEC and FRASER, 1976); VG HB5 (FRASER, 1977); SIEMENS ST 100F (OPPOLZER and KNAUER, 1979); Hitachi H-700 (FIORI and JOY, 1979); Hitachi HU-1000 (ZALUZEC et al., 1978).

Hard x-ray fluorescence of the specimen and its environment has been identified in all of these instruments except the Siemons ST 100F, where the data were insufficient to unambiguously identify the sources. All of the Philips instruments exhibit substantial electron tails unless equipped with spray apertures above and below the variable C2.

Figure 4.2 shows an example of the reduction of the in-hole spectrum to an acceptable level (to be defined) in a modified AEM. The data presented were taken for an extreme case and should not be taken as representative of all microscopes or even newer models of the same manufacturer. The spectra shown in Figs. 4.2 (a) and (b) were obtained from the instrument "as received." The fluorescing source was identified as hard x-rays (ZALUZEC and FRASER, 1976) and satisfactorily reduced with an additional non-beam defining collimator (0.5 mm in diameter, 2 cm long) positioned just above the upper objective lens pole piece. Figs. 4.2 (c) and (d) show the equivalent spectra after modification. The in-hole signal was reduced by nearly two orders of magnitude. A substantial change in the relative Ni/Al K intensity ratio for the two cases is apparent.

The effect of the large in-hole spectrum on quantitative analysis is shown in Fig. 4.3. The measured intensity ratio of the Ni/Al K lines is plotted as a function of specimen thickness for the unmodified instrument (curve A) and for the calculated variation of the Ni/Al K intensity ratio (curve C). These calculations were based purely on electron excitation of the sample. Clearly, in the as-received state, the

thin film approximation is invalid and quantitative analysis is virtually impossible. The initial decrease of the Ni/Al ratio (Curve A) results from the increasing contribution of the electron probe excited
signal with specimen thickness. As this signal begins to dominate and
finally exceed that produced by the uncollimated radiation, the measured
ratio will approach, but not necessarily reach, that corresponding to
electron excitation. The remaining discrepancy will depend on the magnitude of the "in-hole" signal and may be substantial. The ratio eventually goes through a minimum, then begins to increase as absorption effects become important. Analagous trends will be observed if electron
tails were the problem. The results of reduction of the uncollimated

Figure 4.2. Specimen (a) and in-
hole (b) spectra obtained in an
"as received" TEM/STEM JEOL JSEM
200 (1973). Test specimen β-NiAl
disc. Modified stage. STEM mode,
200 keV, probe size 10 nm, specimen 100 nm thick, live time 200
sec. Specimen (c) and in-hole (d)
spectra obtained in an AEM modified to reduce uncollimated fluorescing radiation. Note the
changes relative to (a) and (b).

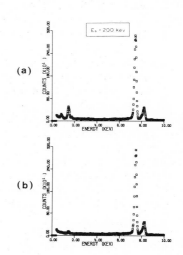

Figure 4.3. The effects of an in-
hole signal on the measured intensity ratio in a NiAl alloy as a
function of specimen thickness.
A - as received instrument; B -
modified instrument; C- calculated
variation based on electron excitation model.

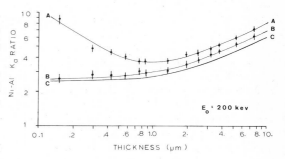

radiation on quantitative analysis is also shown in Fig. 4.3 (curve B).
Agreement with theory is now much better. The discrepancy in the
thinner regions of the foil is related to the detrimental aspects
of contamination. Deviations in the last few data points can be attributed to errors in the thickness measurement and neglect of beam broadening effects in the calculations.

    Ideally, one would like to completely eliminate the in-hole signal.
Realistically this may be extremely difficult. A critical question is
thus: At what level do we reach an acceptable in-hole signal? There is
obviously no general answer to this question, but the following general

criterion seems reasonable. When the intensity of each characteristic x-ray line recorded during an "in-hole" measurement is less than 1% of the integrated intensity (FWHM) measured with the incident electron probe positioned on the thinnest area of the specimen to be studied, then the in-hole signal is deemed acceptable. Both spectra, of course, must be measured under identical operating conditions.

Even when we have obtained an acceptable in-hole signal, it is still necessary to correct all subsequent measurements for the small but detectable remaining spectrum. To good approximation, this may now be accomplished by simple subtraction of the remaining in-hole spectrum from each successive measurement (ZALUZEC and FRASER, 1977). An important assumption in this case is that the in-hole signal essentially remains constant over the same period of machine time as the sample measurement. It is also tacitly assumed that the locally produced radiation resulting from the interaction of the electron probe with the sample itself (see below) are important only to second order, and also that the remotely produced uncollimated radiation is directly related to the emission/probe current, which is taken to be constant. The latter, is a good approximation for thermionic guns; however, for instruments using Field Emission sources, this may be quite inaccurate. In the latter, the acquisition time may be controlled by a measurement of the integrated electron dose (FRASER and WOODHOUSE, 1978).

Four points should be remembered when applying an in-hole subtraction correction. First, the in-hole spectra will change from one specimen to another and thus should be measured for each specimen. Second, for each measurement one should acquire a corresponding in-hole spectrum so that long term changes in the emission/probe current do not invalidate the assumptions. Third, for trace element analysis extreme care must be taken to insure the statistical validity of minor peaks that may be detected both in the specimen and the in-hole data. Finally, it is essential to stress that the in-hole signal must be reduced <u>before</u> subtraction, otherwise additional errors may be introduced. Figure 4.4 is an example of the improper subtraction of the in-hole signal.

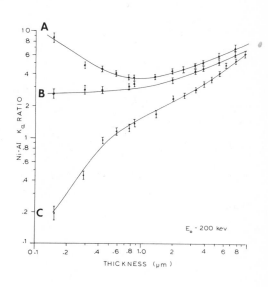

Figure 4.4. Subtraction of the in-hole signal to correct for un-collimated fluorescence. A - as received instrument; B - modified instrument with reduced in-hole signal subtracted; C - erroneous subtraction of in-hole signal without reduction of in-hole signal.

Fluorescence by Uncollimated Radiation: Local Sources

Fluorescence of the specimen and its immediate environment by locally produced radiation is also an important background source. Characteristic x-rays corresponding to the composition of the various support grids, goniometer stage, anti-contamination devices and other materials in the specimen chamber are all contributors. These local sources can sometimes affect quantitative analysis in particularly insidious ways. For simplicity, we will assume in this section that remote sources of uncollimated radiation have been minimized.

The primary source of local radiation is the specimen itself. As the focussed electron probe is scattered within the specimen secondary radiation from a multiplicity of sources is generated. The most significant components of this radiation include: backscattered and diffracted high energy electrons, specimen generated bremmstrahlung and characteristic x-rays.

The most noticable result of local fluorescence is the detection of x-rays which are characteristic of the environment. These x-ray photons will be detected simultaneously with the x-rays which are characteristic of the specimen. Due to the limited energy resolution of solid state x-ray detectors, the combined x-ray spectra may result in overlapping peaks. If the peak overlap problem is not too severe, individual peak intensities can be extracted using computerized data processing techniques. The worst case occurs when an element required in the analysis is also present in the system background. In this event, even quantitation analysis may be difficult and in some cases precluded.

Locally produced radiation can also fluoresce the specimen in the same manner as remotely produced radiation; however, its effects may be more subtle, for example, by selective enhancement of other elements in the specimen. An example is shown in Fig. 4.5 (BENTLEY et al., 1979), where one can immediately observe preferential fluorescence of the elements chronium and iron, relative to the nickel, by the copper K_α line systems peak. Such an effect is quite subtle, considering the apparently small copper systems peaks recorded. When the copper peaks were eliminated (Fig. 4.5a), the fluorescence effects are absent. In general, the specific type of specimen and its composition will influence greatly the magnitude of such fluorescence effects.

Complete elimination of locally generated radiation is virtually impossible; however, characteristic "systems peaks" from measured x-ray spectra can be effectively eliminated by surrounding the specimen with materials which do not generate detectable x-rays. Beryllium is without doubt the best materials choice, except for the potential machining hazards involved. An alternative material is graphite, but it suffers from poor mechanical strength. Regardless of the material chosen, a combination of the following modifications will substantially eliminate characteristic systems peaks:

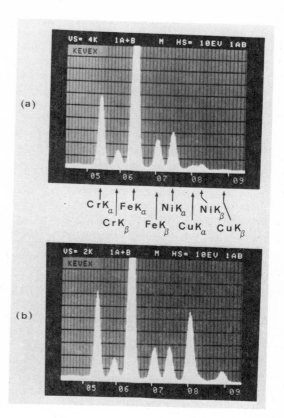

Figure 4.5. Preferential fluorescence of Cr and Fe lines relative to Ni lines by Cu K systems peak. Specimen 316 stainless steel. Same area each spectrum. (a) Be double tilt gomometer (KENIK and BENTLEY, 1979). CuK lines minimized and fluorescent radiation absent. (b) Standard double tilt holder, copper alloy. CuK lines fluorescing Cr and Fe.

1. Low-atomic number specimen stages (SPRYS, 1975; LILJESVAN and ROOMANS, 1976; HREN et al., 1976; ZALUZEC and FRASER, 1976; NICHOLSON et al., 1977; KENIK and BENTLEY, 1979; HEADLEY and HREN, 1978; JOHNSON et al., 1978);

2. Beryllium and/or carbon coated plastic support grids (PANESSA et al., 1978);

3. Low-atomic number substitutes for the anti-contamination devices (HREN et al., 1976; NICHOLSON et al., 1977); and

4. Low-atomic number shields for the objective pole piece as well as modifications to the objective aperture blade (NICHOLSON et al., 1977).

Of these, the greatest improvements result from the use of low atomic number specimen stages and support grids. In any case, these modifications may not substantially affect the detection of systems generated continuum radiation, which may be important if the continuum intensity is monitored for use in quantitative analysis routines. In such cases, it will be necessary to employ a suitable correction scheme (ROBERTSON et al., 1978).

Specimen Contamination

The mechanisms behind the accumulation of contamination as well as procedures for its reduction are important subjects in AEM and will be discussed in Chapter 18. As described there, its detrimental effects are not confined merely to image degradation. For completeness, we summarize here some of the effects of contamination on x-ray microanalysis.

Spatial resolution may be seriously degraded by beam broadening within the contaminant on the entrance surface. Thus, the beam striking the specimen is significantly larger than the microscope conditions would indicate. Space charging of the poorly conducting contaminant may further aggravate this problem (ZALUZEC and FRASER, 1977). The contaminant also contributes to the systems background in the continuum, in any case, and thereby decrease the attainable peak/background ratio. In turn, the minimum detectable limits are adversely affected (ZALUZEC and FRASER, 1977; SHUMAN et al., 1976). If the vacuum fluids contain silicon (or other elements with $Z > 10$), systems peaks may also be added. Finally, preferential adsorption in the contaminant of emitted x-rays from the specimen may occur for energy x-rays and thus directly affect quantitative analysis. On the positive side, contaminant deposits can serve as markers to indicate the approximate region analyzed, help to indicate problems of specimen drift, and may be used to measure the local thickness, since they mark both the top and bottom of the foil (see Chapter 18).

Detector Artifacts

Throughout the preceding discussion it has been tacitly assumed that a suitable solid state Si(Li) x-ray detector has been successfully interfaced to the column of the AEM, and that problems which are specific to such a device are not the limiting factors in analysis. There are, however, a set of artifacts directly linked to the detection and subsequent signal processing of x-ray emission spectra by these solid state detectors. The major artifacts associated with EDS detectors are listed below and the serious reader is referred to the literature for a detailed discussion of their effects and their minimization (FIORI and NEWBURY, 1978; REED, 1976; WOLDSETH, 1973); (1) detection efficiency; (2) generation of silicon escape peaks; (3) peak broadening and distortions; (4) formation of sum peaks due to pulse pile-up; (5) dead-time correction; and (6) microphonics.

4.4 OPTIMUM EXPERIMENTAL CONDITIONS FOR X-RAY ANALYSIS

The ultimate performance of x-ray microchemical analysis in an AEM requires optimization of the specific operating conditions used for the experimental measurements. In general, one wants to maximize the characteristic information measured from the specimen and minimize all other signals. The factors which influence the selection of operating conditions are discussed in this section. In all instances, it will be assumed that the sample is thin enough to avoid complications resulting from electron scattering, x-ray absorption, and x-ray fluorescence. Furthermore, instrumental factors associated with uncollimated fluorescing radiation and sample contamination are assumed to be eliminated.

Detector/Specimen Geometry

Because of the variety of AEM instruments available, it is essential to take into account detector/specimen geometry and its effect on the sensitivity of x-ray analysis. Since all measures of sensitivity are related to the characteristic peak-to-background (P/B) ratio, it is used as a convenient evaluation parameter.

The emission of characteristic x-rays in a thin TEM sample is essentially isotropic. The continuum, on the other hand, is given off highly anisotropically and is strongly influenced by relativistic effects. The intensity of the continuum radiation I_c in the energy range between E_c and $E_c + dE_c$ emitted at an angle Ω with respect to the forward scattering direction (i.e., the incident beam direction) can be written as:

$$I_c(\Omega) \cdot dE_c \cdot d\Omega = \left[I_x \left\{ \frac{\sin^2\Omega}{(1-\beta \cdot \cos\Omega)^4} \right\} + I_y \left\{ 1 + \frac{\cos^2\Omega}{(1-\beta \cdot \cos\Omega)^4} \right\} \right] \cdot k' \frac{dE_c}{E_c} \cdot d\Omega \quad (4.1)$$

where k' is a constant related to the sample thickness and composition and the parameters I_x and I_y are the continuum radiation components resulting from the deceleration of electrons in the specimen (SOMMERFELD, 1931; KIRKPATRICK and WIEDMANN, 1945). A polar plot of the angular distribution of continuum x-rays generated by 100 keV electrons incident on an infinitely thin specimen of nickel is shown in Fig. 4.6.

Figure 4.6. Anisotropic distribution of continuum radiation generated in a thin foil of nickel by a 100 keV electron beam. All intensities normalized to maximum. Forward scattering angle corresponds to observation at zero degrees. E_O - incident electron energy; E - continuum x-ray photon energy; Z - atomic number of specimen.

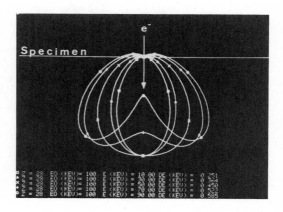

The optimum position of the detector can now be evaluated by calculating the (P/B) ratio as a function of Ω (ZALUZEC, 1978) as plotted in Fig. 4.7 for various incident beam energies. In all cases, the P/B

ratio is maximized when the x-ray detector views the electron entrance
surface at $(\Omega > \pi/2)$. The variation in P/B begins to level out at
angles above 125° (at 100 keV) which is being approached by several new
commercial AEM instruments. The actual inclination of the specimen

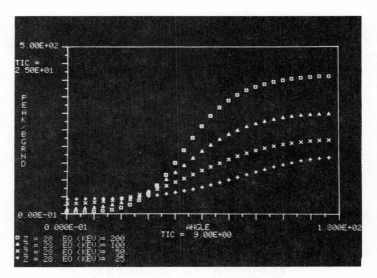

Figure 4.7. Calculated peak-to-background ratio for a thin foil of
nickel as a function of detector observation angle (Ω) at various incident electron beam energies. Z - atomic number of specimen; E - incident electron energy.

relative to the incident electron probe does not affect these calculations, for "thin" films. That is, even though the orientation of the
specimen plane may be oblique to the incident beam the thin area under
the focussed electron probe can be considered to be a segment of a horizontal thin slab. Of course, if the bremsstrahlung generated intercepts
an appreciable mass, absorption effects will influence the calculated
intensity. Such effects have been observed and shown to contribute to
the "systems background" (GEISS and HUANG, 1975).

Detector Collimation

One of the principle advantages of EDS is the potentially large collection angle available. Due to their relatively compact size these detectors can usually be brought within ∿ 2 cm of the specimen and also
results in a corresponding large subtended solid angle. The distinction
between collection and subtending angles rests in the position of the
apex of the respective solid angle cones, Fig. 4.8. The subtending
solid angle is defined with the apex of its cone at the source point of
the characteristic x-rays at the specimen. The collection solid angle

is defined by the penumbra of the detector crystal and any collimators. If the solid state detector is simply mounted on the column of the AEM its large collection solid angle can result in the detection of "systems background" from the specimen environment as well as the specimen related x-rays. Because of this, collimators are usually positioned in front of the detector to restrict its field-of-view. These collimators also serve to reduce the flux of scattered electrons which also impinge on the detector and can introduce spectral artifacts or, if their energy is sufficiently high, actually damage the active area of the silicon crystal.

There is no universal design of collimator and therefore it is necessary to tailor its shape to a specific application. For example, in most studies one usually wants to maximize the specimen signal (thus, the subtending solid angle). On the other hand, the analysis of radioactive specimens requires minimization of the collection solid angle, since the specimen radioactivity can flood the signal processing chain. Most investigators want to maximize the detected specimen signal and minimize the "systems background," unfortuantely these two design requirements are mutually exclusive. A reasonable compromise can be obtained by constructing a collimator which maximizes the subtended solid angle and restricts the penumbra of the collimator cone (i.e., the collection solid angle) to an area at the specimen plane slightly larger than the actual specimen diameter. By using appropriate low-atomic number specimen stages, this relaxed condition on the collection solid angle will not contribute to a detection of characteristic systems peaks and will allow one to maximize the collection efficiency. The relationships between collimator geometry and the respective solid angles are given by STURCKEN, 1976.

Figure 4.8. Schematic illustration of collection solid angle (single hatched region) and subtended solid angle (cross-hatched) of an EDS detector in a TEM/STEM/AEM.

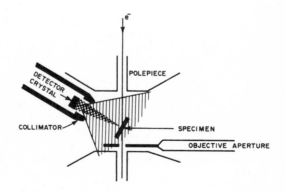

A final consideration is the choice of material to be used in collimator construction. High atomic number materials, such as tantalum or lead, are advantageous since they are effective x-ray shields; but, due to their high atomic number, there is a finite probability that scattering within the bore of the collimator can produce "systems peaks." This problem can be minimized by lining the inner bore with beryllium or graphite.

Selection of Incident Beam Energy and Electron Source

Based on beam broadening effects (Chapter 3), it is apparent that it would be advantageous to increase the incident beam energy to the highest attainable level. Thus, there is considerable interest in performing x-ray microanalysis at energies greater than 100 keV. This becomes even more attractive since the sensitivity of x-ray analysis also increases with increasing energies (JOY and MAHER, 1977; ZALUZEC, 1978). Figure 4.10 plots the calculated P/B ratio generated by the electron excitation of a thin foil as a function of overvoltage (U) of the incident beam. There is a monotonic increase in the P/B ratio with incident electron energy which indicates that the optimum incident electron energy in an ideal AEM may indeed be the highest attainable.

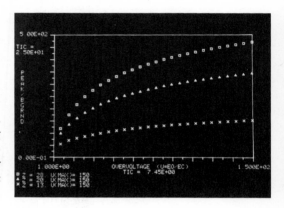

Figure 4.9. Calculated peak-to-background ratio as a function of overvoltage U (= E_O/E_C). E_O - incident electron energy; E_C - x-ray absorption edge energy; Z - atomic number of specimen.

Unfortunately, the problems associates with uncollimated fluorescent radiation in the illuminating system also becomes more acute as the incident beam energy increases (ZALUZEC and FRASER, 1976; ZALUZEC et al., 1978; CLIFF et al., 1978). At accelerating voltages of 200 keV and less, the problems are correctable by the procedures previously outlined. Serious studies using HVEM based microanalysis have only recently been attempted (ZALUZEC et al., 1978, CLIFF et al., 1978). Although the problems are severe, workable solutions are apparently feasible. Further developmental work in the electron-optical system is still necessary in order to produce the required minute high-current density electron probes for HVEM systems to be useful in x-ray analysis.

No electron source will presently satisfy all the requirements of analytical electron microscopy. Each of the parameters which characterize the various emitters (source size, stability, brightness, energy spread, and coherence) must be evaluated and matched to the instrument and the proposed mode of operation. Table 4.1 summarizes representative values of the characteristic parameters for each of the three types of electron sources available: thermionic tungsten, LaB$_6$, and tungsten

Table 4.1

Comparison of Electron Sources

SOURCE		β/Vo	SOURCE SIZE	ENERGY SPREAD (EV)	NOISE	STABILITY	COHERENCY	GUN VACUUM (TORR)
Thermionic	Tungsten Hairpin	1 A/cm^2/sr/eV	50 μm	~2	Low	Good	Low	<10^{-4}
	Pointed Filament	5	~10 μm	~2	Low	Fair	Moderate	<10^{-5}
LaB$_6$	Polycrystalline	10-30	~10 μm	<1	Low	Good	Moderate	<10^{-6}
	Single Crystal	20-50	~5 μm	<1	Low	Good	Moderate	<10^{-6}
Field Emission	Thermally Stabilized	100-500	~10 nm	~0.5	Fair ~1%	Moderate	High	<10^{-8}
	Cold	100-1000	~10 nm	~0.3	Fair ~1%	Fair	High	<10^{-10}

β is the source brightness in A/cm^2/sr, and Vo is accelerating potential in volts

field emission (JOY, 1977). The requirements for thin-foil x-ray microanalysis are quite simple: long and short term stability and maximum
probe current (consistent with the limits set by beam heating and radiation damage). The attainable probe current directly affects the x-ray
count rate. Its dependence as a function of probe diameter and emitter
type are shown in Fig. 4.10. For high spatial resolution, one relies on
the use of small probes and the FEG (field emission gun) appears to be
the optimum choice. However, for TEM/STEM systems, where conventional
transmission electron microscopy is also being performed, LaB_6 filaments
may meet the wider range of desired analytical modes. Both these
sources require better gun vacuums than do conventional tungsten hairpen
filaments, thus increasing both the complexity and cost of the AEM system. Since overall improvements to the vacuum system are necessary to
reduce specimen contamination, this disadvantage may soon disappear.

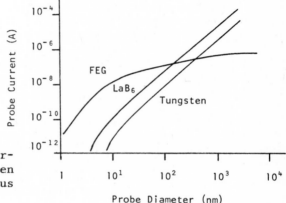

Figure 4.10. Typical probe current variation at the specimen
with probe diameter for various
electron sources.

Imaging and Diffraction Conditions During Analysis

A question often neglected before beginning an x-ray microanalysis
is which imaging mode is most appropriate TEM or STEM? If probe sizes
typical of TEM operation (5.0-0.2 μm) are sufficient to define the region of interest, then there is no need to resort to the potentially
lower current STEM mode. For example, analysis of precipitates extracted on carbon replicas, where the precipitates are thin, large
(200-500 nm), and well separated (spatially) may be advantageously analyzed by TEM instead of STEM. Even if the specimen requires a smaller
probe, it is not always necessary to utilize STEM at its highest resolution.

Diffracting conditions may also play an important role in microanalysis. Both theoretical calculations and experimental measurements
(HASHIMOTO et al., 1962; DUNCUMB, 1962; HALL, 1966; HUTCHINGS et al.,
1978) indicate that anamolous increases in characteristic x-ray production (by as much as a factor of three), can result when thin crystalline
foils are oriented in strongly diffracting conditions. These increases

result from asymmetries in the electron distribution of the Bloch waves and can be phenomenologically explained using a two beam dynamical model. In this approximation, one Bloch wave has nodes at the atom centers, while the node of the second wave will be between the reflecting planes. If the crystal is oriented such that the Bloch wave whose node is at the atom planes is highly excited (corresponding to a negative dynamical deviation parameter) then corresponding increases in characteristic x-ray emission are expected. The calculated variation in the relative probability (Q) of K shell excitation as a function of the deviation parameter is shown in Fig. 4.11 (HALL, 1966).

Such diffraction effects will not affect the validity of analysis if the region of the specimen being investigated is completely homogeneous and a standardless approach is being used. There are experiments where an erroneous analysis may result; for example, in quantitative analysis using thin film standards, when the intensity of either the unknown or the standard is affected. A particularly insidious effect may occur for composition profiles taken across an interface or of a precipitate embedded within a matrix. The diffraction conditions employed for imaging either the interface or the precipitate can be ideal for anamolous x-ray production. Thus, the combination of beam broadening and anamolous x-ray production can invalidate the assumptions of high spatial resolution and the quantitative interpretation of composition profiles.

The possible adverse effects of anamolous x-ray generation during analysis can be avoided if the specimen is oriented with a large positive (or negative) Bragg deviation parameter (w), or at any orientation where no low-order Bragg reflections are strongly excited. Some of the resulting loss in contrast in the images due to the "weak" diffracting conditions can be compensated by electronic signal processing.

Specimen Preparation Artifacts

The preparation of high quality thin specimens for transmission electron microscopy is an art in itself and is obviously important in all aspects of AEM based microanalysis. It is important to insure that these preparation techniques do not introduce artifacts to either the microstructure or the chemistry of the sample. Chemical or electrochemical thinning is potentially detrimental if proper precautions are not exercised. For example, contamination from the polishing solutions may remain on the sample surface and be subsequently detected as trace elements. The formation of oxide films (MORRIS et al., 1977) presents more subtle difficulties , since it may result in the formation of surface layers rich in a particular element. Ion-beam thinning can sometimes be used to remove such surface films, but it may of itself introduce microstructural changes or beam heating effects.

4.5 DATA REDUCTION FOR QUANTITATIVE ANALYSIS

The analysis of x-ray emission data obtained using an AEM based system can become a complex process. In addition to instrumental and

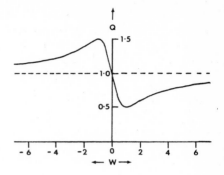

Figure 4.11. Calculated variation in the
relative probability of K-shell excitation
(Q) as a function of the dynamical deviation parameter (ω) (after HALL, 1966).

theoretical correction factors, one must also extract characteristic intensity measurements out of a spectrum of many peaks superimposed upon a
slowly varying background. The major difficulty occurs in multi-element
analysis, where there is spectral interference requiring deconvolution
techniques. It is beyond the scope of this chapter to discuss the details of the various data manipulation techniques used; however, some of
the salient considerations are described.

A general "rule-of-thumb" in choosing a data reduction technique is
"simplest-is-best." Furthermore, there is no substitute for a good,
clean spectrum containing a large number of counts in each channel. In
such a spectrum there will be no need to employ elaborate smoothing functions or peak detection algorithms. If the statistical noise in a measured spectrum requires smoothing, then the operator should consider acquiring more data if possible as a first choice.

The following procedures for obtaining accurate x-ray information
using an EDS system are suggested: (1) Acquire a spectrum from the region of interest. (2) Acquire in-hole data, and correct the sample
spectrum appropriately. (3) Identify all characteristic peaks.
(4) Subtract the background intensity. (5) Using suitable algorithms,
obtain net peak intensities from all characteristic lines.

The use of "automatic" background correction routines is, in general, not recommended. These corrections usually rely on scaling a
"standard" background curve to the measured specimen background. If the
standard background has been produced under a different set of operating
conditions (keV, geometry, thickness, etc.) it will not necessarily have
the same characteristics as the background intensity measured from the
specimen. A generalized polynomial function is adequate and can be
analytically fitted to the background intensities (with appropriate modifications should an absorption edge be encountered). The function need
not be based on any physical model, since we are interested only in obtaining integrated peak intensities. Similar consideration with respect
to obtaining peak intensities by comparison to "standard" spectra also
apply. Unless the standard spectra are acquired under identical operating conditions, they may not accurately duplicate the relevant peak
parameters (position, shape, and width) recorded during the specimen measurements, and thus can result in systematic errors.

It should also be pointed out that the quantitative routines require integrated peak intensities not peak heights. This distinction is important because a solid state detector system translates a fairly well defined x-ray line into a broad gaussian peak the width of which varies with the energy of the incident x-rays. Therefore, two x-ray lines of identical intensity (i.e., number of photons/sec) but of different energy will be recorded as gaussian peaks of equal area but different peak heights. This can be formulated by considering the integral (I) of a gaussian peak of standard deviation σ, and height H· which is given by

$$I = \int_{-\infty}^{+\infty} H \exp\left[-\frac{x^2}{2\sigma^2}\right] dx = \sqrt{2\pi} \cdot \sigma \cdot H \qquad (4.2)$$

The ratio of the integrated intensities of the two different peaks is simply:

$$\frac{I_1}{I_2} = \frac{\sigma_1 \cdot H_1}{\sigma_2 \cdot H_2} \qquad (4.3)$$

Clearly, only when $\sigma_1 = \sigma_2$ does the integrated peak ratio correspond to the peak height ratio. The use of a constant fraction of the total integrated intensity such as the integral over the full width of half maximum (FWHM) will not affect this intensity ratio.

4.6 APPLICATION OF QUANTITATIVE X-RAY MICROANALYSIS:

Parameters of Standardless Analysis

Since instrumental factors play such a large role in determining k it is risky to use data obtained from other instruments. As Goldstein points out in Chapter 3, there is a surprising paucity of published k values, considering popularity of the technique and the potential experimental errors associated with their measurement. Standardless analysis using calculated values of k is, in the author's view, a safer way to proceed. In Chapter 3, Goldstein reviews methods of such analyses following GOLDSTEIN et al. (1977). The present author has refined the selection of parameters in the examples presented at the end of this chapter (ZALUZEC, 1978). A brief summary is presented here.

The number of characteristic x-rays of element A (K_α or L_α) generated by electron excitation of a thin specimen of thickness dt can be written as:

$$\phi_A dt_A = Q_A \frac{N_o C_A}{A_A} \eta \ \omega_A \ a_A \ dt_A \qquad (4.4)$$

Here Q_A is the ionization cross section of the K-shell (or L-shell) of element A; N_o, ρ, C_A, and A_A are, respectively, Avagadro's number, the sample density, the concentration in weight percent, and the atomic weight of A, η the number of electrons bombarding dt, ω_A the fluorescence yield of A and a_A the K_α fraction (e.g., $K_\alpha/K_\alpha + K_\beta$) of the total emission.

The corresponding x-ray intensity (I) which one measures can differ from this generated intensity because of the intrinsic efficiency of the detector. The relationship between the measured intensity (I) and the generated intensity (ϕ_A) is:

$$I_A = \varepsilon_A \phi_A \qquad (4.5)$$

where ε is the efficiency of response of the detector system to the particular characteristic x-ray from element A. For solid state Si(Li) detectors this efficiency is a function of the energy of the given x-ray photon as well as the specific design parameters of the detector. In a general form, this efficiency can be represented by the following equation:

$$\varepsilon_A(E,\alpha) = \Pi_{j\,=\,Be,Au,Si}\left\{\exp\left[-\left(\frac{\mu}{\rho}\right)_j^E \frac{\rho_j t_j}{\cos\alpha}\right]\right\}\left\{1-\exp\left[-\left(\frac{\mu}{\rho}\right)_{Si}^E \rho_{Si}\, t_{Si}^*\right]\right\} \qquad (4.6)$$

The first term in brackets accounts for absorption of the incident x-ray photons before they reach the active region of the Si(Li) crystal and are detected. The product notation Π j=Be, Au, Si indicates that the sources of this absorption result from the cumulative effects of the beryllium window, gold conductive film, and silicon dead layer which spatially precede the active region of the detecting crystal. The term $(\frac{\mu}{\rho})_j^E$ is the mass absorption coefficient of the j element (i.e., Be, Au, Si) for an x-ray photon of energy E, while ρ , and t are respectively the mean density and thickness of each of these absorbers. Should the incident x-ray beam enter the detector inclined at an angle α with respect to its principal axis, then there is an effective increase in the mass thickness of each absorbing medium which is incorporated via the cos α term. If the incident x-ray energy is sufficiently high, then x-rays may be transmitted through the Si(Li) crystal. The corresponding loss in efficiency is accounted for by the second term in brackets, where as previously $(\frac{\mu}{\rho})_{Si}^E$ is the mass absorption coefficient for x-rays of energy E in silicon, ρ is the density of Si, and t* is the thickness of the active region of the Si(Li) detecting crystal.

Combining these last three equations, one obtains the general relationship between the measured x-ray intensity I of a characteristic line of energy E and the local specimen composition:

$$I_A dt_A = \varepsilon_A(E_A,\alpha)\kappa_A C_A \eta_A dt_A \qquad (4.7)$$

with

$$\kappa_A \equiv \frac{Q_A \omega_A a_A}{A_A} \qquad (4.8)$$

An EDS system allows one to measure simultaneously most of the characteristic x-ray photons generated within the region of interest of the specimen and thus the ratio of any two-x-ray lines can be written as:

$$\frac{I_A dt_A}{I_B dt_B} = \frac{\varepsilon_A \kappa_A C_A \eta_A dt_A}{\varepsilon_B \kappa_B C_B \eta_B dt_B} \tag{4.9}$$

Since $\eta_A = \eta_B$ and $dt_A = dt_B$ for a thin film, Eqn. 4.9 simplifies to

$$\frac{I_A}{I_B} = \frac{\varepsilon_A \kappa_A C_A}{\varepsilon_B \kappa_B C_B} \tag{4.10}$$

which is equivalent to the Cliff-Lorimer equation (Chapter 3):

$$\frac{C_A}{C_B} = k_{AB} \frac{I_A}{I_B} \tag{4.11}$$

with

$$k_{AB} = \frac{\varepsilon_B \kappa_B}{\varepsilon_A \kappa_A} \tag{4.12}$$

The x-ray generation constant κ_A may be calculated to good precision for either K_α or L_α from a set of parameterized equations. It may be written in the following terms (for either K_α or L_α):

$$\kappa_A = \frac{Q_A \omega_A a_A}{A_A} \tag{4.13}$$

where A_A is the atomic weight and the other terms will be described below. The ionization cross section Q_A for the K shell is obtained from

$$Q_A(E_o) = \frac{\pi e^4 \cdot Z_A \cdot a(K)}{E_o \cdot E_C} \left\{ \ln\left[b(K) \frac{E_o}{E_C} \right] - \ln(1 - \beta^2) - \beta^2 \right\} \tag{4.14}$$

where

$$a(K) = 0.35, \quad b(K) = \frac{0.2}{U_o \cdot \{1 - \exp(-\gamma)\} \cdot \{1 - \exp(-\delta)\}},$$

$$U_o = E_o/E_C \equiv \text{overvoltage ratio}, \quad \delta = \tfrac{1}{2} \cdot E_C,$$

$$\gamma = \frac{1250}{E_A \cdot U_o^2}, \quad \pi e^4 = 6.4924 \times 10^{20}, \quad \beta = \frac{v}{c}$$

and $E_O + E_C$ are respectively the incident beam energy and the critical excitation energy of the element A. For application to L-shell analysis the constant a(L) is set equal to 0.25 and all other terms are replaced by their L-shell equivalents. When all energies are expressed in keV, the units of Q_A (E_O) are cm^2/atom.

The functions a(K) and b(K) given above were determined analytically (ZALUZEC, 1978) by fitting Eqn. 4.14 to measurements of Q for aluminum, nickel, and silver (HINK and ZIEGLER, 1969; POCKMAN et al., 1947; KIRKPATRICK and BAEZ, 1947). Figures 4.12 and 4.13 compare the predictions of this equation (solid curve) with experiment and with other calculated values.

The fluorescent yield w_A can be calculated from the empirical expression,

$$\left| \frac{w_A}{1 - w_A} \right|^{\frac{1}{4}} = A + BZ + CZ^3 \tag{4.15}$$

in which the coefficients A, B, and C are fitted to experimental measurements (BURHOP, 1955; COLBY, 1968; DYSON, 1973). In this case, the periodic table is broken up into segments and a set of constants is used to determine w for various ranges of Z. Since dedicated minicomputers are normally used for quantitative analysis, it is preferred to use Eqn.

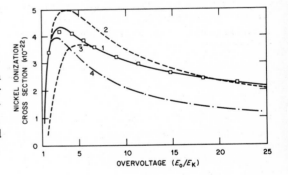

Figure 4.12. Comparison of experimental ionization cross section as a function of overvoltage for nickel (squares) to various calculations. (1) Using Eqns. 4.14; (2), (3) due to POWELL, 1976; and (4) due to WORTHINGTON and TOMLIN, 1956.

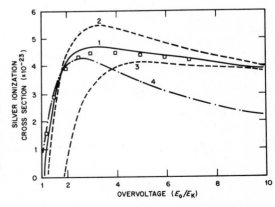

Figure 4.13. Comparison of experimental ionization cross section as a function of overvoltage for silver (squares) to various calculations. (1) Using Eqns. 4.14; (2), (3) due to POWELL, 1976; and (4) due to WORTHINGTON and TOMLIN, 1956.

Figure 4.14. A comparison of experimental measurements with calculations of the constant k_{XXSi} for K_α lines (a) and L_α lines (b) of various elements. Incident electron energy 100 keV.

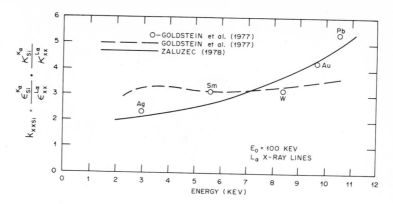

4.15 (or tabulated values) rather than the expression by Wentzel (GOLDSTEIN, Chapter 3). The K_α fraction of the total K-shell emission a_A can be obtained from polynomial fits to experimental data or in tabulated form (SLIVINSKY and EBERT, 1972; HEINRICH, 1976; MCGUIRE, 1970; COLBY, 1968).

Tables 4.1, 4.2 and 4.3 document the calculated x-ray generation constants for K_α and L_α x-rays for incident electron energies of 100, 120, and 200 keV and over the range of atomic numbers $10 \leq Z \leq 92$. The constants for the K_α x-rays for elements with $Z \geq 58$ are omitted since their characteristic energies are greater than 40 keV. Similarly, the constants for L_α x-rays for elements with $Z \leq 30$ are omitted since their energies are less than 1.0 keV and their usage in quantitative analysis is not recommended. Table 4.4 also lists the calculated efficiency of a typical solid state detector between 0.6 and 20.0 keV. The detector parameters used in this calculation were: $t_{Be} = 7.6$ μm; $t_{Au} = 0.02$ μm; $t_{Si(dead\ layer)} = 0.1$ μm; $t_{Si(active\ crystal)} = 3.0$ mm.

There are, of course, alternate formulations which have been used to calculate the constant k_α of Eqn. 4.11. Figure 4.14 compares experimental measurements for 100 keV electrons of this constant (normalized to the silicon K_α line) for K_α (CLIFF and LORIMER, 1975) and L_α lines (GOLDSTEIN et al., 1977) with the predicted k factors using the present formulation (ZALUZEC, 1978) and an earlier parameterization by GOLDSTEIN et al. (1977).

Absorption Correction

The set of equations for the absorption correction in Chapter 3 are presented here in a more generalized parametric form. This formulation is not unique and an alternate formulation may be found in BENTLEY and KENIK (1979).

For the relatively simple geometry found in most electron microprobes, the electron beam strikes the specimen roughly normal to its surface (i.e. the geometry used in Chapter 3). The exiting path length d is then given by

Table 4.2

X-ray generation constant for 100 keV electrons: $\kappa_A = \dfrac{Q_A \cdot w_A \cdot a_A}{A_A}$

Z	κ^{K_α}	κ^{L_α}	Z	κ^{K_α}	κ^{L_α}
10	2.7180E-24	-	51	2.1296E-25	1.2838E-24
11	2.7565E-24	-	52	1.9439E-25	1.2285E-24
12	2.9429E-24	-	53	1.8813E-25	1.2374E-24
13	2.9386E-24	-	54	1.7629E-25	1.1967E-24
14	3.0806E-24	-	55	1.7029E-25	1.1817E-24
15	3.0081E-24	-	56	1.6272E-25	1.1418E-24
16	3.0935E-24	-	57	1.6071E-25	1.1260E-24
17	2.9158E-24	-	58	-	1.1123E-24
18	2.6527E-24	-	59	-	1.1010E-24
19	2.7625E-24	-	60	-	1.0695E-24
20	2.7312E-24	-	61	-	1.0426E-24
21	2.4531E-24	-	62	-	1.0117E-24
22	2.3057E-24	-	63	-	9.9252E-25
23	2.1584E-24	-	64	-	9.5014E-25
24	2.0930E-24	-	65	-	9.3046E-25
25	1.9497E-24	-	66	-	8.9980E-25
26	1.8773E-24	-	67	-	8.7584E-25
27	1.7321E-24	-	68	-	8.5245E-25
28	1.6844E-24	-	69	-	8.3233E-25
29	1.5007E-24	-	70	-	8.0066E-25
30	1.4002E-24	-	71	-	7.7955E-25
31	1.2551E-24	1.4375E-24	72	-	7.5169E-25
32	1.1482E-24	1.4546E-24	73	-	7.2877E-25
33	1.0559E-24	1.4770E-24	74	-	7.0440E-25
34	9.4805E-25	1.4618E-24	75	-	6.8249E-25
35	8.8402E-25	1.5004E-24	76	-	6.5512E-25
36	7.9359E-25	1.4805E-24	77	-	6.3518E-25
37	7.3100E-25	1.49703-24	78	-	6.1266E-25
38	6.6878E-25	1.5013E-24	79	-	5.9367E-25
39	6.1735E-25	1.5170E-24	80	-	5.6990E-25
40	5.6298E-25	1.5121E-24	81	-	5.4648E-25
41	5.1687E-25	1.5148E-24	82	-	5.2626E-25
42	4.6789E-25	1.4935E-24	83	-	5.0906E-25
43	4.2390E-25	1.4707E-24	84	-	4.9396E-25
44	3.8836E-25	1.4611E-24	85	-	4.8135E-25
45	3.5708E-25	1.4531E-24	86	-	4.4345E-25
46	3.2374E-25	1.4208E-24	87	-	4.2970E-25
47	2.9990E-25	1.4146E-24	88	-	4.1248E-25
48	2.7093E-25	1.3685E-24	89	-	3.9931E-25
49	2.5041E-25	1.3485E-24	90	-	3.7966E-25
50	2.2954E-25	1.3115E-24	91	-	3.7048E-25
			92	-	3.4912E-25

Table 4.3

X-ray generation constant for 120 keV electrons $\kappa = \dfrac{Q_A \cdot w_A \cdot a_A}{A_A}$

Z	κ^K_α	κ^L_α	Z	κ^K_α	κ^L_α
10	2.3966E-24	-	51	2.1092E-25	1.1699E-24
11	2.4386E-24	-	52	1.9305E-25	1.1210E-24
12	2.6121E-24	-	53	1.8729E-25	1.1306E-24
13	2.6170E-24	-	54	1.7586E-25	1.0949E-24
14	2.7525E-24	-	55	1.7014E-25	1.0826E-24
15	2.6966E-24	-	56	1.6275E-25	1.0474E-24
16	2.7820E-24	-	57	1.6078E-25	1.0342E-24
17	2.6306E-24	-	58	-	1.0229E-24
18	2.4008E-24	-	59	-	1.0138E-24
19	2.5079E-24	-	60	-	9.8608E-25
20	2.4869E-24	-	61	-	9.6242E-25
21	2.2403E-24	-	62	-	9.3506E-25
22	2.1117E-24	-	63	-	9.1845E-25
23	1.9824E-24	-	64	-	8.8028E-25
24	1.9277E-24	-	65	-	8.6308E-25
25	1.8005E-24	-	66	-	8.3562E-25
26	1.7382E-24	-	67	-	8.1432E-25
27	1.6080E-24	-	68	-	7.9349E-25
28	1.5677E-24	-	69	-	7.7566E-25
29	1.4002E-24	-	70	-	7.4700E-25
30	1.3098E-24	-	71	-	7.2813E-25
31	1.1769E-24	1.2729E-24	72	-	7.0291E-25
32	1.0793E-24	1.2898E-24	73	-	6.8224E-25
33	9.9495E-25	1.3115E-24	74	-	6.6017E-25
34	8.9551E-25	1.2999E-24	75	-	6.4036E-25
35	8.3709E-25	1.3361E-24	76	-	6.1537E-25
36	7.5332E-25	1.3204E-24	77	-	5.9731E-25
37	6.9565E-25	1.3370E-24	78	-	5.7677E-25
38	6.3805E-25	1.3429E-24	79	-	5.5952E-25
39	5.9050E-25	1.3589E-24	80	-	5.3774E-25
40	5.3992E-25	1.3564E-24	81	-	5.1621E-25
41	4.9704E-25	1.3608E-24	82	-	4.9768E-25
42	4.5118E-25	1.3437E-24	83	-	4.8196E-25
43	4.0992E-25	1.3251E-24	84	-	4.6820E-25
44	3.7664E-25	1.3184E-24	85	-	4.5677E-25
45	3.4733E-25	1.3130E-24	86	-	4.2129E-25
46	3.1585E-25	1.2857E-24	87	-	4.0871E-25
47	2.9348E-25	1.2819E-24	88	-	3.9279E-25
48	2.6595E-25	1.2419E-24	89	-	3.8070E-25
49	2.4656E-25	1.2255E-24	90	-	3.6240E-25
50	2.2669E-25	1.1935E-24	91	-	3.5407E-25
			92	-	3.3407E-25

Table 4.4

X-ray generation constant for 200 keV electrons: $\kappa = (Q \cdot w \cdot a)/A_A$

Z	κ^{K_α}	κ^{L_α}	Z	κ^{K_α}	κ^{L_α}
10	1.6659E-24	-	51	1.9705E-25	8.8218E-25
11	1.7089E-24	-	52	1.8131E-25	8.4826E-25
12	1.8455E-24	-	53	1.7672E-25	8.5849E-25
13	1.8640E-24	-	54	1.6659E-25	8.3421E-25
14	1.9765E-24	-	55	1.6165E-25	8.2764E-25
15	1.9519E-24	-	56	1.5491E-25	8.0343E-25
16	2.0299E-24	-	57	1.5313E-25	7.9599E-25
17	1.9346E-24	-	58	-	7.8991E-25
18	1.7795E-24	-	59	-	7.8547E-25
19	1.8732E-24	-	60	-	7.6650E-25
20	1.8718E-24	-	61	-	7.5056E-25
21	1.6988E-24	-	62	-	7.3160E-25
22	1.6133E-24	-	63	-	7.2094E-25
23	1.5256E-24	-	64	-	6.9322E-25
24	1.4943E-24	-	65	-	6.8185E-25
25	1.4057E-24	-	66	-	6.6228E-25
26	1.3667E-24	-	67	-	6.4746E-25
27	1.2732E-24	-	68	-	6.3291E-25
28	1.2500E-24	-	69	-	6.2066E-25
29	1.1242E-24	-	70	-	5.9963E-25
30	1.0589E-24	-	71	-	5.8633E-25
31	9.5813E-25	8.9402E-25	72	-	5.6782E-25
32	8.8471E-25	9.0904E-25	73	-	5.5287E-25
33	8.2126E-25	9.2755E-25	74	-	5.3668E-25
34	7.4433E-25	9.2260E-25	75	-	5.2222E-25
35	7.0063E-25	9.5166E-25	76	-	5.0343E-25
36	6.3494E-25	9.4377E-25	77	-	4.9022E-25
37	5.9047E-25	9.5910E-25	78	-	4.7487E-25
38	5.4541E-25	9.6676E-25	79	-	4.6214E-25
39	5.0836E-25	9.8182E-25	80	-	4.4556E-25
40	4.6813E-25	9.8358E-25	81	-	4.2910E-25
41	4.3404E-25	9.9035E-25	82	-	4.1503E-25
42	3.9682E-25	9.8139E-25	83	-	4.0322E-25
43	3.6311E-25	9.7130E-25	84	-	3.9297E-25
44	3.3602E-25	9.6987E-25	85	-	3.8463E-25
45	3.1207E-25	9.6941E-25	86	-	3.5592E-25
46	2.8578E-25	9.5258E-25	87	-	3.4642E-25
47	2.6737E-25	9.5320E-25	88	-	3.3403E-25
48	2.4392E-25	9.2668E-25	89	-	3.2482E-25
49	2.2760E-25	9.1768E-25	90	-	3.1024E-25
50	2.1056E-25	8.9687E-25	91	-	3.0411E-25

$$d = t \cdot \csc \theta_E \qquad (4.16)$$

where t is the depth of production and θ_E is the take-off angle. Because of the constraints on detector size and position imposed by the objective lens pole piece, the geometry found in an AEM is not so simple. The x-ray geometry found on a general analytical electron microscope is shown in Fig. 4.14 which represents a plane section through the sample containing the incident beam direction and the detector axis. The elevation angle, θ_E, is defined relative to an imaginary plane perpendicular to the incident beam direction and is nominally located at the height of the midplane of the untilted specimen. From this figure, one can show that the relationship between depth of production and exiting path length is now given by:

$$d = t \cdot \frac{\sin \beta}{\cos(\beta - \theta_E)} \equiv t \cdot f(\beta, \theta) \qquad (4.17)$$

For the special case of normal incidence ($\beta = \pi/2$), Eqn. 4.17 reduces to Eqn. 4.16. Care must be taken in using Eqn. 4.17 on samples which are severely bent or having significant thickness changes over the area being analyzed. The angle of electron incidence, β, is also a function of the detector/specimen geometry. It is convenient to plot the various parameters on a stereographic projection from which the value of β can easily be read (ZALUZEC, 1978) but can also be obtained using solid geometry.

The total characteristic intensity (K_α or L_α) emitted in the direction of the detector system, including absorption effects, is given by:

$$\phi_A = {}_0\!\int^{t_o} \phi(t) \left\{ \exp\left[- \left(\frac{\mu}{\rho}\right)^A_{AB} \cdot \rho \cdot d \right] \right\} dt \qquad (4.18)$$

If one assumes $\phi(t)$ is nearly constant, Eqn. 4.18 can be solved to give:

$$\phi_A = \phi_A(0) \left\{ \frac{1 - \exp\left[- \left(\frac{\mu}{\rho}\right)^A_{AB} \cdot \rho \cdot f(\beta, \theta) \, t_o \right]}{\left(\frac{\mu}{\rho}\right)^A_{AB} \cdot \rho \cdot f(\beta, \theta)} \right\} \qquad (4.19)$$

where $\phi(0)$ is the generation function evaluated at zero thickness. One can then easily show that:

$$\frac{C_A}{C_B} = \frac{\varepsilon_B \, \kappa_B \, \delta_B \, I_A}{\varepsilon_A \, \kappa_A \, \delta_A \, I_A} \qquad (4.20)$$

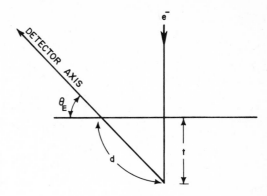

Figure 4.15. Geometry of the absorption correction in an AEM.

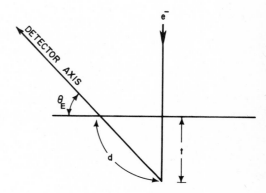

where the ratio δ_B / δ_A simplifies to

$$
\frac{\delta_B}{\delta_A} = \left\{ \frac{\left(\frac{\mu}{\rho}\right)^A_{AB}}{\left(\frac{\mu}{\rho}\right)^B_{AB}} \right\} \cdot \left\{ \frac{1-\exp\left[-\left(\frac{\mu}{\rho}\right)^B_{AB} \cdot \rho \cdot f(\beta,\theta) \cdot t_o \right]}{1-\exp\left[-\left(\frac{\mu}{\rho}\right)^A_{AB} \cdot \rho \cdot f(\beta,\theta) \cdot t_o \right]} \right\} \tag{4.21}
$$

It should be noted that if $\beta = \pi/2$, this equation reduces to the result presented in Chapter 3.

In order to calculate an absorption correction one needs to know the value of the mass absorption coefficient for K_α x-ray photons of element A of the alloy AB which is given by the equation:

$$
\left[\frac{\mu}{\rho}\right]^A_{AB} = \sum_{i=1}^{N} \left[\frac{\mu}{\rho}\right]^A_i C_i \tag{4.22}
$$

In this expression C is the composition in weight percent of the i element in the sample and $(\frac{\mu}{\rho})^A_i$ is the absorption coefficient for the K_α or L_α x-ray of A in the pure element i. Obviously, the composition of the alloy is initially an unknown quantity; and, it is therefore necessary to determine $(\frac{\mu}{\rho})^A_{AB}$ iteratively. As a first approximation, the composition of the alloy is determined ignoring absorption effects. A first-order absorption correction is then calculated and new values of composition are then substituted into the absorption term and the procedure repeated until a suitable convergence is obtained.

These equations are of course only valid for a homogeneous sample of thickness t_o. The application of an absorption correction to an inhomogeneous sample is extremely difficult without some prior knowledge as to the relative composition and thickness of each phase in the excited volume. Thus, in such a situation one usually resorts to analysis of thinner regions of the sample which can be then considered locally homogeneous.

The criterion used to test the validity of the thin-film approximation can be still written in the form suggested by TIXIER and PHILIBERT (1969), that is, for all characteristic x-ray lines from which quantitative results are desired the product of $\chi_A \cdot \rho \cdot t$ must be less than 0.1. In this context χ_A is now defined as

$$\chi_A = \left(\frac{\mu}{\rho}\right)^A_{AB} \frac{\sin\beta}{\cos(\beta-\theta_E)} \qquad (4.23)$$

where all terms in Eqn. 4.23 have their previous definitions. Thus, if the condition

$$\chi_A \cdot \rho \cdot t < 0.1 \qquad (4.24)$$

fails for an x-ray line then the thin-film approximation is invalid and absorption effects must be included within the analysis scheme.

4.7 APPLICATIONS OF STANDARDLESS ANALYSIS

The analysis of homogeneous single phase regions using the standardless approach outlined in the preceding section is a simple procedure. This section will illustrate its application through several examples in an optimized AEM. All data presented in the remainder of this chapter has, therefore, been obtained under low contamination rate conditions and has been appropriately corrected for any residual effects of uncollimated fluorescing radiation. The intensities (or intensity ratios) reported were obtained by integrating gaussian peak profiles (GEISS and HUANG, 1975) over their FWHM after a fitted specimen background curve has been subtracted from all measurements. In all cases the α line intensities reported consists of the sum of $\alpha_1 + \alpha_2$ components.

Table 4.6 compares the determination of the specimen composition using the thin-film standardless technique just outlined with bulk specimen compositions for several well characterized systems. For quantitative analysis using only K shell characteristic x-rays the agreement is excellent. Quantitative analysis using a combination of K and L shell

Table 4.5

Detector Efficiency $\varepsilon(E,\alpha)$

E(keV)	α - 0°	α - 20°	E(keV)	α - 0°	α - 20°
0.60	0.0274	0.0218	10.40	0.9942	0.9939
0.80	0.1945	0.1751	10.60	0.9945	0.9942
1.00	0.4094	0.3866	10.80	0.9948	0.9945
1.20	0.5795	0.5595	11.00	0.9950	0.9947
1.40	0.6973	0.6814	11.20	0.9953	0.9950
1.60	0.7747	0.7621	11.40	0.9954	0.9952
1.80	0.8298	0.8199	11.60	0.9950	0.9954
2.00	0.8159	0.8054	11.80	0.9957	0.9956
2.20	0.8520	0.8433	12.00	0.9917	0.9913
2.40	0.7749	0.7624	12.20	0.9919	0.9916
2.60	0.8158	0.8052	12.40	0.9921	0.9918
2.80	0.8340	0.8243	12.60	0.9921	0.9920
3.00	0.8611	0.8528	12.80	0.9921	0.9921
3.20	0.8677	0.8598	13.00	0.9921	0.9922
3.40	0.8872	0.8804	13.20	0.9919	0.9921
3.60	0.8927	0.8862	13.40	0.9915	0.9920
3.80	0.9071	0.9014	13.60	0.9911	0.9917
4.00	0.9190	0.9141	13.80	0.9887	0.9894
4.20	0.9290	0.9246	14.00	0.9880	0.9890
4.40	0.9374	0.9335	14.20	0.9871	0.9884
4.60	0.9445	0.9411	14.40	0.9848	0.9863
4.80	0.9506	0.9475	14.60	0.9836	0.9855
5.00	0.9558	0.9531	14.80	0.9821	0.9844
5.20	0.9604	0.9579	15.00	0.9805	0.9831
5.40	0.9643	0.9620	15.20	0.9785	0.9816
5.60	0.9677	0.9656	15.40	0.9764	0.9799
5.80	0.9707	0.9688	15.60	0.9740	0.9780
6.00	0.9733	0.9716	15.80	0.9713	0.9758
6.20	0.9756	0.9741	16.00	0.9683	0.9734
6.40	0.9777	0.9762	16.20	0.9651	0.9707
6.60	0.9795	0.9782	16.40	0.9616	0.9678
6.80	0.9811	0.9799	16.60	0.9579	0.9646
7.00	0.9826	0.9815	16.80	0.9539	0.9611
7.20	0.9839	0.9829	17.00	0.9496	0.9574
7.40	0.9851	0.9841	17.20	0.9450	0.9535
7.60	0.9861	0.9853	17.40	0.9402	0.9493
7.80	0.9871	0.9863	17.60	0.9351	0.9448
8.00	0.9880	0.9872	17.80	0.9298	0.9401
8.20	0.9888	0.9881	18.00	0.9243	0.9351
8.40	0.9895	0.9888	18.20	0.9185	0.9299
8.60	0.9902	0.9896	18.40	0.9125	0.9245
8.80	0.9908	0.9902	18.60	0.9062	0.9188
9.00	0.9914	0.9908	18.80	0.8998	0.9130
9.20	0.9919	0.9914	19.00	0.8932	0.9069
9.40	0.9923	0.9919	19.20	0.8863	0.9006
9.60	0.9928	0.9923	19.40	0.8794	0.8942
9.80	0.9932	0.9927	19.60	0.8722	0.8875
10.00	0.9936	0.9931	19.80	0.8649	0.8807
10.20	0.9939	0.9935	20.00	0.8574	0.8738

emission lines is more tenuous, owing to the lack of accurate values for the ionization cross-section in the incident energy range typical of AEM studies. However, the use of the parameterized equations just developed yields results which are remarkably good considering the assumptions made in extending the K-shell formulation to L shell excitation. An example of such K-shell/ L-shell analysis can be seen by reference to the Ni_4Mo results of Table 4.6. In this instance, it was possible to measure the intensities of the Mo K_α, Ni K_α, and Mo L_α x-ray lines simultaneously and although it was necessary to deconvolute the Mo L_α line from the unresolved Mo $L_{\alpha+\beta}$ peak agreement between K-K and K-L analysis is good. Also included in this table is the result of L-shell analysis using only the L lines (Nb and Hf) and once again reasonable agreement is obtained. Two examples of quantitative x-ray analysis using the standardless technique are demonstrated below in greater detail.

Standardless Analysis Using the Thin-Film Approximation: Fe-13r-40Ni

This example will consider the application of standardless analysis in the thin-film approximation (i.e., absorption and fluorescence effects are negligible) to a homogeneous Fe-Cr-Ni specimen. X-ray measurements were made in the STEM mode of operation and a region of the specimen less than 200 nm thick was analyzed using $\sim$ 20 nm diameter probe at an incident energy of 120 keV. After correction for the presence of uncollimated fluorescing radiation, the measured characteristic K_α intensity ratios were:

$$\frac{I_{Fe}}{I_{Ni}} = 1.25 \quad \text{and} \quad \frac{I_{Cr}}{I_{Ni}} = 0.46$$

The efficiency of the solid state x-ray detector used during these measurements can be obtained from Table 4.5 ($\alpha = 0°$), and the values for the x-ray generation constants from Table 4.3. Substituting the values for these parameters into Eqns. 4.11 and 4.12 one obtains the following relative composition ratios:

$$\frac{C_{Fe}}{C_{Ni}} = \frac{\varepsilon_{Ni}}{\varepsilon_{Fe}} \frac{\kappa_{Ni}}{\kappa_{Fe}} \frac{I_{Fe}}{I_{Ni}} = \frac{0.9855}{0.9777} \frac{1.5677 \times 10^{24}}{1.7382 \times 10^{24}} (1.25) = 1.14$$

and

$$\frac{C_{Cr}}{C_{Ni}} = \frac{\varepsilon_{Ni}}{\varepsilon_{Cr}} \frac{\kappa_{Ni}}{\kappa_{Cr}} \frac{I_{Cr}}{I_{Ni}} = \frac{0.9855}{0.9643} \frac{1.5677 \times 10^{24}}{1.9277 \times 10^{24}} (0.426) = 0.354$$

Using the relationship that $\Sigma C = 100$, one solves for the compositions and obtains $C_{Ni} = 400$, $C_{Fe} = 45.7$, and $C_{Cr} = 14.2$ in excellent agreement with the bulk analysis.

Very similar answers are calculated using k_{AB} factors for K_α as discussed in Chapter 3. Using Eqns. 3.9, 3.10 and 3.14, k_{FeNi} and k_{CrNi} at 120 KV are 0.928 and 0.863 respectively. Since no absorption correction is made,

$$\frac{C_{Fe}}{C_{Ni}} = k_{FeNi} \frac{I_{Fe}}{I_{Ni}} = (0.928)(1.25) = 1.16$$

$$\frac{C_{Cr}}{C_{Ni}} = k_{CrNi} \frac{I_{Cr}}{I_{Ni}} = (0.863)(0.426) = 0.354$$

Upon substitution of the values for C_{Fe}/C_{Ni} and C_{Cr}/C_{Ni} as above, one obtains (in wt %): $C_{Ni} = 39.6$, $C_{Fe} = 45.9$ and $C_{Cr} = 14.5$. Both calculation schemes yield compositions which fall well within the known error limits of the measured intensity ratios, I_{Fe}/I_{Ni} and I_{Cr}/I_{Ni}. The agreement between the two formulations is obviously quite reasonable; however, the analysis based on the L-shell differ significantly (see Fig. 4.14).

Standardless Analysis Using the Absorption Correction: β NiA1

The preceding example dealt with quantitative analysis in the thin-film approximation, in this illustration the application of the absorption correction to standardless analysis will be demonstrated. A homogeneous single crystal specimen of β NiAl ($C_{Ni} = 68.5$; $C_{Al} = 31.5$ wt%; $\rho = 5.9$ gms/cm^3) was analyzed at 200 keV in STEM using a probe diameter $\sim$ 20 nm. A characteristic K_α intensity ratio of Ni/Al, of 2.80 was subsequently measured from a region of the specimen $\sim$ 425 nm thick after correction for the residual in-hole spectrum. During these measurements the detector elevation angle θ_E was 0° while the electron incident angle was $\beta \sim 45°$. Using the conventional thin-film analysis equations one obtains a composition ratio.

$$\frac{C_{Ni}}{C_{Al}} = \frac{\varepsilon_{Al}}{\varepsilon_{Ni}} \frac{\kappa_{Al}}{\kappa_{Ni}} \frac{I_{Ni}}{I_{Al}} = (0.7468)(1.491)(2.8) = 3.12$$

giving an apparent composition (wt%) $C_{Al} = 24.3$ and $C_{Ni} = 75.7$. Checking the validity of the thin-film approximation using the Tixier-Philibert criterion, one obtains $\chi \cdot \rho \cdot t \cong 0.02$ for the nickel K_α lines while $\chi \cdot \rho \cdot t \cong 0.9$ for the aluminum K_α line. Clearly, the Al K_α line violates the thin-film criterion and thus the effects of absorption must be included.

One begins the correction procedure by using Eqn. 4.22 to calculate the mass absorption coefficient of the NiAl alloy. From a suitable table of mass absorption coefficients (BRACEWELL and VEIGELE, 1970) one finds that the absorption coefficient for Al K_α x-rays in pure Al and Ni is respectively:

$$\left(\frac{\mu}{\rho}\right)_{Al}^{AlK_\alpha} \cong 410 \ cm^2/gm \ \text{and} \ \left[\frac{\mu}{\rho}\right]_{Ni}^{AlK_\alpha} \cong 4910 \ cm^2/gm$$

while for the NiK_α line the values are

$$\left(\frac{\mu}{\rho}\right)_{Al}^{NiK_\alpha} \cong 63.7 \ cm^2/gm \ \text{and} \left(\frac{\mu}{\rho}\right)_{Ni}^{NiK_\alpha} \cong 60.6 \ cm^2/gm$$

Using the compositions determined from the thin-film model, one obtains a first approximation to the true mass absorption coefficients for the NiAl alloy as:

$$\left(\frac{\mu}{\rho}\right)_{NiAl}^{AlK_\alpha} = (410)(0.243) + (4910)(0.757) = 3817$$

$$\left(\frac{\mu}{\rho}\right)_{NiAl}^{NiK_\alpha} = (63.7)(0.243) + (60.6)(0.757) = 61.4$$

Substitution of these values into Eqn. 4.21 yields a first order absorption correction of:

$$\frac{\delta_{Al}}{\delta_{Ni}} = 0.6486$$

and the resulting composition ratio using Eqn. 4.20 becomes

$$\frac{C_{Ni}}{C_{Al}} = \frac{\delta_{Al}}{\delta_{Ni}} \frac{\varepsilon_{Al}}{\varepsilon_{Ni}} \frac{\kappa_{Al}}{\kappa_{Ni}} \cdot \frac{I_{Ni}}{I_{Al}} = (0.6486)(0.7468)(1.491)(2.8) = 2.02$$

Thus after a first iteration the corrected compositions are:

$$C_{Al}^* = 33.1 \ \text{and} \ C_{Ni}^* = 66.9$$

Using this result and proceeding through a second iteration the new values of mass absorption coefficients subsequently become:

$$\left(\frac{\mu}{\rho}\right)_{NiAl}^{Alk_\alpha} = (410)(0.331) + (4910)(0.669) = 3421$$

$$\left(\frac{\mu}{\rho}\right)_{NiAl}^{Nik_\alpha} = (6.37)(0.331) + (60.6)(0.669) = 61.6$$

This gives a new absorption correction term

$$\frac{\delta_{Al}}{\delta_{Ni}} = 0.6766$$

from which one obtains a calculated composition of

$$C_{Al}^{**} = 32.1 \text{ and } C_{Ni}^{**} = 67.8$$

A third iteration yields

$$C_{Al}^{***} = 32.2 \text{ and } C_{Ni}^{***} = 67.8$$

the ratio eventually converging toward $C_{Al} = 32.1$, $C_{Ni} = 67.9$ a value within 0.6% of the known specimen composition. The convergence in this example was particularly rapid and is a result of the initial thin film estimated composition being within about 7% of the actual composition.

The following results are obtained using the correction schemes outlined by Goldstein in the previous chapter. Using Eqns. 3.9, 3.10 and 3.14, k   at 200 KV is 1.18. If no absorption correction is made,

$$\frac{C_{Ni}}{C_{Al}} = k_{NiAl}\frac{I_{Ni}}{I_{Al}} = (1.18)(2.8) = 3.30$$

giving an apparent composition (wt%) $C_{Al} = 23.3$ and $C_{Ni} = 76.7$. Since an absorption correction is necessary one uses Eqn. 3.19 where

$$\frac{C_{Ni}}{C_{Al}} = k_{NiAl}\left\{\frac{I_{Ni}}{I_{Al}}\frac{\left(\frac{\mu}{\rho}\right)^{Ni}}{\left(\frac{\mu}{\rho}\right)^{Al}}\right\} \cdot \left\{\frac{1-\exp - \frac{\mu}{\rho}^{Al} \cdot \rho \cdot t \cdot \csc\alpha}{1-\exp - \frac{\mu}{\rho}^{Ni} \cdot \rho \cdot t \cdot \csc\alpha}\right\}$$

In this equation α = the effective take off angle = 45° and all $\left(\frac{\mu}{\rho}\right)$ values refer to the specimen. Following the iteration method outlined above, $C_{Al} = 35.5$ wt% and $C_{Ni} = > 64.5$ wt% after the first iteration and $C_{Ni} = 34.0$ wt%, $C_{Ni} = 66.0$ wt% after the third iteration. The convergence is rapid and yields an Al value within 7% relative of the actual specimen composition (31.5 wt%). Such an error in a correction or relative composition shift of almost 50% may be acceptable, since other errors may be of comparable magnitude. For example, if the measured foil thickness is in error by ± 10%, the calculated Al composition will vary by more than ± 1 wt%.

The application of an absorption correction which results in compositions shifts of > 10% should be used with a great deal of caution. Specimen thicknesses sufficient to result in corrections this large will undoubtably have substantial beam broadening effects which can invalidate the geometry used in the formulation of the absorption correction. Furthermore, since the fluorescence correction is directly related to the magnitude of absorbed x-ray flux, working in specimen thickness with absorption corrections of < 10%, for the most part, insures that this correction will be small and can usually be neglected.

4.8 STANDARDLESS ANALYSIS IN COMPLEX SYSTEMS

In this section, the application of x-ray microanalysis to the quantitative interpretation of x-ray intensity measurements in complex multi-component systems is considered. The term complex, in this context, is used to designate any analysis in which the investigator must pay close attention to the specific details of the experimental conditions or data reduction scheme used (beyond those general considerations discussed previously) to obtain accurate quantitative results. Three specific cases will be discussed below: 1. analysis of totally buried peaks; 2. procedures for accurate composition measurements of precipitates and 3. experimental techniques for analysis of radioactive specimens.

Analysis of Totally Buried Peaks

Due to the relatively poor resolution of solid state Si(Li) x-ray detectors, peak overlaps in multicomponent systems are a frequently observed phenomenon. This raises the question; how does one perform quantitative analysis when a peak is totally obscured by neighboring lines?

A particularly relevant example of this problem is the detection of low manganese concentrations in 316 stainless steels. In this situation the Mn K_α line (5.89 keV) is totally overlapped by the Cr K_β line (5.95 keV) while the Mn K_β peak (6.49 keV) is similarly buried under the Fe K_α line (6.50 keV). Fortunately, there is a solution to this apparent dilemma.

The answer is derived from the physics of the x-ray emission process, the relative intensities of all the characteristic x-ray lines from a particular shell are a constant for each element in the periodic table. This constant is, however, a function of atomic number. Typical values for the K_β/K_α ratio are in the range 0.1-0.25, while L shell values are more complex due to the multiplicity of observable emission lines. This variation is related to the intrinsic transition probabilities with a given subshell (DYSON, 1973); however, experimental factors such as self-absorption, detector efficiency, and the mode of excitation also influence the resulting values. Assuming one can obtain accurate values for the appropriate emission lines and the operating conditions employed, then the relative intensity relationships can be used to extract by difference (or deconvolution) the intensity of totally obscured peaks.

Returning to the specific example of Mn analysis, Fig. 4.15 illustrates the detection of Mn using the relative intensity relationship between Cr K_α and K_β lines. Figure 4.16 (a) is an expanded region of the x-ray emission spectrum from a Fe-13Cr-40Ni specimen around the chromium

K-shell lines. Experimental measurements from this specimen yield the following values for Cr, Fe, and Ni K_β/K_α ratios: Cr=0.137±0.001, Fe-0.148±0.001, Ni-0.149±0.001.

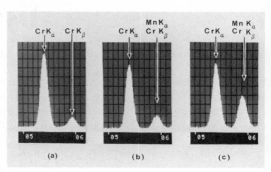

Figure 4.16. Illustration of Cr K_β - Mn K_α complete overlap situations (see text for details) (a) 0% Mn; (b) 1.8% Mn and (c) 8.9% Mn.

Figures 4.16 (b) and (c) show the magnitude of the resulting changes in CrK_β-MnK_α overlap peak which are observed during measurements of Mn rich inclusions in an inhomogeneous austenitic steel. Using the values for Cr K_β/K_α of 0.137 one obtains the local managanese compositions of 1.8 and 8.9 wt% for Figs. 4.16 (b) and (c) respectively. Although this example has dealt specifically with analysis of Mn in steel the technique is clearly applicable to similar overlap problems in any system, assuming of course that accurate measurements of the relative intensity ratios of the respective lines are available.

Quantitative Analysis of Precipitate Phases

Quantitative x-ray microanalysis of second phase precipitation in multicomponent systems is a nontrivial endeavor. Upon observing the presence of second phase regions within the microstructure of a specimen an investigator may attempt to perform analysis using the following procedure: 1. select an approriate imaging mode and probe size; 2. orient the specimen to avoid anamolous x-ray generation effects; 3. measure the characteristic x-ray spectrum from a second phase region in a thin-region of the specimen; and 4. finally convert the relative intensity measurements into composition values using the thin-film equations.

The correct interpretation of the results obtained during such an experiment will depend greatly on: 1. the specific information desired by the analyst; 2. the relative amount of the second phase with respect to the host matrix in the excited volume; and 3. the relationship of the precipitate composition to that of the matrix phase.

The steps outlined above are adequate to provide qualitative or semi-quantitative information; however, quantitative interpretation can be difficult. The complications in this situation arise from the improper application of the thin-film formulation to analysis of the measured x-ray intensities. One should recall, here, that the thin-film equations have been derived assuming that the specimen composition is <u>homogeneous</u> over the analyzed volume. Put in another context, the appli-

Table 4.6

Sample	Incident Energy (keV)	Measured Intensity I_A/I_B	Generated Intensity ϕ_A/ϕ_B	Ratio of X-Ray Generation Constants κ_B/κ_A; $\kappa \equiv \dfrac{Q \cdot w \cdot a}{A}$	Composition Thin Film X-ray Analysis	Bulk Analysis (wt%)
βNiAl	200.	$NiK_\alpha/AlK_\alpha = 2.05$	1.53	1.491	Ni = 69.5 Al = 30.5	Ni – 68.5 Al – 31.5
	100.	$NiK_\alpha/AlK_\alpha = 1.79$	1.34	1.744	Ni = 69.9 Al = 30.1	Ni = 68.5 Al = 31.5
β ZrNb	200.	$NbK_\alpha/ZrK_\alpha = 0.163$	0.163	1.079	Nb = 14.9 Zr = 84.0	Nb = 15.4 Zr = 84.6
β CuZn	100	$CuK_\alpha/ZnK_\alpha = 1.63$	1.634	0.933	Cu = 60.4 Zn = 39.6	Cu = 60.2 Zn = 39.8
Ni₄Mo	120	$MoK_\alpha/NiK_\alpha = 0.113$	0.118	3.475	Mo = 29.2 Ni = 70.8	Mo = 29.0 Ni = 71.0
	120.	$MoL_\alpha/NiK_\alpha = 0.246$	0.323	1.167	Mo = 27.4 Ni = 72.6	Mo = 29.0 Ni = 71.0
NbHf	120.	$NbL_\alpha/HfL_\alpha = 1.30$	1.515	0.516	Nb = 43.9 Hf = 56.1	Nb = 45.9 Hf = 54.1
Fe-13Cr-40Ni	120	$FeK_\alpha/NiK_\alpha = 1.25$	1.26	0.902	Ni = 40.1 Cr = 14.2 Fe = 45.7	Ni = 40.4 Cr = 13.2 Fe = 46.6
	120	$CrK_\alpha/NiK_\alpha = 0.426$	0.435	0.813		
Fe-13Cr-20Ni	120.	$FeK_\alpha/NiK_\alpha = 3.59$	3.619	0.902	Ni = 20.0 Cr = 14.5 Fe = 65.5	Ni = 19.7 Cr = 13.3 Fe = 66.9
	120.	$CrK_\alpha/NiK_\alpha = -0.0880$	0.899	0.813		

cation of the thin-film equations to heterogeneous regions effectively averages their composition over the excited volume. If the object of the experimental measurements is to accurately determine the composition of the precipitated phase, then one must effectively isolate it from the host matrix. This point is most clearly elucidated by reference to the following example.

Needle-like precipitates were observed to form in 316 stainless steel isothermally aged for 10,000 hours at 600½C (MAZIASZ, 1978). Subsequent crystallographic analysis identified these precipitates as having an fcc crystal structure and being consistent with the generic classification $M_{23}C_6$ (space group Fm3m, a $\cong$ 1.06 nm). In order to aid in the understanding of this precipitation process and its effects, accurate determinations of the precipitate composition are required. The apparent composition of this $M_{23}C_6$ phase measured under a systematic set of experimental configurations together with the composition of the aged 316 stainless steel matrix are documented in Table 4.7. The specific examples presented here were chosen to illustrate the potential for systematic errors which may occur during such an analysis.

The matrix composition reported in column 4 of this table was obtained by analyzing a region of the specimen < 200 nm thick approximately equidistant from all surrounding precipitates (the average inter-precipitate spacing was ∿ 500 nm). Column 3 presents the apparent composition observed when a STEM electron probe ∿ 10 nm in diameter (at 120 keV) was focussed on a $M_{23}C_6$ precipitate particle confined within the specimen matrix as shown in Fig. 4.17 (a). Here the precipitate thickness was ∿ 80 nm while the total thickness of the excited volume was on the order of 200 nm. Although the apparent composition is clearly different from matrix, this configuration for analysis is obviously far from optimum. A combination of beam broadening effects as well as the relative precipitate/matrix volume ratio will result in the partial detection of the matrix composition.

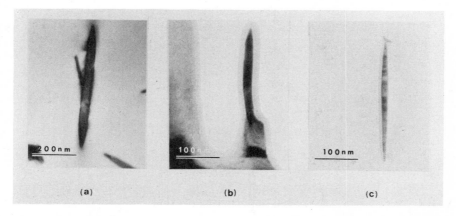

Figure 4.17. Needle-like $M_{23}C_6$ Precipitates in 316 Stainless Steel. (a) Embedded in Matrix; (b) Suspended from Edge of Specimen; (c) Extracted from Matrix.

With a sufficient amount of diligence on the part of the analyst it is usually possible to produce good quality TEM specimens in which the precipitates are partially isolated from the matrix by being suspended out from the foil edge [Fig. 4.17(b)]. Such a configuration improves the relative precipitate/matrix volume ratio and the measured compositions in this case (column 2, Table 4.7) are more representative of the second phase. However, a thin-shell of matrix material may still encapsulate the precipitate. In addition, material in the matrix nearby may be excited by electron scattered by the precipitate and by the contaminant. Thus, the analysis may be still only semi-quantitative. Note that the precipitate shown in Fig. 4.17(b) is clearly also surrounded by a substantial contamination layer produced by prolonged observation in a conventional TEM of relative poor vacuum. Thus, the anamolously high silicon content indicated can be ascribed to the presence of contamination prior to x-ray analysis in the AEM.

The most reliable measurements of the composition are obtained when the precipitate phase is extracted onto a thin carbon replica film using an electrochemical etching process [Fig. 4.17(c)]. The compositions measured here (column 1, Table 4.7) will most closely indicate the actual composition of the three measurements presented.

It is possible, of course, to propose a correction scheme in which one measures the relative volumes of the excited precipitate and matrix regions and then back-calculates the precipitate composition when the second phase is embedded within a matrix. However, not only does this require one to perform the difficult measurement of excited volume but it also requires a prior knowledge of the local matrix composition. Since the most significant precipitation phenomena result in the formation of composition gradients around second phase regions such a correction scheme may be more difficult than the experimental technique of extraction.

Procedures for Analysis of Radioactive Specimens

The final topic considered within this chapter concerns the application of quantitative microanalysis to materials which have been introduced into reactor environments. The observation and characterization of the microstructure resulting from phase transformations produced during irradiation plays an extremely crucial role in understanding the materials subsequent macroscopic properties. Without such information one cannot logically proceed in a systematic development of candidate alloys for the various nuclear energy applications.

After a material has been subjected to a rector environment it generally becomes radioactive, which results in the specimen spontaneously emitting characteristic and continuum radiation to a varying degree. During microanalysis, this radiation will be detected simultaneously with any electron generated x-ray signal. If the specimen is sufficiently active $\gtrsim 10$ μC the intensity of this intrinsic radioactive emission may dominate the electron excited x-ray signal. In fact, due to the large collection solid angle and short specimen-detector distances the

intensity of the intrinsic emission can completely incapacitate the signal processing chain of the detector/multichannel analyzer system due to
count rate limitations. The solution to this predicament is obvious but
not necessarily simple - namely reduce the intensity of the radioactive
emissions bombarding the Si(Li) detector.   This is best accomplished
using complementary approaches.

     First procedure is for the investigator to reduce the excess mass
of material present in the specimen, particularly that surrounding the
rim of the TEM disc.   The majority of the intrinsic emission comes
directly from bulk regions at the periphery of the specimen. Since remote handling of these specimens is sometimes necessary during sample
preparation one can only reduce the specimen size to a certain point.
After this, one must resort to collimation of the x-ray detector system.
The use of high atomic number collimators is particularly relevant here
since the energy spectrum of the irradiated specimen may extend into the
MeV range.   Furthermore, most solid state Si(Li) are enclosed within
relatively thin stainless steel detector heads, and it therefore may be
necessary to wrap lead foil around the immediate vicinity of the
detecting crystal to increase the shielding in this area. The specific
amount of collimation required will obviously depend on the activity of
the specimen being studied. Hopefully, one can by judiciously adjusting
the collection solid angle achieve a condition in which intrinsic radioactive signal allows the analyst to measure in a realistic time frame a
characteristic electron induced x-ray spectrum. An example of such an
analysis is given below.

     A specimen of titanium modified 316 stainless steel was irradiated
in the High Flux Isotope Reactor (HFIR) at ORNL at 600°C to a neutron
fluence producing the following conditions: 30 dpa (displacements per
atom) and 1850 atomic ppm He.   The specimen, after thinning into a 3 mm
diameter TEM self-supporting disc had an activity of $\sim$ 1000 μC and produced a radiation exposure level at contact of $\sim$ 0.5 R/hr.   These
radiations levels were sufficiently high that x-ray analysis in an AEM
optimized for conventional x-ray work was completely prohibited. By constructing a suitable lead collimator adding lead shielding around the
outside of Si(Li) detector, adjusting the collection solid angle to $\sim$ 5
x 10^{-4} sr and reducing the detector preamplifier time constant to 2 μsec
it was possible to attempt microanalysis of this specimen.   After these
modification, the detector deadtime correction in the absence of an electron beam (i.e., the source turned off) was still $\sim$ 40%. Figure 4.18(a)
shows the resulting beam-off spectrum recorded under these conditions,
the characteristic Mn and Cr lines visible resulting from the nuclear
decay process within the specimen (FARRELL, BENTLEY, and BRASKI, 1977;
ENGE, 1966).   The corresponding x-ray spectrum recorded with the electron probe (E = 120 keV) exciting a thin (< 200 nm) region of the matrix of the specimen is shown in Fig. 4.18(b).   Although the electron
generated spectrum is clearly dominated by the intrinsic radioactive
emission, one can still attempt a <u>semi-quantitative</u> analysis of the
matrix composition by subtracting the resulting "in-hole" spectrum
(i.e., spectrum produced by the combined effects of spontaneous emission
and uncollimated fluorescence) from the total spectrum recorded during
electron excitation of the matrix.   The results of such an analysis are

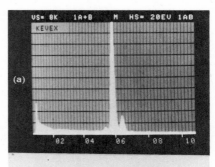

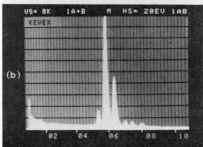

Figure 4.18. EDS Spectrum from Radioactive Ti-Modified 316 SS TEM Specimen. (a) Intrinsic Emission from Specimen; (b) Intrinsic Emmision + Electron Excited X-Rays

given in Table 4.8. Clearly, if the analyst desires more accurate results it will be necessary to reduce the intensity of the intrinsic emission to much lower levels than encountered here. This can be accomplished by altering the specimen preparation technique to reduce the excess sample mass surrounding the electron transparent zone. For example, the use of the window technique (HIRSCH et al, 1977) instead of the jet polishing method will substantially decrease the needless mass of bulk material surrounding the thin region of interest.

SUMMARY

There are a wide variety of instrumental problems which are present to some degree in all AEM instruments. The nature and magnitude of these artifacts can in some instances preclude the simple quantitative interpretation of the recorded x-ray emission spectrum using a thin-film electron excitation model; however, by judicious modifications to the instrument these complications can be effectively eliminated. The specific operating conditions of the microscope necessarily vary from one analysis to another depending on the type of specimen and experiment being performed. In general, however, the overall performance of the AEM system during x-ray analysis is optimized using the highest attainable incident electron energy; selecting the maximum probe diameter and probe current consistent with experimental limitations; and positioning the x-ray detector ina geometry such that it records information from the electron entrance surface of the specimen.

The application of standardless analysis to quantitative composition measurements from areas of a specimen which can be considered to be

Table 4.7

Analysis of $M_{23}C_6$ type precipitate in 316 stainless steel

| | Composition (wt%) | | | |
| | (1) | (2) | (3) | (4) |
Element	Extracted Precipitate	Suspended Precipitate	Embedded Precipitate	Matrix
Si	<.05	2.3*	1.1	1.3
Mo	16.7	14.2	8.7	3.9
Cr	65.9	58.2	30.9	18.4
Mn	<.05	0.2	1.47	1.5
Fe	14.6	21.6	47.9	62.3
Ni	2.8	3.6	10.0	12.6

*Increased silicon content probably due to presence of heavy contamination layer surrounding precipitate. (See figure 4.17b)

Table 4.8

Element	Apparent Matrix Composition (X-ray Microanalysis)	Bulk Analysis Prior to Irradiation (Wet Chemistry)
Si	2.4*	0.4
Mo	2.3	2.5
Ti	n.d**	0.25
Cr	18.5	17.0
Mn	n.d**	0.5
Fe	66.4	67.6
Ni	10.4	12.0
P,S,C	n.d**	0.08

*Specimen contamination due to prolonged observation in a conventional TEM will influence this measurement.

**No detectable peaks present above background after subtracting of in-hole signal.

locally homogeneous is a relatively straightforward process. The accurate analysis of heterogeneous regions in complex alloy systems is on the other hand a more difficult proposal, due in part to the fact that one may not be able to uniquely separate the relative contribution to the measured x-ray spectrum from the different phases in the analyzed volume. Thus, the seemingly simple task of determining precipitate compositions in a thin-foil specimen can become a non-trival exercise if approached incorrectly.

ACKNOWLEDGMENTS

The author would like to acknowledge invaluable discussions and contributions to the work from the following scientists: J. Bentley, P. J. Maziasz, E. A. Kenik, R. W. Carpenter, J. B. Woodhouse, I. D. Ward, and H. L. Fraser; and also to P. A. White, C. McKamey and M. Inman for their assistance in the preparation of this text. This work was supported by the E. P. Wigner Fellowship at ORNL.

REFERENCES

Arthurs, A. M. and Moiseiwitch, B. L., Proc. Roy. Soc. A247, 550 (1958).

Bentley, J., Zaluzec, N. J., Kenik, E. A., Carpenter, R.W., SEM/1979/ Washington, D.C., ed. O. Johari, Chicago Press (1979).

Bentley, J., and Kenik, E. A., Scripta Met., 11, 261-263 (1977).

Bentley, J. and Kenik, E. A., Oak Ridge Natinoal Laboratory, unpublished research, (1979). Available as ORNL TM/6857

Bracewell, B. L.,and Veigle, W. J., Developments in Applied Spectroscopy, 9, ed. A. Perkins, Plenum Press (1971).

Burhop, E. H. S., J. Phys. Radium, 16, 625 (1955).

Burhop, E. H. S., Proc. Cam. Phil. Soc., 36, 43 (1940).

Carpenter, R. W., Bentley, J., SEM/1979/ ed. O. Johari, pub. Chicago Press, Washington, D.C. (1979).

Clarke, D. R., SEM/1978/ 1 ed. O. Johari, pub. Chicago Press, Los Angeles (1978).

Cliff, G., Nasir, M. J., Lorimer, G. W., and Ridley, N., 9th Int. Con. on Electron Microscopy, ed., J. M. Sturgess, pub. Microscopical Society of Canada, Toronto (1978), 540-541.

Colby, J., Advances in X-Ray Analysis, 11, 287 (1968) (Plenum Press).

Dyson, N. A., X-Rays in Atomic and Nuclear Physics, Longman Group, Ltd., London (1973).

Duncumb, P., Phil. Mag., $\underline{7}$ (84), 2101 (1962).

Enge, H., Introduction to Nuclear Physics, Academic Press (1966).

Farrell, K., Bentely, J., and Braski, D., Scripta Met., $\underline{11}$, 243 (1977).

Fink, R. W., Jopson, R. C., Mark, H., and Swift, C. D., Rev. Mod. Phys., 39, 125 (1967).

Fiori, C. and Joy, D. C., National Institute of Health, Bethesda, Md., private communication (1979).

Fiori, C., Newbury, D., SEM/1978/1 ed. O. Johari, pub. Chicago Press, Los Angeles, 401 (1978).

Fraser, H. L., University of Illinois, Urbana, Illinois, private communication (1979).

Fraser, H. L. and Woodhouse, J. B., Electron Microscopy: Proc. of 2nd Workshop, Cornell University, Ithaca, NY, 191 (1978).

Geiss, R. H. and Huang, T. C., X-Ray Spectr., $\underline{4}$ (1975), 196-201.

Goldstein, J. I., Costley, J. L., Lorimer, G. W., and Reed, S. J. B., SEM $\underline{1}$, 315 (1977).

Goldstein, J. I. and Williams, D. B., SEM/1978/I, ed., O. Johari, pub. SEM Inc., AMF O'Hare, IL, 427-434 (1978).

Goldstein, J. I. and Williams, D. B., SEM/1977/1, ed., O. Johari, pub. Chicago Press, IIT Research Institute, Chicago, IL, 651-662 (1977).

Hall, C. R., Proc. Roy. Soc., A$\underline{295}$, 140 (1966).

Hashimoto, H., Howie, A.,and Whelan, M., Proc. Roy. Soc., A$\underline{269}$, 80 (1962).

Headley, T. J. and Hren, J. J., Sources of Background X-Radiation in AEM, ibid., 504-505.

Heinrich, K. F. J., Proc. 11th Conf. Mas (1976), p. 29.

Hirsch, P., Howie, A., Nicholson, R., Pashley, D., Whelan, M., Electron Microscopy of Thin Crystals, Krieger Pub. Co., New York, 1977.

Hink, W. and Ziegler, A., Z. Physik, $\underline{226}$, 222 (1969).

Hutchings, R., Loretto, M. H., Jones, L. P., Smallman, R. E., Analytical Electron Microscopy, 2nd Workshop, Cornell University, Ithaca, New York, 166 (1978).

Hren, J. J., Ong, P. A., Johnson, P.D., and Jenkins, E. J., 34th Proc. EMSE, ed., Bailey, G. W., pub. Claitor's, Baton Rouge, 1976, 418-419.

Johnson, P. F., Bates, S. R., and Hren, J. J., 9th Int. Conf. on Electron Microscopy, ed., Sturgess, pub. Microscopical Society of Canada, Toronto, 1978, 502503.

Joy, D. C.,and Maher, D. M., Analytical Electron Microscopy: Report of a Specialist Workshop, Cornell University (August, 1976), 111113. Materials Science Center, Clark Hall, Cornell University, Ithaca, N.Y. 14850.

Joy, D. C., Maher, D. M., SEM/1977, ed. O. Johari IITRI, Chicago, IL, 325 (1977).

Kenik, E. A., Bentley, J., 35th Ann. Proc. of EMSA, ed., G. W. Bailey, pug. Claitor's, Baton Roughe (1977), 328329.

Lorimer, G. W., Cliff, G., and Clark, J. N., pp. 153159, Developments in Electron Microscopy and Analysis, Academic Press, London, 1976.

Kirkpatrick, P. and Wiedmann, L., Phys. Rev. 67, 321 (1945).

Kirkpatrick, P., and Baez, A. V., Phys. Rev. 71, 521 (1947).

Liljesvan, B., and Roomans, G. M., Ultramicroscopy 2 (1976) 105-107

Lorimer, G. W., Cliff, G., and Clark, J. N., pp. 153-159, Developments in Electron Microscopy and Analysis, Academic Press, London, 1976.

Maziasz, P. J., Proc. of Symposium Metal Physics of Stainless Steel, TRMKAIME Meeting, Denver, 1978.

Morris, P. L., Davis, N. C., and Treverton, J. A., Developments in Electron Microscopy 1977: Proc. of EMAS, ed., D. L. Misell, Pub. Inst. of Phys., Briston, England, 2377380.

Mott, N., and Massey, H., Theory of Atomic Collisions, Oxford Press, London, 1949.

Oppolzer, H., and Knauer, U., SEM/1979, Washington, D.C., ed., O. Johari, pub. Chicago Press.

Perlman, H. S., Proc. Phys. Soc. (London), 76, 623 (1960).

Philibert, J.,and Tixier, R., (1968) B. J. A. P. 1, 685.

Pockman, L. R., Webster, D. L., Kirkpatrick, P., and Harworth, K., Phys; Rev. 71, 330 (1947).

Powell, C. J., Rev. Mod. Phys. 48, 33 (1976).

Reed;, S. J. B., The Electron Microprobe Analysis, Cambridge University Press (1976).

Robertson, B. W., Chapman, J.N., Nicholson, W. A. P., and Ferrier, R. P., 9th Int. Conf. on Electron Microscopy, ed., J. M. Sturgess, pub. Microscopical Society of Canada, Toronto (1978), 550-551.

Shuman, H., Somlyo, A. V., and Somlyo, A. P., Ultramicroscopy 1 (1976) 317-339.

Slivinsky, V. W., and Ebert, P. J., Phys. Rev. A 5, 1581 (1972).

Sommerfeld, A., Ann. d. Physik 11, 257 (1931).

Sturcken, E. F., SEM/1976/I, ed., O. Johari, pub. Chicago Press, IIT Research Institute, Chicago, IL, 247-256 (1976).

Sprys, J. W., Rev. Sci. Instr. 46 [6] (1975) 773-774.

Wentzel, G. Z., Phys. 43, 514 (1927).

Woldseth, R., X-ray Energy Spectrometry, pub. Kevex Corp., (1973), Burlingame, CA.

Worthington, C. R., and Tomlin, S. G., Proc. Phy. Soc. (London), 69, 401 (1956).

Zaluzec, N. J., Ninth Int. Cong. on Electron Microscopy, ed., J.M. Sturgess; pub. Microscopical Society of Canada, Toronto, 1978, Vol. 1, 548-549.

Zaluzec, N. J., Kenik, E. A., and Bentley, J., Analytical Electron Microscopy; Proc. of 2nd Workshop // at Cornell University, July 1978, 179.

Zaluzec, N. J., and Fraser, H. L., 8th Int. Conf. on X-ray Optics and Microanalysis, Boston, 1977, ed., Beaman, Ogilivie, Wittry, pub. Science Press, Princeton, N.Y. (in press).

Zaluzec, N. J., and Fraser, H. L., Analytical Electron Microscopy: Report of a Specialist Workshop, Cornell University (August 1976), 118-120. Materials Science Center, Clark Hall, Cornell University, Ithaca, N. Y. 14850.

Zaluzec, N. J., and Fraser, H. L., 34th Ann. Proc. of EMSA, ed., G. W. Bailey, pub. Claitor's, Baton Rouge (1976), 420-421.

Zaluzec, N. J. and Fraser, H. L., Analytical Electron Microscopy; Proc. of the 2nd Workshop, Cornell University, Ithaca, N.J., July 1978, 122.

Zaluzec, N. J., And Fraser, H. L., Phys. E. 9 (1976) 1051-1052).

Zaluzec, N. J., An Analytical Electron Microscopy Study of the Omega Transformation in a Zirconium-Niobium Alloy Phase, Ph.D. Thesis, University of Illinois, 1978; also published as Oak Ridge National Laboratory Report ORNL/TM6705. Copies available from NTIS, U. S. Department of Commerce, Springfield, VA 22161.

CHAPTER 5

EDS QUANTITATION AND APPLICATION TO BIOLOGY

T.A. HALL and B.L. GUPTA

BIOLOGICAL MICROPROBE LABORATORY, DEPARTMENT OF ZOOLOGY

UNIVERSITY OF CAMBRIDGE, DOWNING STREET, CAMBRIDGE CB2, 3EJ, U.K.

5.1 INTRODUCTION

The quantitative x-ray microanalysis of biological tissue sections differs from that of inorganic specimens strikingly in several ways:

a) Measurements of local concentrations in biological tissue sections may be misleading because the elemental distributions may have been altered inadvertently during the preparation of the specimens -- elements may have been largely lost, or displaced, during preparation.

b) Even if a section as prepared is accepted as a proper object for measurement, the procedures available for measurements in thin organic specimens are subject to systematic errors arising from assumptions about section uniformity or from beam damage.

c) A given procedure may produce a measurement of concentration which is technically correct but biologically irrelevant or ambiguous. For example, in a section stained with a heavy metal, the measurement of amount of element per kg of stained specimen is not in itself useful when much of the mass may consist of the metal stain. A less blatant example: For electrolyte elements which are not bound to organic matrix but are almost entirely "free" in aqueous solution in living tissue, the relevant concentration for the physiologist is millimoles of element per litre of water; if he simply measures mM of element per total mass of tissue, he may get quite different values for regions where there are actually identical values of mM of element per litre of water but different aqueous fractions (i.e. differing values for mass of water/total mass).

d)   Biological tissues are very variable. One must realize that there may be inhomogeneities within a tissue -- gradients for example within the cytoplasm of a single cell; concentrations within an organism may vary with time or physiological condition; there will be differences between individuals of the same species.

Because of these complicating factors, the role of quantitative microanalysis in biology is restricted and the analyst must be clear about the significance of intended measurements.

In this article we shall concentrate on the procedures available for measurements on specimens in the form which is classical for electron microscopy, namely thin or ultrathin sections mounted on thin supporting films or without support. These procedures will be outlined and their points of difficulty and limitations will be discussed. Finally we shall briefly allude to measurements on specimens in other forms, citing references for the reader who needs fuller information about them.

5.2 MEASUREMENTS ON THIN OR ULTRATHIN SECTIONS MOUNTED ON THIN FILMS

Elemental Ratios

The ratio of the concentrations of two elements $\underline{a}$ and $\underline{b}$ can be determined from the equation

$$\frac{C_a}{C_b} = \frac{I_a}{I_b} \left(\frac{I_b}{I_a} \cdot \frac{C_a}{C_b} \right)_r \tag{5.1}$$

where C_a is the mass fraction of element $\underline{a}$ (which can be expressed either as mass of element $\underline{a}$ per unit total mass of specimen, or as millimoles (mM) of element $\underline{a}$ per kg of total specimen mass), I_a is the x-ray peak integral obtained for a selected characteristic emission from the element $\underline{a}$, and the quantity in brackets refers to a thin standard of known composition. The subscript $\underline{r}$ denotes a standard used for ratio measurements.* (The quantities $\underline{I}$ refer to the genuine characteristic signal after correction of the peak integrals for background. The means of background correction, and spectral deconvolution in general, are outside the scope of this paper.)

An increase in section thickness can affect the validity of Eqn. 5.1 in two ways: a correction may be needed for the absorption of soft radiation within the specimen and, because of electron energy-loss, the relative excitation efficiencies for the elements $\underline{a}$ and $\underline{b}$ may not be the same in specimen and standard. For the analysis of inorganic inclusions within sections of soft tissue, the situation is just like the case of

---

*A list of symbols with full definitions is given at the end of this article.

inorganic thin films, while the effects of thickness are actually much less restrictive for the case of elements distributed within the soft tissue matrix. A correction for absorption is outlined below. With respect to excitation efficiency, it is sufficient to ensure that the probe electrons lose on average only a small part of their energy in passing through the section (or standard). This condition is easily satisfied. For example, 50 kV electrons lose on average less than 1 kV while passing through an organic film with density 1 g cm^{-3} and thickness 1 μm.

As thin standards for use with Eqn. 5.1, one may use preparations developed for the analysis of inorganic specimens, for example finely ground pieces of well characterized mineral (CLIFF and LORIMER, 1972) or even slurry-mixes (ROWSE, JEPSON, BAILEY, CLIMPSON and SOPER, 1974). But the effects of absorption and energy degradation are less if one uses one of the organic-matrix standards developed for absolute measurements of concentration (see section on standards below), or a preparation of dried droplets (MORGAN, DAVIES and ERASMUS, 1975). It is also possible to use Eqn. 5.1 without standards, by deriving the bracketed quantity from theory (RUSS, 1974).

While ratio measurements in biological tissue sections are basically like those in inorganic films, and are indeed simpler in practice, generally one cannot deduce absolute concentrations from a set of ratios. In metallic films one can often measure the relative amount of every constituent element, and it is then easy to deduce the absolute concentrations given that they must add up to 100%. A similar calculation is possible in mineralogical specimens when the only missing element is oxygen and its stoichiometry is known. In soft tissues the main constituents are water plus organic material consisting predominantly of the elements C, N. O and H; if absolute concentrations are required, the ratio method cannot suffice.

Millimoles of Element Per Unit Volume

Absolute measurements of elemental mass per unit volume are based on the presence of a standard "peripheral." Prior to sectioning, the tissue must be in contact with a medium of known composition. This medium serves as the standard; it must be sectioned with the tissue and must be present in the section alongside the tissue itself.

Since a characteristic x-ray signal from a thin specimen is proportional to the local amount of the emitting element per unit area, one may write

$$M_a = I_a \, (M_a/I_a)_p \qquad (5.2)$$

where M_a is the local mass of element a per unit area in the analyzed region of the specimen and the subscript p refers to the peripheral standard. It is assumed that the analyzed region of the tissue and the utilized region of the standard are equally thick. If both sides of Eqn. 5.2 are divided by its thickness, we get (since area x thickness = volume)

$$c'_a = I_a \ (C'_a/I_a)p \qquad\qquad (5.3)$$

where C'_g is the mass of element a per unit volume, which is commonly expressed by physiologists in units of mM L^{-1} (millimoles of element per
litre). But it must be noted that the quantity $\underline{C}'$ is mM per litre of
specimen, not mM per litre of water.

Equation 5.3 has been used extensively for the analysis of frozen-
dried sections of quench-frozen epithelial tissues by Dörge and co-
workers (DÖRGE, RICK, GEHRING, MASION and THURAU, 1975; RICK, DÖRGE, VON
ARNIM and THURAU, 1978) and it has been used for measurements on freeze-
substituted epoxy-embedded botanical tissues by PALLAGHY (1973).

Even if the standard contains only one element in known amount per
unit volume, other elements as well can be assayed absolutely in the
specimen by combination with the ratio method. A combination of Eqns.
5.1 and 5.3 gives

$$C'_a = I_a \ \left(\frac{C'_b}{I_b}\right)_p \left(\frac{I_b}{I_a} \ \frac{C'_a}{C'_b}\right)_r \qquad\qquad (5.4)$$

(the ratios C'_a/C'_a and C_a/C_a being of course the same). This equation
gives the concentration of an element a in the specimen, C'_a, in terms of
the concentration of the element b in the peripheral standard and the
relative sensitivity for a and b as determined from a ratio standard or
from theory. PALLAGHY (1973) has used this approach with chlorine in
the epoxy medium as the reference element.

With increasing section thickness, electron path length in the section increases disproportionately (due to electron scatter); excitation
cross sections change due to degradation of the electron energy; the proportionality between I_a and M_a breaks down and Eqn. 5.3 becomes invalid.
The upper limit on thickness set by these factors is not very restrictive for organic material. DÖRGE, RICK, GEHRING and THURAU (1978) have
shown with frozen-dried organic standards that even at a probe voltage
of 15 kV, the proportionality holds up to a "mass thickness" of almost
40 µg cm^{-2}, while at 27 kV the signals are proportional to thickness (except for an absorption effect for soft radiation) up to at least 60 µg
cm^{-2}. On theoretical grounds, the upper limit on thickness should increase approximately in proportion to probe voltage (except for absorption effects).

The method embodied in Eqn. 5.3 is not invalidated in principle by
the use of heavy-metal stains. Of course the use of liquid staining
media during specimen preparation is likely to produce intolerable elemental translocations, but the mere presence of a heavy-metal stain does
not in itself upset the validity of the equation. Thus quantitative
analysis via Eqn. 5.3 may still be possible after staining with osmium
vapor, which may cause only negligible translocations while substantially improving on the poor image contrast which bedevils studies of unstained sections (A. V. SOMLYO, SHUMAN and A. P. SOMLYO, 1977).

The main limitations of the method are:

a)   In the case of cryosections of quench-frozen material, the per-
      ipheral standard and the analyzed region of tissue must be
      equally thick just after the moment of sectioning ("true" sec-
      tioning without chipping, fracturing or edge-compression).
      This condition may be quite difficult to satisfy, depending on
      the tissue and on sectioning technique, and it becomes pro-
      gressively more difficult as lower sectioning temperatures
      come into use in order to guard against elemental diffusion
      (SAUBERMANN, RILEY and BEEUWKES, 1977).

b)   If the cryosections are freeze-dried before analysis, there
      must be no shrinkage during dehydration (or else the peripher-
      al standard and the analyzed tissue must shrink to the same
      extent). Shrinkage changes mass per unit area, and non-
      uniform shrinkage would therefore invalidate Eqn. 5.3.

c)   Uniformity of section thickness and lack of shrinkage are more
      readily achieved in epoxy-embedded sections, but the dangers
      of elemental shifts during embedding (and during freeze-
      substitution) are greater than in the case of cryosections of
      quench-frozen tissues.

d)   The method is generally used with frozen-dried or freeze-
      substituted material. With both of these preparative modes,
      there must be a considerable displacement of elements which
      had been in the aqueous phase in extracellular spaces lacking
      an organic matrix, so that the elemental concentrations in
      such spaces cannot be determined.

Millimoles of Element Per kg of Dried Tissue (Continuum Method)

The requirement for equal thickness of specimen and standard is removed when the analytical method includes the determination of local
variations in "mass thickness" (i.e., mass per unit area). The proportionality between signal I_a and local mass of element $\underline{a}$ per unit area
can be written in the form

$$\frac{I_a}{M\,C_a} = \left(\frac{I_a}{M\,C_a}\right)_o \tag{5.5}$$

where $\underline{M}$ is the local total specimen mass per unit area and the subscript
$\underline{o}$ denotes a thin standard. (Recall that C_a is the mass fraction of element $\underline{a}$ expressed either as mass of $\underline{a}$/total mass or as mM of $\underline{a}$ per kg of
total mass.)

The x-ray continuum can provide the measure of local mass. According to the analysis of KRAMERS (1923), the x-ray continuum intensity
generated in a thin film depends on $\underline{M}$ and on atomic composition in accord with the equation

$$I_w = k \ M \ \sum_i \ (f_i \ Z_i^2 / A_i) \tag{5.6}$$

$$\equiv k \ M \ G$$

Here $\underline{k}$ is a constant, Z_i is the atomic number of element $\underline{i}$, A_i is the atomic weight of element $\underline{i}$, f_i is the mass fraction of element $\underline{i}$ expressed as elemental mass/total mass, the sum is taken over all constituent elements, and I_w is an x-ray continuum count taken in a fixed quantum-energy band (most conveniently a band free of characteristic peaks although such peaks, if present, can be stripped out by deconvolution to give the continuum signal proper). The weighted average of Z^2/A is represented in the second line of the equation, and in the remainder of this text, by the symbol $\underline{G}$.

Substitution of Eqn. 5.6 into Eqn. 5.5 and re-arrangement of terms gives

$$c_a = \frac{I_a}{I_w} \ G \left[\frac{C_a}{(I_a/I_w) \ G} \right]_o \tag{5.7}$$

This equation expresses mass fraction C_a in terms of a ratio (I_a/I_w) which is independent of thickness in thin specimens. The upper limit to thickness is much less stringent than for the method described under Millimoles of Element Per Unit Volume earlier in this section since I and I_w are affected to almost the same extent by the factors which upset the proportionality between I_a and $\underline{M}$. HALL (1979a) has shown that with a probe voltage of 30 kV or more, Eqn. 5.7 may be used for organic sections with mass thicknesses up to at least 200 μg cm^{-2} (given suitable correction, again, for absorption when soft radiations are observed).

The use of Eqn. 5.7 has been discussed in detail in several articles (HALL, 1971; HALL, ANDERSON and APPLETON, 1973; SHUMAN, A. V. SOMLYO and A. P. SOMLYO, 1976; HALL, 1979a). With respect to technical difficulties, we must note the following points here:

a) I_w is the continuum signal generated in the specimen itself. In order to determine this quantity, one should correct the observed continuum count for extraneous contributions from the supporting film and from surrounding bulk material. Continuum from the film can be estimated by putting the beam directly on the bare film. Continuum from bulk surroundings (grid bars or specimen rod) can be greatly reduced if these surroundings are entirely of low atomic number, but this background can probably be accounted for more accurately if there is a heavier element (subscript "e" below) which is present in the bulk material and not in the specimen; one can then use this element as a monitor of bulk-generated continuum and estimate the bulk correction from data obtained with the beam put directly onto the bulk material. Thus, in the correction procedure described here, three runs are involved: the run on the specimen itself, a run with the beam directed onto the surrounding bulk material, and a run with the beam on the supporting film.

The run with the beam on the bulk material establishes a ratio $(I_w/I_e)_b$; here the subscript "b" designates the bulk-run, and I_w and I_e within the brackets are respectively the count in the continuum band and the peak-integral count for the monitor element "e". We assume that the ratio of bulk-generated continuum to the characteristic count I_e is the same whether the beam is directed onto the specimen or onto the bulk material. (The geometry for self-absorption effects may differ in the two cases, and it is therefore best to use a high-energy continuum band and a high-energy monitor line to minimize the effect of absorption on this ratio.) It follows that for the run on the specimen, the bulk-generated continuum background is given by $(I_w/I_e)_b I_e$, where the co-factor I_e is the peak-integral count for the element "e" for the spec-imen run.

When the beam is directed onto the film, the continuum is generated within the film, and also within the bulk surroundings by electrons scattered from the film. Hence the continuum generated within the film itself is given by $[I_w - (I_w/I_e)_b I_e]_f$, where the subscript f denotes the run with the beam directed onto the film, and the second term rep-resents the bulk contribution during this run.

The corrected signal I_w, representing continuum generated within the specimen itself, is hence given by

$$I_w = I_{wt} - (I_w/I_e)_b I_e - [I_w - (I_w/I_e)_b I_e]_f \qquad (5.8)$$

where I_{wt} is the uncorrected continuum count for the specimen run, and the film run is done with live-time and beam current matching the specimen run.

The significant bulk background may come from either a grid or a specimen rod. For cases where both contribute significantly, a similar though longer equation is readily formulated.

b) Equation 5.7 requires knowledge of the quantities Z^2/A (or "G"). This is not a problem in the case of the standard, where the composition is known, but an estimate must be made for the specimen. The specimen must be regarded as a composite of organic matrix, whose elemental com-position cannot be fully known, plus heavier elements whose characteris-tic radiations are detected. The problem is simply to guess the value $\underline{G}$ for the organic matrix; once this value is postulated, Eqn. 5.7 can be solved iteratively with successive corrections to the overall specimen $\underline{G}$ to take account of the effect of the heavier elements (or one may use the closed form given by HALL (1971, page 238). The concentrations reached in this way are a consistent set, but they are still erroneous if the assumed value of $\underline{G}$ for the organic matrix is wrong. Fortunately $\underline{G}$ does not vary much among organic compounds (HALL 1979a). A typical value for protein (exluding sulphur, which is assayed by its x-ray emis-sion) is $\underline{G=3.2}$. Uncertainty over the real $\underline{G}$ value of the organic matrix in an analyzed region of dehydrated soft tissue introduces a probable error of the order of a few per cent.

c) Beam damage seriously complicates the use of Eqn. 5.7. Discussion
of this problem is deferred to Section 5.4 below.

Apart from technical problems, we must note two basic limitations
in the use of the continuum method for measurements of mM of element per
kg in dehydrated specimens:

a) Measurement of mM of element per kg of "dry weight" has an obvious
biological significance for elements which are mainly bound to the organic matrix. For elements which were predominantly in aqueous solution
prior to dehydration, the interpretation is less direct and the significance is less definite (as noted above in Section 5.1).

b) Whatever the method of analysis, one must expect dislocation of the
elements in matrix-free extracellular spaces when dehydration is part of
the preparative procedure. Measurements in such spaces in frozen-dried
sections are therefore dubious.

The continuum method has been extensively used for the microanaly-
sis of frozen-dried sections by A. P. Somlyo and co-workers (A. V.
SOMLYO et al., 1977; A. P. SOMLYO, SHUMAN and A. V. SOMLYO, 1979) and by
Roomans and co-workers (ROOMANS and SEVÉUS, 1976; ROOMANS, 1978).

Millimoles of Element Per kg Wet Weight (Continuum Method, Frozen-Hydrated Sections)

The continuum method and Eqn. 5.7 may be applied to frozen-hydrated
sections for the direct determination of mM of element per kg "wet
weight." The method requires the use of a cold-stage to maintain the
hydration even during analysis. The major advantages of this difficult
technique are the capability for the analysis of small matrix-free extracellular spaces, and the capability of measuring local "dry-weight
fraction."

A complication arises in the estimation of the value G for the
matrix, since the tissue matrix in frozen-hydrated material is a composite of frozen water (with $G = 3.67$) and organic material (G approximately 3.2). This is not a serious problem in practice, since the organic ("dry-weight") fraction in most soft tissue is only 10 - 30%, and
a fairly accurate match in matrix G is provided by a standard consisting
of a frozen aqueous solution containing organic material at a level of
about 20%. If the local dry-weight fraction in an analyzed region is
actually determined, as described in the next section, then a value for
the composite matrix can actually be worked out from the observations.

Discussion of the major problem of beam damage is again deferred to
Section 5.4.

The continuum method has been used extensively for the microanaly-
sis of frozen-hydrated sections by Gupta and co-workers (GUPTA, HALL and
MORETON, 1977; GUPTA, BERRIDGE, HALL and MORETON, 1978a; GUPTA, HALL and
NAFTALIN, 1978b; GUPTA and HALL, 1979).

Dry-Weight and Aqueous Fractions

A local dry-weight fraction can be measured in a frozen-hydrated section by observing the continuum intensity, dehydrating the section within the microscope column by removal from the cold-stage, and observing the continuum intensity from the same region again after the dehydration (GUPTA, HALL and MORETON, 1979). If there were no shrinkage during dehydration, and the continuum signal were simply proportional to mass per unit area, the dry-weight fraction F_d would be given simply by

$$F_d = \frac{I_{wd}}{I_{wh}} \qquad (5.9)$$

where I_{wh} is the continuum signal in the hydrated state and I_{wd} is the continuum signal after dehydration. In the authors' experience Eqn. 5.9 is unreliable because dehydration is often accompanied by substantial shrinkage which tends to increase mass per unit area and hence increase the continuum signal. Since shrinkage must affect elemental and continuum signals in the same way, one can correct for shrinkage by establishing a shrinkage factor $\underline{S}$,

$$S = \frac{I_{ah}}{I_{ad}} \qquad (5.10)$$

where I_{ah} and I_{ad} refer to the characteristic intensities for an element $\underline{a}$ before and after dehydration. The version of Eqn. 5.9 corrected for shrinkage is

$$F_d = \frac{I_{wd}}{I_{wh}} \, S \qquad (5.11)$$

Fully detailed illustrations of the use of Eqn. 5.11 have been presented by GUPTA (1979), and biologically significant applications have appeared in two studies (GUPTA et al., 1978 a,b).

Equation 5.11 neglects the fact that $\underline{G}$, and hence the continuum signal per unit mass thickness, are different in the hydrated and the dehydrated states. Actually it is not difficult to take account of this fact and a straightforward derivation leads to an improved equation*

$$F_d = \frac{S}{S + Y[(I_{wh}/I_{wd}) - S]} \qquad (5.12)$$

---

*Equation 5.12 has not been published before. The derivation is given in Appendix I.

where $\underline{Y}$ is the value of $\underline{G}$ in the dried matrix divided by the value of $\underline{G}$ for water. While Eqn. 5.12 must be more accurate than Eqn. 5.11, in practice they give quite similar results. As a typical example, if $I_{wd}/I_{wh} = 0.20$, $S = 1$ and $Y = 3.2/3.67$, then Eqn. 5.11 gives $F_d = 0.20$ and Eqn. 5.12 gives $\underline{F_d} = 0.22$.

GUPTA et al. (1979) find that it is better to make the measurements before and after dehydration on adjacent fields rather than on the identical field. The initial field is prone to severe beam damage on re-irradiation after dehydration, due perhaps to the expression of latent beam damage when the specimen temperature is raised for the dehydration.

The aqueous fraction F_a is, of course, simply

$$F_a = 1 - F_d \qquad (5.13)$$

Conversion to mM of Element Per Litre of Water

In the case of physiological studies involving electrolyte elements, one generally wants to know their concentrations in the aqueous phase, i.e. mM of element in a given volume of water. In cases where it may be shown or assumed that an electrolyte element is virtually entirely in the aqueous phase (i.e., that it is negligibly bound), one can deduce local mM per litre of water from the measurement of local mM per kg of tissue in the frozen-hydrated state ($\underline{C}$) and the measurement of the local aqueous fraction F_a. Given negligible binding, the concentration of an element $\underline{b}$ in the aqueous phase, C''_b, is simply

$$C''_b = \frac{C_b}{F_a} \qquad (5.14)$$

Equation 5.14 has been used to compare microprobe results with aqueous-phase measurements by ion-selective microelectrodes (GUPTA et al., 1978a).

Absorption Corrections

Because the mean atomic number and the x-ray absorption cross sections in soft tissues are low, absorption in thin sections is generally negligible. When a correction is necessary for a low-energy radiation, it may be calculated from a simple expression based on the assumption that x-ray generation is uniform throughout the depth of the section. The formula derived by HALL (1971, p. 240) is

$$\frac{I'_a}{I_a} = \frac{P\,M\,\csc\theta}{1 - \exp(-P\,M\,\csc\theta)} \qquad (5.15)$$

where I_a is the observed peak integral for the characteristic radiation of element a, I_a' is the integral that would be obtained if there were no absorption in the specimen, P is the effective x-ray absorption coefficient in units $cm^2 g^{-1}$, M is local section mass per unit area in units $g cm^{-2}$ and α is the takeoff angle. The effective coefficient P is obtained by summing over the constituent elements i:

$$P = \sum_i (f_i P_i) \qquad (5.16)$$

In order to use Eqn. 5.15, one must have a value for the local mass thickness M. This can be measured by comparison with the continuum signal from a mass standard, a thin organic film of known thickness and composition, via the equation

$$M = M_o \frac{I_w}{I_{wo}} \frac{G_o}{G} \qquad (5.17)$$

the subscript o referring here to the mass standard. For this standard the authors use polycarbonate film ("Makrofol," supplied by Siemens), thickness 2 μm, mass thickness 240 $\mu g cm^{-2}$, composition $C_{16}O_3H_{14}$ and G-value 3.08. In order to avoid beam damage, the continuum signal from the mass standard should be recorded with the beam spread enough to keep the integrated beam loading below 10^{-10} Coulomb per μm^{-2}.

Equation 5.15 also requires assumptions about specimen composition in order to estimate P and G. If one calculates absorption for sodium x-rays in a section of ice with $M = 100$ $\mu g cm^{-2}$ (thickness 1 μm and density 1 $g cm^{-3}$), with a takeoff angle of 35°, the result is a correction factor I_a'/I_a equal to 1.3. It is clear from this that absorption is generally quite negligible for ultrathin sections of organic material, that it is generally unimportant as well for more penetrating radiations in 1-μm sections, and that the requisite assumptions need not be very accurate when the correction has to be made. When peripheral standards are used, even for sodium the absorptions in standard and tissue may be similar enough to make an explicit correction unnecessary. Absorption corrections based on an equation similar to (15), but in the context of inorganic thin films, are discussed in greater detail by Goldstein in Chapter 3 of this book.

Standards

To minimize corrections, the most suitable standards for the analysis of sections of soft tissue are standards of similar thickness and composition. In the case of epoxy-embedded tissue sections, there are good ways to prepare epoxy standard-sections "doped" with known concentrations of many cations and anions (SPURR, 1974; SPURR, 1975; ROOMANS and VAN GAAL, 1977). In the case of cryosections of quench-frozen tissue, the standards are generally cryosections of quench-frozen aqueous solutions containing known concentrations of salts and some 10 - 20% of organic macromolecular material. Typical macromolecular additives are albumin (DÖRGE et al., 1975), gelatin (ROOMANS and SEVÉUS, 1977),

high-molecular-weight Dextran (GUPTA et al., 1977) and polyvinylpyrroli-
done ("PVP") (SAUBERMANN and ECHLIN, 1975). ROOMANS and SEVÉUS (1977)
have compared several of these materials as to their suitability for
quantitative standards.

With cryosections, the aqueous standards are best used in the peripheral form, i.e. tissue and standard are in contact and are quench-
frozen and sectioned together. The standard then remains hydrated
during the analysis of frozen-hydrated tissue sections and will dehy-
drate with the tissue when the section is to be analyzed frozen-dried.
For physiological studies, the peripheral standard is actually the extra-
cellular medium within which the functioning and composition of the
tissue are to be determined. It is then essential to establish that the
macromolecular component does not interfere with physiological performance. Albumin, Dextran and PVP have all been used in peripheral standards, but tolerance for them varies with the type of tissue (ECHLIN,
SKAER, GARDINER, FRANKS and ASQUITH, 1977).

Correct preparation of aqueous solutions for peripheral standards
involves a minor point which generally escapes attention. When the
method of measurement is to provide results in terms of mM of element
per unit volume (see above), the standard solution should be prepared
with measured amounts of element per unit volume of solution; when the
method will measure mM of element per kg wet weight (see above), the
solution should be prepared with measured amounts of element per unit
total mass of solution. (Either procedure is all right for measurements
of mM of element per kg dry weight since the essential datum in that
case is mM of element per kg of non-aqueous material.)

Another minor difficulty stems from the change of volume on
freezing, which has a confusing effect on the measurement of amount of
element per unit volume. This effect will not be considered here.

5.3 EFFECTS OF CONTAMINATION WITHIN THE MICROSCOPE COLUMN

Contamination deposited on the section within the microscope may
include elements of analytical interest or may simply add to local specimen mass. The extent of the problem depends on the nature of the microscope vacuum and also on the thickness of the analyzed sections.

Contamination seems to be insignificant in microscopes operated at
ultra-high vacuum (e.g. Philips 400, or field-emission STEM's).

For microscopes run at ordinary vacuum, in the authors' experience
sulphur and chlorine may occasionally deposit onto specimens mounted on
a cold-stage, but this contamination can be reduced by regular out-
gassing of the foreline trap in the roughing-pump line, and the effect
can then be prevented by use of a protective cold-cap fitted closely
above the specimen.

"Mass" contamination -- carbon or organic debris -- affects the
methods of quantitation which rely on the measurement of local mass,
specifically the continuum method. With an ordinary vacuum system, the

rate of contamination is usually not high enough to add significantly to the mass of a 1-μm section during analysis, but the effect may be large for an ultrathin specimen exposed to a finely focussed beam. Deposition rates are notoriously unreproducible and the dependence of rate on beam current is not monotonic, but under poor conditions local mass in ultra-thin specimens may be increased on the order of 50% during a 100-sec ana-lytical run. Hence the application of the continuum method to the quan-titative analysis of ultrathin specimens requires either ultra-high vacuum or a very effective anti-contamination shield.

5.4 EFFECTS OF BEAM DAMAGE

Beam-induced loss of "mass" (material of low atomic number) does not affect the analytical scheme outlined in Section 5.2 but it does pose a serious problem in the use of the continuum method. The beam breaks chemical bonds, leading to the escape of volatile fragments and volatile reaction products. In uncooled organic films, typically 20 - 30% of the mass is lost under a beam loading of the order of 10^{-10} Coulomb μm^{-2} and the specimen then becomes fairly stable against further loss (STENN and BAHR, 1970). A dose of 10^{-10} C μm^{-2} is almost invariab-ly exceeded many-fold during microanalysis.

One approach to the problem of mass loss is to assume that specimen and standards are affected to the same degree. I_W and I_{WO} would then be changed by the same factor and the basic Eqn. 5.7 would remain valid. In practice this assumption may be all right most of the time, but it is risky and must be false at times, in view of the fact that the end-point for mass loss actually varies from about 10% to about 90% in a range of organic films (BAHR, JOHNSON and ZEITLER, 1965).

A more promising approach to the control of mass loss is the use of low temperature. It is difficult to determine accurately the relation-ship between temperature and beam-induced loss of mass, probably because of the problem of establishing the temperature of the specimen itself, but it is known that stage-cooling by liquid nitrogen substantially re-duces and/or slows the loss (HALL and GUPTA, 1974; SHUMAN et al., 1976), and loss is slight or nil near the temperature of liquid helium (RAMAMURTI, 1977; RAMAMURTI, CREWE and ISAACSON, 1975; DUBOCHET, 1975).

In the authors' laboratory, where 1-μm frozen-hydrated or frozen-dried sections are analyzed on a stage held at -170°C, mass loss during analysis occurs inconsistently. We attribute the inconsistency to vari-ability in the temperature of the specimen, due to uneven contact be-tween the sections and the conducting support film. However, one can tell after an analytical run whether substantial loss has occurred. Mass loss makes the analyzed region appear relatively "bleached" against a darker surround in the scanning transmission image. Losses of only a few per cent produce a visible bleaching. Another sign of loss is a fall in the continuum count rate during the analysis. On the basis of these criteria, we judge that analyses can be done with only slight mass loss, albeit inconsistently, under our conditions of operation (50-kV beam, beam currents near 2 nA, analyzed areas 0.1 - 1 μm^2, real time of run 100 - 200 sec, sections frozen-hydrated or frozen-dried and 1-μm

thick, mounted on thin nylon film coated with 15 nm of evaporated aluminum).

There is also a danger that the beam may remove certain elements,
generally by volatilization. The literature is too fragmentary and dispersed to be reviewed here, but losses of Na, S, Cl and K have been observed. The process of elemental loss is usually slow enough to be
followed as a progressive drop in characteristic-line intensity during
analysis. Since mass loss and elemental loss are both manifestations of
the mobility of small volatile molecules, conditions which permit loss
of mass are likely to lead to elemental losses as well.

5.5 SPECIMEN PREPARATION

In the preparation of tissue sections for electron-microscopical
morphology, liquid media are commonly used for fixation, dehydration,
staining, precipitation of soluble components of interest and/or embedding. In quantitative analytical microscopy, such media are generally avoided because of the risk of altering the distributions of the elements of interest, and cryopreparative techniques are preferable.

The cryopreparation of tissue sections begins with the quench-
freezing of a small piece of tissue (for a summary of modes of quench-
freezing, see HALL 1979b). The pathway then follows the alternative
routes of freeze-substitution or cryosectioning.

Freeze-substitution entails replacement of the ice by a cold solvent, replacement of the solvent by embedding medium, and finally sectioning. A great deal of experimentation has still to be done to establish just how successful this technique can be in preserving elemental
distributions. The effect of infiltrating with the liquid embedding
medium is particularly suspect. But several workers have developed and
used freeze-substitution for analytical microscopy (LÄUCHLI, SPURR and
WITTKOPP, 1970; FORREST and MARSHALL, 1976; VAN ZYL, FORREST, HOCKING
and PALLAGHY, 1976; HARVEY, HALL and FLOWERS, 1976; BUROVINA, GRIBAKIN,
PETROSYAN, PIVAROVA, POGORELOV and POLYANOVSKY, 1978). LECHENE and
WARNER (1977) and CHANDLER and BATTERSBY (1979) have assessed the potential of freeze-substitution less optimistically.

The cryosectioning route entails sectioning of the frozen block,
and then freeze-drying or else transfer of the hydrated section onto a
cold-stage in the analytical microscope for analysis in the hydrated
state. The combination of quench-freezing and cryosectioning is in principle the simplest and "cleanest" preparative procedure for analytical
microscopy but again a great deal remains to be learned about the requisite conditions, especially with regard to temperatures. We do not know
how much diffusion may occur in a block of frozen tissue at the usual
sectioning temperatures (which range from -30°C in some laboratories to
-110°C in others), and displacements also occur above -70°C in some
freeze-drying routines (BARNARD and SEVÉUS, 1978).

Another area of relative ignorance is the subject of ice-crystal
formation. Quench-freezing should be as rapid as possible to minimize

ice-crystal formation, which may shift diffusible elements and limit the analytical spatial resolution, but the poor thermal conductivity of tissues restricts rapid freezing to the neighborhood of the surface when a piece of tissue is quench-frozen (SJÖSTRÖM, 1975). Cryoprotectants like glycerol can inhibit crystal formation but are not used as they penetrate cells and lead to elemental shifts. However the addition of certain non-penetrating macromolecules (e.g. Dextran, PVP) to the extra-cellular medium can inhibit ice-crystal formation even within cells, to a distance of 50 - 100 μm from the surface (FRANKS, ASQUITH, HAMMOND, SKAER and ECHLIN, 1977). The explanation of this inhibition is not known.

There is an enormous literature on the tricky subject of cryoprep-aration. Detailed descriptions of the techniques used in some labora-tories are given by DÖRGE et al. (1978), A. V. SOMLYO et al. (1977) and GUPTA et al. (1977), and cryopreparation was the main subject of a re-cent conference (ECHLIN, 1979).

5.6 SPECIMENS OTHER THAN SECTIONS MOUNTED ON THIN FILM

Quantitative x-ray microanalysis is applied to many types of bio-logical specimen other than tissue sections mounted on thin films. Such analyses are mostly outside the scope of "analytical microscopy," but the interested reader can refer to the following papers:

Tissue sections on bulk supports. The method of COLBY (1968) has been adapted for biological use by WARNER and COLEMAN (1974, 1975).

Fluids obtained by micropuncture: INGRAM and HOGBEN (1967), LECHENE (1974), MOREL (1975), ROINEL (1976), QUINTON (1978), GARDLAND, BROWN and HENDERSON (1978). (The paper by Quinton describes the energy-dispersive analysis of droplets dried onto bulk supports, by means of diffracting spectrometers.)

Tissue homogenates: MORGAN et al. (1975) describe the analysis of sec-tions of tissue homogenates using a cobalt internal standard. A known concentration of cobalt is added to the homogenate, and the other elements in the sections are then assayed by means of the ratio method with cobalt as the reference element.

LIST OF SYMBOLS USED IN THIS ARTICLE

A_i Atomic weight of element i.

C_a Mass fraction of element a (units can be either mM of element per kg of specimen or g of element per g of specimen, to suit conven-ience).

C'_a Amount of element a per unit volume of specimen.

C''_a Amount of element a per litre of tissue water.

f_a Mass fraction of element $\underline{a}$, units g of element per g of specimen.

F_d Local dry-weight fraction (non-aqueous mass/total mass in the hydrated tissue).

F_a Local aqueous fraction ($F_a = 1 - F_d$).

G The weighted average of Z^2/A, $\sum_i f_i Z_i^2/A_i$, for a specimen or for any multi-element component such as water or organic matrix.

I_a Spectral peak integral for characteristic radiation of element $\underline{a}$ (except with subscript $\underline{w}$, which refers to continuum radiation).

I'_a Peak integral after correction for self-absorption in the specimen (see Eqn. 5.15).

I_{ah} Value obtained for I_a in a hydrated section.

I_{ad} Value obtained for I_a in the section after dehydration.

I_w Quantum count generated within the specimen itself, taken within some fixed quantum-energy band of the x-ray continuum, excluding any contribution from characteristic radiations.

I_{wh} Value obtained for I_w in a hydrated section.

I_{wd} Value obtained for I_w in the section after dehydration.

I_{wt} Total continuum count before correction for extraneous sources.

k A constant.

M Local total specimen mass per unit area (area normal to the electron beam). This quantity is the same as the quantity $\underline{pt}$ defined by Goldstein in Chapter 3 of this book.

M_a Local mass of element $\underline{a}$ per unit area.

M_a Local mass of water per unit area in a hydrated section (used only in Appendix I).

M_d Local mass per unit area in the section after dehydration.

P_i x-ray absorption coefficient for (absorbing) element $\underline{i}$, $cm^2\ g^{-1}$. This quantity is the same as the quantity $(\mu/p)_{\underline{i}}$ defined by Goldstein in Chapter 3 of this book.

P Effective X-ray absorption coefficient (Eqn. 5.16).

S Shrinkage factor (Eqn. 5.10).

Y $\underline{G}$(dried tissue)/$\underline{G}$(water).

Z_i Atomic number of element i.

α Takeoff angle.

SUBSCRIPTS

b Bulk material in the neighborhood of the specimen.

e Monitor element in the bulk material.

f Specimen supporting film.

o Thin standard (general).

p Thin standard, peripheral.

r Thin standard used for elemental ratios.

REFERENCES

Bahr, G. F., Johnson, F. B., and Zeitler, E., 1965, Lab. Invest., 14, 1115.

Barnard, T., and Sevéus, L., 1978, J. Microscopy, 112, 281.

Burovina, I. V., Gribakin, F. G., Petrosyan, A. M., Pivovarova, N. B., Pogorelov, A. G., and Polyanovsky, A. D., 1978, J. comp. Physiol., 127, 245.

Chandler, J. A., 1977, X-Ray Microanalysis in the Electron Microscope (Amsterdam: Elsevier/North-Holland Biomedical Press).

Chandler, J.A., and Battersby, S., 1979, Microbeam Analysis in Biology, edited by C. Lechene and R. Warner (New York: Academic Press), in press.

Cliff, G., and Lorimer, G. W., 1972, Proc. 5th European Congress on Electron Microscopy, no editor cited (Bristol: The Institute of Physics), 140.

Colby, J. W., 1968, Advances in X-Ray Analysis, edited by J. B. Newkirk, G. R. Mallett and H. G. Pfeiffer, Vol. 11 (New York: Plenum Press), 287.

Dörge, A., Rick, R., Gehring, K., Mason, J., and Thurau, K., 1975, J. de Microscopie et Biologie Cell., 22, 205.

Dörge, A., Rick, R., Gehring, K., and Thurau, K., 1978, Pflügers Arch., 373, 85.

Dubochet, J., 1975, J. Ultrastruc. Res., 52, 276.

Echlin, P. (editor), 1979, Low Temperature Biological Microscopy and
    Analysis (Oxford: Blackwell). (Articles from J. Microscopy, 111,
    Part 1 and 112, Part 1.)

Echlin, P., and Galle, P. (editors), 1975, Biological Microanalysis
    (J. de Microscopie et Biologie Cell., 22, no. 2-3) (Paris: Société
    Francaise de Microscopie Électronique).

Echlin, P., and Kaufmann, R. (editors), 1978, Microprobe Analysis in
    Biology and Medicine (Microscopica Acta, Suppl. 2) (Stuttgart:
    Hirzel Verlag).

Echlin, P., Skaer, H. leB., Gardiner, B. O. C., Franks, F., and Asquith,
    M. H., 1977, J. Microscopy, 110, 239.

Erasmus, D. A. (editor), 1978, Electron Probe Microanalysis in Biology
    (London: Chapman and Hall).

Forrest, Q. G., and Marshall, A. T., 1976, Proc. 6th European Congress
    on Electron Microscopy, edited by Y. Ben-Shaul, Vol. II (Jerusalem:
    Tal Publishing Co.), 218.

Franks, F., Asquith, M. H., Hammond, C. C., Skaer, H. leB., and Echlin,
    P., 1977, J. Microscopy, 110, 223.

Garland, H. O., Brown, J. A., and Henderson, I. W., 1978, Electron
    Probe Microanalysis in Biology, edited by D. A. Erasmus (London:
    Chapman and Hall), 212.

Gupta, B. L., 1979, Microbeam Analysis in Biology, edited by C. Lechene
    and R. Warner (New York: Academic Press), in press.

Gupta, B. L., and Hall, T. A., 1979, Fedn. Proc. Fedn. Am. Socs. exp.
    Biol., 38, 144.

Gupta, B. L., Hall, T. A., and Moreton, R. B., 1977, Transport of Ions
    and Water in Animals, edited by B. L. Gupta, R. B. Moreton, J. L.
    Oschman and B. J. Wall (London: Academic Press), p. 83; 1979,
    X-Ray Optics and Microanalysis - Eighth International Conference,
    edited by D. Beaman, R. Ogilvie and D. Wittry (Princeton: Science
    Press), in press.

Gupta, B. L., Berridge, M. J., Hall, T. A., and Moreton, R. B., 1978a,
    J. exp. Biol., 72, 261.

Gupta, B. L., Hall, T. A., and Naftalin, R. J., 1978b, Nature (London),
    272, 70.

Hall, T. A., 1971, Physical Techniques in Biological Research, edited by
    G. Oster, 2nd edition, Vol. IA (New York: Academic Press), p. 157;
    1979a, Microbeam Analysis in Biology, edited by C. Lechene and R.
    Warner (New York: Academic Press), in press; 1979b, J. Microscopy,
    in press.

Hall, T. A., Anderson, H. C., and Appleton, T. C., 1973, J. Microscopy, 99, 177.

Hall, T. A., Echlin, P., and Kaufmann, R. (editors), 1974, Microprobe Analysis as Applied to Cells and Tissues (London: Academic Press).

Harvey, D. M. R., Hall, J. L., and Flowers, T. J., 1976, J. Microscopy, 107, 189.

Ingram, M. J., and Hogben, C. A. M., 1967, Analytical Biochemistry, 18, 54.

Kramers, H. A., 1923, Phil. Mag., 46, 836.

Läuchli, A., Spurr, A. R., and Wittkopp, R., 1970, Planta (Berlin), 95, 341.

Lechene, C., 1974, Microprobe Analysis as Applied to Cells and Tissues, edited by T. Hall, P. Echlin and R. Kaufmann (London: Academic Press), 351.

Lechene, C.P., and Warner, R.R., 1977, Annual Review of Biophysics and Bioengineering, 6, 57.

Morel, F., 1975, J. de Microscopie et Biologie Cell., 22, 479.

Morgan, A. J., Davies, T. W., and Erasmus, D. A., 1975, J. Microscopy, 104, 271.

Pallaghy, C. K., 1973, Austral. J. biol. Sci., 26, 1015.

Quinton, P. M., 1978, Micron, 9, 57.

Ramamurti, K., 1977, Proc. 35th Annual Meeting of the Electron Microscopy Society of America, edited by G. W. Bailey (Baton Rouge: Claitors), 560.

Ramamurti, K., Crewe, A. V., and Isaacson, M. S., 1975, Ultramicroscopy, 1, 156.

Rick, R., Dörge, A., von Arnim, E., and Thurau, K., 1978, J. Membrane Biol., 39, 313.

Roinel, N., 1977, Proc. 35th Annual Meeting of the Electron Microscopy Society of America, edited by G. W. Bailey (Baton Rouge: Claitors), 362.

Roomans, G. M., 1978, Interaction of Cation and Anion Transport in Yeast (doctoral thesis), (Nijmegen, The Netherlands: Stichting Studentenpers).

Roomans, G. M., and Sevéus, L., 1976, J. Cell Sci., 21, 119; 1977, J. submicr. Cytol., 9, 31.

Roomans, G. M., and Van Gaal, H. L. M., 1977, J. Microscopy, 109, 235.

Rowse, J. B., Jepson, W. B., Bailey, A. T., Climpson, N. A., and Soper, P. M., 1974, J. Physics E: Sci. Instri., 7, 512.

Russ, J. C., 1974, Microprobe Analysis as Applied to Cells and Tissues, edited by T. Hall, P. Echlin, and R. Kaufmann (London: Academic Press), 269.

Saubermann, A. J., and Echlin, P., 1975, J. Microscopy, 105, 155.

Saubermann, A. J., Riley, W. D., and Beeuwkes, R., 1977, J. Microscopy, 111, 39.

Shuman, H., Somlyo, A. V., and Somlyo, A. P., 1976, Ultramicroscopy, 1, 317; 1977, Scanning Electron Microscopy), p. 663.

Sjöström, M., 1975, J. Microscopy, 105, 67.

Somlyo, A. P., Shuman, H., and Somlyo, A. V., 1979, Frontiers in Biological Energetics: Electrons to Tissues, edited by T. S. Leigh, P. L. Dutton and A. Scarpa (New York: Academic Press), in press..

Somlyo, A. V., Shuman, H., and Somlyo, A. P., 1977, J. Cell Biol., 74, 828.

Spurr, A. R., 1974, Microprobe Analysis as Applied to Cells and Tissues, edited by T. Hall, P. Echlin and R. Kaufmann (London: Academic Press), p. 213; 1975, J. de Microscopie et Biologie Cell., 22, 287.

Stenn, K., and Bahr, G. F., 1970, J. Ultrastruc. Res., 31, 526.

Van Zyl, J., Forrest, Q. G., Hocking, C., and Pallaghy, C. K., 1976, Micron, 7, 213.

Warner, R. R., and Coleman, J. R., 1974, Microprobe Analysis as Applied to Cells and Tissues, edited by T. Hall, P. Echlin and R. Kaufmann (London: Academic Press), p. 249; 1975, Micron, 6, 79.

MAIN LITERATURE REFERENCES

Quantitative x-ray microanalytical microscopy

   Shuman et al. (1976). Continuum method.
   Shuman, A. V. Somlyo and A. P. Somlyo (1977). Continuum method.
   Hall (1971). General (pre-dates the energy-dispersive era).

Books on x-ray analytical microscopy in general

   Chandler (1977).
   Erasmus (1978).

Proceedings of conferences on microprobe analysis in biology

Hall, Echlin and Kaufmann (1974).
Echlin and Galle (1975).
Echlin and Kaufmann (1978).

APPENDIX I
Derivation of Equation 5.12 for Dry-Weight Determination

Notation: M_a = mass of water per unit area before dehydration

M_d = organic mass per unit area after dehydration

The organic mass per unit area before dehydration must be $(S M_d)$, and the dry-weight fraction F_d must be given by

$$F_d = \frac{S M_d}{S M_d + M_a} = \frac{S}{S+(M_a/M)} \tag{5.18}$$

(The contribution of heavy elements to the continuum is neglected in this formulation.)

From Eqn. 5.6 the continuum signal before dehydration is

$$I_{wh} = k \left[G(water) M_a + G(organic) S M_d \right] \tag{5.19}$$

and the continuum signal after dehydration is

$$I_{wd} = k \, G(organic) \, M_d \tag{5.20}$$

Division of Eqn. 5.19 by Eqn. 5.20 and re-arrangement gives

$$\frac{M_a}{M_d} = Y \left(\frac{I_{wh}}{I_{wd}} - S \right) \tag{5.21}$$

where $Y = G(organic)/G(water)$. Substitution of Eqn. 5.21 into Eqn. 5.18 gives

$$F_d = \frac{S}{S + Y\left(\dfrac{I_{wh}}{I_{wd}} - S\right)}$$

APPENDIX II
Sample Calculations

What follows is an example of one set of microprobe data worked out in different ways. In order to bring out the differences clearly, the data are hypothetical and idealized; the numbers are related as one would expect in the absence of statistical fluctuations. For a detailed examination of actual experimental data from the authors' laboratory, the reader may refer to GUPTA and HALL (1979).

Table 5.1 presents the hypothetical experimental data, plus a final column giving the continuum counts corrected for extraneous background be means of Eqn. 5.8. For calculations of mM of element per litre it is assumed that there was a peripheral standard ("periph" in the Table), made by sectioning a frozen solution containing 150 g of Dextran, 140 mM of Na, 150 mM of Cl and 10 mM of K per litre of solution; for calculations of mM per kg, it is assumed that the standard solution was made up to contain these amounts in an aliquot with a total mass of 1 kg. The hypothetical ratio standard is a frozen section of saline solution containing 15% Dextran and 100 mM KCl. The mass standard is a polycarbonate film, $C_{16}H_{14}O_3$, with mass per unit area $2.4 \cdot 10^{-4}$ g cm^{-2}. Live time per run and probe current are assumed constant (except for reduced current to avoid overload when the beam is directly on the holder, "bulk" in the Table). The holders are Dural, containing Al plus a few per cent Cu. The tissue section with its peripheral standard film is nylon, cast in the laboratory and coated with evaporated aluminum. It is assumed, in the first calculations below, that the runs are done first with the specimen frozen-hydrated, and that the section is then dehydrated in the microscope column preceding the final set of runs.

TABLE 1

Initial Data, Plus Corrected Continuum Count

Field	Peak Integrals				I_{wt}	I_w
	Na	Cl	K	Cu		
ratio standard		2000	2100			
mass-standard				300	6500	6000
mass-bulk				6000	10000	
Hydrated section						
periph	560	5400	378	75	3650	3000
film				50	600	500
cytoplasm	51	1800	2600	60	3320	2700
bulk				5000	10000	
Dehydrated section						
periph	724	6000	420	30	1022	462
cytoplasm	62	1800	2600	40	1327	747

Calculations

1. <u>Dry-weight fraction of the standard</u>.

Equation 5.12 is used. The shrinkage factor $\underline{S}$ for the peripheral standard is best calculated from the chlorine data, since there may be a substantial absorption effect for sodium and the potassium peak-to-background ratio in a 10-mM hydrated standard is too low to permit good statistical accuracy. The chlorine data and Eqn. 5.10 give $\underline{S = 5400/6000 = 0.9}$. $\underline{G}$ for Dextran is about 3.34 (taking a composition $C_6H_{12}O_5$), so $\underline{Y}$ for the standard, in Eqn. 5.12, is 3.34/3.67. These values inserted into Eqn. 5.12 give $F_d = .150$. This equals the nominal 15% dry-weight, showing that the material had not lost water during handling prior to the deliberate dehydration.

2. <u>Local dry-weight fraction in the cytoplasm</u>.

(a). The chlorine counts in the cytoplasm are the same before and after dehydration, indicating no local shrinkage, so $\underline{S = 1}$. The tissue is regarded as a composite of water and proteinaceous matter, and the composition for the latter is taken to be (by weight) .07 H, .50 C, .16 N, $\underline{.25\ O}$ and $\underline{.02\ S}$. The $\underline{G}$-value for this composition is $\underline{3.28}$, so $\underline{Y = 3.28/3.67}$. The data table gives $I_{wh}/I_{wd} = 2700/747$. Insertion of these values into Eqn. 5.12 gives $\underline{F_d = 0.30}$.

(b). If one uses the simpler and less accurate Eqn. 5.11 , the result for local F_d is $\underline{1 \times (747/2700) = 0.28}$, which is not much different.

3. <u>mM of element per kg wet weight</u>.

The calculation is based generally on Eqn. 5.7.

(a). Approximation assuming $\underline{G(specimen) = G(standard)}$, and neglecting absorption:

$$C_a \cong (I_a/I_w)\ C_{ao}\ (I_w/I_a)_o \tag{5.22}$$

$$C_{Na} = (51/2700)\ 140\ (3000/560) = 14.2\ \text{mM per kg.}$$

$$C_{Cl} = (1800/2700)\ 150\ (3000/5400) = 55.6\ \text{mM per kg.}$$

$$C_K = (2600/2700)\ 10\ (3000/378) = 76.4\ \text{mM per kg.}$$

(b). Calculation of $\underline{C_K}$ by ratio method: The potassium count of 378 from the 10 mM of K in the hydrated standard would be subject in practice to a large probable error. While the numbers in the present example have been contrived as if statistical fluctuations did not exist, in practice it would be better to calculate $\underline{C_K}$ through a use of the ratio method, with Cl as the reference element. Instead of Eqn. 5.4, which was derived under the assumption of uniform section thickness, we may use the readily-derived formula

$$c_K = D_{Cl} \, (I_K/I_{Cl}) \, [(I_{Cl}/I_K) \, (c_K/c_{Cl})]_r \qquad (5.23)$$

Substitution of the data and the calculated value of C_{Cl} gives

$$c_K = 55.6 \, (2600/1800) \, (2000/2100) \, (100/100) = 76.4 \text{ mM per kg.}$$

(c). More accurate calculation using G-values: If we regard the cyto-plasm as consisting of water plus proteinaceous matter, with dry-weight fraction F_d, we may write

$$G(\text{cytoplasm}) = (1-F_d) \, G(\text{water}) + F_d \, G(\text{proteinaceous})$$

$$= (1-F_d) \, 3.67 + F_d \, 3.28$$

$$= 3.67 - .39 \, F_d$$

while $G(\text{peripheral}) = (.85) \, 3.67 + (.15) \, 3.34 = 3.62$

If we use the measured value of .30 for F_d, then in Eqn. 5.7

$$\frac{G(\text{cytoplasm})}{G(\text{periph.})} = \frac{3.67 - .30 \, (.39)}{3.62} = \frac{3.55}{3.62} = .98$$

and all of the concentrations above should be corrected by this factor, so the preferred values become

$$C_{Na} = .98 \, (14.2) \; = \; 13.9 \text{ mM per kg,}$$

$$C_{Cl} = .98 \, (55.6) = 54.5 \text{ mM per kg}$$

and $C_K = .98 \, (76.4) = 74.9 \text{ mM per kg.}$

(If measurements were not made on the dehydrated specimen, there would be no measured value of local F_d available, but one could still make a correction by guessing a value. Values of F_d less than 0.3 imply an even closer match between G(cytoplasm) and G(periph.), so the uncertain-ty connected with the G-value correction is slight.)

A complete formalism should also include the effect of the heavier elements on the G values. This correction is not difficult to formu-late, but is left out here for the sake of simplicity.

(d). Absorption correction for sodium: The absorption correction is calculated from Eqn. 5.15. We have to work out the G-values, the total mass per unit area M and the effective absorption coefficients P for the analyzed field and the peripheral standard.

As worked out above, in this case G(cytoplasm) = 3.55 and G(periph.) = 3.62. For the mass standard, the composition $C_{16}H_{14}O_3$ gives a G-value 3.08. Then Eqn. 5.17 gives

$$M(\text{cytoplasm}) = 2.4 \; 10^{-4} \, (2700/6000)(3.08/3.55) = .937 \; 10^{-4} \text{ g cm}^{-2}$$

and $M(\text{periph.}) = 2.4 \; 10^{-4} \, (3000/6000)(3.08/3.62) = 1.02 \; 10^{-4} \text{ g cm}^{-2}.$

The absorption coefficients are worked out from Eqn. 5.16, using published coefficients for the constituent elements. Table 5.2 gives the effective coefficients $\underline{P}$ for water, dehydrated tissue and Dextran.

TABLE 5.2

Absorption coefficients of water, tissue and Dextran for sodium radiation

(see Eqn. 5.16).

Absorbing element	P_i	Water f_i	Water f_iP_i	Dry tissue f_i	Dry tissue f_iP_i	Dextran f_i	Dextran f_iP_i
H		.11	-	.07	-	.07	-
C	1534	0	0	.50	767	.44	675
N	2450	0	0	.16	392	0	0
O	4109	.89	3652	.25	1027	.49	2013
S	2103	0	0	.02	42	0	0
Total			3652		2228		2688 $cm^2\ g^{-1}$.

The effective cross-sections for the tissue and for the peripheral standard are calculated from these component coefficients, using $F_d = .3$ for the tissue and $F_d = .15$ for the peripheral standard.

$$P(\text{cytoplasm}) = .3(2228) + .7(3652) = 3225,$$
$$P(\text{periph.}) = .15(2688) + .85(3652) = 3507.$$

The example is worked out here for a takeoff angle of 35°. Table 5.3 summarizes the other values inserted into Eqn. 5.15 and gives the resulting correction factors. The Table also gives the values worked out in the same way for the dehydrated section.

TABLE 5.3

Values of the main quantities in the absorption correction for sodium.

Field	$10^{-4}\ M$	P	Correction Factor I_a/I_a'
hydrated tissue	.937	3225	.77
hydrated periph.	1.02	3507	.75
dehydrated tissue	.280	2228	.94
dehydrated periph.	.153	2688	.97

The application of the absorption corrections to the sodium counts from the hydrated tissue and standard changes the calculated sodium concentration from 13.9 to 13.9 (0.75/0.77)=13.5 mM per kg.

With respect to this correction, the following points should be noted:

With a suitable standard, it is seen that the correction almost cancels out. If the local value of $\underline{F_d}$ were not measured, it would be adequate for most purposes to estimate the correction using a guessed value for $\underline{F_d}$.

In the hydrated section, tissue and standard both consist predominantly of water so similar absorption can be expected. Tissue and standard are not so similar after dehydration, as is reflected in Table 5.3, but of course there is much less absorption after dehydration.

The calculations presented here are for section thicknesses of approximately 1 μm. It is clear that for ultrathin sections (100-nm thick), the absorption correction is usually negligible.

Again, the heavier elements have been left out of the example for the sake of simplicity, but they can be included readily in an iterative routine.

<u>mM of element per litre of water:</u>

If it is assumed that the elements Na, Cl and K are entirely in the aqueous phase in the analyzed region of tissue, then the values of mM per kg wet weight may be converted to mM per litre of water by dividing by the local aqueous fraction $\underline{F_a = 1 - F_d}$.

In the present example this gives

C''_{Na} 13.5/0.7 = 19.3 mM per litre of water,

C''_{Cl} 54.5/0.7 = 77.9 mM per litre of water and

C''_{K} 74.9/0.7 = 107. mM per litre of water.

<u>mM per kg dry weight:</u>

Only a few laboratories are analyzing frozen sections in the hydrated state. More often the sections are dehydrated by freeze-drying before they are put into the analytical microscope. Measurement by the continuum method then gives concentrations in terms of mM per kg of dry weight. As an example we take Table 5.1 again, simply omitting the runs on the section prior to dehydration. The data for the dehydrated section can be processed directly by Eqn. 5.7. We now have a dehydrated peripheral standard containing 140 mM of Na, 150 mM of Cl and 10 mM of K per 150 g of Dextran; the concentrations work out to be Na 881, Cl 944 and K 62.9 mM per total kg. <u>G(cytoplasm)</u> is now taken to be 3.28 and <u>G(periph)</u> to be 3.34. In order to show the computation more neatly, Eqn. 5.7 is written in the form

$$C_a = C_{ao} \ (I_a/I_w) \ (I_{wo}/I_{ao}) \ (G/G_o).$$

Substitution of the values above and of the observed counts gives

$$C_{Na} = 881 \ (62/747) \ (462/724) \ (3.28/3.34) = 45.8 \ \text{mM per kg dry},$$

$$C_{Cl} = 944 \ (1800/747) \ (462/6000) \ (3.28/3.34) = 172 \ \text{mM per kg dry, and}$$

$$C_{K} = 62.9 \ (2600/747) \ (462/420) \ (3.28/3.34) = 236 \ \text{mM per kg dry}.$$

With respect to this calculation a few fine points should be noted:

(a). Again, one may correct for sodium absorption by the method given in Section 5.3d above. The calculated correction factors are included in Table 5.3.

(b). We have already noted that in measurements by the continuum method, there is an uncertainty associated with the limited knowledge of the local value of $\underline{G}$ in the specimen. The quantity $\underline{G/G_0}$ is quite near unity in frozen-hydrated material where water is the main constituent of both specimen and standard; the incomplete knowledge of the composition of the organic matrix leads to a greater uncertainty of the value of $\underline{G/G_0}$ in frozen-dried material.

(c). Although one cannot know the $\underline{G}$-value for the organic matrix precisely, one $\underline{can}$ correct for the effect on $\underline{G}$ of the measured heavier elements, and this should be done at least for dehydrated specimens. For this correction, the concentrations are expressed as mass fractions $\underline{f}$ and the values of $G = \Sigma(f_i Z_i^2/A_i$ are re-calculated. In the present case the mass fractions are as shown in Table 5.4.

TABLE 5.4

Mass-fractions f in the dehydrated material

$$f_a \ (g/g) = 10^{-6} \ A \ C_a \ (mM/kg).$$

	"Periph."		Cytoplasm	
	C_a	f_a	C_a	f_a
Na	881	.0203	45.8	.0011
Cl	944	.0335	172	.0061
K	62.9	.0024	236	.0092
Dextran		.9438		
"protein"				.9836

The corrected G-values are then:

G(periph.)   =   .9438(3.34)+.0203(121/23)+.0335(289/35.5)+.0024(361/39)
=3.55, and G(cyto.) = .9836(3.28)+.0011(121/23)+.0061(289/35.5)+.0092
(361/39) = 3.37.

   The corrected ratio G/G_0 is thus $3.37/3.55 = .949$ in place of $3.28/3.34 = .982$ in Eqn. 5.7. The resulting corrected concentrations are C_{Na} = 44.3, C_{Cl} = 166, and C_K = 228 mM per kg dry weight.

mM per unit volume of specimen.

   Equation 5.3 can be used to measure mM of element per unit volume in both frozen-hydrated and dehydrated sections.

   When the equation is applied to the data for the frozen-hydrated section in Table 5.1, the results are

$$C'_{Na} = 51 \ (140/560) = 12.7 \text{ mM per litre,}$$

$$C'_{Cl} = 1800 \ (150/5400) = 50.0 \text{ mM per litre   and}$$

$$C'_K = 2600 \ (10/378) = 68.8 \text{ mM per litre.}$$

   When the equation is applied to the data obtained from the same section after dehydration, the results are

$$C'_{Na} = 62 \ (140/724) = 12.0 \text{ mM per litre,}$$

$$C'_{Cl} = 1800 \ (150/6000) = 45.0 \text{ mM per litre, and}$$

$$C'_K = 2600 \ (10/420) = 61.9 \text{ mM per litre.}$$

   These two sets of results should be compared with the results from the continuum method applied to the same hydrated specimen (Section 5.3c): C_{Na} = 13.9 mM per kg wet; C_{Cl} = 54.5 mM per kg wet; C_K = 74.9 mM per kg wet.

   The numerical differences have nothing to do with the difference in units (mM per litre vs. mM per kg). The difference in results for the hydrated section arises from an assumption which was built into the hypothetical Table 5.1, namely that the section thickness is not quite uniform. This is manifest in the value I_W = 2700 compared to I_{WO} = 3000. If such a non-uniformity exists, Eqn. 5.7 is still valid but an error arises from the use of Eqn. 5.3.

   A further source of error has been introduced into the hypothetical data for the dehydrated sections: for K and Cl, the tissue counts in the dehydrated state are the same as in the hydrated state, but dehydration has led to changes in these counts in the peripheral standard, indicating non-uniform shrinkage during dehydration (the sodium count changes in the tissue as well, but only because of reduced absorption). If non-uniform shrinkage occurs on dehydration, an error results when Eqn. 5.3 is applied to the analysis of frozen-dried sections.

(The actual occurrence of non-uniform section thickness or shrinkage must depend on the techniques of sectioning, mounting and dehydration, and on the nature of the tissue.)

Summary.

The calculations above show how various results may be obtained from different formulations, all applied to one body of data. (Here we neglect the possible difference in the standard prepared volumetrically or gravimetrically.). The results are summarized in Table 5.5.

TABLE 5.5

Summary of the Results of the Calculations

Specimen	Equation numbers	Concentrations			F_d	Units
		Na	Cl	K		
hydrated	5.3	12.7	50.0	68.8		mM per litre of tissue
dehydrated	5.3	12.0	45.0	61.9		mM per litre of tissue
combined data	5.12				.3	
hydrated	5.7	13.5	54.5	74.9		mM per kg wet tissue
dehydrated	5.7	44.3	166.	228.		mM per kg dry tissue
combined data	5.7,12	19.3	77.9	107.		mM per litre of water

CHAPTER 6

MONTE CARLO SIMULATION IN ANALYTICAL ELECTRON MICROSCOPY

DAVID F. KYSER

IBM RESEARCH LABORATORY

SAN JOSE, CALIFORNIA 95193

6.1 INTRODUCTION

Spatial resolution in Analytical Electron Microscopy (AEM) is ultimately limited by electron scattering within the foil target. The degree of electron scattering is determined by experimental variables such as incident beam voltage, foil thickness, foil density, and tilt angle of the foil with respect to the incident electron beam. In order to predict the optimum experimental conditions to achieve a desired resolution, we must have a reliable theoretical model for calculating the effects of these experimental parameters on spatial resolution. One model, based on a single scattering event at the middle of the foil, has been proposed by GOLDSTEIN et al. (1977) and some typical results are given in Chapter 3 of this book. Another model, based on Monte Carlo simulation of electron scattering throughout the foil, has been proposed by KYSER and GEISS (1977). Additional results obtained with a sophisticated Monte Carlo model have been given by GEISS and KYSER (1979) and NEWBURY and MYKLEBUST (1979). Finally, FAULKNER et al. (1977) and FAULKNER and NORRGARD (1978) have used a simplified Monte Carlo model to estimate electron scattering and spatial resolution on AEM.

The purpose of this chapter is to provide an elementary, complete description of (a) the basic physical concepts used in Monte Carlo simulation of electron trajectories, (b) implementing a Monte Carlo program on a digital computer, and (c) some results obtained with Monte Carlo calculations of spatial resolution and x-ray production. The motivation for this work arose from the technical discussions reported at the Cornell Specialist Workshop on AEM (1976) and subsequent interaction with some of the participants.

6.2 BASIC PHYSICAL CONCEPTS IN MONTE CARLO SIMULATION

The Monte Carlo simulation of electron trajectories and x-ray production is similar to that utilized previously by KYSER and MURATA (1974a) for chemical microanalysis of thin films on substrates, except that the substrate is absent in the present calculations. The Monte Carlo model utilizes a true "single-scattering" approach wherein the primary electrons are elastically scattered along their trajectory by the atomic nuclei they encounter at the end of a mean free path, and they lose energy continuously (i.e. linearly with path distance) between successive scattering points due to electron-electron scattering. This electron-electron scattering (inelastic) is assumed to be very small angle scattering in comparison to the electron-atom scattering. The production of x-rays (or some other interaction) is determined by the energy-dependent cross-section for ionization of that particular x-ray (or cross-section for some other particular interaction). The separate parts of the model will now be treated in more detail. An excellent series of papers on alternate approaches to Monte Carlo models of electron scattering are contained in the NBS Monte Carlo Workshop Proceedings (1975). In addition, a new approach for Monte Carlo simulation which directly includes the inelastic electron scattering has been discussed by SHIMIZU et al. (1976). The Monte Carlo model used in the present work is based on a continuous-slowing-down approximation for energy loss due to inelastic electron scattering, and neglects the small angular scattering associated with energy loss.

Electron Scattering

The scattering model of MURATA et al. (1971, 1972) is utilized to predict the spatial trajectory of each incident primary electron. Rutherford elastic scattering of the negatively charged electron by a "screened" positively charged atomic nucleus is assumed to control the angular distribution of scattering. The Rutherford equation for the differential elastic scattering cross-section is:

$$\frac{d\sigma(\theta)}{d\Omega} = \sigma'(\theta) = \frac{Z(Z+1)e^4}{4E^2(1-\cos\theta+2_\beta)^2} \qquad (6.1a)$$

where Z is the atomic number of the scattering atom, e^- the electron charge, E the kinetic energy of the scattered electron, θ the scattering angle, and β the screening parameter to account for electrostatic screening of the nucleus by orbital electrons. The total cross-section σ is obtained by integrating $\sigma'(\theta)$ over the range $\theta = 0$, π and also over the range $\phi = 0$, 2π in $d\Omega = \sin\theta d\theta d\phi$. The result is:

$$\sigma = \frac{\pi e^4 Z(Z+1)}{4E^2\beta(\beta+1)} \qquad (6.1b)$$

The distance along the trajectory between elastic scattering events is given by the mean free path for scattering:

$$\Lambda(cm) = 1/n\sigma = A/N_A\rho\sigma \qquad (6.2)$$

where n is the volume density of atoms (cm^{-3}), σ the total cross-section for scattering (cm^2/atom), N Avagadro's number (atoms/mole), A the atomic weight of the scattering atom (gm/mole), and ρ the mass density of the target (gm/cm^3). For a mixed-element target composed of i elements, each with weight fraction C_i,

$$\rho\Lambda(gm/cm^2) = 1/N_A\Sigma(C_i\sigma_i/A_i) \qquad (6.3)$$

There is no statistical distribution assumed for Λ about its value calculated from Eqn. 6.2 or 6.3. Note that in Eqn. 6.3, the mass density of the alloy $\rho = \Sigma C_i\rho_i$.

The probability $P(\theta)d\Omega$ of scattering into the solid angle $d\Omega = \sin\theta d\theta d\phi$ is given by

$$P(\theta)d\Omega = \frac{\sigma'(\theta)}{\sigma} d\Omega \qquad (6.4)$$

If $F(\theta)$ is the indefinite integral of $P(\theta)d\Omega$ over θ, then the directional cosine ($\cos\theta$) for a particular scattering event can be chosen by generating a uniformly distributed random number R (between 0 and 1) and setting $F(\theta) = R$. Then $\cos\theta$ is deduced from the relation

$$\cos\theta = 1 - [2\beta R/(1+\beta-R)] \qquad (6.5)$$

The atom specie i which scatters the incident electron in a mixed target is also chosen in a Monte Carlo fashion by generating another uniformly distributed random number between 0 and 1. The probability that the electron will be scattered by atom i is simply the fractional cross section

$$p_i = (C_i\sigma_i/A_i)/\Sigma(C_i\sigma_i/A_i) \qquad (6.6)$$

Hence, if a generated random number R is in the range $(0-p_i)$ the electron is assumed to be scattered by that specie of atom. If R is in the range (p_i-1), scattering is caused by another specie of atom. The actual specie of atom that scatters the electron is determined by comparing the generated R with the ranges, 0 to p_1, p_1 to $(p_1 + p_2)$, $(p_1 + p_2)$ to $(p_1 + p_2 + p_3)$, etc. In this manner, the atom specie that scatters is determined in a random fashion but is weighted by its fractional cross-section.

Energy Loss Between Elastic Scattering Events

The energy loss along the trajectory, between each scattering event, is approximated by the Bethe continuous-slowing-down equation

$$\frac{dE}{ds} = \frac{- 2\pi e^4 \rho N_A}{E} \sum_i \frac{C_i Z_i}{A} \ln \frac{\gamma E}{J_i} \qquad (6.7)$$

where $\gamma = 1.166$ for relativistic energy electrons and J is the mean ionization energy for atom specie i. Numerical values for J can be calculated from an approximation due to BERGER and SELTZER (1964):

$$J_i = [9.76Z_i + 58.8/Z_i^{0.19}] \qquad (6.8)$$

A somewhat different set of values for J have been proposed by DUNCUMB and REED (1968), especially for low Z . Fortunately this uncertainty in J enters via a logarithm term in Eqn. 6.7, and hence the sensitivity of dE/ds to this uncertainty is reduced accordingly. The numerical values of J used in the present calculations are calculated with Eqn. 6.8. For targets with only one element, Eqn. 6.7 becomes

$$\frac{dE}{ds} \text{ (KV/cm)} = - 7.85 \times 10^4 \frac{\rho Z}{A} \frac{1}{E} \ln \frac{1.166E}{J} \qquad (6.9)$$

Note that Eqns. 6.7 and 6.9 are only valid for $\gamma E/J > 1$. Hence, the electron trajectory can only be simulated down to a minimum energy E = J/γ when using Bethe's equation for energy loss. The value of E will be limited by the largest value of Z in a mixed element target.

Sequence of Calculations

The sequence of events in a trajectory simulation is illustrated in Fig. 6.1. An electron with energy E_0 is impinged at the origin and at 90° to the surface of a semi-infinite target, the x-y plane being the surface. The incident angle can be varied. The first scattering event is assumed to occur at the origin. The scattering angle θ_0 and step length Λ_0 is calculated by Monte Carlo techniques, as described previously. Since Rutherford scattering is axially symmetric about the incident direction of the electron being scattered, a uniformly generated random number must be used to assign a value to the azimuthal angle ϕ_0. With Λ_0, θ_0, ϕ_0 determined, then the spatial position 1 of the next scattering event is determined with respect to 0. The electron energy at point 1 is then calculated by decrementing the energy with respect to its value at point 0 via Eqn. 6.7. At point 1 the sequence is repeated, using E_1 to calculate Λ_1 and θ_1. Another random number is used to generate ϕ_1; and, hence, the spatial position of point 2 is found. The sequence is repeated continuously until the electron energy has decreased to some chosen value near to, but greater than, J /γ. The sequence is terminated there because Eqn. 6.7 will become indeterminate at lower energies. In addition, the step length becomes very small, and the number of steps increases significantly as the electron energy decreases. If the electron escapes the surface as a back-scattered electron, the energy and direction is saved in the computer program. In order to represent the averaging effects of a real electron beam that contains many incident electrons, a large number of electron trajectories must be simulated. There will not be any identical electron trajectories because of the large number of random numbers and step lengths used. An example of the electron trajectories simulated for a particular case is shown in Fig. 6.2. The complex nature of electron scattering is obvious in such trajectory plots.

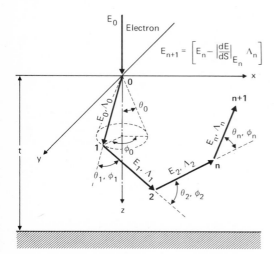

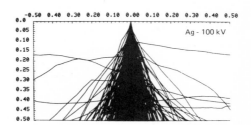

Fig. 6.1. Geometry of the initial steps of electron scattering and energy loss in a Monte Carlo simulation.

Fig. 6.2. Projection of 400 electron trajectories in a 0.5 μm Ag foil for a 100 kV point source. θ = 0°. The scales are in μm, and are equal.

Spatial Distribution of Energy Loss and X-ray Production

In order to deal quantitatively with the results of such Monte Carlo simulation of electron trajectories, we must form histograms in the variables of interest. For example, if we are interested in the energy-distribution $N(E)$ or angular distribution $N(\theta)$ for the transmitted electrons, then we must count up the number of electrons within each energy increment ΔE, or count up the number of electrons within each angular increment $\Delta\theta$ and plot $N(\theta)$ versus θ with a histogram resolution of $\Delta\theta$. The precision of the histogram (i.e. its fluctuation about a mean value) will be controlled by the number of electron trajectories within that increment. For higher precision, we must simulate more trajectories and/or increase the histogram increment (i.e. decrease histogram resolution). This is simply a consequence of the statistical nature of the trajectory simulation. It is at this point where Monte Carlo calculations and closed-form analytic calculations diverge in practice.

The statistical uncertainty of a Monte Carlo calculation is analogous to x-ray emission statistics, i.e. the standard deviation S =

$(N)^{1/2}$ where N is the number of trajectories simulated. Hence, the relative standard deviation $S = S/N = (N)^{-1/2}$. For $N = 10^4$, $S = 1\%$. This assumes that all N electrons have passed through every histogram cell. Normally this is not the case, and hence N should be the number of electrons which have passed through a particular cell if spatial resolution is the application involved. In that case, the standard deviation will be different for each cell, and unique for each particular combination of variables in the calculation. Sometimes only 1-dimensional histograms are desired, e.g. in the depth-distribution of x-rays produced. Or sometimes the total interaction of the electron-target configuration is summed to provide just a single quantitative number. In each particular case, however, the statistics are always related to the total number of electrons simulated by $S\alpha(N)^{1/2}$.

The spatial distribution of energy loss and x-ray production can be calculated by forming a 2-dimensional histogram in radius r and depth z with resolution Δr and Δz, respectively for a point source beam incident. This concept is illustrated in Fig. 6.3 for a foil target. The histogram cell position in the radial direction is defined by the index J, and the cell position in the z direction is defined by the index I. The energy deposited in cell (I,J) by any and all electrons which travel through it is calculated and saved for subsequent use. The cell position refers to the midpoint of the cell. For radial symmetry, the cells can be described as concentric donut-shaped volumes. For non-radial symmetry, conventional cartesian coordinates and square cells can be utilized, either in 2 or 3 dimensions. The spatial distribution of energy deposited by electron beams in polymeric films on thick substrates has been a valuable aid in understanding the complex nature of electron beam lithography technology (see KYSER and MURATA, 1974b).

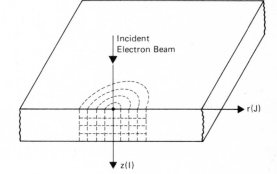

Fig. 6.3. Donut-shaped cells used to form a 2-dimensional Monte Carlo histogram with radial symmetry.

For quantitative calculations of the relative (or absolute) characteristic x-ray production in the foil, another equation must be introduced to describe the ionization rate for a particular atomic level along the electron path. Along the step length ds between scattering events, the electron is assumed to have a constant energy E. Then the number of ions dn created of specie i is

$$dn_i = [C_i \rho N_A / A_i] \ Q_i(E)ds \qquad\qquad (6.10)$$

where Q (E) is the energy-dependent cross-section for ionization and ds = Λ. The number of x-ray photons produced is then proportional to dn through the fluorescent yield ω; and the contribution from all step lengths is summed, as well as that from all electron trajectories. In the present calculation, the energy E is evaluated at the beginning of the step rather than the middle of the step. Since the cross-section Q changes slowly with E at large E, this difference is negligible. In similar fashion to the spatial distribution of energy deposition described previously, the spatial distribution of x-ray production can be calculated by forming a 2-dimensional histogram in r and z. The characteristic x-rays produced in cell (I, J) by any and all electrons which travel through it are summed. An example of such a distribution if shown in Fig. 6.4. Smooth lines have been drawn through the histogram points to enable simple display of the results. Note that at the higher beam voltage, the distribution is more concentrated.

The choice for an expression to define the cross-section for ionization Q (E) is still of some concern. There are a variety of expressions proposed, and these have been discussed recently by POWELL (1975) and KYSER (1976) and GOLDSTEIN et al. (1977). Fortunately in most AEM applications the mean energy loss rate is very small, typically less than 1 eV/Å. Hence, x-ray production in thin foils at high beam voltages can be considered to occur almost monoenergetically. However, with the present Monte Carlo program, we automatically include this small effect, and have chosen to use the expression for Q (E) proposed by REUTER (1971) for the energy-dependence only. For mixed-element targets, some choice must be made for the absolute value of Q (E) also, especially when K, L, and M lines are mixed together. This will be particularly important in quantitative chemical microanalysis with x-ray measurements.

For application to spatial resolution calculations in chemical microanalysis with x-ray measurements, the Monte Carlo simulation model has some unique advantages relative to existing alternative models:

1. Can simulate any foil thickness
2. Can simulate any beam distribution incident
3. Can simulate any beam angle incident
4. Can simulate any x-ray takeoff angle
5. Self absorption correction included automatically
6. Can simulate any compound element target
7. Atomic number correction included automatically

Some of these advantages will be obvious in the subsequent sections of this chapter. In addition, there is an economical advantage for foil calculations in contrast to bulk calculations. In Monte Carlo calculations for bulk targets, a large part of the computational time is allocated to the later stages of the electron trajectory where the energy is low and the step lengths are short and numerous. For foil targets, the electrons traverse the foil relatively quickly and there is much less computer time spent (per electron) than for bulk targets. Hence, it is economically more feasible to calculate electron trajectories for foil targets, and this can be used to advantage to improve the Monte Carlo statistics for foil targets, even at high beam energies.

6.3 DESIGN, IMPLEMENTATION, and OUTPUT OF A MONTE CARLO PROGRAM

Due to the variety of approaches to computer programming, there is not a unique computer program for Monte Carlo simulation of electron scattering and energy loss in solid targets. A brief review of the various approaches taken is described by BISHOP (1975). Because a detailed listing of our computer program and its associated input/output details is inappropriate here, we will instead refer to several different examples which are available in the literature. A very simplified program, along with a detailed explanation of its use, has been published by CURGENVEN and DUNCUMB (1971). A more sophisticated program has been published in the appendix of an excellent paper by SHIMIZU (1972). In addition, a very detailed report on a Monte Carlo program, including the computer code and some examples, has been published by HENOC and MAURICE (1975). Based on the information contained in these three papers, along with the references of MURATA et al. (1971, 1972), the typical analytical group with access to a computer facility should be able to start up their own Monte Carlo program for electron trajectory simulation. However, there are some topics which are often neglected in the description of such programs. We will now discuss some of those which will aid the typical user in implementing a Monte Carlo program.

Computer Generation and Utilization of Random Numbers

In computer programs which utilize Monte Carlo models, there is usually a need for a relatively large number of "random" numbers. These numbers are used in various ways, e.g. in Eqn. 6.5 to pick a particular value of $\cos \theta$ or to pick a particular azimuthal angle $\phi = 2\pi R$ in Fig. 6.1 at each scattering point. The term "uniformly distributed" random number refers to the concept that there is equal probability for any particular number to be generated, within a particular interval of numbers. Usually the interval is between 0 and 1. There are many different ways to generate a "uniformly distributed" random number between 0, 1 and to test for its randomness. The full treatment of this subject is beyond the scope of this chapter. However, one of the most useful methods for Monte Carlo simulation, and the one employed in the present work, is that of the congruential method for generating pseudorandom numbers. The details of this method can be found in HAMMERSLEY and HANDSCOMB (1964) and in KNUTH (1969). In such methods, one starts with an "initial" random number which is subsequently utilized in a computer subroutine program to trigger the calculation of a sequence of additional random numbers. The randomness and length of the random number list is controlled by the input parameters in the subroutine. The list of random numbers is subsequently used in simulation calculations as described. Interestingly enough, the list of random numbers can be duplicated at will by simply using the same initial random number as the trigger. Hence, Monte Carlo calculations can be repeated exactly if desired, with the identical statistics, etc. Neverless, the results of the calculation are still based on randomness and statistics. The point is that with the congruential method, the randomness can be repeated exactly. This appears to be a dichotomy at first, but is only a detail which is of no concern for the present applications. In practice, one normally uses a series of initial random numbers and their associated lists and divides up the electron trajectories such that a fraction of the total electron trajectories come from only one list. In this way some additional randomness can be incorporated into the calculations. The choice of an "initial" random number is not critical, and any simple convenient scheme can be used such as throwing dice, darts, or guessing at a useful sequence of digits to form a number. There are also some types of electronic calculators which can generate a random number upon command.

In calculations of spatial resolution and other applications which involve electron beams, there is often a need to include within the Monte Carlo simulation the finite size of the incident beam. If the distribution of the incident beam current density is known in practice, this can be simulated by having each successive electron in the Monte Carlo calculation enter the target surface such that the distribution of all the electrons together is the same as that desired. If the distribution is Gaussian in radius, then one can either utilize a "Gaussian" random number generator subroutine or simply generate a "Gaussian-distributed" random number from a "uniformly-distributed" one (R_U). The radial distance of each electron starting position is then

$$r_G = (-2\sigma^2 \ln R_U)^{\frac{1}{2}} \qquad (6.11)$$

where σ = standard deviation of desired Gaussian distribution. If cartesian coordinates are used to describe the initial scattering point, then $x_0 = r_G \cos(2\pi R)$ and $y_0 = r_G \sin(2\pi R)$ where R is another random number between 0, 1. With subsequent electrons, different values of R_U and R would be generated and used. Hence, for a large number of electrons, the distribution will be Gaussian and practically continuous with radius.

Computational Time and Its Control

One common objection to Monte Carlo calculations is their alleged high cost due to long computations on a digital computer. However, the actual cost is completely within the control of the user, and is primarily determined by the model used, the target configuration, and the number of electrons simulated N. For simple calculations, the model of CURGENVEN and DUNCUMB can be used (1971). As described previously, the foil application is more economical than the bulk application. Finally, the desired statistical certainty sets the value for N. Since Monte Carlo calculations are not meant to be used for rapid on-line analysis or control, there can be a reasonable time delay between program submission to a computer and the output calculations. Many programs can wait until low-load computer times to be run. For debugging a program, only a few trajectories need to be simulated to check the operation and output calculations for reliability. A high-precision Monte Carlo program should only be run after establishing its reliability. The output of any Monte Carlo calculation can be saved for subsequent use in other applications. However, one must then decide in advance the output variety desired and anticipate the need.

As a typical example, the curves shown in Fig. 6.4 were generated with 100,000 electrons and with a cell size of 25 Å in order to get good histogram resolution. This was accomplished on an IBM System 360/Model 195 computer in about 10 minutes of CPU (central processing unit). Hence, the value of about 10,000 trajectories/1 CPU minute is obtained. For thicker foils and lower resolution (larger cell size), much fewer trajectories are needed. Most of our calculations are done with N = 10,000 and require about 1 CPU minute.

Condensation and Output of Results Obtained

The variety of output data available from a Monte Carlo simulation of electron trajectories and energy loss is quite large. It includes the following:

1. electron trajectories plotted for qualitative distribution
2. backward scattered and forward scattered electron yield
3. backward scattered and forward scattered energy distribution
4. backward scattered and forward scattered angular distribution
5. spatial distribution of energy deposition
6. spatial distribution of x-ray production

Of course, the quantitative accuracy of such calculations depends intimately on the accuracy of the physical model employed. This is true of any model, and the Monte Carlo method does not avoid this necessity. However, the Monte Carlo model is a very intuitive and easy model to simulate physical processes with, once the output desired is identified. The angular, energy, and number yield of backward scattered electrons from foil targets is of some interest, and has been discussed recently by NIEDRIG (1977). The spatial distribution of energy disposition and its relevance to electron beam lithography has been discussed by KYSER and MURATA (1974b). The value of Monte Carlo calculations for spatial resolution and quantitative microanalysis has already been discussed. In cases 5 and 6, the output histogram can be in 1, 2, or 3 dimensions as desired. The Monte Carlo model is not limited to flat surfaces or homogeneous targets. Specific boundary conditions on the target can easily be accommodated and incorporated, e.g. lateral discontinuities to approximate an inclusion. In some cases, just a simple graphical plot of the electron trajectories will suffice to provide some qualitative or semi-quantitative information; and some examples of this are presented in the next section of this chapter.

6.4 APPLICATIONS TO X-RAY MICROANALYSIS

One particularly important application of Monte Carlo simulation is for x-ray microanalysis. The complex nature of electron scattering and

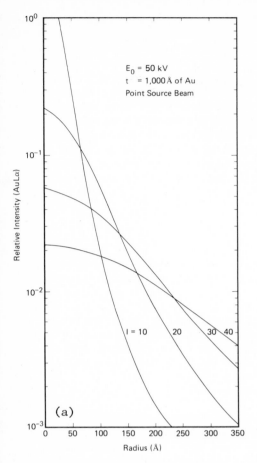

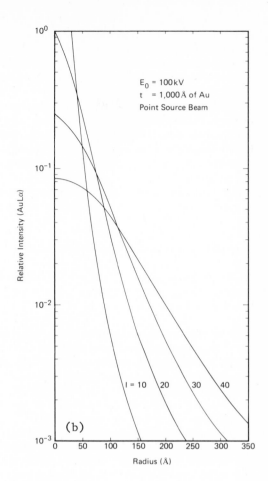

Fig. 6.4. The spatial distribution of AuLα x-ray production in a 0.1 μm Au foil for (a) 50 kV and (b) 100 kV point source beam. The index I = 10, 20, 30, 40 corresponds to a cell thickness of 25 Å at depths of 250, 500, 750, and 1000 Å, respectively in the Monte Carlo calculations.

energy loss in solids requires a reliable, accurate model to describe the importance of such variables as beam voltage, film thickness, film material, and tilt angle on the spatial distribution of x-ray production. An early application of Monte Carlo simulation for quantitative analysis of very thin films on thick substrates was made by KYSER and MURATA (1974a). Successful analyses were made of both the composition and thickness of alloy films at beam voltages encountered in an SEM or an electron microprobe. More recent work by REUTER et al. (1978) has helped to confirm the accuracy of Monte Carlo simulation in this lower voltage regime by careful comparison with experimental results. In this section, we will present some applications of the Monte Carlo model to calculations of the 1-dimensional depth distribution and the 2-dimensional radial distribution of x-ray production in thin foils at high beam voltages such as that encountered in AEM. Fig. 6.5 illustrates the geometrical problem encountered when the incident electrons are able to penetrate through the target without complete attenuation. For quantitative analysis, we must know the depth-distribution of x-ray production, assuming that the foil is homogeneous in composition. For spatial resolution, we must know the radial distribution of x-ray production. The following sections present some examples of this.

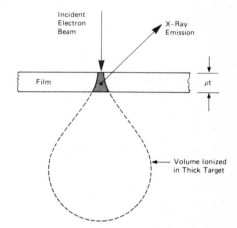

Fig. 6.5. A qualitative display of the volume analyzed in a foil, relative to that in a thick target.

Depth Distribution of X-ray Production

Figure 6.6 shows the Monte Carlo results obtained for the depth distribution of AuLα generated in a 1,000 Å Au foil at 50 kV and 100 kV. The vertical scale is not an absolute scale, but the relative values at 50 kV and 100 kV are absolute. This immediately indicates the higher efficiency (per electron) for AuLα generation at 50 kV. The Monte Carlo data plotted in Fig. 6.6 is the same as that used to form Figs. 6.4(a), (b). From Fig. 6.4, the intensities at the same depth histogram index I have been summed over the radial histogram, and a weighting factor included for the donut-shaped volumes. The statistical fluctuation in the histogram is apparent, and the smooth line is drawn afterward. It is obvious that even for this extreme case, there is very little change in the depth distribution for 100 kV from its value at the surface.

Figure 6.7 shows the depth distribution of x-ray production in 4,000 Å foils of Al, Cu, and Au at 100 kV for comparison. Again the change in the depth distribution is very little over the foil thickness, and suggests that a reasonable approximation for $\Phi(\rho z)$ is that $\Phi(\rho z)$ = constant K.

The depth distribution of x-ray production is an important concept in understanding the magnitude of the self-absorption or attenuation of the x-rays as they emerge from the foil. In conventional electron probe microanalysis of thick targets at lower beam voltages, the absorption correction can be a significant one for accurate quantitative analysis. If we define $f(\chi)$ as the fraction of x-rays produced which escape the

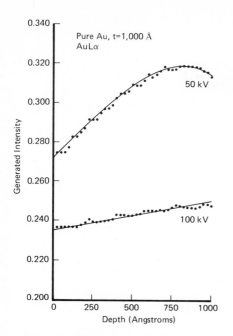

Fig. 6.6. The depth distribution of AuLα produced in a 0.1 μm Au foil with a 50 kV and 100 100kV beam, based on Monte Carlo calculations.

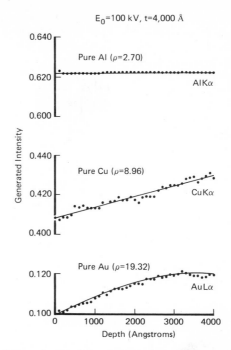

Fig. 6.7. The depth distribution AlKα, CuKα, and AuLα, in 0.4 μm pure foils at 100 kV, based on Monte Carlo calculations.

target at a specific angle ψ, then

$$f(\chi) = \Sigma_0^{\rho t} \; \Phi(\rho z)e^{-\chi \rho z}d\rho z / \Sigma_0^{\rho t} \; \Phi(\rho z)d\rho z \qquad (6.12)$$

where $\chi = (\mu/\rho)\csc\psi$ and μ/ρ = mass absorption coefficient. If $\Phi(\rho z) = K$, then

$$f(\chi) = \frac{1-e^{-\chi \rho t}}{\chi \rho t} \qquad (6.13)$$

This is the absorption correction in the thin-film approximation, and is the same result as that given by PHILIBERT and TIXIER (1975) in a different form. It is easily seen that as χ becomes vanishingly small (i.e. low μ/ρ) then f(χ) becomes 1, and as χ becomes infinitely large (i.e. high μ/ρ) then f(χ) becomes close to 0. Because of uncertainties in the values of μ/ρ for specific x-rays and absorbers, it is helpful to determine the effect of an uncertainty Δχ on the uncertainty in f(χ). This if found by taking the differential

$$\frac{df(\chi)}{d\chi} = \frac{2e^{-\chi \rho t}-1}{\chi} \qquad (6.14)$$

If $\chi\rho t \ll 1$ and $\Delta\chi/\chi \ll 1$, then $\Delta f \cong \Delta\chi/\chi$ and so the uncertainty in f is proportional to the uncertainty in χ. Figure 6.8 shows the dependence of $f(\chi)$ upon $\chi\rho t$ for the thin film approximation. Note that $\chi\rho t$ is dimensionless, and so is $f(\chi)$. If the mass thickness ρt and absorption coefficient μ/ρ is known, then the curve in Fig. 6.8 can be used to quickly calculate the magnitude of the absorption, or difference in absorption, for any x-rays. For long wavelength x-rays such as AlKα in a matrix such as Cu, the correction is important for quantitative analysis.

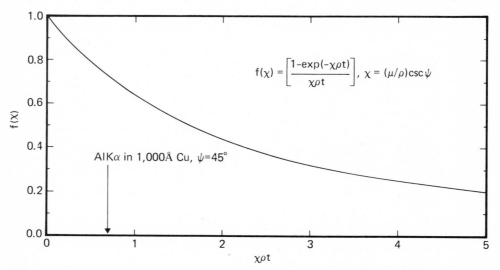

Fig. 6.8. Universal curve for the fraction of total x-ray production which escapes the foil at angle ψ due to selfabsorption, based on the thin-film approximation.

Total X-ray Production in Foils

For the total x-ray production in a foil, the radial and depth histograms must be summed over all values of I and J. Then the spatial distribution is not important, and the relative signals from different elements in the foil can be calculated, or the voltage and/or thickness dependence can be calculated for a particular element. One particularly interesting application of the latter type was made by GEISS and KYSER (1979) to predict the voltage-dependence of the characteristic peak intensity P with respect to the continuous background intensity B (at the peak spectral position). To calculate the background B, an appropriate cross-section equation was used within the Monte Carlo simulation instead of $Q(E)$ for the characteristic x-ray, and then the beam voltage was varied systematically to provide P/B versus beam voltage E_0. Separate Monte Carlo calculations were made for P and B at the same value of E_0. The theoretical, and subsequent experimental results with an Fe foil, showed that P/B increases continuously with increasing E_0. However, the absolute value of P (per electron) decreases with increasing E_0. Fortunately, B decreases <u>faster</u> than P, and so P/B increases with E_0. This contradicts the results of some experiments reported in the literature, and specifically contradicts the conclusions reached by FAULKNER and NORRGARD (1978) regarding the voltage-dependence of P/B.

Experimentally, it is very important to eliminate the stray electron and x-ray bombardment of the foil, or else the quantitative analysis will not be reliable. A simple test for this can be the observed voltage-dependence of P/B. If P/B goes through a maximum value and then decreases, the x-ray production is not entirely due to the primary electron beam at a point on the foil. A more detailed analysis of the various sources of spectral contamination in AEM is presented by GOLDSTEIN and WILLIAMS (1978).

If the foil is very thin, then we can estimate the voltage-dependence of P/B in the following manner. The peak intensity $I_P \propto Q_P$ and the background intensity $I_B \propto Q_B$. Let $Q_P \propto (1/U_0) \ln U_0$ where $U_0 = E_0/E_c$ and E_c is the critical excitation energy for the characteristic x-ray. From KRAMERS (1923) we can deduce that $Q_B \propto 1/U_0$ in the isotropic approximation (see REED, 1975) and for $E_c \ll E_0$. Hence in the thin film approximation,

$$\frac{P}{B} = \frac{I_P}{I_B} \propto \ln U_0 \qquad (6.15)$$

and thus P/B increases continuously with E_0. This is also confirmed with Monte Carlo calculations for finite thickness foils. Note also that P/B (observed) = P/B (generated) since the depth distributions; and, hence, the absorption corrections are essentially identical for both I_P and I_B.

Radial Distribution of X-ray Production

If we were to neglect the effect of electron scattering within the foil, then the radial distribution of x-ray production would be the same as the radial distribution of the incident electron beam. If the beam current density distribution is Gaussian and described by

$$J(r) = \frac{I_0}{2\pi\sigma^2} \exp \frac{-r^2}{2\sigma^2} \qquad (6.16)$$

then the fraction of the total x-ray production volume in the foil without scattering of electrons is

$$f(r) = \frac{t \int_0^r J(r)\, 2\pi r\, dr}{t \int_0^\infty J(r)\, 2\pi r\, dr} = 1 - \exp \frac{-r^2}{2\sigma^2} \qquad (6.17)$$

Figure 6.9 shows a plot of Eqn. 6.17 for standard deviation values of σ
= 25, 50, and 100 Å. For such a Gaussian beam, the "diameter" is FWHM =
2.36σ. At r = FWHM/2, f(r) = 0.50 and at r = FWHM, f(r) = 0.94. Since
the definition of resolution is somewhat arbitrary, we will arbitrarily
use the value of f(r) = 0.90 to determine spatial resolution in the presence of electron scattering.

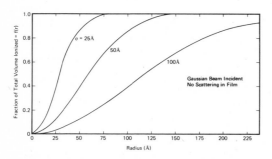

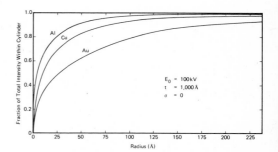

Fig. 6.9. Analytical results for
the fraction of total x-ray production versus radius of right cylinder
and a Gaussian beam incident, without electron scattering in the foil.

Fig. 6.10. Monte Carlo results
for the fraction of total x-ray
production versus radius of
right cylinder and a point
source beam (σ = 0) in 0.1 μm
foils of Al, Cu, and Au at
100kV, θ = 0°.

With electron scattering included, the curves of Fig. 6.9 will be
changed, depending on the particular conditions employed. Figure 6.10
shows the Monte Carlo results of a simulation for a point source beam (σ
= 0) and 1,000 Å foils of Al, Cu, or Au at 100 kV. These curves are generated by summing the x-ray production in all histogram cells from r = 0
to r = r and from z = 0 to z = t. Then that intensity is normalized
with the total intensity from all cells, and plotted versus radius r.
Hence, the volume is defined as a right circular cylinder, and the resolution for microanalysis can be defined as the value of d = 2r which
gives a particular value of f(r).

Figure 6.11 shows the same case as Figure 6.10, except that σ = 50
Å to illustrate the sensitivity of d to σ. The spatial distribution of
x-ray production with a finite value for σ was calculated by distri-
buting successive electron trajectories about r = 0 with a Gaussian distribution, as discussed previously.

Figure 6.12 shows a composite result of Monte Carlo calculations
for a 1,000 Å Cu foil with E_0 = 50, 100, 200 kV and σ = 0, 50, 100 Å for
comparison. The implications of beam voltage and of beam diameter
desired on instrument design for high-resolution microanalysis have been
discussed by GEISS and KYSER (1979). It was concluded that it will be
better to utilize a 200 kV field emission (FE) gun source, since the improvement in AEM resolution capability with FE versus thermionic W at
100 kV is very little.

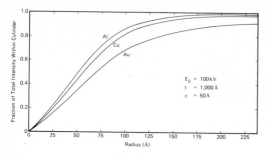

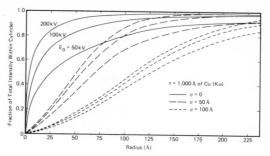

Fig. 6.11. Monte Carlo results for same foils as Fig. 6.10, but with a Gaussian beam (σ = 50 Å) incident, θ = 0°.

Fig. 6.12. Composite of Monte Carlo results for 0.1 μm foils of Cu at various beam voltages and Gaussian beam diameters (FWHM = 2.36σ), θ = 0°.

By utilizing curves such as those in Figs. 6.10-6.12, we can tabulate the spatial resolution for various conditions with a common definition for resolution. A summary of the Monte Carlo calculations is given in Table 6.1 for 100 kV and σ = 0, 50 Å in C, Al, Cu, and Au foils with various thicknesses. All of these results are for normal beam incidence. When foil thickness is very small, then the incident beam size is an important contribution to the spatial resolution obtained, especially for low atomic number foils. At larger values of thickness, the beam size is less important (in the range of σ = 50 Å) because then the resolution is limited by electron scattering within the foil. The actual contribution from each source depends on the particular case investigated.

The values for spatial resolution given in Table 6.1 agree fairly well with the values given by GOLDSTEIN in Table 3.5, Chapter 3, of this book for the valid ranges of the non-Monte Carlo approach. However, the values in Table 6.1 do not agree with some of the values given by NEWBURY and MYKLEBUST (1979), especially for thick Au foils where they differ by more than a factor of 3. The source of this difference is not apparent at the present time. The values of d in Table 6.1 for a thin Al foil (e.g. 1000-2000 Å) agree with experimental tests of spatial resolution in microanalysis of Si_3N_4 and Al_2O_3 foils presented by GEISS and KYSER (1979) with a small beam diameter. The values of d in Table 6.1 for thin Cu foils (1000-2000 Å) also agree with the summary of measurements in ferrous materials (Table 3.7) given by GOLDSTEIN in Chapter 3 for σ = 50 Å.

Electron Trajectory Plotting

In many cases, a simple graphical plot of the electron trajectories can provide the necessary qualitative or semi-quantitative information for AEM applications. While the trajectories are simulated in 3 spatial dimensions, they can only be easily plotted in projection upon a 2-dimensional plane, e.g. the x-z plane. In the following discussion, we

TABLE 6.1

Summary of Monte Carlo calculations defining the cylinder diameter d ($\overset{\circ}{A}$) which contains 90% of the total x-ray production generated within the foil for E_0 = 100 keV, σ = 0. The values in parentheses are for σ = 50 $\overset{\circ}{A}$.

Film material and x-ray	Film thickness ($\overset{\circ}{A}$)			
	400	1000	2000	4000
C (Kα)	-- --	50 (200)	130 (230)	250 (300)
Al (Kα)	20 (195)	90 (220)	200 (250)	460 (460)
Cu (Kα)	40 (205)	160 (250)	550 (550)	1400 (1400)
Au (Lα)	80 (220)	380 (400)	1300 (1300)	4500 (4500)

will show a variety of trajectory plots for special cases and indicate the kind of information which can be deduced.

Figure 6.13 shows 400 electron trajectories in 0.5 μm foils of Al, Cu, and Au at 100 kV with normal beam incidence. Note that with increasing atomic number Z of the target, the spatial distribution of the trajectory plot broadens. This is in agreement with the quantitative results given in Table 6.1 for x-ray microanalysis resolution, and it is interesting to compare the actual distribution with the resolution definition used to calculate d in Table 6.1. In Table 6.1, d is the diameter of a right circular cylinder within which 90% of the total x-ray production is contained. Figure 6.13 shows how the electron trajectories are actually distributed in r and z. For application to thinner foils, it is a good approximation to simply truncate the trajectories at depth z = t where t is the foil thickness of interest. Hence, fig. 6.13 is also appropriate for foils with t < 0.5 μm after truncation. Figure 6.14 shows 400 electron trajectories in 0.5 μm foils of Al, Cu, and Au at 200 kV. At this high voltage, the trajectory distributions are narrower than those in Figure 6.13, as expected. Note that in Figs. 6.13 and 6.14, the horizontal scale has been expanded by 2X the vertical scale in order to display the results more clearly.

Figure 6.15 shows 400 electron trajectories in 0.2 μm foils of Cu and Au at 100 kV. Note that the horizontal scale is now expanded by 4X the vertical scale, in order to display the scattering effects clearly. Figure 6.16 shows the same target as Fig. 6.15 except at 200 kV. The spatial distribution of trajectories in 0.2 μm of Cu at 100 kV [Fig. 6.15(a)] is about the same as that in 0.2 μm of Au at 200 kV [Fig. 6.16(b)].

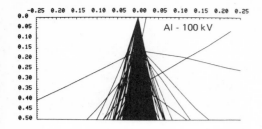

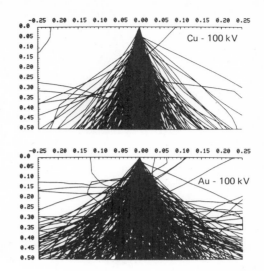

Fig. 6.13. Projection of 400 electron trajectories in 0.5 μm foils of (a) Al, (b) Cu, and (c) Au for a 100kV point source beam, θ = 0°. The scales are in μm, but unequal by 2X.

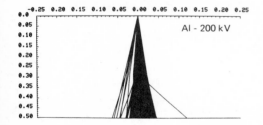

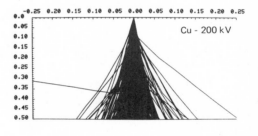

Fig. 6.14. Projection of 400 electron trajectories in 0.5 μm foils of (a) Al, (b) Cu, and (c) Au for a 200 kV point source beam, θ = 0°. The scales are in μm, but unequal by 2X.

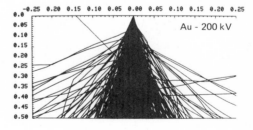

At fixed beam voltage, the P/B ratio of x-ray signals is dependent on the tilt angle of the foil with respect to the beam. This effect has been discussed by GEISS and HUANG (1975) who found that P/B goes through a maximum around 30° tilt. Our Monte Carlo program has the capability to simulate any tilt angle, and Fig. 6.17 shows the same cases as Fig. 6.15 except that θ = 45° tilt and the scales are now the same. Note that there is now severe degradation of the spatial resolution, and it becomes difficult to actually define the resolution in such cases. Figure 6.18 shows the same cases as Fig. 6.16 also except for θ = 45° tilt.

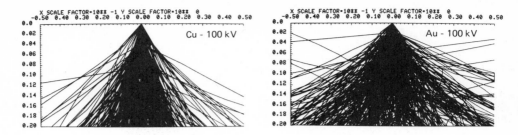

Fig. 6.15.   Projection of 400 electron trajectories in 0.2 μm foils of
(a) Cu, and (b) Au for a 100 kV point source beam, θ = 0°. The scales
are in μm, but unequal by 4X.

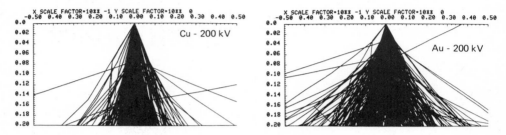

Fig. 6.16.   Projection of 400 electron trajectories in 0.2 μm foils of
(a) Cu and (b) Au for a 200 kV point source beam, θ = 0°. The scales
are in μm, but unequal by 4X.

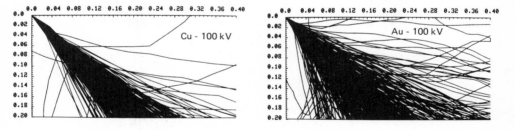

Fig. 6.17.   Projection of 400 electron trajectories in 0.2 μm foils
of (a) Cu and (b) Au for a 100 kV point source beam, θ = 45°. The
scales are μm, and are equal.

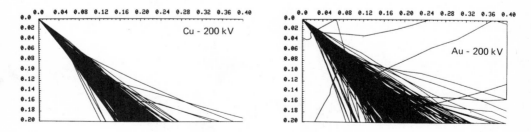

Fig. 6.18.   Projection of 400 electron trajectories in 0.2 μm foils
of (a) Cu and (b) Au for a 200 kV point source beam, θ = 45°. The
scales are in μm, and are equal.

From trajectory plots such as those given in Figs. 6.13-6.18, several conclusions can be reached. (1) at fixed beam voltage, thinner foils and lower Z foils give higher spatial resolution for microanalysis, (2) at fixed foil thickness, higher beam voltage and lower Z foils give higher resolution for microanalysis (3) at fixed voltage and thickness, lower tilt angles and lower Z foils give higher resolution for microanalysis. The quantitative aspects of spatial resolution must be treated as discussed previously, and shown in Table 6.1. The quantitative aspects of chemical analysis can also be treated with Monte Carlo simulation as discussed in section 6.4, but a simpler and more practical method is discussed by GOLDSTEIN in Chapter 3 of this book for very thin foils. When that method cannot be relied upon, e.g. when the foils are very thick, then Monte Carlo simulation must be utilized to account for the electron scattering.

6.5 SUMMARY

Monte Carlo simulation of electron scattering and energy loss in AEM is a very powerful tool for both qualitative and quantitative design and interpretation of experiments. The accuracy of the Monte Carlo calculation is dependent on the accuracy of the physical approximations used in its design, and there is a continuing research effort to identify the areas which need better approximations. However, the Monte Carlo method is a very "physical" one in which the basic concepts are easily grasped and interpreted, and there is a wealth of output available in the form of electron energy, angular, and number distributions for absorbed, backscattered, and transmitted electrons in AEM applications. The calculation of spatial resolution in x-ray microanalysis has enabled the identification of those variables which control and ultimately limit the maximum resolution attainable. The depth distribution of x-ray production has also been calculated, and the validity of the thin-film approximation for the x-ray absorption correction has been evaluated. For most situations, it appears that the depth distribution is approximately constant, and this leads to a simple analytic equation for the absorption correction.

Several versions of a Monte Carlo program are available in the literature, and the serious user of AEM is encouraged to start a project on Monte Carlo simulation and tailor it to the application desired. Many applications demand a good model for interpretation and experimental design, and Monte Carlo methods are probably the most flexible and adaptable, especially for quantitative interpretation. The technical insight and subsequent rewards to be gained are very worthwhile.

ACKNOWLEDGMENTS

The author gratefully thanks Dr. Kenji Murata (Osaka University, Japan) for assistance in preparing a Monte Carlo computer program for simulation of electron scattering in solid targets, Dr. Roy Geiss (IBM Research Division) for many stimulating and encouraging discussions and collaboration on AEM research, and Mr. Richard Pyle (IBM Systems Product Division) for providing the computer software support for interactive computer graphics.

REFERENCES

Berger, M. J. and Seltzer, S. M., National Academy of Sciences-National Research Council, Publ. 1133 (Washington, D. C., 1964), p. 205.

Bishop, H. E., NBS Special Publication 460 (Washington, D. C., 1975), p. 5.

Curgenven, L. and Duncumb, P., Tube Investments Res. Lab. Report 303 (Essex, England, 1971).

Duncumb, P. and Reed, S. J., NBS Special Publication 298 (Washington, D. C., 1968), p. 133.

Faulkner, R. G., Hopkins, T. C., and Norrgard, K., X-Ray Spect. 6, p. 73 (1977).

Faulkner, R. G. and Norrgard, K., X-Ray Spect. 7, p. 184 (1978).

Geiss, R. H. and Huang, T. C., X-Ray Spect. 4, p. 196 (1975).

Geiss, R. H. and Kyser, D. F., Ultramicroscopy 3, p. 397 (1979).

Goldstein, J. I., Costley, J. L., Lorimer, G. W., and Reed, S. J., SEM/1977/I, p. 315 (1977).

Goldstein, J. I. and Williams, D. B., SEM/1978/I, p. 427 (1978).

Hammersley, J. M., and Handscomb, D. C., Monte Carlo Methods (Methuen and Co. Ltd., London, 1964).

Henoc, J. and Maurice, F., C.E.N.S. Report CEA-R-4615 (French Atomic Energy Commission, 1975).

Knuth, D. E., The Art of Computer Programming - Seminumerical Algorithums, Vol. 2 (Addison-Wesley Publ. Co., 1969), Ch. 3.

Kramers, H. A., Phil. Mag. 46, p. 836 (1923).

Kyser, D. F. and Murata, K., IBM J. Res. Dev. 18, p. 352 (1974a).

Kyser, D. F. and Murata, K., Proc. 7th Int. Conf. Electron and Ion Beam Science and Technology (The Electrochemical Society, 1974b), p. 205.

Kyser, D. F., Proc. 11th Annual Conf. of the Microbeam Analysis Society (Miami Beach, 1976), p. 28A.

Kyser, D. F. and Geiss, R. H., Proc. 12th Annual Conf. of the Microbeam Analysis Society (Boston, 1977), p. 110A.

Murata, K. Matsukawa, T., and Shimizu, R., Jap. J. Appl. Phys. 10, p. 678 (1971).

Murata, K., Matsukawa, T., and Shimizu, R., Proc. 6th Int. Conf. X-Ray Optics and Microanalysis (Univ. of Tokyo Press, 1972), p. 105.

Newbury, D. E., and Myklebust, R. L., Ultramicroscopy 3, p. 391 (1979).

Niedrig, H., SEM/1978/I, p. 841 (1978).

Philibert, J. and Tixier, R., Physical Aspects of Electron Microscopy and Microbeam Analysis (Wiley, 1975), p. 333.

Powell, C. J., NBS Special Publication 460 (Washington, D. C., 1975), p. 97.

Reed, S. J., Electron Microprobe Analysis (Cambridge Univ. Press, 1975), Ch. 19.

Reuter, W., Kuptsis, J. D., Lurio, A., and Kyser, D. F., J. Phys. D. 11, p. 2633 (1978).

Reuter, W., Proc. 6th Int. Conf. X-Ray Optics and Microanalysis (Univ. of Tokyo Press, 1972), p. 121.

Shimizu, R., Quantitative Analysis With Electron Microprobes and Secondary Ion Mass Spectrometry (Nuclear Research Institute, Julich, Germany, 1972), p. 156.

Shimizu, R., Kataoka, Y., Ikuta, T., Koshikawa, T., and Hashimoto, H., J. Phys. D. 9, p. 101 (1976).

Workshop Proc. on Monte Carlo Calculations in Electron Probe Microanalysis and Scanning Electron Microscopy, ed. by K. F. Heinrich, D. E. Newbury, and H. Yakowitz (Nat. Bur. of Stds., 1975) - NBS Spec. Publ. 460.

Workshop Proc. on Analytical Electron Microscopy (Cornell Univ., 1976).

CHAPTER 7

THE BASIC PRINCIPLES OF ELECTRON ENERGY LOSS SPECTROSCOPY

DAVID C. JOY

BELL LABORATORIES

MURRAY HILL, NEW JERSEY 07974

7.1 WHAT IS ELECTRON ENERGY LOSS SPECTROSCOPY?

Electron Energy Loss Spectroscopy (EELS) is the study of the energy distribution of electrons which have interacted with a specimen. In the next four chapters of this book we will be examining the way in which this technique can be used, in conjunction with an electron microscope, to provide a powerful microanalytical method giving detailed quantitative information about the chemical, physical and electronic nature of our sample. In this chapter the basic physics of EELS will be discussed and the important concepts will be defined and described. The following chapters will then take these ideas and apply them in the contexts of biological, materials science and solid state studies.

7.2 WHAT IS REQUIRED?

The three components of any EELS experiment are (1) a source of electrons, (2) a device (the spectrometer) for analyzing the energy of the scattered electrons and (3) a suitable sample of the material that we wish to study. In general no restrictions need be placed on any of these components, but for our purposes we will limit the discussion to an experimental arrangement which is compatible with the operation of a conventional transmission (or scanning transmission) electron microscope. In this case the source of electrons will be the gun of the microscope, and the incident electrons will be assumed to have an energy in the range of 50 to 200 keV. Because we are using a microscope the incident beam will be capable of being focussed on to a specimen area, whose size and position can be controlled, and which can be imaged using the normal optical system of the instrument. The specimen itself will be a "thin" section of the material to be examined. A discussion of how thin this section must be will be given later, but as a general guide it is adequate to say that a sample which is thin enough to give a good conventional TEM image, at the chosen accelerating voltage, should also give usable EELS data. The spectrometer is then placed after the specimen to analyze the transmitted electrons, as shown in Fig. 7.1.

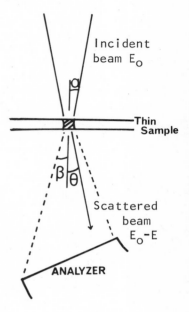

Fig. 7.1. The geometry of energy loss spectroscopy in the electron microscope illustrating the scattering angle θ, the spectrometer acceptance angle β and the incident beam convergence angle α.

    All the information that can be obtained about the sample is contained in the angular and energy distribution of the electrons that have passed through the specimen. By studying these distributions, which result from interactions between the electrons and the specimen, and analyzing them in terms of an appropriate model, the required information can be found. The most general way to characterize these interactions is to measure the momentum changes suffered by the electrons as they pass through the sample. This is done by measuring both the angle θ through which an electron is scattered, and E its change in energy relative to its incident energy E_0. For some types of studies this procedure is necessary and examples of this approach are given in Chapter 10 by Silcox. Usually, however, it is sufficient to collect all the transmitted electrons lying within a cone of some width β about the incident beam direction, and then to analyze these for their energy loss. This destroys the information about the momentum transfer, since the spectrometer is now integrating information from the whole angular range which it is accepting, but because it allows us to use simpler designs of spectrometers, and as the information produced can be readily related to the important properties of the material by simple mathematical models of the interactions, this approach is the one most generally used. The result that we obtain from such an experiment is then an Electron Energy Loss Spectrum in which we plot the transmitted signal intensity I(E) as a function of the energy loss E for all the electrons scattered within the angular cone β accepted by the spectrometer.

7.3 DESCRIBING THE ENERGY LOSS SPECTRUM

Figure 7.2 shows a typical electron energy loss spectrum recorded from a thin foil of silicon using 100 kV incident electrons and with $\beta = 3.10^{-3}$ radians. The signal intensity I(E) is plotted in the vertical direction and the energy loss E on the horizontal axis with the loss increasing from left to right, with zero-loss then indicating that the energy of the electron is the same as its incident energy. A large maximum in the intensity centered at E = 0 is the most prominent feature. This is usually called the "zero-loss peak", and it contains both unscattered electrons (those which have suffered no interactions while passing through the sample and so retain their original energy and direction of motion) and electrons which have interacted with the sample but without losing significant amounts of energy. In any specimen from which a useful EEL spectrum can be obtained this zero-loss peak will always be the largest single component. The remainder of the spectrum then contains the "inelastically scattered" electrons, those which have suffered both a deflection and a loss of energy as the result of some interaction with the sample. The most obvious characteristic of this part of the spectrum is the rapid fall of the average signal intensity as E increases. In fact, as can be seen, by the time an energy loss of 50 eV is reached on this specimen the signal has fallen by a factor of about 50 times, and a change in the gain of the recording system is required to keep the detail on the spectrum visible. This dramatic fall in intensity after the zero-loss peak is a feature of all energy loss spectra and present difficulties in both the recording and the analysis of the spectra. Superimposed on this falling intensity are various features deriving from specific interactions and it is these, rather than the overall shape of the spectrum, that contains the information that we wish to obtain.

Fig. 7.2. Typical ELS data recorded from a thin silicon foil at 100 kV with an acceptance angle of 3 m. rad. Note the zero-loss peak, the plasmon peaks, then a gain change of 50 followed by the SiL_{23} edge.

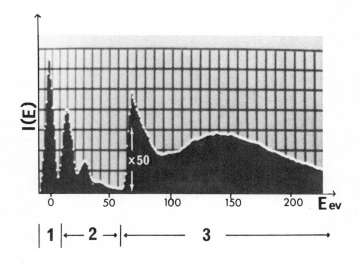

Having looked at the major features of the spectrum it is now necessary to see how we can make use of it for micro-analytical purposes. To
do this it is convenient to introduce some mathematical concepts. In
our EELS experiment we measure the intensity I(E) of the transmitted
signal at some energy loss E for a measurement made over a solid angle
dΩ about some angle θ from the original direction. If the intensity of
the incident beam is I, then I(E,θ)/I is the fraction of the incident
electrons which have been deflected into our measuring device at θ while
losing energy E. This fraction is directly proportional to the physical
quantity called the "doubly differential cross-section" $d^2\sigma(E,\theta)/dE \cdot d\Omega$
which is simply the probability that any incident electron will suffer
an energy loss E while being scattered into the solid angle dΩ about θ.
The constant of proportionality is N, the number of atoms per unit area
in the volume examined. Thus,

$$\frac{I(E,\theta)}{I} = N \cdot \frac{d^2\sigma(E,\theta)}{dE \cdot d\Omega} \qquad (7.1)$$

This cross-section, which is what we in effect measure in an EEL experiment, can be related to the properties of our sample in two different
but equivalent ways. This first of these is the "Macroscopic" approach.
In this we ignore all the microscopic details of the specimen such as
its chemistry and crystallography and instead describe it only in terms
of its dielectric constant ε, which is the physical parameter which measures the response of the material to an electromagnetic wave (such as
electrons or light). The dielectric constant ε is a complex number
having both a real part (which describes the refraction of the wave in
the material) and an imaginary part (which describes the absorption of
the wave). The relationship between $d^2\sigma(E,\theta)/dE \cdot d\Omega$ and ε can be shown
to be

$$\frac{d^2\sigma}{dE \cdot d\Omega} = \frac{(Im(-1/\varepsilon))}{2\pi^2 a_0 \, nE} \cdot \frac{1}{\theta^2 + \theta_E^2} \qquad (7.2)$$

where a_0 is the Bohr atomic radius and n is the density of atoms/ cm^3.
θ_E is the "characteristic inelastic scattering angle", defined as

$$\theta_E = \frac{E}{2E_0} \qquad (7.3)$$

where as before E is the energy lost by the electron relative to its initial energy E_0. Using this model our EELS experiment thus actually measure $Im(-1/\varepsilon)$. Once that has been determined the real part $Re(1/\varepsilon)$ can
also be obtained by a mathematical manipulation called the Kramer-Kronig
transform (e.g. see DANIELS, et al. 1970) and hence the complete complex
dielectric constant of the specimen can be found, and from that all the
optical properties.

An advantage of this technique for determining ε is that it can be
done for any value of θ (i.e. at a finite momentum transfer) whereas an
optical measurement can only be done at θ = 0. However, while the

dielectric constant is an important physical guide to a material its use-fulness is limited to rather specialized applications. For our purposes therefore the second description, the "Microscopic" approach is more valuable. In this the cross-section is related to specific types of in-teractions between the specimen and the incident electron in terms of a mathematical model involving quantities which are of micro-analytical interest. To employ this approach it is useful to divide the spectrum up into three sections, as shown in Fig. 7.2, and to then consider each portion separately to see the physical processes occurring there and thus how a measurement of the EEL spectrum can yield useful analytical data.

7.4 THE MICRO-ANALYTICAL INFORMATION IN THE EEL SPECIMEN

Region 1 - Around Zero-Loss

The "zero-loss" peak has already been referred to as a dominant fea-ture of the spectrum and so it demands some attention even though it is of limited use in micro-analysis. It is firstly clear that the defini-tion of "zero-loss" needs to be qualified because the peak has a finite energy width. There are two reasons for this. First, any spectrometer has a definite energy resolution which restricts its ability to distin-guish between electrons of different energy if this difference is below a specified value. Typically a resolution of from 1 to 5 electron volts is found in practical systems. Second, the beam of incident electrons has an energy spread, both from the inherent energy width of the elec-tron source and from instabilities in the accelerating potential, which can range from 1.5 eV for a thermionic tungsten filament down to 0.3 eV for a cold field emitter. Together these two effects give an energy re-gion several volts wide within which we cannot distinguish electrons which have lost no energy from those which have lost a small amount.

With this proviso the zero-loss peak is then made up from three groups of electrons. First, there are those which are "unscattered", having passed through the sample within being involved in any interac-tion. Since they have not interacted with the sample they cannot carry any significant information about it. However, they do measure the "total scattering power" of the sample since the unscattered signal I_{un} is given by

$$\frac{I_{un}}{I} = \exp\ [-N\sigma_T(\theta)] \qquad (7.4)$$

where $\sigma_T(\theta)$ is the total cross-section for scattering outside the angle θ. The angular distribution of the unscattered beam is of course the same as that of the incident beam (Fig. 7.3(a)), a cone typically 10^{-3} rads. wide. Second, there are the "elastically scattered electrons" which have been deflected by the nuclear charge of an atom. The cross-section for elastic scattering is given by

$$\frac{d\sigma}{d\Omega} = \frac{4}{a_0^2} \cdot \frac{Z^2}{(q^2+1/p^2)^2} \tag{7.5}$$

where $q = 4/\lambda\pi\sin(\theta/2)$, λ is the electron wavelength and $p = a_0/Z^{1/3}$ where Z is the atomic number of the scattering nucleus. Note that this cross-section contains no E dependence since no energy loss is involved. The cross-section for elastic scattering into the acceptance angle of the spectrometer, found by integrating Eqn. 7.5, is of the order of 10^{-18} cm²/atoms for a light element such as Si at 100 kV. A more visual way of expressing this is to define a "mean free path" L, where

$$L = \frac{1}{n\sigma} \tag{7.6}$$

is the average distance the incident electron travels between elastic interactions. Thus for Si, $L \sim 1000$ Å, and so in an average TEM specimen of about 500 Å thickness most electrons will experience one, or more, such events. The angular distribution of these electrons has the form

$$\frac{I_{EL}(\theta)}{I_{EL}(0)} = [\theta^2+\theta_0]^{-2} \tag{7.7}$$

where $\theta^0 = (\lambda/2 \ a \ Z^{-\frac{1}{4}})$. This is a broad distribution, typically 30 x 10^{-3} rad. at half height (see Fig. 7.3(b)), compared to the incident beam. (Note that in the case of a crystalline sample the elastic scattering peaks at those angles satisfying Bragg's condition $\theta \sim \lambda/d$, where d is the lattice spacing, so the peak form is different but its width is similar.) The significance of the elastic scattering is that it affects our ability to efficiently collect the inelastically scattered electrons since, in a thick sample, there is a chance that the electron will undergo both an elastic and an inelastic event. Because the elastic scattering angle is large compared to typical spectrometer acceptance angles, this will reduce the collected signal.

The elastic electrons are not of much use in micro-analysis in general but it can be shown (e.g. ISAACSON, 1978) that if we collect over a large enough angular range that the elastic cross-section $\sigma_{EL} = k_1 \cdot Z^{3/2}$, whereas the total inelastic cross-section $\sigma_{IE} = k_2 \cdot Z^{1/2}$. A ratio of the elastic and inelastic signals is therefore proportional to Z, the mean atomic number of the sampled volume of the specimen. After calibration on a suitable pure specimen this technique can then be used for micro-analysis (EGERTON, PHILIPS and WHELAN, 1975). This result is also used for the imaging and identification of single atoms as described elsewhere in this book by Isaacson.

The third group of electrons present are those which have generated, or been scattered by, "phonon excitation." Phonons are vibrations of the atoms in the sample and, although present in all materials, are of most significance in crystals. The energy loss associated with phonon scattering is small, typically 0.1 eV or less so these losses are indistinguishable from the other components of the zero-loss peak. The angular distribution of the phonon scattered electrons is a broad peak (5 m.rad or so wide) centered around each Bragg direction. Since these electrons carry no useful analytical, or imaging, information and only serve to reduce the transmitted intensity, they will not be considered further.

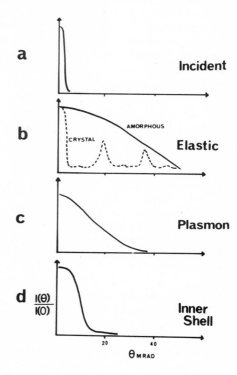

Fig. 7.3. The angular distribution of (a) the incident beam, (b) elastically scattered electrons, (c) plasmon excitations and (d) inner shell losses.

Region 2 - The Low-Loss Region

The portion of the spectrum extending from the edge of the zero-loss peak out to about 50 eV can conveniently be called the "low-loss" region. It can be seen from Fig. 7.2 that the signal intensity in this region is comparable to that in the zero-loss peak. Since about 50% of the electrons passing through a typical sample are either unscattered or elastically scattered, regions 1 and 2 of the spectrum account between them for all but a few percent of the intensity in the spectrum.

The energy losses in this region are due to interactions in which valence or conduction electrons are excited. The best known of these events is the "Plasmon" excitation which, historically, was the first energy-loss process used for micro-analysis (see WILLIAMS and EDINGTON, 1976). Plasmon excitations occur in metals and alloys which have a large number of free electrons, such as aluminum. These electrons behave like a gas, maintaining an equilibrium electron density because of the coulomb repulsion between them. A fast electron incident on the sample disturbs the equilibrium and the gas is set into oscillation. Because many valence electrons are involved this is known as a "collective excitation."

The frequency of the oscillation w_p is proportional to $(n_E)^{1/2}$ where n_E is the number of free electrons/cm^3 involved in the oscillation. To generate this requires an energy $E_{pl} = hw_p$ where h is Planck's constant, so the incident electron suffers an energy loss of E_{pl} if it excites a plasmon. For typical metals $w_p \sim 10^{-16}$ rad/sec, so E_{pl} is about 20 eV. The angular distribution of plasmon losses (see Fig. 7.3(c)) is again broad, the half width being 10 to 20 m.rads. The plasmon mean free path L_p is usually between 500 and 1500 Å at 100 kV, so that in an average specimen a large fraction of the electrons transmitted will have lost energy by creating a plasmon. If the specimen is thick enough the electron may excite another plasmon so that its total energy loss becomes $2E_{pl}$, and this process can be repeated. The characteristic EEL spectrum from a metal is thus that of the zero-loss peak followed by several equally spaced plasmon peaks of decreasing amplitude. Because the plasmon energy loss depends on the free electron density it can be used to identify the material. Unfortunately, all materials showing good plasmon peaks have about the same value of E_{pl} so that the measured value is not unique to any one metal or alloy. However, shifts in the plasmon peak position can be used to monitor the change in concentration of one element in an alloy, once a calibration is obtained, and this technique has been successfully applied to a variety of metallurgical problems (see WILLIAMS and EDINGTON, 1976). Nevertheless, the limited range of materials to which this technique can be applied make it less important than the innershell losses described in the next section.

Before leaving the plasmons it is worth noting that they provide an accurate way of measuring the sample thickness. The probability P(m) of an electron exciting m plasmons, and losing $m \cdot E_{pl}$, is simply

$$P(m) = \frac{1}{m!} \left(\frac{t}{L_p}\right)^m \exp(-t/L_p) \qquad (7.8)$$

where t is the sample thickness. Thus the ratio of the probabilities of exciting zero and one plasmon, P(0) and P(1), is

$$\frac{P(1)}{P(0)} = t/L_p. \qquad (7.9)$$

On the spectrum this is just the ratio of the intensity of the first plasmon peak to the zero-loss peak and hence t can be found when L_p is known (e.g. from references such as JOUFFREY, 1978). This is a convenient way of assessing the sample thickness for, as is discussed later, if $t \geq L_p$ then the sample is probably too thick for microanalysis using the inner-shell losses.

While plasmon losses are important for metals, in most materials the valence electrons are not free enough to participate in such a collective oscillation. The energy loss structures seen in the low loss regions in these cases are mainly due to the excitation, or ionization, of electrons from various bound states. The lowest energy losses (below 15 eV) come from the excitations of electrons in molecular orbitals. Each molecular type has a characteristic spectrum in this range which, although too complex to be interpreted directly, can be used as a "fingerprint" (HAINFIELD and ISAACSON, 1978). Work on this aspect of EELS is only just beginning but it appears promising despite the fact that the molecular groups which give these features are readily destroyed by radiation damage and this will limit the sensitivity of this technique.

Above 15 eV the spectrum shows structures which are mostly due to valence shell excitations. It is noticeable that there is often a peak which resembles the plasmon peaks found in a metal. For example, most biological sections show a broad peak centered around 20 eV loss even though few, if any, free electrons are present. One suggested explanation (ISAACSON, 1972) is that it comes from the excitation of electrons in the $\pi \rightarrow \pi^*$ orbitals. Much work remains to be done on this important part of the spectrum.

Region 3 - Higher Energy Losses

The final region to be considered is that lying above 50 eV. As shown in Fig. 7.2 it has the general form of a smoothly falling "background" on which are superimposed the inner-shell "ionization edges." It is this portion of the EEL spectrum which is most important for micro-analysis as the ionization energy of an edge is a unique property of the element from which it came. In addition to this quantitative chemical analysis, a study of the edges also yields electronic and structural data about the sample.

The "background" contains no microanalytical information but is the dominant feature of the spectrum. It comes from a variety of effects such as the excitations from valence states to vacuum, multiple plasmon losses and the tails of features at lower energy losses. In a pure element the background before the first ionization edge is mostly due to valence excitations and plasmon losses, but after the first edge it is mainly the tail of the preceding edge. In a multi-element system, each successive edge will contribute to the background intensity of the edges that follow at higher energy losses. Because of this complexity it is impossible to calculate the background from first principles, but it has been found experimentally (EGERTON, 1975; MAHER, JOY, EGERTON, MOCHEL, 1979) that the energy differential cross-section for the background has the form

$$\frac{d\sigma}{dE} = A \cdot E^{-r} \qquad (7.10)$$

where E is the energy loss and A and r are parameters depending on the acceptance angle β of the spectrometer. For a fixed value of β the background then falls as E^{-r}, where r is between 3 and 5. The characteristic edges of interest are riding on this rapidly changing background which must thus be stripped from the spectrum before progress can be made. The angular distribution of the background is forward peaked, but fairly broad, so that for small collection angles β the intensity is low but as β is increased the background intensity rises monotonically up to the largest values of β (> 50 m.rad).

The "characteristic edges" contain the analytical information. Each edge represents the energy loss associated with the ionization of an electron from an inner shell of an atom. The energy loss at which the edge starts is the classical ionization energy of the atom and uniquely characteristic of that element. A simple energy loss measurement is therefore sufficient to identify the chemical components of the sample. Figure 7.4 shows the major edges occurring in the loss range

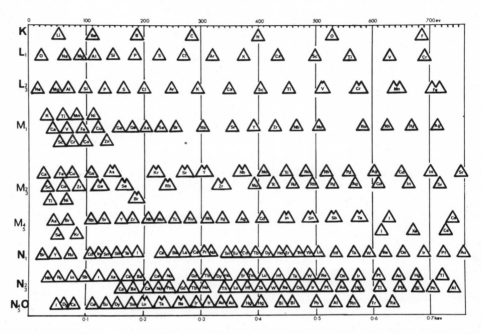

Fig. 7.4. The characteristic energy losses for inner shell ionizations in the range 0-700 eV arranged by shell type K, L, M, etc.

0-700 eV, arranged by the shell (K,L,M, etc.) from which the electron is removed. When used in this way the EELS technique is similar to energy dispersive x-ray spectroscopy (EDXS) and, in fact, the ionizations observed in EELS are the first stage of the events that produce characteristic x-rays. By analogy from Eqn. 7.1 the EELS signal I_E from an ionization edge is

$$I_E = I \cdot N \cdot \sigma_E^\beta \qquad (7.11)$$

where σ_E^β is the appropriate cross-section for the edge and for the angle accepted by the spectrometer. The x-ray signal I_X detected in an EDXS measurement under the same condition would be

$$I_X = I \cdot N \cdot \sigma_E \cdot \omega \cdot F \qquad (7.12)$$

where σ_E is the x-ray ionization cross-section, ω is the fluorescent yield and F is the collection efficiency of the x-ray detector. It is obvious that the EELS measurement should be more efficient because we

are using a primary effect (the ionization) rather than the ensuing decay, and this is correct. Although the EELS cross-section is smaller
than the x-ray cross-section (since the spectrometer only collects some
of the electrons which have caused ionizations) this difference is more
than compensated by the collector efficiency factor in the EDXS case.
Even with the best modern detectors in a TEM/ STEM system F is only the
order of 0.02 at best and this value falls drastically as the energy of
the x-ray quanta decreases below about 1.5 keV. The detection of x-rays
from light elements is therefore very inefficient, and this is further
worsened by the fluorescent yield factor which is close to unity for a
heavy element but only about 10^{-2} for magnesium and falling as about Z^4
for lighter elements. In the EELS case on the contrary the performance
actually improves as Z falls because the ionization cross-sections rise
as the energy loss at which they occur falls. For the study of light
elements EELS therefore has a clear advantage over EDXS. For heavier
elements the EELS advantage is lessened because of poor signal-noise
ratios at high energy losses, and although lower energy edges (L, M,
etc.) could be used these then may now be masked by the strong K-edges
from light elements. The ideal arrangement is to have both techniques
available. Substituting values in Eqn. 7.11 we find that the minimum
detectable mass, ignoring the noise contribution of the background,
should approach 10^{-20} grams for values of I consistent with electron
microscope operation and current published work has demonstrated figures
in this region (ISAACSON, 1978; COSTA et al., 1978).

The most important edges for micro-analytical purposes are the
K- and the L-edges, with the K-edges being especially significant as
they cover the elements Li to F which cannot be detected using EDXS.
Figure 7.5 shows a carbon K-edge, illustrating the usual triangular
shape which becomes more evident when the background is stripped as
shown in Fig.7.5(b). The differential cross-section for an edge has the
form

$$\frac{d^2\sigma}{d\Omega \cdot dE} = \frac{4a_0^2 \cdot f(E,\theta)}{(E_o/R)(E/R)} \cdot \frac{1}{\theta^2 + \theta_E^2} \qquad (7.13)$$

for E greater than E_k the edge energy, where $\theta_E = E/2E_0$ and R is
Rydbergs constant (13.6 eV). The quantity $f(\theta,E)$ is called the "oscillator strength" and represents the number of electrons per atom that take
part in the particular energy loss process. Its value is approximately
constant for small values of θ, but falls to zero for large values. It
can thus be seen from Eqn. 7.13 that the scattering is forward peaked
(Fig. 7.3(d)) with a width varying as θ_E. Putting $E = E_k$ we find that θ_E
is about 2 m.rads at 100 kV. The value of $f(\theta,E)$ falls to zero at about
$\theta = \sqrt{2\theta_E}$ which is typically about 100 m.rads at 100 kV.

The total cross-section for a K-shell ionization is found by integrating Eqn. 7.13 from $E = E_k$ to E_0 and for all values of θ. This total
cross-section σ_k is the same as the x-ray value, and is of the order of
10^{-20} cm² per atom for a light element at 100 kV. This is two orders of
magnitude smaller than the elastic or plasmon cross-sections and ex-
plains the weakness of the edge signal. Looked at another way, the mean

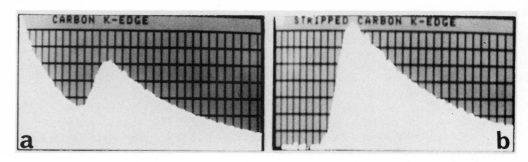

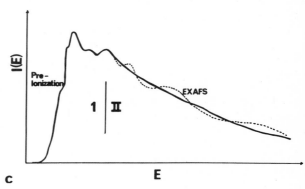

Fig. 7.5. A carbon K-edge stored in MCA memory (a) in the "as-recorded" form (b) after modelling and then stripping the background from beneath the edge. The sketch in (c) shown schematically the pre-ionization and extended fine-structures present on a K-edge.

free path, L_k, is several microns so that in our normal 500 Å specimen only a few electrons interact to produce the ionization.

In practice our spectrometer can only accept electrons lying within a cone angle of β and, because of the necessity of stripping the background from beneath the edge, we can only integrate the signal over a limited energy "window" Δ from E_k to $E_k+\Delta$. We therefore need the appropriate cross-section $\sigma_k(\beta,\Delta)$ for these conditions. The ratio of $\sigma_k(\beta,\Delta)$ to σ_k in effect represents the efficiency with which our system is collecting the ionization loss events from the sample, and it can be shown (ISAACSON and JOHNSON, 1975) that this can be written to a good approximation as

$$\frac{\sigma_k(\beta,\Delta)}{\sigma_k} = \eta_\beta \cdot \eta_\Delta \tag{7.14}$$

so that the effects on the collection efficiency of the acceptance angle and the energy window can be treated independently. From Eqn. 7.13 the "angular efficiency factor" η_β is found to be

$$\eta_\beta = \frac{\ln(1+\beta^2/\theta_E^2)}{\ln(2/\theta_E)} \tag{7.15}$$

where as usual $\theta_E = E/2E_0$. Figure 7.6 shows a plot of η_β for different values of β and for a range of values of the energy loss E, for $E_0 = 100$

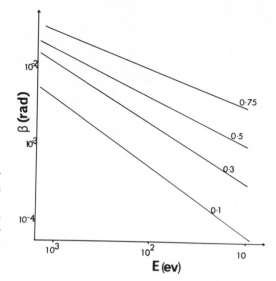

Fig. 7.6. The detection of inelastically scattered electrons as
a function of the collection angle.
Each line on the graph shows the
fractional collection efficiency
achieved with a spectrometer acceptance angle β at some energy loss
E.

keV. For typical values of β, (5 to 10 m.rads), and energy losses in the
range 100 to 1000 eV we see that we can expect to collect between 20 and
50% of the inelastically scattered electrons, a figure which is to be
compared with the 1% efficiency of an EDXS detector. Strictly speaking
some allowance should be made for the fact that, when performing microanalysis, our beam of incident electrons is focussed into a cone of
width α so as to get the maximum current in a small area. This has the
effect of making the efficiency rather smaller (ISAACSON, 1978) but as
long as $\beta > \alpha$ this correction is not of much concern.

The "energy efficiency" factor can be calculated by noting that the
energy dependent part of Eqn. 7.13 can be written, for $E > E_k$, as

$$\frac{d\sigma}{dE} = B \cdot E^{-s} \qquad (7.16)$$

where B and s are parameters dependent on β. The shape of the edge is
thus similar to that of the background (Eqn. 7.10). This gives

$$\eta_\Delta = 1 - \left(1 + \frac{\Delta}{E_k}\right)^{1-s} \qquad (7.17)$$

so that for $\Delta = 0.5\ E_k$ and $s \sim 4$ we can have an "energy efficiency" of
70%.

The K-edge region contains much micro-analytical data in addition
to a simple indication of the elemental composition. By measuring the
intensity of the edge signal and applying Eqn. 7.1, as described later,
this analysis can be made quantitative. Two other pieces of information
are also available as indicated in the sketch Fig. 7.5(c). Just around
the edge energy E_k there is "pre-ionization fine structure" the form of
which is a function of the density of unoccupied bound states into which

the ionized electron can be placed. This structure varies depending on the chemical and crystallographic state in which the atom finds itself, so for example quite marked changes can be seen in the shape of a carbon edge observed in graphite, diamond and silicon carbide (EGERTON and WHELAN, 1974a). A second effect is the "Extended Fine Structure" (often called EXAFS by analogy with an equivalent effect observed in x-ray absorption spectra) which is a weak oscillatory structure extending for many hundreds of electron volts away from the edge. These oscillations occur because the electromagnetic wave produced by the ionization for some energy loss E has a wavelength $\lambda = 12.25/(E-E_k)^{1/2}$Å. At a loss of 100 eV above an edge this is a wavelength of 1.2 Å which is of the same order as an atomic spacing. Consequently, the wave can be diffracted by neighboring atoms and return to interfere with the outgoing wave. An analysis of these EXAFS provides important information about atomic arrangements and this topic is discussed in more detail in Chapter 9 by Maher.

L-edges (and the M, N etc. edges) are more complex than K-edges and of more limited use so they will not be discussed in detail here. However, all the information contained in a K-edge is also available from these other edges, though their shape is often very different, and the analysis is usually more difficult.

7.5 COLLECTING THE ENERGY LOSS SPECTRUM

It is now appropriate to see how an EEL spectrum can be collected from the specimen in our microscope. The spectrum is obtained by passing the transmitted signal into the spectrometer which disperses it into its various energy components. This can be done by passing the electrons through a magnetic, or electrostatic, field or a combination of both and successful analyzers based on each of these principles have been constructed. But for the purposes of micro-analysis in an electron microscope the magnetic spectrometer is the most convenient approach, and the rest of this discussion will concentrate on these devices.

Figure 7.7 shows the ray paths through such an analyzer, often called a "magnetic prism" both because of its shape and its similarity to the action of a glass prism on white light. Electrons leaving the specimen are bent through a radius R in the prism and brought to a focus again. If all the electrons were of the same energy then the point on the specimen would be imaged to a point in the focal plane of the spectrometer. But if there is a spread of energies in the transmitted beam then those electrons which have lost energy will be bent through a smaller radius and come to a focus at a point displaced from the zero-loss focus. If the energy difference of the two rays is ΔE, and their eventual separation is Δx, then the quantity $D = \Delta E/\Delta x$ is called the "dispersion" of the spectrometer. For a magnetic prism analyzer

$$D = 2R/E_0 \qquad\qquad (7.18)$$

so that for a typical radius of 20 cm, D is 4 microns/eV at 100 kV.

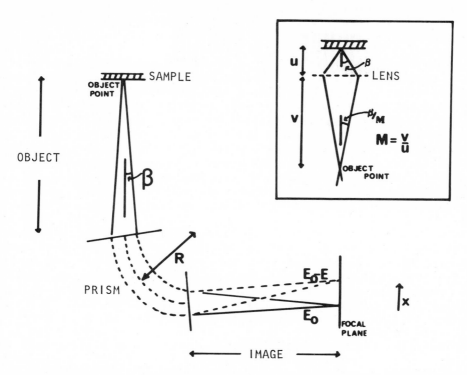

Fig. 7.7. The electron ray paths through the magnetic prism spectrometer showing the object and image planes and the direction of dispersion. The inset shows how the use of a lens to couple the analyzer to the microscope reduces the angular divergence of the beam passing into the spectrometer.

If a slit is placed in the image plane of the prism then electrons of any given loss can be selected, or a spectrum can be formed by scanning the dispersed electrons over the slit. The "resolution" of the spectrum would then be $\delta = S/D$ where S is the width of the slit. In principle this seems to imply that any desired resolution could be achieved by closing down the slit spacing (subject to the limitations set by the energy spread of the incident beam of course). This is not in fact the case because it can be demonstrated (e.g. ISAACSON, 1978) that the relation

$$\Omega \cdot \frac{E}{\delta E} \sim 1 \qquad (7.19)$$

(where δE is the energy resolution, and Ω is the solid angle accepted by the spectrometer, here equal to $\pi\beta^2$) is obeyed. It has already been shown in Fig. 7.6 that we would like β to be large if possible, but choosing a value of say 10 m.rads for and substituting into Eqn. 7.19 gives a resolution value at 100 keV of 30 eV, which may be higher than we wish. The reason for this is that the spectrometer, like any "lens" suffers from aberrations. This aberration is proportional to β^2, so that each point on the object becomes a disc, in the plane of the slit, of size $C\beta^2$ and the resolution is consequently limited to $C\beta^2/D$, where C is the aberration constant.

The limitation set by Eqn. 7.19 can be overcome in several ways. One is to slow the electrons before analyzing them using a "retarding field analyzer", since reducing E_0 will allow any given spectrometer to achieve a better resolution. Another is to try to reduce the aberration constant C of the spectrometer in some way. This can be done by modifying the shape of the magnetic pole faces (e.g. CREWE et al., 1971) and more than an order of magnitude improvement in C is possible. Or we can change the way in which the spectrometer is coupled to the microscope. If a lens is placed between the specimen and the spectrometer object point (see Fig. 7.7) then for a magnification of M, the entrance pupil into the analyzer is β/M for a collection angle at the specimen of β. Thus the constant of unity on the right hand side of Eqn. 7.19 is increased by M^2, and a substantial improvement in resolution can be obtained for any given value of β. However, there is an upper limit to what can be achieved because the lens will also magnify the area W on the specimen from which the electrons are coming to give an object size M.W, which will be imaged through the analyzer to give a resolution limit of M.W/D. There may thus be an optimum magnification for coupling the spectrometer to the microscope. If the dimension W is very small, as in STEM where it could be the size of the scanned probe ~ 100 Å, then M can be made very large before the effect of source size becomes significant. But in TEM, where the dimension W is perhaps a few microns, only a limited magnification can be used before the source size, rather than the angular aberration, becomes dominant. Even in the STEM there will be a limit to the field of view over which the beam can be allowed to scan during analysis unless the beam is "unscanned" to bring it back on to the axis before it enters the spectrometer.

By correctly choosing the coupling optics a resolution of 1 or 2 eV for an acceptance angle of 10 m.rads is possible, even with simple analyzer designs (e.g. EGERTON, 1978a, JOY and MAHER, 1978a). No higher resolution is possible because of the inherent spread of energies in the incident beam, but at this level most fine structure and low-loss structures can be resolved. For the study of EXAFS a resolution of 3-5 eV is adequate, while chemical micro-analysis requires only 10-20 eV. Using a higher resolution than is necessary may lead to a fall in the signal collected and, since the current on to the specimen is limited if high spatial resolution is required, and because many samples of interest are highly sensitive to electron bombardment, it is usually more important to operate the spectrometer so as to achieve the best signal/noise rather than the highest resolution. In the case of the inner shell losses this is helped by choosing the optimum value of β (JOY and MAHER, 1978b). Although both the edge signal and the background rise with β, the loss signal saturates for large while the background continues to increase. Experimentally a value of $\sim E_k/E_0$ is found to give the best results.

7.6 RECORDING AND ANALYZING THE DATA

In order to make full use of the information available from the EEL spectrum it is necessary to record it in a form suitable for subsequent processing. Originally spectra were recorded by placing a photographic plate in the position now occupied by the slit. While this procedure had the advantage that the whole spectrum was recorded in parallel, the numerical data could only be obtained by performing micro-densitometry on the plate. In all current systems the signal is detected by a scintillator/photomultiplier, or a solid-state diode, placed after the defining slit. The signal can be collected and handled either in an analog form, or digitally including counting the electron signal. Properly adjusted either type of detector and either method of signal handling will give comparable results, the advantages of the digital approach for very low signals being balanced by the ease with which the analog method can cope with the zero-loss region. Although the spectrum can be recorded for analysis by displaying it on a chart recorder, the best technique is to store the spectrum in a multi-channel analyzer (MCA) such as is used for EDXS work (MAHER et al., 1978). Figure 7.8 shows one way in which this can be done. The analog output voltage from the photomultiplier detector is passed into a voltage to frequency converter which produces an output train of pulses whose instantaneous repetition rate is proportional to the signal level. Each channel of the MCA is accessed in turn for a "dwell-time" period of between 10 and 200 milli-seconds, during which all the pulses produced are stored in that channel. A scan ramp both sweeps the spectrum over the slit and sequentially steps through the MCA addresses so that the complete spectrum is built up with a known energy increment per channel (typically 1 eV/ channel). Because the spectrum is scanned over the slit, the time spent

Fig. 7.8. A block diagram of a data collection system for ELS based on a multichannel analyzer of the type used for EDS operation. The computer is used both for manipulation of the spectra and for permanent storage of data on floppy discs.

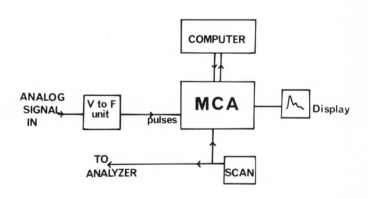

accumulating data at any particular energy loss is small compared with the total recording time and so signal/noise ratios are worse than they would be if a parallel recording scheme could be devised. Although a variety of electronic parallel recording systems have been tried none has yet been proved to be acceptable.

The MCA is well suited to storing the EEL spectrum because it can accommodate a very wide dynamic range, and because once the spectrum has been collected this way it can then be processed either directly or with a computer. The most important data processing that is required is that for quantitative micro-analysis. The formula for finding N, the number of atoms contributing to the observed intensity in an inner-shell edge, follows directly from Eqn. 7.1, thus

$$N = \frac{I_p(\beta,\Delta)}{I \cdot \sigma(\beta,\Delta)} \qquad (7.20)$$

where $I_p(\beta,\Delta)$ is the intensity under the edge and above the background measured over the energy window Δ and in the angular range β, and $\sigma(\beta,\Delta)$ is the appropriate partial cross-section. In order to account for such effects as plasmon losses, back-scattering etc. it can be shown (EGERTON and WHELAN, 1974b) that I should be replaced by $I_\ell(\beta,\Delta)$ which is the signal received through the spectrometer under the zero-loss peak again for the collection angle β and the window Δ (see Fig. 7.9) so that

$$N = \frac{I_p(\beta,\Delta)}{I_\ell(\beta,\Delta) \cdot \sigma(\beta,\Delta)} \qquad (7.21)$$

I_ℓ can be obtained from the MCA memory directly by integrating over the required number of channels, but to obtain I_p it is first necessary to strip the background from beneath the edge. This can be done by using a computer to fit the expected background for $A \cdot E^{-r}$ (see Eqn. 7.10) to a

Fig. 7.9. A sketch of the low-loss and inner-shell excitation regions of a spectrum showing the intensities required for quantitation. In both cases the intensities are measured over the same energy window Δ using the same spectrometer acceptance angle β.

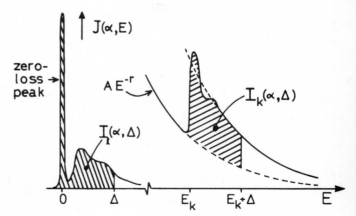

region of the spectrum prior to the edge. When the best fit has been achieved the model background can then be extrapolated beneath the edge for the required window Δ, and subtracted from the spectrum to give the real value of I_p. Chapter 9 by Maher looks in detail at these processes for quantitation and their accuracy.

7.7 THE EFFECTS OF SPECIMEN THICKNESS

The assumption implicit in all of the discussion in this chapter has been that each electron is only scattered once by the specimen before it leaves the sample. When the sample is thin this is a reasonable assumption, but as the sample becomes thicker an electron can be "multiply scattered." That is to say that having produced a K-shell ionization it can subsequently undergo an elastic scattering, which will deflect it and possibly cause it not to be collected by the spectrometer, or another inelastic collision which will leave it with an additional energy loss and deflection. On the other hand, increasing the sample thickness increases the number of atoms which contribute to the loss signal and hence the signal goes up. When both of these effects are considered we find that as the sample thickness is increased the characteristic edge signal rises to a maximum and then falls away, while the background rises steadily. Beyond the thickness at which the edge signal is a maximum, the shape of the edge also changes as the multiple scattering causes additional energy losses to the electrons leaving the sample (see Fig. 7.10). This corresponds to a specimen thickness of about one mean free path for inelastic scattering, which is about 1000 Å for carbon, or 800 Å for iron, at 100 kV. Since the plasmon mean free path is close to the inelastic mean free path this critical specimen thickness can be monitored by observing the plasmon peaks, as discussed above.

The best peak to background, and hence the optimum micro-analytical sensitivity, will be obtained for samples below this thickness as the background falls more rapidly than the peak signal. An ideal biological sample thickness would thus be 300-500 Å, which is thinner than conventional "thin" sections. The problems raised by this requirement are discussed in more detail in Chapters 8 and 9 by Johnson and Maher. It is this sensitivity of the shape of the spectrum to the sample thickness

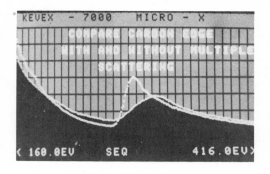

Fig. 7.10. Two spectra recorded from a carbon film illustrating the effect of multiple scattering on the carbon K-edge. Note that both the shape of the edge and its peak-to-background ratio are affected.

that causes most of the problems in EELS interpretation, and ultimately sets a limit to what can be achieved. In principle the multiple-scattering spectrum from a thick sample can be "deconvoluted" to give the simple single-scattering profile, but the mathematical manipulation is time consuming and cannot yet be incorporated as a routine part of a micro-analysis procedure. The onus is thus on the operator to provide samples thin enough to give interpretable spectra.

7.8 TO SUMMARIZE

In this chapter the basic concepts of EELS have been discussed. The following chapters now develop these ideas in their application to the biological and materials sciences. In addition to demonstrating the abilities of the technique some of the current problems will also be considered, in particular the topics of sample thickness, contamination and radiation damage. Even though this technique is still in the early days of its development it represents the most sensitive analytical procedure that is compatible with the requirements of a high resolution electron microscope and it can already be regarded as a routine tool for many purposes.

REFERENCES

In addition to papers referred to in the text, this list of references also contains a few "classic" papers which deserve to be read by anyone interested in this topic.

Costa, J.L., Joy, D.C., Maher, D.M., Kirk, K. and Hui, S. (1978), "A Fluorinated Molecule as an Ultramicroscopic Tracer," Science 200, 537-9.
Describes an application of EELS to biology.

Crewe, A.V., Isaacson, M., and Johnson, D.E. (1971), "A High Resolution Electron Spectrometer for Use in Transmission Scanning Electron Microscopy," Rev. Sci. Instrum. 42, 411-20.
Valuable paper describing techniques for improving spectrometers.

Daniels, J., Festenberg, C.V., Raether, H. and Zeppenfeld, D. (1970), "Optical Constants of Solids by Electron Spectroscopy," Springer Tracts in Modern Physics (Berlin: Springer-Verlag) 54, 77-135.
Use of Kramer-Kronig transform on EEL spectra to derive dielectric constant of specimen.

Egerton, R.F. (1975), "Inelastic Scattering of 80 keV Electrons in Amorphous Carbon," Phil. Mag. 31, 199-215.
The first paper to put the study of inner shell losses on a firm foundation.

Egerton, R.F. (1978a), "A Simple Electron Spectrometer for Energy Analysis in the Transmission Microscopy," Ultramicroscopy 3, 39-47.
Build it yourself analyzer.

Egerton, R.F. (1978b), "Quantitative Energy-Loss Spectroscopy," Proc. 11th Annual SEM Symposium (Chicago: SEM Inc.) 1, 13-23.
Good review of the techniques of quantitation now available.

Egerton, R.F., Philips, J. and Whelan, M.J. (1975), "Applications of Energy Analysis in a TEM," Developments in Electron Microscopy and Analysis (ed. J. Venables) (London: Academic Press), 137-41.
Demonstrates several interesting uses for EELS.

Egerton, R.F. and Whelan, M.J. (1974a), "Electron Energy Loss Spectra of Diamond, Graphite and Amorphous Carbon," J. Electron. Spectr. Rel. Phen. 3, 232-6.
Studies of pre-ionization structures. See also Egerton, R.F. and Whelan, M.J. (1974), "The Electron Energy Loss Spectrum and Band Structure of Diamond," Phil. Mag. 30, 739.

Egerton, R.F. and Whelan, M.J. (1974b), "High Resolution Micro-Analysis of Light Elements by Electron Energy Loss Spectrometry," Proc. 8th Int. Cong. on Electron Microscopy (Canberra: Australian Acad. Sciences) 1, 384-5.
Discussion of quantitation theory and applications.

Ferrier, R.P. and Hills, R.P.T. (1974), "Selected Area Electron Spectroscopy," Advances in Analysis of Microstructural Features by Electron Beam Techniques (London: Metals Society), 41-66.
Excellent general discussion of EELS and useful estimates of sensitivity for various elements.

Hainfeld, J. and Isaacson, M. (1978), "The Use of Electron Energy Loss Spectroscopy for Studying Membrane Architecture - A Preliminary Report," Ultramiscroscopy 3, 87-95.
Studies using low-loss structures.

Hiller, J. and Baker, R.F. (1944), "Microanalysis by Means of Electrons," J. Appl. Phys. 15, 663-76.
The classic, original, paper on EELS. A fine example of creative research and clear writing. The spectra are so good that it is not obvious what real progress has been made in the 35 years since then.

Isaacson, M. (1972), "Interactions of 25 keV Electrons With the Nucleic Bases Adenine, Thymine and Uracil: II. Inner Shell Excitations and Inelastic Cross Sections," J. Chem. Physics 56, 1813-18.
Demonstrates pre-ionization structures in biological materials.

Isaacson, M. (1978), "All You Wanted to Know About ELS and Were Afraid to Ask," Proc. 11th Ann. SEM Symposium (Chicago: SEM Inc.) 1, 763-76.
Excellent and concise survey of EELS.

Isaacson, M. and Johnson, D. (1975), "The Microanalysis of Light Elements Using Transmitted Energy Loss Electrons." Ultramicroscopy 1, 33-52.
Basic reference for much work on inner-shell losses.

Jouffrey, B. (1978), "Electron Energy Losses with Special Reference to HVEM," Short Wavelength Microscopy (New York: N.Y. Acad. Sci.) 29-46.
Good discussion of electron physics.

Joy, D.C., and Maher, D.M. (1978a), "A Practical Electron Spectrometer for Chemical Analysis," J. Microscopy 114, 117-30.
More do-it-yourself spectrometry.

Joy, D.C. and Maher, D.M. (1978b), "The Choice of Operating Parameters for Micro-Analysis by Electron Energy Loss Spectroscopy," Ultramicroscopy 3, 69-74.
Optimization of spectrometer conditions.

Maher, D.M., Joy, D.C., Egerton, R.F. and Mochel, P. (1979), "The Functional Form of the Energy Differential Cross-Sections for Carbon," J. Appl. Phys. (to be published) and also see Proc. Specialist Workshop on Analytical Electron Microscopy, Ithaca, N.Y. (Cornell University), Materials Research Laboratory Report 3082 (1978), 236-41.
Analytical forms for cross-sections of K-edges and backgrounds.

Maher, D.M., Mochel, P. and Joy, D.C. (1978), "A Data Collection and Reduction System for Energy-Loss Spectroscopy," Proc. 13th Ann. Conf. Microbeam Analysis Society (Ann Arbor: Kyser), 53A-G.
Use of an MCA to store EEL spectra.

Williams, D.B. and Edington, J.W. (1976), "High Resolution Microanalysis in Materials Science Using Electron Energy Loss Measurements," J. of Microscopy 108, 113-45.
Good general review with emphasis on plasmon losses.

CHAPTER 8

ENERGY LOSS SPECTROMETRY FOR BIOLOGICAL RESEARCH

DALE E. JOHNSON

CENTER FOR BIOENGINEERING RF-52
UNIVERSITY OF WASHINGTON
SEATTLE, WASHINGTON 98195

8.1 INTRODUCTION

Electron Energy Loss Spectrometry as a technique for the study of biological specimens is still very much in the early stages of development. Because of its high information gathering efficiency and high energy resolution (see Ch. 7, Joy) the potential for the use of ELS in the microanalytical study of biological thin specimens seems clear. However, it is also appreciated that certain problems (e.g. background intensity, radiation damage) are fundamental limitations to the full exploitation of this ELS potential. The current experimental situation is one in which both a variety of applications are being evaluated and, at the same time, the fundamental characteristics of the technique are being investigated, hopefully leading to optimum experimental parameters for application of the technique.

It is our purpose in this chapter to review for biological specimens: (1) the kinds of information available, (2) some possible applications, and (3) the limitations to the effective use of ELS in the study of thin specimens. Because of the nature of the technique, we consider only thin specimens, and we will also only consider the use of ELS combined with electron optical instrumentation capable of producing at least moderate spatial resolution. For the basic principles of ELS, the reader is referred to Ch. 7, Joy.

8.2 CHARACTERISTICS OF A TYPICAL SPECTRUM

The general features of energy loss spectra characteristic of biological specimens are basically the same as those described in detail in Ch. 7, (Joy). Several points, however, should be made concerning these biological spectra.

Consider first the low-loss region (0-50 eV) due to excitation of valence shell electrons. Below about 10 eV the energy loss events reflect transitions between molecular energy levels and will show structure very similar to UV and vacuum UV spectra for the same molecules. Between about 10 eV and 50 eV is almost invariably found a broad peak centered about 25 eV with some structure possibly present on the low energy side of the peak. This energy loss peak is due to excitation and ionization of lower lying valence shell electrons modified to some extent by collective effects (see Ch. 7, Joy). These loss events are clearly not typical plasmon events and the degree of collective effect is far from clear [JOHNSON (1972)]. This peak is broad, large, and generally non-characteristic.

In the higher energy loss region (E> 50 eV) absorption edges found on the smoothly decreasing background are generally due to the excitation and ionization of K shell electrons of low atomic number elements. The edges themselves reflect the elemental composition of the specimen while fine structure near the edge may reflect chemical bonding states of the atoms and structure away from the edge may reflect the spatial arrangement of atoms surrounding the excited atom (EXAFS-see Ch. 7, Joy and Ch. 9, Maher).

To illustrate these features, a typical spectrum is shown in Fig. 8.1. The fractions of total cross section (σ/σ_T) and total energy loss (E/E_T) indicated in each region are only approximate due to the difficulty in extrapolating measured spectra to higher energies. They do serve, however, to indicate the predominance of low energy loss events but with significant energy deposition found in the higher energy loss region. This particular aspect has relevance in the understanding of radiation damage mechanisms [ISAACSON, JOHNSON and CREWE (1973)]. (Ch. 16, Glaeser).

Fig. 8.1. A typical energy loss spectrum of a biological material - in this case, a thin ($\sim$ 400 Å) sublimed film of cytosine. Indicated are the fractions of the total cross section (σ/σ_T) and total energy loss E/E_T found in the regions of valence shell and inner shell excitations. The spectra were taken at 25 kV using a scanning microscope with a field emission source [see JOHNSON (1972) for complete experimental details].

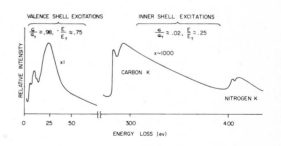

One instrumental point should be made regarding the energy resolution required in the study of various parts of the energy loss spectrum. For simply the detection of inner shell ionization edges, 10 to 20 eV resolution may be sufficient. However, to adequately resolve fine structure in the low lying (0-10 eV) region and at the onset of inner shell excitations, ~0.5 eV resolution is probably necessary and would require the use of a field emission electron source.

In summary, information is available in the energy loss spectrum regarding elemental, chemical, and molecular composition of biological specimens, and this information can be obtained from spatially localized areas in either conventional or scanning microscopes.

8.3 SENSITIVITY OF ELS TECHNIQUES

Elemental Microanalysis

Certainly the most detailed investigations of the sensitivity of energy loss techniques has been in the area of low atomic number elemental analysis [ISAACSON and JOHNSON, (1975)] The absolute accuracy of theoretical calculations has been somewhat limited by an inexact knowledge of background intensities (both in magnitude and angular distribution) but a clear picture still emerges of a technique with a significant improvement in sensitivity over x-ray techniques for low Z analysis. The particular atomic number element below which, ELS is more sensitive than EDS, is of course, instrument dependent and is still a subject of experimental investigation. It probably lies, however, between Na and Ca for K excitation.

As an illustration of an experimental comparison of ELS and EDS sensitivity in a typical instrument, we have plotted experimental spectra for Al K excitations in Fig. 8.2. Al was used because of ease of specimen preparation as a thin film, and because it lies within the atomic number range of interest. All the experimental parameters are given in the figure caption and the relative heights of the two peaks will change somewhat as a function of these parameters. From this figure, two important points should be made. First, the background beneath energy loss peaks is generally higher than the corresponding EDS background simply because electron scattering events producing this ELS background are more probable than Bremsstrahlung events of the same energy loss. Second, both spectra were taken for the same accumulation time but, while each channel of the EDS spectrum was counting for this time, each channel of the ELS spectrum was counting for only 1/100th of this time due to the serial nature of the ELS data collection system. Thus, if a parallel data collection system had been used, the ELS spectrum would have had 100 times more counts/channel than shown. This example indicates the crucial role that the development of a parallel data collection system will play in the application of ELS to elemental microanalysis.

Chemical and Molecular Microanalysis

As a tool for the study of both chemical bonding states and molecular energy levels, ELS is unique in that it is a volume (as opposed to surface) technique and the volume studies can be localized with high spatial resolution. No other potentially competing technique (e.g. UV absorption, ESCA, Auger) offers this combination and thus cannot be compared directly in sensitivity.

Fig. 8.2. ELS and EDS spectra
gathered simultaneously from a
thin ($\sim$ 500 Å) Al film evaporated
onto a formvar substrate. Incident energy = 100 kV, accumulation
time = 100 s, channels full scale
= 100, beam current = 4 x 10^{-9}Å.
The energy loss spectrometer was a
straight edge magnetic sector
operated with $\sim$5MR acceptance
angle. ELS data collection-serial
with energy electing slit, scin-
tillator, photomultiplier and
counting circuitry. The x-ray
detector was a 30 mm^2 Kevex detector at 2 cm detector-specimen
distance.

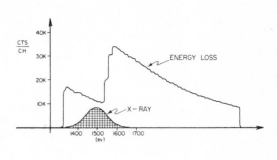

NOTE: Because of the serial ELS data collection, the accumulation
time/channel in the ELS spectrum was only 1 sec/channel.

Extended Fine Structure (EXAFS) Studies

A comprehensive and detailed comparison of the sensitivity of ELS
and photon beam systems for microanalytical purposes has been given by
ISAACSON and UTLAUT, (1978). Their study takes into account a number of
factors, including radiation damage. For the particular case of EXAFS
studies from spatially well resolved areas, their conclusion is that ELS
should have significantly increased sensitivity over the alternative
synchrotron radiation technique.

8.4 APPROACHES TO THE QUANTITATIVE USE OF ELS

Elemental Microanalysis

The use of ELS to determine quantitatively, the relative or absolute amount of an element present in a sample is rapidly increasing.
The principal technique employed to date is described in Ch. 9 (Maher),
and involves the use of calculated or measured cross sections and collection efficiencies, and also the use of standards along with the measured
energy loss spectra.

The use of standards to eliminate the need for explicit knowledge
of cross sections and collection efficiencies is a very sucessful technique in thin film EDS work [SHUMAN, SOMLYO, and SOMLYO, (1976)] and for
ELS application would require the measurements of ELS peak intensities
from standards of known and similar content to the specimen studied. At
the same time, some quantity proportional to total mass thickness of the
specimen is measured for both specimen and standard. This could be, for
example: low lying energy loss peaks, specimen scattering, or even EDS
Bremsstrahlung measurements. These measurements would then be used to

calculate a ratio of the concentration of an element in the unknown to the concentration in the standard. In this method, of course, the energy loss peak integration window would have to be large enough to integrate out any variation in peak shape due to differences in: (1) chemical bonding state of the element, or (2) multiple scattering, between the specimen and the standard.

Chemical Microanalysis

The point has previously been made that ELS, with its high energy resolution, can provide information on chemical bonding states. In principle, this information can be used to calculate not only the amount of an element present in a specimen but also the amounts in various chemical bonding states. For example, near edge (Carbon K) fine structure, such at that shown in Fig. 8.3, could be correlated with number of C atoms in each bonding state. This very exciting prospect is unfortunately limited severely by radiation damage (see Sect. 8.6 and Ch. 16, Glaeser) but still needs to be mentioned. One clear application in biological microanalysis would be in the determination of bound to free ion ratios (e.g. for Na, K, Ca).

Molecular Microanalysis

Analogous to the above discussion, it is possible to determine the amounts of particular molecules in a specimen from characteristic energy loss spectra. This is also subject to a severe radiation damage limitation and for the situation of several molecular species present, would require careful application of deconvolution techniques on, for example, rather complicated low lying (0-10 eV) spectra to separate particular molecular components.

Dielectric Constant Determination

It is also possible to characterize in a quantitative way the response of a biological specimen to the passage of a fast electron by use of the dielectric constant formalism [DANIELS, et al., (1970)] The experimental energy loss spectra is used to calculate the energy loss function (Im $1/\varepsilon$) (see Ch. 7, Joy) and this in turn is used to calculate the optical constants (ε_1 and ε_2) of the material. This can be particularly useful in attempts to understand the nature of particular energy loss events [JOHNSON, (1972)].

8.5 EXAMPLES OF TYPICAL EXPERIMENTAL RESULTS

Experimental Spectra

As an example of the kinds of experimental results obtained from biological materials with ELS, we have grouped together in Fig. 8.3 data from the energy loss spectrum of a thin film of the nucleic acid base,

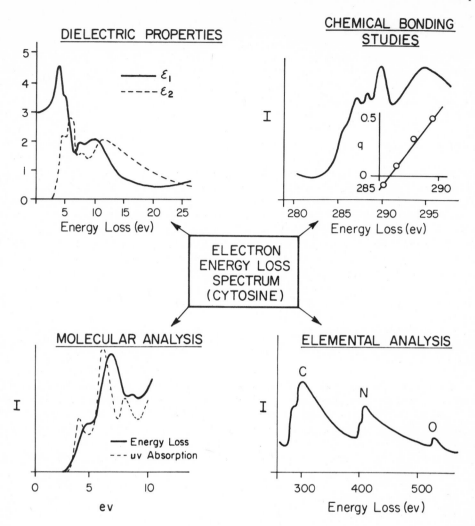

Fig. 8.3. A selection of ELS information experimentally obtained from a thin (~400Å) sublimed film of cytosine. Shown are: the low lying spectrum compared with UV absorption, the dielectric constants calculated from the energy loss function, inner shell ionization edges, and near edge fine structure at the carbon K ionization edge. The spectra were taken at ~25 kV in a scanning microscope with a field emission source and a magnetic sector analyzer. [For complete experimental details, see Johnson (1972)].

cytosine. Shown are: (1) the low lying spectrum (0-10 eV) compared with UV absorption data, (2) the optical constants calculated from the energy loss function, (3) the K shell energy loss peaks reflecting the elemental composition, and (4) fine structure at the carbon K edge reflecting varying amounts of charge on the individual C atoms within this aromatic hydrocarbon. The insert plots calculated charge on each atom vs. energy of the individual peaks, [JOHNSON, (1972)].

Low Z Elemental Mapping

Mapping, as a special case of elemental microanalysis (namely the determination of a two-dimensional distribution of elemental composition) is a technique where the increased sensitivity of ELS is particularly useful. Basically, the technique involves using an energy loss spectrometer system to select electrons from energy loss regions characteristic of a given element, and then to use this characteristic signal, to form the contrast in either a conventional transmission or a scanning transmission image. If the effects of mass thickness variation within the specimen are corrected for (see Sec. 8.6), the images then "map" the distribution of that element since the image intensity is directly correlated with elemental distribution. An example of the approach is the work by COSTA et al., (1978) in the mapping of fluorine labeled serotonin in air-dried platelets, with the conclusion that the serotonin was localized in the dense bodies of the platelets. The reader is referred to the original work for micrographs and discussion.

Molecular Species Mapping

Due to radiation damage and spectral complexity, the ELS mapping of molecular species using low lying energy loss events (0-10 eV) appears to be a difficult task. Preliminary results have been reported by HAINFIELD et al., (1978) in an attempt to use ELS in localizing various components of biological membranes. Their results demonstrate clearly the problems involved and also show the important role that mass thickness effects can play in ELS mapping studies (see Sect. 8.6).

Extended Fine Structure (EXAFS)

In Fig. 8.4 we have plotted the low lying and K shell excitation spectra for a thin film of graphite, illustrating extended fine structure which can be used in determining interatomic distances. These spectra were obtained in an instrument with only moderate energy resolution on ($\sim$3-4 eV) but with good spatial resolution ($\sim$100 Å). A detailed analysis of such spectra resulting in the calculation of C-C spacings in graphite has been reported by KINCAID, et al., (1978). Their spectra were obtained in a high energy resolution spectrometer system not designed for spatial resolution.

In an analogous fashion to the analysis of extended x-ray absorption fine structure, [COSTA, (1978)] extended fine structure past an energy loss ionization edge can be used to provide information on distance, number and type of atoms surrounding the excited atom. Basically, the analysis involves a fourier transform of the momentum spectrum of the ejected electron with peaks in the fourier transform corresponding to various interatomic distances. The analysis techniques are well developed for x-ray absorption studies and directly applicable to energy loss data. Some care must be taken using ELS to either correct for multiple scattering events or to demonstrate that they are negligible for sufficiently thin specimens.

Fig. 8.4. Low lying and carbon K edge spectra from a thin film of graphite. Incident energy = 100 kV, accumulation time = 100s, beam current = 10^{-9}A ($\div$ 3 x 10^3 for low lying spectra). Spectra were taken in a JEOL 100 C with a straight edge magnetic sector analyzer and $\sim$1.5MR acceptance angle. The fine structure shown past the K ionization edge (after correction for multiple scattering) can be used to determine the arrangement of atoms about the excited atom (EXAFS). This type of analysis has been accomplished by Kincaid et al., in a special purpose-large beam spot electron spectrometer system [KINCAID, et al., (1978)].

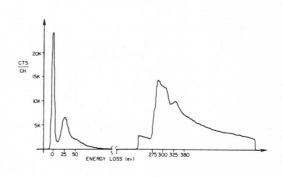

The extent to which this technique can be useful in biological microanalysis is not clear. This is principally because it requires high signal levels ($\sim10^4$-10^5 electrons/data point) to adequately resolve the low contrast structure and at the same time, the structure itself will be changed by radiation damage of the specimen. The ability to extract spatially resolved information about the local environment of particular elements in biological materials is sufficiently attractive, however, that at least a few attempts are being made to obtain EXAFS data using electron microscope techniques on biologically interesting material.

8.6 PRACTICAL LIMITATIONS

Radiation Damage

The general subject of radiation damage in the environment of the electron microscope is covered in detail in Ch. 16 (Glaeser) and we will discuss only two points briefly here.

i Mass Loss

In elemental microanalysis using characteristic energy loss electrons or x-rays, mass loss is the principle deleterious effect of radiation damage.

The problem is certainly no more severe with energy loss techniques than with x-ray techniques and, to the extent that the energy loss technique is more sensitive, the amount of mass loss may be reduced with

ELS. However, it is also clear that mass loss is very much element dependent and in low Z elemental microanalysis, for which ELS is particularly suited, some of the elements (e.g. F and O) may also be particularly prone to this type of damage.

Overall, mass loss is a problem for all types of electron beam induced elemental microanalysis and for certain elements, both total counts and count rates should be observed to detect and correct for mass loss. In addition, steps may be taken, for example, cooling and coating of the specimen, in an effort to reduce the rate and extent of this mass loss.

ii Bond Scission

For certain types of information made available by the high energy resolution of ELS, (e.g. energy loss spectra related to molecular composition or chemical bonding states) radiation induced bond scission is a severe and fundamental limitation. In essence, the same inelastic events, which with some probability result in characteristic energy loss electrons are also with some probability capable of permanently destroying the electronic structure which makes possible the characteristic spectra.

The important parameters in assessing the limitation imposed by bond scission in any particular application are: (1) The ratio of the probability of a characteristic event to the probability of a damaging event and (2) the initial number of undamaged molecules within the irradiated area.

If we assume a simple model of the radiation damage process, namely that a single damaging event produces a damaged molecule, then the number of undamaged molecules at a dose D is given by:

$$N = N_o \exp(D/D_{1/e}) \tag{8.1}$$

This has been shown to be a reasonable approximation for radiation damage in nucleic acid bases [ISAACSON, JOHNSON, AND CREWE (1973)] A decreasing number of undamaged molecules with increasing dose implies a limit to the number of characteristic energy loss events obtainable from a given irradiated area. The number of characteristic energy loss events ($\equiv$C) obtained for a dose D is, through simple integration, given by:

$$C = (\sigma_C/\sigma_D) \cdot N_o [1-\exp(-D/D_{1/e})] \tag{8.2}$$

with: σ_D = Damage cross section (= $1/D_{1/e}$)

σ_C = Characteristic excitation cross section

and the maximum number of characteristic events obtainable is given by:

$$C_{max} = (\sigma_C/\sigma_D) \cdot N_o \tag{8.3}$$

A limitation on spatial resolution follows directly from the above relation. A minimum number of characteristic events required corresponds to a minimum number of molecules originally in the beam and thus, for a given specimen, to a minimum area irradiated.

As an example, consider the use of the characteristic energy loss spectra of the nucleic acid bases. From ISAACSON, JOHNSON, and CREWE (1973) we have, for an incident energy of 25 keV:

$$\sigma_C \text{ (0-10 eV)} \cong 2 \times 10^{-2} \text{ Å}^2/\text{Molecule}$$
$$\sigma_C \text{ (Carbon K fine structure-estimated)} \cong 2 \times 10^{-4} \text{ Å}^2/\text{Molecule} \,.$$
$$\sigma_D \text{ (Base Molecule in DNA)} \cong 10^{-1} \quad \text{Å}^2/\text{Molecule}$$

So that the above relation would estimate for the measurement of characteristic base spectra in a DNA specimen:

$$C_{max} \cong 2 \times 10^{-1} \quad N_o$$

and for the measurement of carbon K fine structure from base molecules in a DNA specimen:

$$C_{max} \cong 2 \times 10^{-3} N_o$$

Thus for example, if we require (for adequate statistics) that $\overline{C}_{max} >$ 10^4, then we should have initially at least 5×10^4 base molecules within the beam for measurement of characteristic base spectra and at least 5×10^6 base molecules initially within the beam for measurement of carbon K fine structure. If we then assume a 500 Å thick specimen and $\sim$ 1 base molecule/ 200Å^3, we will estimate a minimum beam diameter (necessary to include N_o) of $\sim$150Å for measuring characteristic base spectra and $\sim$150Å for measuring characteristic base spectra and $\sim$1500Å for measuring near edge fine structure. Since both σ_C and σ_D scale with the incident energy, the estimate will be independent of incident energy.

The above analysis is only an approximate one, and ignores, for example, the increasing background due to the damaged molecules.

The ratio of σ_C/σ_D in biological materials can range over several orders of magnitude [ISAACSON, JOHNSON, AND CREWE, (1973)] and should be estimated or measured in planning any such application of ELS.

Specimen Thickness

i Effect on Background and Peak Heights

Since in ELS, we are dealing with a technique which relies on electron transmission, it is obvious that specimen thickness will be an important factor. The exact nature of the thickness effects are still a subject of experimental investigation but two main points are clear. First, the amount of non-characteristic background ($\equiv$B) underlying a characteristic energy loss will increase with thickness. This may not be a fundamental limitation, however, since in many cases (e.g. uniform concentration of an element through the section thickness) the characteristic peak ($\equiv P$) increases more rapidly with thickness than the noise in the background.

Second, there is a thickness for which the probability is a maximum that an electron will undergo <u>only</u> the energy loss scattering event of interest. This is a desirable condition since additional elastic scattering events will remove electrons from the spectrometer acceptance angle and additional inelastic events will remove electrons from the characteristic peak. The probability of such a single characteristic energy loss event ($\equiv P_1$) as a function of thickness ($\equiv T$) can be written as the product of the probability of a single event (T/λ) and the probability of no other events ($= \exp{-T/\lambda_o}$):

$$P_1(T) = \frac{T}{\lambda} \exp{-T/\lambda_o} \tag{8.4}$$

With: λ = mean free path for the characteristic event

λ_o = mean free path for all other scattering events

The assumptions are made that $\lambda << \lambda_o$ and that all elastic events are scattered outside the spectrometer acceptance angle.

From this simple expression, $P_1(T)$ is a maximum for $T \sim \lambda_o$. For amorphous carbon, this would correspond to $T \sim 600$ Å at 100 kV (10). [BRUNGER and MENZ, (1965)]. The above expression is only an approximate guide to the optimum thickness and the maximum in $P_1(T)$ is a fairly broad one.

ii Specimen Mass Thickness Effects in Mapping

To discuss the effects of varying specimen mass thickness on characteristic energy loss mapping, consider the simple situation represented in Fig. 8.5. Assume we have a region of the specimen (0) which has icreased mass density over the background region (B) and that also region 0 contains a mass fraction ($\equiv f = m_z/\rho t$) of element Z. If we wish to produce a map showing the location of element Z by using characteristic energy loss events of element Z, the increased mass density of region 0 will have two main effects. First, more electrons will be effectively removed from the beam by elastic scattering out of the aperture and other inelastic events in region 0, and second, the probability for background events in the region of the characteristic peak is increased in region 0. If we define C_2 to be the ratio of the window 2 energy loss signal for region 0 ($\equiv I_{02}$) to the window 2 signal for region B ($\equiv I_{B2}$), we can write approximately:

$$C_2 = \frac{I_{02}}{I_{B2}} = \frac{\exp(-k\rho t\sigma_T) \cdot (\sigma_{B2} k\rho t + \sigma_{Z2} f k\rho t)}{\exp(-\rho t\sigma_T) \cdot (\sigma_{B2} \rho t)} = \left[1 + f \frac{\sigma_{Z2}}{\sigma_{B2}}\right] \cdot k\exp[-(k-1)\rho t\sigma_T] \tag{8.5}$$

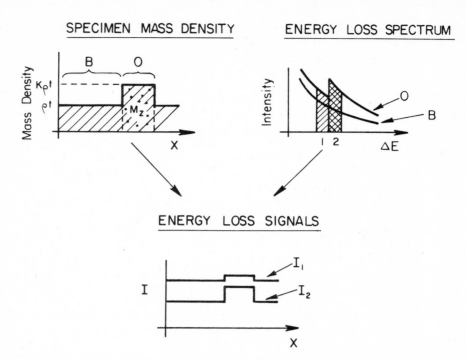

Fig. 8.5. A schematic diagram of the use of ELS to produce a one dimensional map of the distribution of element Z. The specimen consists of a background region ($\equiv$B) with a region of increased mass density ($\equiv$O) containing a mass of element Z (=m). Two energy windows are shown overlaid on spectra from regions O and B and the energy loss signals from these windows are plotted. In the text, it is shown that I_2 cannot be used along to unambiguously locate element Z.

where σ_T = cross section for elastic scattering out of the aperture and other inelastic events.

σ_{B2} = cross section for background events within the aperture and within window 2.

σ_{Z2} = cross section for characteristic energy loss events within the aperture and within window 2.

It is clear from this expression that C_2 is a function not only of the concentration of the element Z but also of the mass thickness ratio (k). A similar expression for window 1 is given by:

$$C_1 = \frac{I_{O1}}{I_{B1}} = \frac{\exp(-k\rho t\sigma_T)\cdot(\sigma_{B1}k\rho t)}{\exp(-\rho t\sigma_T)\cdot(\sigma_{B1}\rho t)} = K\exp[-(k-1)\rho t\sigma_T] \; . \qquad (8.6)$$

Indicating that the object (O) will show contrast even for a window not containing the characteristic peak (window 1), and that this contrast will be a function of the mass thickness ratio (k). However, we look at the ratio:

$$C \equiv \frac{C_2}{C_1} = (1+f\frac{\sigma_{Z2}}{\sigma_{B2}}) \qquad (8.7)$$

We have a quantity which depends only on the mass fraction f and the relative cross sections for characteristic and background energy loss events at the characteristic peak.

From the above simple and approximate analysis, it is clear that contrast in a single map using an energy window at the characteristic peak of element Z is neither necessary nor sufficient to indicate the presence of detectable amounts of element Z. Such contrast is not necessary since mass thickness effects may mask the presence of the element and such contrast is not sufficient, since mass thickness effects may mimic the presence of the element. However, by taking the ratio of the contrast obtained using energy windows containing and not containing the characteristic peaks, the effects of mass thickness variation can be removed. The above approach is certainly not the only way to eliminate these effects [JEANGUILLAUME et al., (1978)] but is given as an example and to indicate the importance of correcting for mass thickness effects in energy loss elemental mapping.

8.7 SUMMARY

In spite of the relatively small amount of experimental ELS results available on biological specimens, the potential of the technique seems sufficiently clear that, as the instrumentation becomes more and more available, ELS should begin to play a significant role in the analytical electron microscopy of biological materials. There are certainly limitations (in some cases severe) to the application of the technique in biological research and these need to be taken into account in the planning and execution of any ELS study. The most likely role for ELS in biological research will be as a valuable complement to EDS, providing in certain cases, data either not available or available only with less sensitivity using EDS.

REFERENCES

Brunger, W., and Menz, W., 1965, Zeit. Fur Phys., 184, 271.

Costa, J.L., Joy, D.C., Maher, D.M., Kirk, K., and Hui, S., 1978, Science, 200, 537.

Daniels, J., Festenberg, C.V., Raether, H., and Zeppenfeld, D., 1970, Springer Tracts in Modern Physics, (Berlin: Springer-Verlag), 54, 77.

Hainfeld, J., and Isaacson, M., 1978, Ultramicroscopy, $\underline{3}$, 87.

Isaacson, M.S., Johnson, D.E., and Crewe, A.V., 1973, Rad. Res., $\underline{55}$, 205.

Isaacson, M.S., and Johnson, D.E., 1975, Ultramicroscopy, $\underline{1}$, 33.

Isaacson, M.S. and Utlaut, M., 1978, Optic, $\underline{50}$, 213.

Jeancuillaume, C., Trebbia, D., and Colliex, C., 1978, Ultramicroscopy, $\underline{3}$, 237.

Johnson, D.E., 1972, Rad. Res., $\underline{49}$, 63.

Kincaid, M.M., Meixner, A.E., and Platzman, P.M., 1978, Phys. Rev. Let., $\underline{40}$, 1296.

Shuman, H., Somlyo, A.V., and Somlyo, A.P., 1976, Ultramicroscopy, $\underline{1}$, 317.

Stern, E.A., Contemp. Phys. $\underline{4}$, 289. (1978).

CLASSIC REFERENCES

See Ch. 7 (Joy) for a complete list.

CHAPTER 9

ELEMENTAL ANALYSIS USING INNER-SHELL EXCITATIONS: A MICROANALYTICAL TECHNIQUE FOR MATERIALS CHARACTERIZATION

DENNIS M. MAHER

BELL LABORATORIES

MURRAY HILL, NEW JERSEY 07974

9.1 INTRODUCTION

Experiments based on transmission electron microscopy play an extremely important role in materials characterization and diagnostics. The high resolution which can be achieved by modern commercial instruments is being used routinely to derive structural and crystallographic information from both the image and diffraction pattern. In materials diagnostics, these two capabilities are enhanced greatly by the ability to obtain <u>direct</u> elemental information at a comparable spatial resolution (i.e. $\lesssim 10$ nm) and thereby place this elemental information in the context of the microstructure and micro-crystallography of the specimen. This combination of techniques, in part, has been the goal of analytical electron microscopy. The principle of microarea analyses is to probe a small volume of a specimen and to detect the many signals which are generated as a result of the interaction between the incident-electron beam and this volume. The desired elemental information is carried either: i) in the secondary emission of X-rays or Auger electrons which occur during the decay of the primary excitation process; or ii) in the transmitted-electron energy-loss spectrum which reflects the primary excitations (i.e. plasmons, valence-shell electrons and inner-shell electrons). Since the preliminary work of WITTRY, FERRIER and COSSLETT (1969) was reported, there has been considerable interest in the use of inner-shell excitations for direct elemental analysis and it is this aspect of electron energy-loss spectroscopy that will be detailed here. Those interested in the analysis techniques and materials applications of plasmon excitations should see a recent review by WILLIAMS and EDINGTON (1976).

In order to place this chapter in perspective, it must be realized that although a large number of fundamental studies which concern the physics of inner-shell losses has been reported, relatively little quantitative work has been done in materials science, per se. On the other hand, numerous papers dealing with elemental identification have appeared recently. Since even elemental identification, in general, may require some form of quantitation or data processing, the subject matter will be approached with this in mind. Moreover, the discussion assumes that a magnetic-prism spectrometer is used and that energy-loss spectra are recorded and processed directly in a multichannel analyzer whose core memory is interfaced to a small computer, as described in the chapter by JOY (1979).

9.2 BASIC CONSIDERATIONS

In this section four basic aspects of electron energy-loss spectroscopy are considered in quite general terms. These are: (1) the spectrum; (2) the spectral dynamic range; (3) spectral background; and (4) edge shapes. To some degree this information has been covered in the chapter by JOY (1979). However, it seems appropriate to elaborate on these aspects of the subject in order to establish a suitable framework for subsequent sections on quantitation, elemental identification and detection limits.

Spectrum

For elemental analysis the energy-loss spectrum is recorded in the forward direction (i.e. $\theta \to 0$, where θ is measured with respect to the incident-beam direction) and within a semi-angle β subtended at the specimen. Therefore, the resulting spectrum is a plot of electron intensity versus the energy lost, i.e. $I(E)$, where E is the energy loss relative to the incident-beam kinetic energy E_0. A typical spectrum exhibits a sharp peak at $E = 0$ (i.e. the zero-loss peak) followed by one or more broader peaks in the range up to $E \sim 50$ eV. These low-loss peaks are due to the interaction of transmitted electrons with the valence or conduction electrons, resulting in plasmon and/or single-electron excitations. At higher energy losses, the excitation of inner-shell electrons to various unoccupied states above the Fermi level is reflected by a rise in intensity (i.e. an "edge") just above $E = E_k$, where E_k is the appropriate binding energy for an atomic shell k (k representing the type of shell: K, L, M etc.). Since each element has unique inner-shell binding energies (e.g. LARKIN, 1977), it is possible to study a particular element (atom) in an environment of many elements (atoms) by electron energy-loss spectroscopy using "characteristic" edges. For immediate reference the K-shell binding energies of elements z = 3 through 15 and L_{23}-shell binding energies for z = 11 through 30 are given in Table 9.1.* As can be seen from the tabulation, these elements are all accessible to analysis using an electron spectrometer which can record losses out to ~ 2000 eV. Moreover, in principle, the remaining elements in the periodic table are similarly accessible using M- and N-shells. For the reader's convenience, the correspondence between the spectral notation of edges, as used in this chapter, and the electronic configuration of atoms is summarized in Table 9.2.

---

* Tables 9.1 - 9.4 are at the end of this chapter.

Dynamic Range

The cross-sections for inner-shell excitations are small compared to those associated with the features in the low-loss region of a spectrum and therefore by comparison edge signals are lower (possibly several orders of magnitude). In addition edges are superimposed on a continuously decreasing background whose mean-signal level at any loss may, in general, be high compared to the net edge intensity. Therefore, it is usually necessary to increase the detector gain at various intervals when recording a spectrum in order to resolve edge structures. This is illustrated clearly in Fig. 9.1 which is an energy-loss spectrum recorded over the modest energy range 0-200 eV from a silicon crystals. The spectrum can be divided into two regions, namely below and above 50 eV. In the region below 50 eV, the zero-loss peak dominates and two plasmon peaks (at 17 and 34 eV) are observed with a low detector (photomultiplier) gain whereas a gain change of $\sim$ 50X is required to observe the L_{23}-edge structure (E > 99 eV). In the case of a high-energy K edge (e.g. the silicon K-edge at 1839 eV), a dynamical range of 10^5 to 1 relative to the low-loss region may be required to observe edge structure above background. One's ability to record and computer process both the low-loss region and edges of interest at an appropriate gain condition will be limited, in general, by either the recording chain (including detector) or computer memory. Since at present small computers (e.g. PDP 11/03) are being used for data processing, the memory capacity of the computer is the limiting factor. Therefore, depending on the situation, it may be necessary to record spectra at various gains, G, and then splice these regions together using calibrated gain factors, GF = $1/\Delta G$. The importance of this in quantitation will become obvious in subsequent sections.

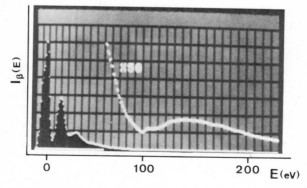

Fig. 9.1. Energy-loss spectra that illustrate the dynamical range required to observe both the low-loss (shaded black) and inner-shell (unshaded) regions from a silicon crystal. The zero-loss peak and plasmaloss peaks (at 17 eV and 34 eV) are observed clearly with a low photomultiplier gain whereas a gain change of $\sim$ 50X is required to observe the L_{23}-edge structure (> 99 eV). These spectra were recorded into a multichannel analyzer operating in the sequential mode with E_0 = 100 keV, probe diameter $\sim$ 10 nm, β = 3 mrad, spectrometer resolution $\sim$ 5 eV and 200 msec dwell time per channel.

Spectral Background

As can be seen clearly from the higher gain spectrum in Fig. 9.1, edges ride on top of a decreasing background and, unlike x-ray peaks, edges are of indefinite extent (i.e. the edge profile and its associated background converge only at $E \rightarrow E_0$). For the L_{23}-edge in pure silicon (i.e. Fig. 9.1) the background is due to the tail of the plasmon excitations and non-characteristic single valence-electron excitations. However, in the general case, the background at any edge threshold will be due to the sum of the "tails" from all lower lying edges together with those contributions from the low-loss region, as described above for silicon. This situation has several important consequences. Firstly, to obtain the true edge shape, as well as net edge intensity, the background must be extrapolated under the edge of interest. This is inherently less accurate than the interpolation procedures used for x-ray analysis. Secondly, it might appear that different mathematical models for the background would have to be used each time a higher energy-loss edge in a spectrum is processed. In this regard nature has been kind to us, since the functional form of an edge for $E \gg E_k$ is the same as for plasmon tails plus non-characteristic single-electron excitations. Therefore, for edges which are reasonably well separated in E, the same functional form is applicable throughout the spectrum and only the fitted parameters change. Lastly, it should be clear that the range of energy losses over which the background can be extrapolated must be limited and this range is referred to as the energy window Δ.

Because of the complexity of this situation, it is not possible to calculate the background at an arbitrary point in the spectrum from first principles and therefore one proceeds empirically. It has been found that, over a moderate energy range, the background prior to the onset of an edge can be represented by an expression of the form $A \cdot E^{-r}$, where as before E is the energy loss, A and r being constants. Since the spectrum is stored electronically in a micro- or mini-computer, the values of A and r can be found from a digital fit to the background over an energy range of typically 50 to 100 eV preceding the edge. To achieve this result, the appropriate signal intensity and energy-loss values are converted to logarithms and a linear least-squares criterion is used to obtain the desired constants. The modeled background is then extrapolated beyond the edge for a specified value of Δ and the true-edge profile obtained by stripping the extrapolated background intensity from the total signal intensity. These procedures are illustrated in Fig. 9.2 for the case of a K-edge recorded from an amorphous carbon film. In practice it has been found that the modeled background and experimental data prior to the edge differ by less than the channel-to-channel noise fluctuations. The values of A and r depend sensitively on the material, its thickness t, β and E . Typical fitted values of r vary from 2.5 to 4.5 and therefore can deviate significantly from the expected theoretical value of 4 (e.g. ISAACSON and JOHNSON, 1975).

After the background has been stripped, the limiting profile of an edge (i.e. for $E \gg E_k$) again can be described by an inverse power law which shall be denoted as $B \cdot E^{-s}$, where B and s are constants. For the

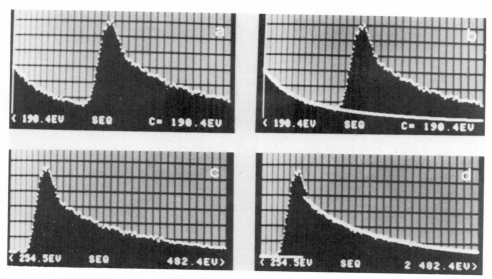

Fig. 9.2. Illustration of the basic steps required to characterize an edge: a) raw spectral data of a K-edge (E = 284 eV) from an amorphous-carbon specimen ~ 20 nm thick; b) exponential background fit and extrapolation (solid curve); c) edge profile after stripping off the extrapolated background; and d) exponential fit to the limiting portion (i.e. $E > E_k+30$) of the stripped edge (solid curve).

case of the K-edge from amorphous carbon, measurements have been made of the exponent s as a function of β (MAHER, JOY, EGERTON and MOCHEL, 1979) and the experimental results compared with calculated values derived from a hydrogenic model of the inner-shell ionization process (EGERTON, 1979). The agreement between the two is within 5% which clearly lends support to the validity of the empirical background extrapolation procedure outlined above. An example of the edge fit for amorphous carbon is shown in Fig. 9.2(d).

Edge Shapes

The probabilities that primary electrons of kinetics energy E_0 will lose a given amount of energy depend on the elemental constituents and thickness of the specimen. These probabilities are expressed in terms of the cross sections for each type of energy-loss event, as discussed by JOY (1979). The expected spectral intensity, $I(E,\theta)$ is then directly proportional to the double-differential cross section (i.e. the energy, solid-angle differential cross section)

$$\frac{d^2\sigma(E,\theta)}{dE\ d\Omega}$$

which is the probability that a primary electron will undergo an energy-loss E and be scattered into an angle θ with a solid-angle dΩ. The

energy-loss spectra considered here are recorded with a circular
aperture centered about the incident-beam axis and subtending a half-
angle β at the specimen. Therefore, the measured intensity I (E) is
directly proportional to an energy-differential cross section given by

$$\frac{d\sigma(E)}{dE} = \int_0^{\theta=\beta} \frac{d^2\sigma}{dEd\Omega} \ 2\pi \sin\theta \ d\theta. \qquad (9.1)$$

where θ = 0 corresponds to the incident-beam direction.

In the microscopic description $d^2\sigma(E,\theta)/ded\Omega$, the probability of an
energy-loss event is proportional to the differential oscillator
strength

$$\frac{df(E,q)}{dE}$$

where the momentum transfer q is related to θ and the incident-electron
momentum k_0 by

$$q^2 = k_0^2(\theta^2 + \theta_E^2) \qquad (9.2)$$

with $\theta = E/2E_0$. The quantity $df(E,q)/dE$ which is referred to as the
generalized oscillator strength is a measure of the number of electrons
per scattering center (i.e. atom) that contribute to any given energy
loss (INOKUTI, 1971; MANSON, 1972). As a consequence of this description one obtains

$$\frac{d^2\sigma(E,\theta)}{dEd\Omega} = \frac{4a_0^2R^2}{E_0} \cdot \frac{1}{\theta^2 + \theta_E^2} \cdot \frac{df(E,q)}{EdE} \qquad (9.3)$$

for E > E , where a_0 is the Bohr radius of the hydrogen atom (=
0.529x10^{-8} cm) and R is Rydberg's constant (= 13.6 eV).

For particular inner-shell excitation the generalized oscillator
strength is the property of a specific atom and it can be calculated
without regard to solid-state effects from a knowledge of the initial-
wave function (i.e. continuum state). Once the energy and momentum dependence of df/dE is known, the edge shape can be calculated, e.g. by
integrating Eqn. 9.1 after making the appropriate substitution from Eqn.
9.3. The result is obviously a single-scattering profile of an edge
without consideration of background effects. The theoretical shape of
the K-edge for carbon has been calculated using a hydrogenic model of
the inner-shell ionization processes (EGERTON, 1979) and the shapes of
K-, L_{23}- and M_{45}- edges have been calculated for boron, magnesium and
molybdenum, respectively, by REZ and LEAPMAN (see LEAPMAN, 1979). The
latter authors used a full quantum mechanical treatment to derive the
appropriate wave functions and their results are shown in Fig. 9.3. The
calculated boron K edge exhibits a characteristic "saw-tooth" shape and
can be compared directly with the stripped carbon K edge shown in Fig.
9.2(c), whereas , the L_{23} edge of magnesium exhibits a characteristic
"sleeping whale" shape which is even more exaggerated for the molybdenum
M_{45} edge. These delayed maxima for L_{23} and M_{45} edges arise because of
the centrifugal barrier that must be overcome by an electron excited
from a non-spherically symmetric shell. This phenomenon is well known

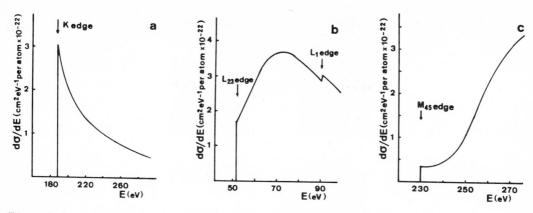

Fig. 9.3. Theoretical edge profiles for the boron K edge (a), the magnesium L_{23} and L_1 edges (b) and the molybdenum M_{45} edge (c). In each
case E_0 = 80 keV and β = 10 mrad (from Leapman (1978).

in atomic physics. The absence of a sharp threshold at E and corresponding delayed maximum make it more difficult to uniquely identify L
and M, as well as N, edges for the matter in a spectrum. For comparison
to theory, raw spectral data and stripped edges are shown in Fig. 9.4
for silicon L_{23}, antimony M_{45} and lead N_{45}. These spectra were recorded
from single-crystal films using a 10 nm dia probe. The general form of
these edges is as predicted from theory. Clearly the detailed shapes of
L, M and N edges will depend on both the spectrometer resolution δ and,
for example, the separation between the L_2 and L_3 thresholds. Therefore, for the case of titanium where this separation is 6 eV (see Table
9.1), two maxima may be resolved at threshold (COLLIEX et al., 1976) and
the edge profile will differ significantly from that shown in Figs.
9.3(b) and 9.4(b). For $\delta \lesssim 2eV$ fine-structure effects (see SILCOX,
1979) just above the threshold energy for an inner-shell excitation will
be resolved clearly and the edge shape in this region will be altered
markedly from the predicted by the atomic model (see SILCOX, 1979). The
importance of edge shapes in elemental identification will be discussed
and illustrated by example at a later stage of this chapter.

9.3 PROGRESS IN QUANTITATION

Over the past several years it has been demonstrated experimentally
that electron energy-losses caused by inner-shell excitations and detected in the transmission mode provide a sensitive method of microanalysis especially for elements in the first two rows of the periodic
table (e.g. ISAACSON and JOHNSON, 1975). Unlike Auger microanalysis,
the energy-loss method can detect elements distributed within the in-
terior of a specimen, provided the latter is sufficiently thin. x-ray
and Auger techniques measure secondary processes which involve both electrons in higher level shells and the inner-shell vacancies created by
the incident electrons, where electron energy-loss spectroscopy measures
the inner-shell excitation directly through its effect on the transmitted electrons. Therefore, relatively simple equations can be used
for quantitative analysis and corrections for absorption or yield of the

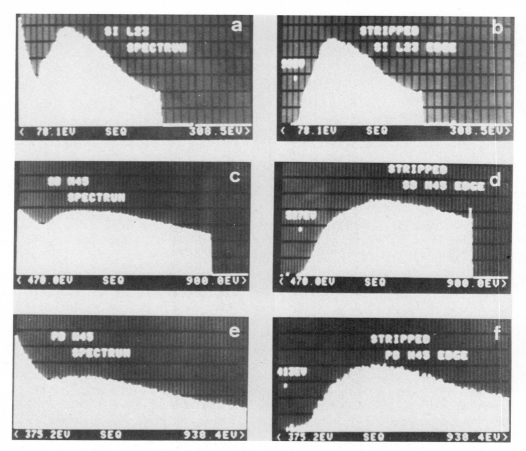

Fig. 9.4. Experimental spectrum and stripped edge profiles from single crystals of silicon, antimony and lead: a-b) silicon L_{23} edge recorded with β = 3 mrad, δ = 5 eV and ~ 200 msec per eV; c-d) antimony M_{45} edge recorded with β = 7.5 mrad, δ = 10 eV and ~ 300 msec per eV; and e-f) lead N_{45} edge recorded with β = 3 mrad, δ = 10 eV and ~ 200 msec per eV. In all cases E_0 = 100 keV and probe diameter measurement requires that one record the low-loss region of the spectrum in addition to the regions containing characteristic edges and that an appropriate <u>partial</u> cross section is known for the inner-shell excitation.

secondary process are not required. Electron energy-loss spectroscopy can be employed as a standardless technique - that is, the absolute amount of a particular element within a given region of a specimen (defined, for example, by the incident-electron beam) is measured.

In a recent paper by JOY, EGERTON and MAHER (1979), three quantitation approaches were discussed, namely methods based on <u>efficiency factors</u>, <u>calculated partial cross sections</u> and the use of <u>standards</u>. Results obtained from these three methods were examined in terms of their stability, relative accuracy and absolute accuracy. Lastly, errors were established for each criterion. These studies were initiated in order to experimentally test many of the assumptions involved in quantitation and to assess the possible influence on quantitation of operating procedures and instrumental factors. It is felt that this work summarizes the present position of quantitation and therefore in this section the methods, results and conclusions are redocumented. More recent results are also included.

Analysis Methods

The quantity N of a particular element contributing to the measured intensity in an edge can be obtained to a good approximation, say $\lesssim 5\%$ (EGERTON, 1978a), from the equation

$$N \doteq \frac{I_k(\Delta,\beta)}{I_1(\Delta,\beta)} \cdot \frac{1}{\sigma_k(\Delta,\beta)} \cdot \qquad (9.4)$$

Here $I_k (\Delta,\beta)$ is the integrated intensity under the edge and above the background measured over an energy-loss window Δ and for a semi-scattering angle β; $I_e(\Delta,\beta)$ is the integrated intensity under the low-loss region of the spectrum measured from the zero-loss peak up to Δ and for the same angle β; and $\sigma (\Delta,\beta)$ is a partial cross section for excitations to continuum states for shell k which cover a range Δ of energy losses and a range of scattering angles from zero to β. The angle β is normally determined by an aperture placed between the specimen and electron spectrometer. If $\sigma (\Delta,\beta)$ is expressed in cm^2 per atom, then Eqn. 9.4 gives N in atoms per cm^2 of surface within the incident-electron beam for STEM or within an apertured selected area for CTEM.

In the data systems presently in use, the values of I_k and I_e are are determined through interactive computer programs which read the spectrum from memory of the multi-channel analyzer and allow regions of interest (i.e. low-loss, background and edge regions) to be defined, for example, by painting them with a cursor controlled by the operator.

The first step in the quantitation of a spectrum is to subtract off the detector dark current. This contribution is recorded both at the beginning and end of each spectral measurement (as depicted schematically in Fig. 9.5). The next step is to obtain I_k in a form which can be related directly to I_e. Since each channel in the memory of the multi-channel analyzer corresponds to a known increment of energy-loss (e.g. 1 eV per channel), the relative position of an edge, i.e. E_k, and the energy window Δ over which an integration is to be carried out both can be defined directly from the edge spectral profile. Therefore, with reference to Fig. 9.5, one specifies E_k, the energy range over which the background is to be modeled (E_{BG}) and the end point for background extrapolation ($E_k + \Delta$). The computer then fits an inverse power law $A \cdot E^{-r}$ to the background, extrapolates this fit to $E_k + \Delta$ and strips the extrapolated intensity from the total intensity. The value of $I_k(\Delta)$ is then obtained by summing the net number of counts in each channel over the appropriate energy range. The edge region may or may not be recorded at the same detector gain as I_e (or for that matter a second edge of interest). Therefore, the appropriate integral for quantitation may be I (Δ) multiplied by a gain factor, GF = $1/\Delta G$, which is read into the program.

Gain factors must be known accurately and in practice they have been determined from measurements of the integrated edge intensity (e.g. the K-edge from an amorphous carbon film) as a function of the detector gain. Finally the low-loss region given in $-\delta/2 \leq E \leq \Delta - \delta/2$ (again see

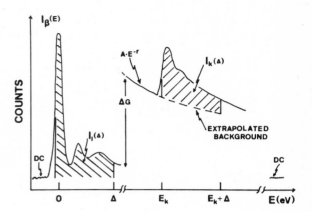

Fig. 9.5. Schematic electron energy-loss spectrum of the signal intensity I (E) versus energy loss E where DC signifies the dark current of the detector and all other quantities are defined in the text.

Fig. 9.5) must be defined and the integral $I_e(\Delta)$ calculated directly from the contents of memory.

The time required to model and extrapolate the background under an edge, as well as derive values of $I_e(\Delta)$ and $I_k(\Delta)$, is about one minute, the limiting factor being the time it takes to mark the regions of interest and input relevant experimental parameters, i.e. E_0, β, dark currents, gain factor and Z. The last step is to calculate $\sigma(\Delta,\beta)$ and three methods are now described.

Method 1: Efficiency Factors

The partial cross-section $\sigma(\Delta,\beta)$ in Eqn. 9.4 must be used because only electrons scattered through angles equal to, or less than, β are collected by the spectrometer and only electrons suffering energy losses between the threshold energy E_k and $E_k+\Delta$ are made use of in the analysis. However, electrons are scattered through all angles up to π, and they can lose any amount of energy between E_k and E_0. The cross-section $\sigma(\Delta \to \inf, \pi)$ which takes account of these effects is the total cross-section σ_T and therefore is the same quantity as the ionization cross-section used in quantitative x-ray analysis. If we restrict our discussion to K-edges then it can be shown (ISAACSON and JOHNSON, 1975) that the partial and total ionization cross-sections are related, to a good approximation, by

$$\sigma(\Delta,\beta) \doteq \sigma_T \cdot \eta_\Delta \doteq \eta_\beta \quad . \tag{9.5}$$

That is the variables Δ and β can be treated separately in their action on σ. If appropriate forms for the functions η_Δ and η_β can be found then $\sigma(\Delta,\beta)$ can be calculated because σ is available from standard x-ray references (for a comprehensive review, see POWELL, 1976). The quantities η_β and η_Δ are usually referred to as "efficiency factors" since η_β is the fraction of electrons collected by the spectrometer as a result of its finite acceptance angle β compared to the total inelastic signal, and η_Δ is the fraction of the signal lying in the energy window Δ compared to the true integral which stretches to E_0. Analytical forms can be found for both efficiency factors if K-edges are used and some reasonable approximations are made. η_β is obtained by noting that the inelastically scattered electrons have an angular distribution $I(\theta)$ about the incident-beam direction ($\theta=0$) of the form

$$I(\theta)/I(0) = (\theta^2 + \theta_E^2)^{-1} \tag{9.6}$$

where θ is the characteristic scattering angle for an energy loss E, and as given previously equals $E/2E_0$. Using this expressing one readily finds that the fraction η_β of the total signal intensity lying within the angle β is

$$\eta_B = \frac{\ln[1+(\beta^2/\vec{\theta}_E^2)]}{\ln(2/\vec{\theta}_E)} \tag{9.7}$$

where $\vec{\theta}$ is equal to $(E_k+\Delta/2)/2E_0$, that is the average scattering angle over the energy window from E_k to $E_k+\Delta$. Measurements show that expression 9.7 does represent the variation of the collected signal with β quite well (EGERTON and JOY, 1977). Since the quantities E_0, E_k, β and Δ are all fed into a computer program, the value of η_β can be calculated directly when required.

The value of η is found by noting that, after stripping away the background, the K-edge intensity I (E) has the form

$$I_K(E) = B\cdot E^{-s} \quad (E > E_K) \tag{9.8}$$

where B and s are constants, as discussed in the above section. The ratio of η_Δ of the signal integrated over the window Δ to that extending over the energy-loss range from E_k to E_0, then is found simply by integrating Eqn. 9.8 and the result is

$$\eta_\Delta = 1 - \frac{E_K}{E_k + \Delta}^{s-1} \tag{9.9}$$

For given values of E_k, Δ and s, η can be calculated. As in the case of the background fit, the portion of the edge to be modeled can be specified, e.g. by <u>painting</u> the region of interest using a cursor. The computer then reads the appropriate intensity values and calculates the best fit value of s and then η_Δ. For that part of the edge profile immediately following its onset at E_k, the shape is very dependent on such variables as the specimen thickness and even the chemical bonding of those atoms which are being probed. As a result the accuracy of the fit in this region is not very good. However, from 20 eV or more away from the edge, the stripped edge is accurately described by $B\cdot E^{-s}$, as can be seen from Fig. 9.2(d). Consistent values of the parameter s are obtained for windows of 80 to 100 eV. The value of $\sigma(\Delta,\beta)$ now can be found using the computed values of η_β and η_Δ and the appropriate total cross-section σ_T. In the program used by JOY et al., 1979, the value of σ is computed using the Bethe cross-section (see MANSON, 1972) but with the constants selected in accord with the recommendations of POWELL (1976). The accuracy of σ_T for light elements at 100 kV is not particularly good and the literature contains other values for σ_T which can differ by up to a factor of 2 in either direction. Until better experimental and theoretical data are available, it is necessary to treat these computed σ_T values with some caution in as far as their absolute value is concerned, although the ratios of the σ_T for different elements is expected to be fairly accurate.

Method 2: Calculated Partial Cross Sections

A second approach of obtaining the required partial cross-section $\sigma(\Delta,\beta)$ is to calculate it directly from first principles. The ionization cross-section can be computed using quantum mechanical methods (MANSON, 1972 and LEAPMAN, REZ and MAYERS, 1978) and this has been shown to give results which are in detailed agreement with the best experimental data, although considerable computing time is required. However, the simplest procedure is to approximate the initial-state and final-state wave functions appropriate for the ionization process by Coulombic wave functions, so that the theoretical situation becomes similar to that of the hydrogen atom, for which the generalized oscillator strength (for a given energy and momentum transfer) is known analytically (BETHE, 1930). For the conditions normally used in transmission electron spectroscopy, this hydrogenic approximation gives results which are in good agreement with experimental data for K-shell losses (EGERTON, 1979). With some correction in the region of the excitation threshold, a similar approximation can be used for L-shells (EGERTON, unpublished). Therefore values of $\sigma(\Delta,\beta)$ can be calculated for K- and L-edges in less than 1 sec by means of a short Fortran program, using the experimental values of Δ and β.

A complete quantitation program incorporating all the required integration, background stripping and computational steps can be put into about 13 kilo-byte of store on a conventional 16 bit word mini-computer. This is about the same storage as is required for the program described in the above section.

The only inputs required for the hydrogenic cross-section programs (called SIGMAK and SIGMAL) are the threshold energy of the edge E_k, the energy window Δ, the accelerating voltage E_0, the scattering angle β and the atomic number Z. The required partial cross-section is then produced directly and so it can be inserted straight into Eqn. 9.4. Because this method does not rely on an edge fitting procedure, smaller energy windows can be used without loss of accuracy. This is of special importance when close-lying edges must be analyzed in a spectrum.

Method 3: Standards

A third method of performing a practical quantitation based on Eqn. 9.4 is to use a separate experiment to find $\sigma(\Delta,\beta)$. This implies the use of a standard, a procedure which clearly introduces a number of additional problems into the quantitation since effects like the specimen thickness and orientation are liable to have an influence on the observed form of the edge. However this method has the compensating advantage that it can be applied to any edge for which a suitable standard can be found. In order to illustrate this approach, L-edges are considered, in particular the silicon L_{23} at 99 eV.

Figure 9.4(b) shows such an edge after the background has been stripped, as described in the above section. The variation of the integral under the edge now can be found as a function of the energy window Δ. If these integrals, in turn, are divided by the corresponding low-loss integrals evaluated over the same energy window, then the result for each Δ is directly proportional to the partial cross section $\sigma(\Delta,\beta)$. As can be seen from Eqn. 9.4, the proportionality constant is N. Therefore, the problem is to find a method of determining an accurate value of N from, say, a single crystal and then back calculate $\sigma(\Delta,\beta)$ from the appropriate spectral values of $I_k(\Delta,\beta)$ and $I_e(\Delta,\beta)$. Figure 9.6 shows the experimental variation of $I_k(\Delta,\beta)$ normalized to $I_k(100,\beta)$ for a silicon

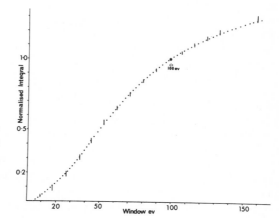

Fig. 9.6. Normalized silicon L-edge intensity, i.e. $I_k(\Delta, 3$ mrad$)/I_k(100$ eV, 3 ,rad$)$, versus the energy window (from Joy et al. 1979).

L-edge with β = 3 mrad and E_0 = 100 keV. This curve represents the function f (Δ) where to a good approximation*

$$f_L(\Delta) = \sigma_L(\Delta,3 \text{ mrad})/\sigma_L(100 \text{ eV}, 3 \text{ mrad}) \qquad (9.10)$$

---

*A more rigorous approach would be to plot $(I_k(\Delta,\beta)/I_k(100,\beta) \cdot I_e(100,\beta)/I_e(\Delta,\beta)$ against Δ to evaluate $f_k(\Delta,\beta)$.

Once an absolute value for σ (100) at 3 mrad is known, the partial cross section for any other energy window can be determined. The function f (Δ) can be read either from a graph or calculated from a polynomial fit to the appropriate curve and this analytical form incorporated into the computer program, so that one then evaluates σ (Δ) = σ (100)·f (Δ) for a given Δ. Certainly f (Δ) can be determined at other values of β that are appropriate for a particular microscope/spectrometer configuration and this variation built into the program.

An absolute value for the partial cross section at β = 3 mrad and Δ = 100 eV can be obtained by simultaneously taking an energy-loss spectrum and a convergent-beam diffraction pattern from the same area of a crystal. This thickness t of the crystal can be derived directly from the convergent-beam diffraction pattern to an accuracy of about 3% (e.g. AMELINCKX, 1964) and then the number N of atoms-cm^{-2} calculated from t and the bulk density ρ of silicon*. Since N is now known, the partial cross section can be found using Eqn. 9.4 and the intensity data from the energy-loss spectrum. (The result for silicon was that σ (100 eV, 3 mrad) = 5.23x10^{-20} cm^2 per atom at E_0 = 100 keV.)

The spirit of this method is in essence the same as that described in the above section, except that the correction (or efficiency) factors are evaluated differently. The method has considerable applicability in materials science, and is very easy to carry out in practice since the necessary correction with Δ can be found graphically or analytically. Although the experimental form of the result may be slightly dependent on the thickness of the crystal, the net effect is small for energy windows greater than ∿ 30 eV. Apparently this is because the major consequence of thickness is to redistribute the intensity only immediately after the edge.

In the absence of an on-line computer capable of performing cross-section calculations of the type described in the above sections, this approach would appear to be the only way to carry out L- and M-edge analyses. Furthermore cross-sections obtained in this way can be compared directly to those derived from model calculations.

Tests of Analysis Methods

Because the techniques of data collection and analysis in electron energy-loss spectroscopy are relatively new, it is important to test the results as fully as possible. This implies not only comparing results obtained from energy-loss measurements with those obtained from other techniques (e.g. energy dispersive x-ray spectroscopy), but also checking the results from energy-loss measurements for self-consistency and even performing "round-robin" type of experiments in which the same sample is examined in various microscopes by different workers in an attempt to look for instrumental effects.

---

*N = ρt/m where m is the mass per atom (i.e. the atomic weight divided by Avogadro's number.

With this in mind three test criteria, in ascending order of severity, were considered by JOY et al. (1979) in order to examine the quality of the quantitation routines described in the above section. These criteria were:

(1) N for an element of interest must remain constant within limits when the experimental parameters of the system (i.e. STEM vs. CTEM, E_0, Δ or β) are varied. This is a stability test of the entire analysis method;

(2) Measurement of the composition of compounds of known stoichiometry (such as BN, MgO, etc.) should agree with the expected results and likewise should be independent of the operating conditions. This is a test of the relative accuracy of a quantitation method when applied to two or more elements; and

(3) The value of N should agree with measurements made on the same material by independent quantitation methods and with similar measurements performed on different instruments. This is a test of the absolute accuracy of the procedures employed and of the effects of instrumental artifacts on the data.

If all three of these criteria are satisfied then the method is an absolute, quantitative mecroanalytical technique. If only the first two conditions are met then although the method will be useful for compositional studies of compounds etc., standards will be necessary in order to obtain absolute numbers.

Stability of Quantitation Methods

A basic requirement of any quantitation method is that the result be independent of system variables. In electron energy-loss spectroscopy and for a given spectrometer resolution, these variables are the mode of operating the microscope (i.e. STEM or CTEM), E_0, β and Δ. In this section the results which are discussed reflect all of these variables, however, the one of interest is Δ. This quantity can be varied over a wide range when data are being analyzed and therefore it is very important that the stability of N as a function of Δ be evaluated when the remaining variables are typical for one's own system.

This aspect of quantitation has been examined in considerable detail using methods 1 and 2 and when applied to K-edges, e.g. from amorphous carbon. Figure 9.7 shows typical results using these two methods to quantitate a spectrum obtained from a carbon film $\sim$ 40 nm in thickness, at $E_0 = 100$ keV in the STEM mode. For discussion purposes the results in Fig. 9.7 are normalized to the value of N predicted at $\Delta = 300$ eV and, with reference to this figure, the following general remarks can be made.

a. The efficiency factor method gives values of N which vary by about 25% for Δ in the range 50 to 100 eV. The results for $\Delta < 50$ eV (not shown) vary discontinuously because the value of the exponent s, which is derived from a fit to stripped edge profiles, changes

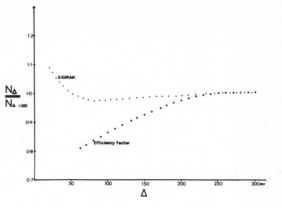

Fig. 9.7. Representative results
of an N determination as a function of Δ for an amorphous-carbon
specimen ~ 40 nm thick using the
efficiency-factor method (Δ) and
calculated cross-section method
(o): STEM mode; E_0 = 100 keV; and
β = 3 mrad (from Joy et al. 1979).

rapidly in this energy region and hence the calculated energy-
efficiency factors change accordingly. Beyond $\Delta \sim$ 50 eV, however,
the N values are smoothly varying out to large windows. Above Δ =
175 eV the stability is better than 5%. Clearly the results will
be less sensitive to spectral noise when the window is large. In
the case of thick films where substantial multiple scattering (e.g.
the convolution of low-loss excitations with inner-shell excitations) occurs, the shape of a K-edge can be altered significantly
and this method breaks down. For carbon this conditions has been
found to set in at thicknesses greater than about 150 nm. In summary, this approach satisfies the stability criterion provided that
a sufficiently large energy window can be chosen and multiple scattering is minimal.

b.   The stability of N values derived using the calculated (SIGMAK)
       cross-section method is seen to be excellent even for relatively
       narrow windows. From about 30 to 300 eV the variation is within
       5%, and some of the variation that is observed for large Δ may be
       due to cumulative errors in the integral as a result of inaccura-
       cies in the background extrapolation. This hypothesis still must
       be checked. Since the integrated edge intensity is relatively in-
       sensitive to the exact shape of the edge profile, acceptable stabil-
       ity (i.e. 5%) with this method has been obtained for spectra ex-
       hibiting some degree of multiple scattering (i.e. t > 150 nm).

       In order to test the stability of N using the calculated L cross-
section method (SIGMAL), the L-edge from copper evaporated onto a carbon
film was studied (JOY et al., 1979). The variation of N with Δ and at
two values of β was determined when operating in the CTEM mode at E_0 =
80 keV. Typical results are shown in Fig. 9.8 where it can be seen that
for Δ in the range 50 to 500 eV, the total variation in N is less than ~
15%. As mentioned before, some of this variation may be due to cumulative errors in the background extrapolation technique.

       The stability of the standards method has not been tested fully as
of yet, but preliminary data obtained for the L_{23} edge from silicon suggest a stability of 10% or better. Since the experimental correction

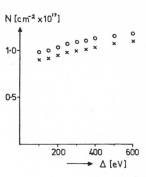

Fig. 9.8. Variation of N for copper (evaporated onto carbon) with Δ and β fron an L-loss analysis using hydrogenic cross sections with a modified correction for outer screening: CTEM mode; E_0 = 80 keV; and o is = 8.6 mrad and x is 4.6 mrad (from Joy et al. 1979).

curve (e.g. Fig. 9.6) is derived from an edge integral, the method should be applicable even in the presence of some multiple-scattering contributions. However, more data are required to substantiate this conclusion.

In summary, all three methods have been shown to be stable within 10% or better when applied under properly chosen conditions. Clearly it is important that stability tests be carried out as a first step when doing quantitative studies.

Relative Accuracy of Atomic Ratios

Quite often the relative amount, N_1/N_2, of two elements is required, rather than the absolute value of N. In this case the ratio method can be employed, as is most often done in x-ray microanalysis. If the edges of interest are measured and quantitated under the same conditions, then $I^1(\Delta,\beta)$ and $I^2(\Delta,\beta)$ in Eqn. 9.4 cancel when taking the ratio and one obtains

$$\frac{N_1}{N_2} = \frac{\sigma_k^2(\Delta,\beta)}{\sigma_k^1(\Delta,\beta)} \cdot \frac{I_k^1(\Delta,\beta)}{I_k^2(\Delta,\beta)} \qquad (9.11)$$

where the inner-shell k of the two edges need not be the same. Thus, the K-edge intensity can be evaluated for the lowest atomic-number element and, for example, the L- or M-edge intensity evaluated for the higher atomic-number element for which the K-edge intensity may be quite weak. This approach has been employed in the case of a number of precipitate structures which are important to the metallurgist (LEAPMAN and COSSLETT, 1977; LEAPMAN and WHELAN, 1977; FRASER, 1978a; and LEAPMAN, SANDERSON and WHELAN, 1978) and errors ranging from 10 to 30% have been reported.

When a quantitation method is used to measure the atomic ratio of a known compound, one, is, in effect, is testing the relative accuracy of the partial cross-section for each of the elements concerned. Several results can be obtained from a test of this kind: the atomic ratio could vary with the choice of Δ, or the ratio could be reasonably constant but not in agreement with the expected value, or it could be both constant and correct within an acceptable experiment error.

One of the problems with this type of experiment is to find test samples, since many compounds containing light elements with two suitable K-edges tend to suffer differential mass-loss under the electron beam. However, two compounds have been found to be satisfactory, namely BN and MgO, both of which are readily available in a pure form suitable for microanalysis (JOY et al., 1979). An additional difficulty in these experiments is that two (or more) edges in the same spectrum may be affected differently by multiple scattering, i.e., the lowest energy-loss edge will be more severely affected than will a higher energy-loss edge (see Fig. 9.9, for example). This means that the specimens should be as thin as possible and that the methods used for analysis should be able to deal effectively with some multiple scattering. In the limit of thick specimens it will be necessary to derive the single-scattering intensity from the spectrum by a deconvolution process (see JOY, 1979) before any analysis is possible.

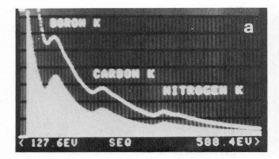 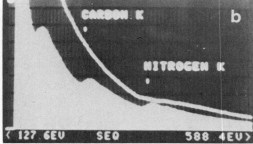

Fig. 9.9. Spectra from a BN crystal supported on an amorphous-carbon substrate showing in (a) the attenuation of the K-edge intensity from boron with increased thickness (shaded spectrum is from the thinner crystal) and in (b) the attenuation of K-edges from carbon and nitrogen with increased thickness (shaded spectrum is from the thinner crystal): STEM mode; $E_0 = 100$ keV and $\beta = 3$ mrad.

The efficiency-factor method has been used to evaluate the B to N ratio from BN (STEM mode, $E_0 = 100$ keV and $\beta = 3.0$ mrad) and typical results from such an experiment are summarized in Table 9.3. In general, the errors relative to the expected value were <10% whereas the stability was better than a few percent. In these experiments it was found that the value of N for each element varied with the choice of Δ in the same way as shown in Fig. 9.7 for carbon. It must be recalled that for this method the total cross sections σ are determined by the constants inserted into the Bethe formula. Although the absolute value of σ for any element can vary over a range in excess of 2:1 by the choice of the constants published in the literature, the ratio of the values obtained is surprisingly constant indicating that the functional variation of any of the standard x-ray cross-section formulae must be correct.

Method 2 has been found to give slightly better results when analyzing compounds. Figure 9.10 shows typical results when using the SIGMAK routine to analyze MgO (CTEM mode, E_0 = 80 keV and β = 8.6 mrad). Both the N values for Mg and O remain stable with Δ, and the resultant atomic ratio Mg:O = 0.93 ± 0.03 is in good agreement with the expected value. This indicates that the SIGMAK calculation correctly accounts for the variation of the partial cross section with edge energy. It does not, of course, prove that the absolute value is correct since the multiplying constants will cancel in any ratio. Equally good results have been obtained on the BN system using SIGMAK (STEM mode, E = 100 keV and β = 3 mrad). A ratio of B:N = 0.90 ± 0.03 has been found for energy windows in the range 40 to 100 eV. Although the absolute value of the ratio is in 10% error with the expected value, the agreement is to be considered satisfactory at this stage. It is also worth noting that in the case of BN, a carbon K-edge (from the support film) is interposed between the two edges used for the quantitation. The success of the programs in dealing with this situation indicates that the background stripping routines are reasonably accurate.

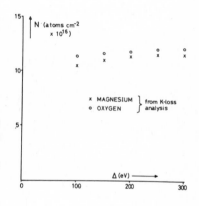

Fig. 9.10. Compositional analysis of MgO showing the variation of N with Δ using the calculated cross-section method for K edges: CTEM mode; 80 keV; β = 8.6 mrad; and Mg:O = 0.93±0.03 (from Joy et al. 1979).

The third quantitation method is just now being tested for various compounds and the results are too preliminary to report. Clearly the relative accuracy of atomic ratios derived from this approach will be governed by the accuracy of the original partial cross sections and experiments are in progress to assess this accuracy.

In summary, atomic ratios using K-edge intensities have been obtained with a relative accuracy of 15% or better using both methods 1 and 2. Therefore, either method may be considered suitable for routine microanalysis.

Absolute Accuracy of Quantitation

A final test of any quantitation method is that a calculated N value agrees, in absolute terms, with an independently determined value. This is not a pre-requisite for a microanalytical technique since, in general, it is atomic ratios that are most often required. Thus, energy dispersive x-ray spectroscopy is widely used for microanalysis even though the result is not an "absolute" measurement. However, because

the quantitation relation is so simple in electron energy-loss spectroscopy, there is a strong incentive to achieve data-reduction programs
and analysis methods which do provide correct, absolute values of N.

JOY et al. (1979) examined this aspect of methods 1 and 2. Results
from the same specimen area were compared using two independent methods
to obtain N and also results from the same specimens but obtained in different microscopes and under different modes of operation were compared.

The specimens chosen for this purpose were a series of amorphous-
carbon films (prepared by E. J. Fulham, Inc.) with nominal values of
thickness ranging from 20 nm to 160 nm. These films were examined both
in the as-received condition and after an anneal at 300°C. Although no
elements other than carbon were detected in these films, considerable
changes in the characteristics of the films were observed after annealing, indicating the possible presence of hydrocarbons. The experiments were carried out independently at two locations* and were conducted on two different microscopes; the first was a JEOL 100B, operated
at 80 kV in the CTEM mode and equipped with a magnetic prism spectrometer (EGERTON, 1978b); the second was a JEOL 100B, operated at 100 kV
in the STEM mode and also using a magnetic prism spectrometer (JOY and
MAHER, 1978a). For both situations the spectrometer resolution was between 5 and 10 eV, as measured from the zero-loss peak. Spectra were
computer processed using routines of the type described above.

Independent microscopic determinations of N are possible using several techniques. The best known of these is that based on the Plasmon
Peaks. Provided that the mean-free path for a plasmon excitation λ is
known, then the thickness of the specimen can be estimated from measurements of the integrated intensities in the low-loss region of the spectrum. Strictly speaking in the case of amorphous carbon the prominent
loss peak observed at 24 eV is not definitely classified as a plasmon
loss but also reflects some contribution from single-electron excitations. However, for our purpose it can be treated in an identical way
to a normal plasmon. The thickness can be derived from several relationships and the one chosen by JOY et al. gives

$$t/\lambda_p = \log_e(I_T/I_o) \tag{9.12}$$

where I_0 is the integrated intensity under the zero-loss peak, I_T is the
total spectrum intensity (measured by integrating the spectrum over an
energy window of 200 eV or more) and t is the specimen thickness. This
relationship was expected to be more reliable than those based on the
ratio of the areas of individual plasmon peaks and the zero-loss peak
(e.g. JOUFFREY, 1978). However, more recent theoretical work (FARROW,
JOY and MAHER, unpublished) has indicated that under the correct conditions, the two relationships should be equivalent. Nevertheless, it can
be seen from Eqn. 9.12 that if one knows λ then t can be calculated
from a simple determination of two integrated intensities. Two clarifi-
cations regarding the accuracy of this approach, particularly for the

---

*The University of Alberta and Bell Laboratories.

case of amorphous carbon, need to be made. Firstly, the λ values tabulated in the literature (EGERTON, 1975, and KIHN, SEVERLY and JOUFFREY, 1976) are experimental values calculated from specimens whose thickness was assumed to be known and secondly, the thickness uniformity of the films usually is not given. Therefore, the literature values must be treated with caution. An estimate of 10% probable error would seem likely. Once the film thickness is estimated, its density ρ is required in order to obtain N. Densities quoted for carbon films vary between 1.8 and 2.2 gm/cm^3, the exact value certainly being sensitive to the prior history of the film. Thus, a further uncertainty of at least 10% in the absolute determination of N is introduced.

The results obtained from quantitation methods 1 and 2 were compared to the values of N derived from Eqn. 9.12 and assuming the density of a thin amorphous-carbon film to be 2 gm/cm^3. The values of N, as a function of Δ, for one set of experiments are shown in Fig. 9.11 and the results in the limit of large Δ are summarized in Table 9.4. It can be seen that the values of N derived from the integrated intensities in the plasmon portion show good agreement ($\sim$ 10%) and this was considered a justification for using this approach as a "standard" for comparison.

The value of N obtained from the K-edge at 80 kV (with β = 8.6 mrad) was 1.22 ± 0.02 x 10^{17} atoms/cm^2 over the range 30 eV < Δ < 250 eV using the SIGMAK routine. This compares with a K-edge value at 100 kV (for β = 3 mrad) of 1.74 ± 0.03 x 10^{17} atoms/cm^2 for 30 eV < Δ < 250 eV probably within ± 15% of the "true" value. Since in the STEM case only also using the SIGMAK routine. Analyzing the 100 kV data using the efficiency-factor method, resulted in a value of 1.48 ± 0.05 x 10^{17} atoms/ cm^2 (200 eV < Δ < 250 eV).* Although not shown on this figure, N values were obtained at both kV's for other values of β. The measured spread in N was of the order ± 5%. Similar experiments were done on thicker films and in every case the comparative values of N exhibited the same relative order as those given in Table 9.4. After annealing the carbon films, the values of N determined under all conditions decreased by about 40%, with a small reduction in the spread between the methods used to evaluate N.

Fig. 9.11. Quantitative analyses of an amorphous-carbon specimen ($\sim$ 20 nm thick): 1) CTEM mode, E_0 = 80 keV, K-loss analysis using SIGMAK, β = 8.6 mrad; 2) STEM mode, E_0 = 100 keV, K-loss analysis using SIGMAK, β = 3 mrad; 3) STEM mode, E_0 = 100 keV, K-loss analysis using efficiency-factor method, β = 3 mrad; 4) STEM mode, E_0 = 100 keV, plasmon analysis, β = 3 mrad; and 5) CTEM mode, E_0 = 80 keV, plasmon analysis, β = 8.6 mrad (from Joy et al. 1979).

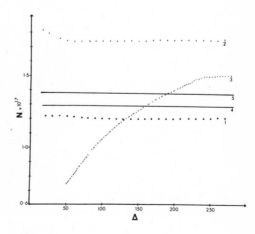

---

*See Appendix for worked examples when using methods 1 and 2.

If an error of ± 10% is assumed for the determination of N from the "standard", then the 80 kV SIGMAK value and the limiting value from the 100 kV efficiency-factor method, are in good agreement. However, the 100 kV SIGMAK value is about 25% higher than these. Since this ordering is maintained over the entire range of film thicknesses studied, the 100 kV SIGMAK value is clearly showing some systematic discrepancy compared to the other data. This general level of agreement was considered encouraging, particularly bearing in mind the uncertainties in the determination of N from the "standard." The implications of this are illustrated by the fact that if the density of a carbon film had been taken as 2.2 gm/cm^3 rather than 2 gm/cm^3, the errors would have been reversed, with the 100 kV value being in better agreement with the "standard" than the 80 kV values. Thus, for any operating condition, on either microscope and using either data reduction system, the value of N obtained is a few tens of thousands of carbon atoms were being analyzed, these results are certainly encouraging.

Future Considerations

Under suitably chosen conditions, it should be possible at present to carry out quantitative analyses with a relative accuracy of 10% or better and an absolute accuracy of ± 20% or better. It is felt that these limits are now fixed by partial cross-section determinations and/or specimen related variables. Clearly there is a need for uniformly-thin standards which can be quantitatively characterized by independent techniques. Ideally <u>one</u> of these standards should be suitable for both energy-loss analysis and energy-dispersive x-ray analysis. Several approaches to obtaining standards are being pursued and among these approaches are ion implantation (MAHER, JOY and MOCHEL, 1978) and diode sputtering (MAHER, JOY and SCHMIDT, unpublished). Systematic experiments which utilize method 3 should be extremely beneficial, particularly in the area of metallurgy where pure-metal single crystals can be used for standards.

Clearly variables other than those mentioned up until now must be considered in future quantitation studies and a brief discussion of a few of these is given below.

(a) Beam Convergence - α_0: The convergence of the incident beam α_0 is convoluted with the scattered-electron distribution and when $\alpha_0 \geq \beta$ this convolution results in a smaller effective value of β. In the case where $\alpha_0 \sim \beta$, this effect can produce an error of up to 30% in $\sigma(\Delta,\beta)$ unless a correction is made (ISAACSON, 1978). If a quantitative analysis is made in the STEM mode at near maximum resolution (i.e. point-to-point) then one will usually be in the regime where $\alpha_0 \geq \beta$ (since $\alpha_0 \sim 10$ mrad for this case) and proper account will have to be taken of the convolution between the incident beam and the scattered-electron distribution.

(b) Acceptance Semi-Angle of Detector - β: The quantity β must be known to about 5% if significant errors are to be avoided in the calculation of the partial cross sections (e.g. using SIGMAK or SIGMAL). Therefore, it is important to keep in mind that this accuracy can only be maintained in practice if the variation of β with objective lens current or specimen height is carefully calibrated, e.g. from a diffraction pattern.

(c) Specimen Thickness - t: The present rule of thumb is that in quantitative analysis t should be less than the total mean-free path for inelastic scattering (i.e. $\sim$ 100 nm at E_0 = 100 keV). Equation 9.4 for obtaining N is not exact, in that it takes account of elastic and inelastic multiple scattering only to first order (EGERTON, 1978a). Similarly Eqn. 9.11 for obtaining N_1/N_2 is not exact and mixed scattering (i.e. primary electrons which excite both an inner-shell electron and one or more valence-shell electrons) can effect $I_k^1(\Delta,\beta)$ and $I_k^2(\Delta,\beta)$ differently, depending on the E 's and thickness. Therefore, the accuracy of both equations decreases with increasing thickness. In order to put explicit limits on the acceptable thickness required to achieve a given accuracy using Eqns. 9.4 and 9.11, spectra taken as a function of thickness must be deconvoluted in order to obtain the single-scattering case and the results obtained then compared to the predictions from the raw spectral data.

(d) Specimen Orientation: For highly-disordered specimens, the orientation is of little importance but for crystalline materials, Eqns. 9.4 and 9.11 will be less accurate when strong Bragg reflections lie inside or just outside the detector aperture (EGERTON, 1978a). Given our present limited understanding of this situation, it is best to orient crystalline specimens so as to minimize the intensity in the diffracted beams (i.e. use quasi-kinematical conditions) when recording spectra. The present rule of thumb here is that β should be less than the Bragg angle θ but greater than the mean-angle $\bar{\theta}_E$ for valence-electron scattering. For E_0 = 100 keV this implies that β should be in the range 3 to 10 mrad.

9.4 ELEMENTAL IDENTIFICATION

Over the last seven years a significant number of inner-shell spectra from a variety of materials have appeared in the literature. One concludes from these results that K, L_{23}, M_{23}, M_{45} and N_{45} edges will be the most useful in elemental identification (e.g. see LEAPMAN, 1979). Other edges, such as L_1, M_1, etc., will be extremely difficult to detect because of their low cross sections compared to those for K, L_{23}, M_{23} etc. which are typically 5 x 10^{-18} to t x 10^{-22} cm^2 per atoms at E_0 = 100 keV. According to the Bethe model (POWELL, 1976), the total ionization cross section for K and L edges is given approximately by

$$\sigma_k(\Delta\to\infty,\pi) \doteqdot \frac{CZ_1}{E_k E_0} \quad B_k \ \ln \frac{A_k E_0}{E_k} \tag{9.13}$$

where C, B and A are constants and Z_1 is the number of electrons associated with the excited subshell (i.e. n(1) ^{1}, as in Table 9.2). This expression illustrates that $\sigma(\Delta\to\infty,\pi)$ is proportional to Z_1, and neglecting the ln term, E^{-1}. Therefore, as Z decreases, both K and L cross sections increase and this is why the energy-loss technique is sensitive to low Z elements ($3 \leq Z \leq 11$) when using K edges and intermediate Z elements ($11 \leq Z \leq 30$) when using L edges. The significance of this situation is given by the following (over simplified) example. If one can detect a K edge from a light element (i.e. $3 \leq Z \leq 11$) at E (i.e. $1s^2$), then it certainly should be possible to detect a L_{23} edge at $E_{L_{23}} = E_K \pm 300$ eV (i.e. $2p^6$), say, for the case where N_1/N_2 is in the range 0.5 to 1.5 which may be typical of a metal, or semiconductor, carbide, nitride or oxide. "Should be possible" is the key phrase in elemental identification and there are a number of factors which need to be considered.

Threshold Energy

In principle, elemental identification with electron energy-loss spectroscopy is as straight forward as with energy-dispersive x-ray spectroscopy. This can be illustrated by an example taken from a diagnostics problem in silicon device processing. Spectra in the energy-loss range 0 to 400 eV were recorded from two areas, one believed to be "good" and one "defective" (see Figs. 9.12(a) and 9.12(b), respectively). The first spectrum exhibits silicon (L_{23}) and boron (K) edges, as expected from the known doping of the device at the point of analysis. However, the second spectrum exhibits a carbon (K) edge, indicating that carbon contamination has occurred during some processing step of this device. In a spectral analysis of this type, elemental identification can be obtained by measuring the threshold energy of an observed edge and comparing this with binding energies organized as charts of the type shown by JOY (1979) or with binding-energy tables (e.g. LARKINS, 1977). In the near future, it is expected that most commercial data-collection systems will include edge markers, as commonly used in energy-dispersive x-ray analysis.

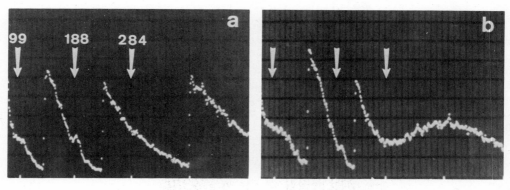

Fig. 9.12. Spectra recorded from two different areas of a sectioned silicon device illustrating edge identification from the threshold energies for silicon (E = 99 eV), boron (E = 188 eV) and carbon (E = 284 eV). In (a) the analyzed volume was from a good region of the device and in (b) the analyzed volume was from a defective region of the device: STEM mode; $E_0 = 100$ kV; and $\beta = 7.5$ mrad.

In many practical situations the multiplicity of edges which are observed can be considerable. The ability to combine electron energy-loss and energy-dispersive x-ray techniques has obvious advantages in these situations. When analyzing for light elements in a higher average Z matrix, this combination makes it possible, in principle, to use the x-ray results to identify the matrix edges which usually will be the L, M etc. type.

Shape Analysis

In many complex situations and especially when the edge signal to the background signal (S_k/S_{BG}) is low, it is clearly important to accurately identify threshold energies. To achieve this accuracy the spectrometer resolution δ should be 5 eV or better and the background prior to the edge should be modeled, extrapolated, and stripped from the presumed edge intensity. Threshold structure can be identified more readily from this result and E_k accurately established. The "true"-edge shape which is the result of applying these operations then can be compared to expected shapes, as discussed in prior sections. The elemental origin of the presumed edge then obtained from E and the true shape relative to the expected shape for k should be in agreement, taking due account of solid-state effects and multiple scattering. To illustrate the utility of this approach which shall be referred to as shape analysis, typical raw spectral data of a silicon L_{23} edge from SiO_2 is shown in Fig. 9.13(a) and this result should be compared to the pure-silicon spectrum in Fig. 9.4(a). At first it was difficult to rationalize the marked differences in the two results. However, after the background is stripped from the SiO_2 data, the "true"-edge shape is revealed and it has the expected L_{23} form and compares quite closely to the stripped pure-silicon edge shown in Fig. 9.4(b). The threshold energy can be defined accurately from this result and shows a chemical shift of about 4 eV relative to pure silicon.

This analysis method has been applied in a number of control studies and the results have confirmed its general utility in elemental identification.

Elemental Maps

Spectral analysis for elemental identification is powerful and ultimately necessary. However, at present, spectra are sequentially recorded and the time required to determine the spatial distribution of several elements (known or unknown) may be substantial. This problem can be overcome by forming an image with those electrons which have lost an increment of energy $E_k + \delta$ where E_k corresponds to the binding energy of a specified element. If the energy dispersion of the spectrometer is adjusted so that the energy loss E + δ is passed through the selecting slit of the analyzer, then the video signal will vary depending on whether or not a detectable amount of that element is present in the volume examined. In the absence of the element the signal will correspond to the local background intensity for that energy loss, while in the presence of the element the signal will be increased by an amount depending on the quantity of the element in the irradiated volume. Therefore in a STEM with a magnetic prism spectrometer, a video display for the selected energy can be a first order indication of the elemental concentration. Such an elemental "map" is shown in Fig. 9.14. Care must be exercised interpreting such images, since local changes in mass thickness will also cause variations in image intensity (e.g. JEANGUILLAUME, et al., 1978). The simplest distinction between elemental and mass-thickness variation may be obtained by comparing images in energy increments just before and just after the edge of interest. The elemental information will then exhibit either enhanced contrast or a reversal of contrast, going from dark (before the edge) to bright

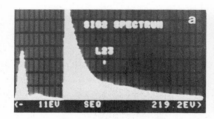 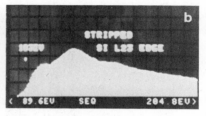

Fig. 9.13. Illustrating shape analysis applied to an SiO_2 spectrum: (a) low-loss region and the silicon L-edge region after the change, (b) the stripped L edge. STEM mode; E_o = 100 keV; β = 3 mrad.

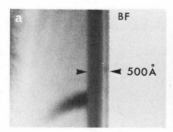 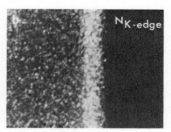

Fig. 9.14. Illustration of how elemental maps can be used to obtain the spatial distribution of a specific element. Bright-field STEM (a) and elemental map (b) images showing the distribution of nitrogen ion implanted into silicon. The specimen was prepared such that the thin section was parallel to the ion implanted direction and the implanted zone was ~ 50 nm thick. The elemental map was recorded just above the K-edge of nitrogen (E = 402 eV) and in the scanning transmission mode (E_0 = 100 keV, probe diameter ~ 10 nm, β = 3 mrad, spectrometer resolution ~ 10 eV and 50 sec exposure time).

Fig. 9.15. Energy dispersed image (scanning transmission mode) of an amorphous carbon film showing the variation of intensity which occurs on crossing the K-edge (E = 284 eV). The amount of energy loss increases left to right and the experimental conditions were as in Fig. 9.14).

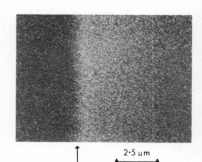

(after the edge). This effect can be illustrated quite clearly by an energy dispersed, scanning transmission image of an amorphous carbon film (Fig. 9.15). In this case, the film is of uniform thickness ($\sim$ 160 nm) and the variation in energy loss across the image gives rise to a sharp enhancement of the contrast at the carbon K-edge (284 eV), the contrast going from black to white and then slowly dropping back to its original level with increasing energy loss.

This imaging technique has been applied in a number of studies related to materials science (e.g. JOY and MAHER, 1977, WILSON et al., 1977, JOUFFREY, 1978, and FRASER, 1978b). In practice, we have used the technique to obtain a rapid <u>indication</u> of certain elemental distributions. Image features showing the expected elemental contrast are identified and then spectral analyses carried out on these features.

9.5 DETECTION LIMITS

In prior sections it was assumed that the systems variables were optimized for maximum sensitivity. This statement must be qualified because the coupling between systems variables and constraints set by analysis methods or sensitivity maximization can be in conflict. For example, if a spectrometer resolution of δ is required for a particular experiment, then the quantities which can be varied to achieve maximum sensitivity are the incident-beam current density J (electrons-cm^{-2} sec^{-1}), t, β, the dwell time τ (sec. per channel) of the multichannel analyzer and Δ. The available current density will be limited by the source. On the other hand, the usable current density (and/or dwell time) may be fixed by specimen related variables (e.g. radiation damage, heating, mass loss or mass gain). For obvious reasons it may not be possible to vary t and the range over which β can be varied may be fixed by α_o, the microscope optics, θ_B or δ. Therefore, in practice the coupling between these variables is crucial in determining one's ability to record an edge with the best possible sensitivity or statistics. Here the measure of "best" is given by the signal-to-noise ratio which may be expressed as S_k/S_{BG}, $S_k/\sqrt{S_{BG}}$ or $S_k/\sqrt{S_k + S_{BG}}$.* The expression which is applicable will depend on the nature of the specimen and objectives of the experiment. For specimens of interest to materials science, $S_k/\sqrt{S_{BG}}$ is the important quantity in elemental identification (i.e. "seeing" an edge), whereas $S_k/\sqrt{S_k + S_{BG}}$ is what governs the minimum detectable limits.

Importance of β

The dominant feature of an energy-loss spectrum for E > 50 eV is the background. In order to achieve the optimum sensitivity and statistics it is necessary to maximize the edge intensity with respect to the background intensity. As discussed by JOY (1979), the edge intensity is strongly peaked in the forward direction (i.e. $\theta \to 0$) whereas the background has a broader angular distribution which is of low intensity in the forward direction and rises for increasing θ. The measured situation for the K edge from amorphous carbon is shown in Fig. 9.16.

---

*$S_k = I_k$ (Δ,β) and S_{BG} is the extrapolated background intensity, I (Δ,β).

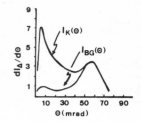

Fig. 9.16. Experimental curves of the angular distribution of the background intensity I (θ) under the K-edge and the stripped K-edge intensity I (θ) from an amorphous-carbon specimen (Colliex et al. 1976).

Because of this difference in angular distribution, $S_k/\sqrt{S_{BG}}$ and $S/\sqrt{S_{BG}+S_k}$ exhibit a maximum for some value of β ≤ 40 mrad. The effect of varying β on the K edge visibility for carbon in silicon is shown clearly in Fig. 9.17. Detailed measurements have shown that $S/\sqrt{S_{BG}}$ for the K edges from C, N, O, F, Na, Al and Si (JOY and MAHER, 1978b) exhibit a broad maximum at β ∼ $θ_E$ (see Table 9.1 for 100 keV values) where as $S_k/\sqrt{S_{BG}+S_k}$ for K edges from B, C and O exhibit a broad maximum at β ∼ 10 mrad (EGERTON, ROSSOUW and WHELAN, 1975). Although similar results have not been reported for L, M and N edges, it is expected that there will be an optimum value of β for which the signal-to-noise ratio is maximized for these edges. The practical significance of these results is that, when performing elemental analyses on an unknown, it may be necessary to record spectra at several values of β in order to unambiguously identify a weak edge. Moreover only by choosing a detector aperture which maximizes $S_k/\sqrt{S_{BG}+S_k}$ for a particular k, will it be possible to approach the minimum detectable limits of the technique.

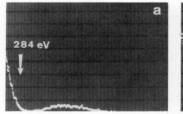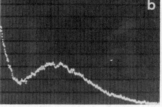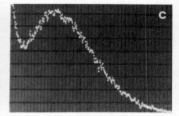

Fig. 9.17. Spectra from silicon containing carbon showing the change in the carbon K-edge visibility (i.e. $S_k/\sqrt{S_{BG}}$) with detector collection angle: a) β = 0.3 mrad; b) β = 0.75 mrad; and c) β = 1.5 mrad. These results were obtained in the STEM mode at E_0 = 100 keV and with δ = 20 eV.

Minimum Detectable Limits

The mathematical framework for estimating the minimum detectable mass (MDM) and minimum mass fraction (MMF) has been given by ISAACSON and JOHNSON (1975). One of the important points to be learned from their work is that to obtain the minimum detectable limit both the conditions of minimum mass and MMF must be satisfied. The reasons for this are that: 1) in order to exceed a minum required count rate (R_{min}) for detection, the volume analyzed must contain a minimum mass of the element which is of interest; and 2) in order to exceed the confidence limit set by $S_k/\sqrt{S_k+S_{BG}}$ a minimum mass fraction must be present in the same volume. Therefore, one can see that a given atomic fraction of a particular element may be detectable in a 100 nm thick crystal but the same concentration could go undetected when t ∿ 50 nm because the mass of the element within the analyzed volume is too small for R_{min} to be exceeded. In the cases where R_{min} is satisfied, the detection limit is set by $S_k/\sqrt{S_k+S_{BG}}$ and hence the background under the edge of interest. As a result of this situation, electron energy-loss spectroscopy is not a trace-element technique when employed in the microarea mode.

Calculations of the minimum detectable limits can be done rigorously only for specific situations where the systems variables (J, t, β, σ, τ, Δ and R_{min}) are specified and assumptions are made concerning: 1) the origin and functional form of the background; and 2) the ratio of I_R (E) to I_{BG} (E) and the functional form of the edge for E >> E_k.

The most direct approach to estimating the MDM and MMF is from experiment. If it is assumed that the measured quantity N, or N_1/N_1+N_2, is linearly proportional to the minimum detectable quantity then

$$\frac{M_{mea}}{(S_k/\sqrt{S_{BG}}+S_k)\ mea} \doteq \frac{MDM}{3} \qquad (9.14)$$

or

$$\frac{(MF)_{mea}}{(S_k/\sqrt{S_{BG}}+ S_k)_{mea}} \doteq \frac{MMF}{3} \qquad (9.15)$$

where M_{mea} is the measured mass of an element (i.e. N·m·analyzed area), $(MF)_{mea}$ is the measured atomic fraction (i.e. N_1/N_1+N_2) of an element in the analyzed volume and the confidence limit is $S_k/\sqrt{S_{BG}}+S_{BG} \geq$ 3. This approach assumes that the matrix mass remains constant and the MF of the analyzed element decreases (i.e. S_{BG}∿ constant).

Results derived from experiments where the host lattice is silicon have been treated within these approximations and the extrapolated results are that the MDM of carbon is ∿ 5 x 10^{-17} gms and the MMF of nitrogen is ∿ 0.1 at %. These numbers are in general agreement with those predicted from rigorous calculations (JOY and MAHER, unpublished) and of those reported throughout the literature when using a standard heated filament and operating in the STEM mode with a 10 nm dia probe. Certainly an improvement can be expected for higher brightness sources.

9.6 SUMMARY

In this chapter the use of inner-shell excitations in elemental analysis has been discussed. It is expected that the fundamental understanding and the instrumental developments in the field will progress rapidly. It is especially important that the number of scientific and technological problems which are tractable to quantitation be explored fully during this same period. Clearly the useful specimen thickness is an important factor in carrying out analyses on specimens of practical interest. Therefore, the need for higher accelerating voltages is obvious. For specimens of the same thickness now studied at 100 kV, one could achieve higher precision and increased sensitivity at, say, 200 kV. Moreover, the same precision and sensitivity now achieved at 100 kV would be possible for thicker specimens at the higher accelerating voltage. Hopefully, experiments at 200 kV will be possible in the near future so that these predictions can be tested.

REFERENCES

Amelinckx, S., 1964, Direct Observation of Dislocations (London: Academic Press) p. 193.

Bethe, H., 1930, Ann. Phys., 5, 325.

Colliex, C., Cosslet, V. E., Leapman, R. D. and Trebbia, P., 1976, Ultramicroscopy, 1, 301.

Egerton, R. F., 1979, Ultramicroscopy (in press).

Egerton, R. F., 1978a, Ultramicroscopy, 3, 243.

Egerton, R. F., 1978b, Ultramicroscopy, 3, 39.

Egerton, R. F. and Joy, D. C., 1977, Proc. 35th Ann. Meeting EMSA, (Baton Rouge: Claitors Press), p. 252.

Egerton, R. F., Rossouw, C. J. and Whelan, M. J., 1976, Developments in Electron Microscopy and Analysis, J. Venables ed. (London: Academic Press), p. 129.

Fraser, H. L., 1978a, Proc. 9th Int. Congress on Electron Microscopy, (Toronto: Imperial Press), 1, p. 552.

Fraser, H. L., 1978b, Proc. 11th Ann. SEM Symposium (Chicago: SEM, Inc.), 1, p. 627.

Inokuti, M., 1971, Rev. Mod. Phys. 43, 297.

Isaacson, M., 1978, Proc. 11th Ann. SEM Symposium (Chicago: SEM, Inc.), 1, 763.

Isaacson, M. and Johnson, D., 1975, Ultramicroscopy, 1, 33.

Jeanguillaume, C., Trebbia, P. and Colliex, C., 1978, Ultramicroscopy, 3, 237.

Joy, D. C., 1979, this book.

Joy, D. C. and Maher, D. M., 1978a, J. of Microscopy, 114, 117.

Joy, D. C. and Maher, D. M., 1978b, Ultramicroscopy, 3, 69.

Joy, D. C. and Maher, D. M., 1977, Developments in Electron Microscopy and Analysis (Briston: The Institute of Physics), Ser. No. 36, p. 357.

Joy, D. C., Egerton, R. F., and Maher, D. M., 1979, Proc. 12th Ann. SEM Symposium (Chicago: SEM, Inc.) in press.

Jouffrey, B., Short Wavelength Microscopy (New York: N. Y. Acad. of Sci.) p. 29.

Larkins, F. P., 1977, Atomic Data and Nuclear Data Tables, 20, 312.

Leapman, R. D., 1979, Ultramicroscopy, 3, 413.

Leapman, R. D. and Cosslett, V. E., 1977, Vacuum, 26, 423.

Leapman, R. D. and Whelan, M. J., 1977, Developments in Electron Microscopy and Analysis (Briston: The Institute of Physics), Ser. No. 36, p. 361.

Leapman, R. D., Rez. P. and Mayers, D., 1978, Proc. 9th Int. Cong. on Electron Microscopy (Toronto: Imperial Press), 1, p. 526.

Leapman, R. D., Sanderson, S. J. and Whelan, M. J., 1978, Metal Sci., 12, 215.

Maher, D. M., Joy, D. C. and Mochel, P., 1978, Proc. 9th Int. Congress on Electron Microscopy (Toronto: Imperial Press), 1, p. 528.

Maher, D. M., Joy, D. C., Egerton, R. F., and Mochel, P., 1979, J. of Appl. Phys. (in press).

Manson, S. T., 1972, Phys. Rev., 16, 1013.

Powell, C. J., 1976, Rev. Mod. Phys., 48, 33.

Silcox, J., 1979, this book.

Williams, D. B. and Edington, J. W., 1976, J. of Microscopy, 108, 113.

Willson, C. J., Batson, P. E., Craven, A. J. and Brown, L. M., 1977, Developments in Electron Microscopy and Analysis (Bristol: The Institute of Physics) Ser. No. 36, p. 365.

Wittry, D. B., Ferrier, R. P. and Cosslett, V. E., 1969, Brit. J. Appl. Phys., 2, 1967.

APPENDIX

In order to give some indication of the numbers which are derived from a quantitative analysis, two examples are presented of a N determination using methods 1 and 2. The results are for a K-edge analysis from an amorphous-carbon specimen ($\sim$ 40 nm thick) and the operating conditions were: STEM mode; E_0 = 100 keV; α_0 = 2.0 mrad; probe diameter $\sim$ 10 nm; J $\sim$ 10 amp-cm^{-2}; β = 3.0 mrad; and τ = 200 msec per channel at 1 eV per channel.

METHOD 1 - Efficiency Factors

For this method N is given by

$$\frac{I_K(\Delta,\beta)}{I_\ell(\Delta,\beta)} \cdot \frac{GF}{\sigma_T \cdot \eta_\Delta \cdot \eta_B}$$

and the results of the analysis are as follows:

Input -   Output -

E_0 = 100 keV;   r = 3.72
β = 3 mrad;   s = 4.58
E_K = 284 eV; η_Δ = 0.85;
Δ = 200 eV; and η_β = 0.2215;
GF = 0.046 I_K = 45,507 counts;
                                                 I_ℓ = 1,536,779 counts;
                                                 σ_T = 2.29 x 10^{-20} cm^2 per atoms;
                                                 and N = 3.26 x 10^{17} atoms-cm^{-2}.

METHOD 2 - Calculated Cross Section (SIGMAK)

For this method N is given by

$$\frac{I_K(\Delta,\beta)}{I_\ell(\Delta,\beta)} \cdot \frac{GF}{\sigma_K(\Delta,\beta)}$$

and the results of the analysis are as follows:

Input -   Output -

E_0 = 100 keV; I_K = 34,178 counts;
β = 3 mrad; I_ℓ = 1,535,877 counts;
E_K = 284 eV; σ_K = 2.17 x 10^{-21} cm^2 per atom;
Δ = 100 eV;   and N = 4.72 x 10^{17} atoms-cm^{-2}.
GF = 0.046; and
Z = 6.

If one assumes that the specimen is 40 nm thick, as was predicted from the evaporation conditions and measured using a quartz crystal technique, then*

$$N = \frac{t \cdot \rho}{m} \doteq \frac{4 \times 10^{-6} \cdot 2}{2 \times 10^{-23}} \rightarrow$$

$$N \doteq 4 \times 10^{17} \text{ atoms-cm}^{-2}.$$

Therefore the two methods predict values of N which are within ± 20% of the <u>expected</u> value derived from macroscopic techniques.

---

* $\rho = 2$ gm-cm ^{3} and $m = \dfrac{\text{atomic weight (12)}}{\text{Avogadro's number } (6 \times 10^{23})}$.

Table 9.1a. Binding Energies (E_K) and Characteristic Scattering Angles ($\theta_E = E_K/2E_0$) for Elements $3 \leq Z \leq 11$

Z	ELEMENT	E_K (eV)*	θ_E (mrad)[+]
3	Li	54	0.27
4	Be	112	0.54
5	B	188	0.94
6	C	284	1.42
7	N	402	2.01
8	O	532	2.66
9	F	685	3.43
10	Ne	870	4.35
11	Na	1072	5.36
12	Mg	1305	6.53
13	Al	1560	7.80
14	Si	1839	9.20
15	P	2146	10.73

---

*From Larkins (1977)
[+] $E_0 = 100$ keV

Table 9.1b. Binding Energies (E_{L_2}, E_{L_3}) and Characteristic Scattering Angles $(\theta_E = E_{L_{23}}/2E_0)$ for Elements $11 \leq Z \leq 30$.

Z	ELEMENT	E_{L_2} (eV)*	E_{L_3} (eV)*	θ_E (mrad)[+]
11	Na	31	31	0.16
12	Mg	51	51	0.26
13	Al	73	73	0.37
14	Si	100	99	0.50
15	P	136	135	0.68
16	S	165	164	0.82
17	Cl	202	200	1.01
18	Ar	250	249	1.25
19	K	296	294	1.48
20	Ca	350	346	1.74
21	Sc	407	402	2.02
22	Ti	462	456	2.30
23	V	521	513	2.59
24	Cr	584	575	2.90
25	Mn	651	640	3.23
26	Fe	721	708	3.57
27	Co	794	779	3.93
28	Ni	872	855	4.32
29	Cu	951	931	4.71
30	Zn	1043	1020	5.16

*From Larkins (1977)

[+] $E_0 = 100$ keV and where appropriate the average value (i.e. $\overline{E}_{L_{23}}$) of θ_E is given.

Table 9.2. Spectral Notation and Its Correspondence to the Electronic Configuration of Atoms

EDGE	SHELL			k_i n(ℓ), where $\ell = 0,1,2...n-1$		
k	n	k_1, n(0)	k_{23}, n(1)	k_{45}, n(2)	k_{67}, n(3)
K	1	K, 1(s)			
L	2	L_1, 2(s)	L_{23}, 2(p)		
M	3	M_1, 3(s)	M_{23}, 3(p)	M_{45}, 3(d)	
N	4	N_1, 4(s)	N_{23}, 4(p)	N_{45}, 4(d)	N_{67}, 4(f)
O	5	O_1, 5(s)	O_{23}, 5(p)	O_{45}, 5(d)	O_{67}, 5(f)
P	6	P_1, 6(s)	P_{23}, 6(p)	P_{45}, 6(d)	P_{67}, 6(f)

DEFINITIONS

Spectral Notation E_{k_i}:

k   denotes the inner-shell edge K,L, M...

i   denotes the energy level from which the electron is excited within an inner shell.

Electronic Configuration of Atoms $n(\ell)^{Z_\ell}$:

n   denotes the shell (i.e. principal quantum number of one orbital).

ℓ denotes the subshell (i.e. azimuthal quantum number) and takes values $0,1,2...n-1$ where $\ell = 0$ is the s subshell, $\ell = 1$ is the p subshell, $\ell = 2$ is the d subshell and $\ell = 3$ is the f subshell.

Z_ℓ denotes the number of electrons in a subshell where the maximum number is $2(2\ell+1)$.
     See equation 13 in Section 4.

CORRESPONDENCE

k   is equivalent to n.

i   is related to ℓ through the usual selection rules which govern angular momentum. The result is that for all $n(\ell)$ such that $\ell = 0$ there is only one energy level (i.e. $K, L_1, M_1, ...$) and for each energy level $n(\ell)$ such that $\ell > 0$ there are two sublevels (i.e. L_2 and L_3, M_2 and M_3, M_4 and $M_5...$).

Table 9.3. Atomic ratios for BN Using Efficiency Factor Method (STEM mode, $E_0 = 100$ keV and $\beta = 3.0$ mrad).

ENERGY WINDOW	B:N
100	1.03
78	0.79
46	0.78
38	0.76

Table 9.4.   Summary of Quantitation Results from Amorphous Carbon

Method	$N(\text{atoms-cm}^{-2}\times10^{17})$	
	80 kV (CTEM)	100 kV (STEM)
PLASMON SPECTRUM	$1.38;=0.14$[a]	1.27 ± 0.13[b]
EFFICIENCY FACTORS		1.48 ± 0.05 $(\beta = 3.0 \text{ mrad})$
SIGMAK	1.22 ± 0.02 $(\beta = 8.6 \text{ mrad})$	1.74 ± 0.03 $(\beta = 3.0 \text{ mrad})$

_____

a) $\beta = 8.6$ mrad, $\lambda_p = 720$ Å and $\rho = 2$ gm-cm^{-3}.

b) $\beta = 3.0$ mrad, $\lambda_p = 900$ Å and $\rho = 2$ gm-cm^{-3}.

CHAPTER 10

ANALYSIS OF THE ELECTRONIC STRUCTURE OF SOLIDS

JOHN SILCOX

PROFESSOR OF APPLIED AND ENGINEERING PHYSICS
CORNELL UNIVERSITY
ITHACA, NEW YORK 14853

10.1 INTRODUCTION

In this chapter, we will be concerned with some of the fine detail of energy loss spectroscopy and the means in which it can can be exploited to provide valuable and detailed information on electronic structure. In many respects this work complements and has similarities to electronic information available through the use of photoelectron spectroscopy, inelastic x-ray scattering, optical reflectivity and Auger spectroscopy. Combined with the small probe capability of modern electron microscopy, this becomes a valuable probe of materials. In addition, there are unique features which make energy loss spectroscopy valuable. In the following pages, we will indicate how and where these features are apparent. Our endeavor is also to provide a road map giving directions to various features in the spectrum and to give an appreciation of the underlying physics so that the terrain in general can be appreciated. Finally, we hope also to identify snags that might give rise to roadblocks and outline possible routes circumventing these snags so that an evaluation of the effort necessary might be made.

In chapter 7 Joy presents an outline of the scattering physics along with the definitions of quantities such as cross-section, dielectric function, $\varepsilon(q,\omega)$ used in describing the scattering. We will expand some of these ideas here and introduce one or two others. Joy also gives a description of some of the scattering processes that we will be discussing along with a description of elastic scattering. This background will be assumed in the following discussion. We will not discuss apparatus since this question too has been discussed earlier but we will be concerned somewhat with questions of data analysis.

10.2 SCATTERING KINEMATICS

In Fig. 10.1, we outline the simple kinematics associated with plane wave inelastic scattering. In this model, the high energy electron is in a plane wave state e^{ikR} where R is the position vector of the high energy electron of wave vector k and energy

$$E_1 = \frac{\hbar^2 k^2}{2m} \tag{10.1}$$

upon interaction with the specimen, the electron undergoes inelastic scattering to a new energy state

$$E_2 = \frac{\hbar^2 k^2}{2m} \tag{10.2}$$

with a momentum transfer of

$$\hbar q = \hbar\ (k \to k')$$

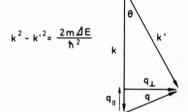

$$k^2 - k'^2 = \frac{2m\,\Delta E}{\hbar^2}$$

Fig. 10.1. Illustrating the kinematics of plane wave inelastic scattering.

For small energy losses, $(E = E_1 - E_2)$ and small scattering angles $(\sin\theta \cong \theta)$, the parallel and perpendicular components of momentum transfer $(q_{\parallel} + q_{\perp})$ can be written respectively as

$$q_{\parallel} = k\theta_E \text{ and } q_{\perp} = k\theta, \text{ where } \theta_E = \frac{E}{2E_o} \tag{10.3}$$

Under these conditions,

$$q^2 = k^2(\theta^2 + \theta_E^2) \tag{10.4}$$

A measurement of the energy loss E therefore gives θ_E and the measurement of θ completes the evaluation of q. The experiment outlined in Fig. 10.1 therefore gives the proportion of the incident electrons which transfer to the specimen an energy $E = \hbar w$ and momentum $\hbar q$. This proportion will be determined by the ability of the specimen to absorb that energy and momentum and, in turn, this will be determined by the energy levels associated with the electrons and the nuclei forming the specimen. To give a simple example, an electron bound in the L-shell of aluminum can be put into the conduction band by the transfer of at least 72 eV. If 172 eV is transferred then in the excited state, the excited electron will have a kinetic energy of 100 eV.

The number of fast electrons scattered in this way will be determined by the cross-section for that particular process. Before discussing the factors entering into establishing the cross-section, it is of some value to discuss the underlying physics of scattering in simple terms so that the concept of a cross-section might become more familiar. We draw an analogy with elastic scattering from a crystal. The atoms, consisting of positively charged nuclei and negatively charged electrons, are arranged in space at regular intervals. A useful description of this condition is to describe a crystal as the superposition of waves of charge density with the wavelengths of the charge density fluctuations representing the periodicities of the crystal. The elastic interaction of the high energy electron with a static wave of charge density provides an electron optical diffraction pattern analogous to the light optical diffraction pattern formed by a grating. In an inelastic process, energy is transferred to the crystal and can be absorbed in a number of ways. To consider the case of the aluminum L-shell electron given above, a positive hole is left in the L-shell and a negative electron has 100 eV of kinetic energy and is propagating away from the hole. This can be regarded as a dyanmic fluctuation of the charge density with an angular frequency, ω, given by the quantum condition $E = \hbar\omega$. The wavelength of the charge density flucutation is now not restricted to the crystal periodicity and the fast electron is diffracted accordingly. Thus, inelastic scattering can be regarded as creating dynamic fluctuations in the charge density with a frequency ω and a wave vector q. It remains to relate the amplitude of the charge fluctuation to the electronic structure.

As indicated above the frequency of the charge fluctuation reflects the energy difference between the energy levels of the solid. We note also that the wave number reflects the spatial dependence of the wave functions associated with the electrons in the solid. A central quantity in the theory of the processes is the matrix element $\rho_q{}^{if} = <f\,e^{iqr}\,i>$ which reflects a bound electron going from the initial state $|i>$ to the final state $<f|$. The operator e^{iqr} arises from the interaction of the high energy electron with the bound electron.

Same feeling for the significance of this quantity might come from consideration of the same quantity for elastic scattering in which the final state is identical with the initial state. For this situation, we have

$$\rho_q^{ii} = <i\left|e^{iqr}\right|i> = \int\psi_i^*(r)e^{iqr}\psi_i(r)dr$$

$$= \int\psi_i^2(r)e^{iqr}dr$$

$$= \int\rho(r)e^{iqr}dr \qquad (10.6)$$

where $\psi_i(r)$ is the wave function of an electron in the ground state i
and $\rho_i(r)$ is then the probability density of the electron in that state.
In this picture, ρ_q^{ii} becomes the scattering amplitude due to a wave of
probability density of wavelength q. For elastic scattering ρ_q^{ii} is non-
zero only at values of q equal to reciprocal lattice vectors, i.e. at
points corresponding to Bragg reflections. F is inelastic scattering, ρ_q^{if}
can be regarded as the scattering amplitude of the dynamic charge fluctu-
ation of wave vector q produced by exciting a <u>bound</u> electron from $|i>$ to
$<f|$. This picture is adequate for energy absorption processes that involve just one bound electron undergoing a transition from one electron
state to another.

There are other mechanisms whereby the specimen can absorb energy,
however, in which the electrons interact strongly with each other and in
which many electrons participate. These interactions are described in
general as collective excitations. Perhaps the simplest of these in
terms of the above discussion is the bulk plasmon. This can be regarded
as a charge density fluctuation involving many electrons analogous to
the sound waves that pass through a fluid medium. This would involve a
substantial volume of the sample in comparison with, for example, the
inner-core excitation discussed above which is likely to be a much more
localized fluctuation.

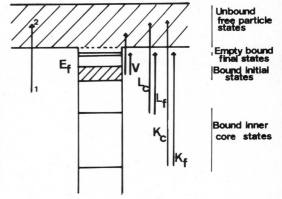

Fig. 10.2. Possible electronic
structure states and transitions
that might give observable spectra. K_F, L_F inner core losses
giving fine structure, K_c, L_c
inner core losses giving continuum
spectra, V, Valence spectrum $1 \rightarrow 2$
Excited electron with kinetic
energy.

Given the above discussion, we can now look at the factors in the
cross-section and identify the ways in which the loss processes can be
identified. The most straight forward version and the most convenient
for inner-shell excitations is the differential cross section in energy
and solid angle Ω which can be put in the form

$$\frac{\partial^2 \sigma}{\partial E \partial \Omega} = \frac{1}{a_0^2} \frac{\left| \rho_q^{if} \right|^2}{q^4} \tag{10.7}$$

where a_0 is the Bohr radius. The final state wave functions $<f|$ due to
enter into ρ_q^{if} have been normalized to unit energy. For present purposes, it is enough to note that the q^{-4} term arises from the Coulomb
interaction and that the ρ_q^{if} contains the information relevant to the

electronic structure. The analogy with elastic scattering does persist in that the structure factor terms relevant to electron diffraction take the form $\rho_G^{i}{}^{'}/G^2$ at reciprocal lattice points and need to be squared to appear in an intensity calculation as in the above.

Other forms of this cross-section appear valuable. For small angles of scattering, the matrix element can be given in an approximate form

$$q <i|\varepsilon q \cdot \underline{r}|f> = q d^{if} \tag{10.8}$$

where εq is a unit vector parallel to q. Here d^{if} is identical with the dipole matrix element which determines the response of the system to an electromagnetic wave polarized along εq. Thus, at small scattering angles the response function becomes comparable to that relevant for optical properties. Two generalizations appear. First, it is common to introduce an optical oscillation strength in such studies. In this case a generalized optical oscillation strength can be introduced, defined as

$$\frac{df}{dE} = \frac{2mE}{\hbar^2}\left|\rho_q^{if}\right|^2 \tag{10.9}$$

At small angles of scattering, this agrees with the dipole form but it differs at large angles. The second generalization, of more detailed concern here, is the introduction of the frequency and wave vector dependent dielectric function $\varepsilon(q,\omega)$. Since both electrons and electromagnetic fields interact with the sample through an electric field, we should not be too surprised at this. For electron losses, the cross-section can be put in the form

$$\frac{\partial^3\sigma}{\partial E\,\partial^2 q} = \frac{1}{2\pi^2}\cdot\frac{1}{a_o nE_o}\cdot\frac{1}{q^2}\ \text{Im}\ -\frac{1}{\varepsilon(q,\omega)} \tag{10.10}$$

$$= \frac{1}{2\pi^2 a_o nE_o}\cdot\frac{1}{q^2}\ \frac{\varepsilon_2}{\varepsilon_1^2 + \varepsilon_2^2}$$

where $\varepsilon(q,\omega) = \varepsilon_1 + i\varepsilon_2$ gives the real, ε_1, and imaginary, ε_2, parts of the dielectric function n is the electron density. At $q \to 0$, these will tend to the appropriate optical equivalents. We emphasize that this is an equivalent form to that given above but which tends to be convenient for a number of purposes. In discussing plasma losses, for example, the real part of the dielectric function for simple metals like aluminum will go through zero at the plasma frequency, ω_p. Then the energy loss function, $\text{Im}-\frac{1}{\varepsilon}$, goes through a maximum with a peak value of ε_2^{-1}. At high energy losses $\varepsilon_1 \to 1$ and $\varepsilon_2 \ll 1$. Then the energy loss function tends to $\varepsilon_2(q,\omega)$, i.e. it matches the optical absorption. We note that this is a very general formulation and, for some situations, the optical constants as computed or measured can be used to generate the electron loss profiles. The effect of damping, ε_2, is sometimes sufficient to move the peak in the energy loss function considerably away from the condition $\varepsilon_1 = 0$. Similarly, not infrequently, peaks arise in Im-ε^{-1} even though ε_1 does not go through zero but is in a rising pattern with frequency. Such effects do represent collective responses of the solid, but are superimposed upon a background response that inhibits the spontaneous response represented by $\varepsilon_1 = 0$. In general, the complex range of responses is possible of which the bulk plasmon and single particle excitations are two extreme representatives.

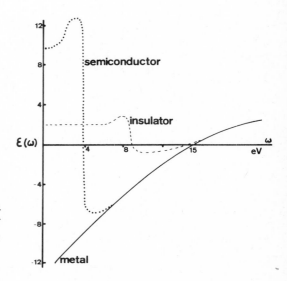

Fig. 10.3. A sketch of common dielectric functions $\varepsilon(\omega)$ for typical materials, e.g. Al, SiO$_2$ and Si.

10.3 INNER-CORE EXCITATIONS

We turn now to consideration of the electronic structural information available from inner-core excitations. In general, the unfilled energy levels can be either discrete (corresponding to unfilled band states) or a continuum (corresponding to free particle states) in terms of the energy dependence. Thus, at the onset of an edge, such simple considerations suggest that fine structure can arise reflecting the discrete levels. We note the importance of this in that since a particular edge of an atomic species has been identified then exploration of the energy levels in the immediate vicinity of that atom is feasible. At small angles of scattering, the cross-section as indicated above will reflect the particular directionality of the q-vector and will probe certain directions in the way that polarized electromagnetic radiation does. Thus, in a crystal, the final states might well have a directionality injected by the crystal structure and a probe of the angular dependence of the cross-section will reveal that directionality. Selection rules provide another useful separating key in identifying such states. Essentially, at small angles, dipole selection rules will apply and these can be used to label components parts of the final state wave function as s, p, d, etc. Fine structures could also arise from structure in the initial state wave function. Clearly in all cases, small changes can be expected from changes in the local chemical environment and thus the local chemistry can be probed. At large scattering angles, different selection rules come into play and it is reasonable to expect to see different transitions arise.

Fig. 10.4. Sketch of core edge structure
A) Fine structure arising from transitions to unoccupied bound states.
B) Excitation to continuum states. Low
intensity modulations (~ 5% intensity) in
this region are the EXAFS structure.

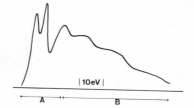

| 10eV |

A   B

All of the above considerations reflect features within the vicinity of the core edge where the discrete bound levels are expected to dominate the final state wave function. In principle a number of factors
will play a role in the exact details. The joint density-of-states and
the matrix elements ρ_q^{if} are clearly very important. However, other questions might also be significant in a given case. The excited electron
leaves behind a positive hole. In some cases, it seems possible that
the final state might be a band state not included directly in the band
structure but resulting from the Coulomb attraction between the positive
hole and the electrons and at an energy below the appropriate bound
levels. This type of effect is often termed an exciton or an excitonic
effect. Complete identification will often entail a substantial calculation. Another effect which may reduce the intensity at the onset of
the appropriate edge associated with the other electrons in the solid is
the centrifugal barrier. In some cases, this can lower the intensity
considerably over a distance of fifty volts or so above the edge and
make it difficult to study the features of interest.

A final component of the edge spectra of some considerable current
interest is the possibility of observing extended x-ray absorption fine
structure. This does not strictly represent electronic structure but
rather atomic structure surrounding the atom in question. The observation is of an intensity modulation superimposed on the slowly decaying
intensity after the edge. The modulation has a magnitude of perhaps 5%
of the background intensity with wavelengths on the order of 20-50 eV.
The interpretation of the modulation rests on the behavior of the excited electron in the final state. This electron can be scattered from
the neighboring atoms giving spherical waves reflected back to the
origin. These effects must be included in the final state wave function. When these waves interfere constructively or destructively in the
vicinity of the initial state then maxima or minima result in the appropriate cross-section as reported. Analysis of this modulation is not
straight forward. First the background has to be stripped and the resultant spectra, suitably weighted, has to be Fourier transformed.
Good statistics (i.e. high accuracy) appear to be necessary, since the
modulation initially is no more than 5% of the background. Typically,
the x-ray spectroscopists look for an accuracy of order ~ 0.01%, although data of lower precision has been recorded and used. Electron experiments, for various reasons, appear not to have been taken to comparable levels although it is not clear that there are any good reasons why
this is the case. A particularly valuable experiment at present seems
to be a comparison of an analysis for both x-rays and electrons.

10.4 VALENCE ELECTRON EXCITATIONS

We turn now to a somewhat more complex area at low energy loss. The more valuable form of the cross-section is probably the dielectric formulation although a number of effects involving single particles are seen. The dominant excitation at low energies is the bulk plasmon, normally occurring when the real part of the dielectric function is zero. It has long been a feature of inelastic electron scattering. Recently, interest has been renewed in details of the plasmon dispersion, particularly as a means of study of the damping mechanisms in different directions in a single crystal. Current understanding suggests that this is due to single electron excitations across band gaps and it appears possible that these measurements can be used to give an indication of the values of these parameters, in particular crystalline directions. At the large angles of scattering, conservation of energy and wave vector make it possible for the plasmon to give up its energy directly to excitation of electron-hole pairs across the Fermi surface, a mechanism known as Landau damping. In principle, the shape and wave-vector dependence in this region can be used as an indication of exchange and correlation effects in the electron gas. It seems clear, however, that some considerable effort will be necessary before this is solidly established.

Somewhat weaker features in the spectrum at low energies are somewhat more rewarding. As indicated earlier, dipole selection rules are applicable at small angles of scattering. At large angles, divergences from the selection rules can occur resulting in the possible observation of optically forbidden transitions. Such effects have now been seen and appear clearly as a different way to probe the band structure. Comparison with the inner core would be a valuable adjunct. The factors determining the occurrence of peaks include the joint density-of-states and matrix elements and, in some instances, these two will show a dependence on momentum transfer (i.e. the angle of scattering). Where this is possible, it then becomes feasible to trace out the location of these maxima as a function of orientation.

In anisotropic crystals, the above techniques become particularly valuable. In many respects, the information might parallel similar information available through optical reflectivity, but the electron loss data appears in many respects to complement the optical reflectivity. The ability to change both the incident orientation and the angle of scattering appears to be particularly valuable at very small scattering angles ($< 10^{-4}$ radians). Again, the data obtained under these conditions is equivalent to the uses of polarized optical radiation, but retrieval of the necessary optical constants seems to be more straight forward in the electron case. Complications do arise in the low loss regime due to two effects that can be exploited. These are losses due to surface modes and to Cerenkov radiation. Surface losses appear in a number of guises. Perhaps the best known is the surface plasmon which comprises charge oscillations on an interface between two media (e.g. metal-oxide) with an exponentially damped electromagnetic field into either medium. It can be identified through the dependence on the angle of incidence with respect to the surface and gives more information with respect to the dielectric function. In a second version of this excitation, the electromagnetic field inside the specimen can be sinusoidal instead of exponentially damped. Such excitations now are electromagnetic waves propagated inside the slab (e.g. oxide-semiconductor-oxide) with the surfaces acting as a wave guide. Excitations of this nature are different from bulk plasma oscillations in that they include this transverse electromagnetic wave propagating inside the specimen.

Losses due to the creation of electromagnetic radiation propagating inside the crystal also arise as a result of retardation effects. An electron travelling faster than the speed of light in a given medium will be accompanied by the emission of electromagnetic radiation. If the electron speed is v and velocity of light in vacuo is c_0 then the condition becomes $v > c_0 \sqrt{E(g,w)}$ or $E(g,w) > \beta^2$.

$$v > \frac{c_0}{\sqrt{\varepsilon(q,\omega)}} \quad \text{or} \quad \varepsilon(q.w) > \beta^2 \qquad (10.11)$$

For 75 keV electrons, for example, $\beta = 1/2$ and the condition becomes $\varepsilon(q,w) > 4$.

10.5 FINAL COMMENTS

The bulk of this chapter will have given the reader an impression of the wide variety of electronic structure information available through inelastic electron scattering. Some fairly obvious comments can be made. First, within an electron microscopy context this complements the microstructural information and vice versa. The momentum dependence of inelastic scattering is unique, providing valuable insights through both identification of optically forbidden transitions and through orientation dependence. The range of losses is broad and covers a useful range between 0 and 1 keV. It provides an exploration of the bound unoccupied states. All of these are unique capabilities that provide a valuable complementary role for this technique in parallel with other techniques such as photoelectron spectroscopy, optical reflectivity and Auger spectroscopy, many of which are surface tools.

We end on a cautionary note. Multiple scattering can cause an occurrence of extra peaks in positions in which they do not belong. Similarly, contamination or alteration of the specimen by the instrument can provide a major disturbance of the observed spectrum and great care has to be excercised (see Chapters 16, 17, and 18).

ACKNOWLEDGMENTS

Various of my colleagues have contributed over the years to the development of my understanding of this topic. Research support by the N.S.F. through the Materials Science Center at Cornell and through DMR7809204, is also gratefully acknowledged.

REFERENCES

Chen, C.H., Silcox, J., and Vincent, R., "Physical Aspects of Electron
    Microscopy and Microbeam Analysis," ed. Siegel, B., and Beaman.
    publ. John Wiley & Sons, p. 303 (1975).

Colliex, C., Cosslett, V.E., Leapman, R.D., and Trebbia, P., Ultramicro-
    scopy 1, 301 (1976).

Daniels, J., Festenberg, C.V., Raether, H., and Zeppenfeld, K. Springer
    Tracts in Modern Physics. Springer-Verlag, Berlin 54, 77 (1970).

CHAPTER 11

STEM IMAGING OF CRYSTALS AND DEFECTS

C.J. HUMPHREYS

DEPARTMENT OF PHYSICS, ARIZONA STATE UNIVERSITY

TEMPE, ARIZONA 85281 U.S.A. *

11.1 INTRODUCTION

Crystals are like people: it is the defects in them which tend to make them interesting! This chapter describes the use of STEM imaging for the structural characterization of crystalline materials, perfect and imperfect. The object of the chapter is to describe basic principles as clearly as possible, using a minimum of mathematics.

Section 11.2 discusses in some detail the very useful and elegant principle of reciprocity. Although the principle is very simple, its application to practical microscopy often requires considerable care. It is shown that, correctly applied, reciprocity must always be valid for elastic scattering but that invoking reciprocity is not appropriate for those situations in which it cannot usefully be applied.

In Section 11.3 the important problem of the signal-to-noise ratio in STEM images is discussed, and the implications this has for the application of reciprocity. Also image recording techniques of particular interest to materials scientists are mentioned, for example the application of the so-called Z-contrast method of Crewe to crystalline materials. Sections 11.4 and 11.5 discuss in some detail STEM imaging of defects, using reciprocity to interpret STEM images by reference to the corresponding CTEM images which would be formed under reciprocally related conditions.

Section 11.6 considers finer points of interpretation. In particular the failure of the column approximation for the interpretation of some STEM images, and methods of calculating STEM images if it is not appropriate to apply reciprocity. The important and topical question of

---

*On leave from Department of Metallurgy and Science of Materials, University of Oxford, Oxford, England.

the possible advantages of STEM relative to CTEM for the penetration of thick specimens is discussed in detail in Section 11.7. The article concludes with a brief outline of some current developments in STEM.

11.2 PRINCIPLE OF RECIPROCITY IN STEM AND CTEM

The reciprocity principle was originally proved by Helmholtz for the case of the propagation of sound waves. The principle essentially states that if a certain signal is detected at a point A when a source is placed at another point B, then the same signal in amplitude and phase (and hence also in intensity) would be detected at B if the source were placed at A. The principle is a consequence of time reversal symmetry and is therefore applicable to many different problems. In particular VON LAUE (1935) proved the above statement for the case of electron diffraction and elastic scattering, and POGANY and TURNER (1968) extended the theory to include inelastically scattered electrons with small energy loss, for which case there is is an approximate reciprocity of intensities rather than of amplitudes.

This elegant theorem forms the basis of the interpretation of many STEM images. The reason is that under the appropriate conditions STEM images may be shown by reciprocity to be similar to CTEM images (COWLEY, 1969, CREWE and WALL, 1970, ZEITLER and THOMSON, 1970). Hence the very large amount of knowledge that exists concerning the interpretation of CTEM images may be "taken over" and used, in many cases, as the basis for interpreting STEM images.

However, it must be emphasized that the simplicity of the reciprocity principle can constitute a trap for the unwary. Although the principle must always be valid for elastic scattering, in practice it is not always easily applicable, nor is it always appropriate. We consider below aspects of reciprocity of particular relevance to crystals and defects. For further discussion of reciprocity the reader is referred to Chapter 2 by Cowley.

Reciprocity for Electron Microscopes

Consider first the idealized microscopes shown schematially in Fig. 11.1. For the idealized CTEM (Fig. 11.1(a)) electrons emitted from a point source placed at B travel through the lenses, apertures and the specimen and are detected at a point A. We now apply reciprocity (i.e. time reversal) to the complete electron-optical system of Fig. 11.1(a) and we obtain Fig. 11.1(b) as follows: electrons are now emitted from A, which becomes the point source. The electrons in the windings of the electromagnetic lenses in Fig. 11.1(b) are time-reversed relative to Fig. 11.1(a), thus the magnetic fields are reversed in sign, hence in Fig. 11.1(b) the probe lens focuses the electrons onto the specimen as required (note the importance of applying reciprocity to the complete system). The electrons propagate through the specimen and are focused by the imaging lens to a point detector at B. Provided that the electron-optical system in Fig. 11.1(b) is identical to that in Fig.

11.1(a) then using reciprocity the signal at A in Fig. 11.1 (a) must be identical to the signal at B in Fig. 11.1(b). The following points should be noted:

(1) As regards the incident electrons the "top" (entrance surface) of the specimen in Fig. 11.1(a) is the "bottom" (exit surface) of the specimen in Fig. 11.1(b) (see below).

(2) A reciprocity of apertures is necessary. Thus, the angle of incidence, $2\alpha_C$, in Fig. 11.1(a) is equal to the angle of exit, $2\beta_S$, in Fig. 11.1(b): the angle of exit, $2\beta_C$, in Fig. 11.1(a) is equal to the angle of incidence, $2\alpha_S$, in Fig. 11.1(b). (We identify α and β

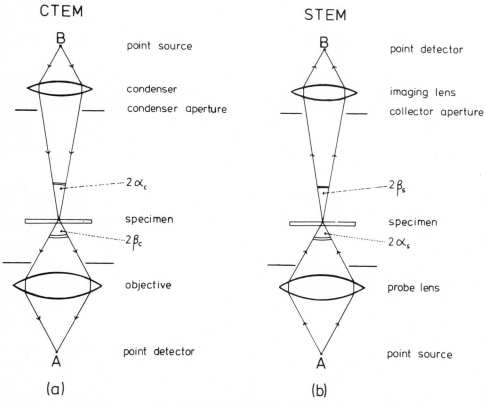

Fig. 11.1. Schematic diagram illustrating the principle of reciprocity applied to CTEM and STEM. (a) Electrons from a point source B in CTEM form an incident cone, angle $2\alpha_C$, on the specimen. The objective aperture allows an exit cone of electrons, angle $2\beta_C$, to be focused by the objective lens to a point detector A. (b) The reciprocity related STEM diagram, obtained by applying time reversal to the CTEM system of (a). In (b) electrons from a point source A are focused by the probe-forming lens into a cone of angle $2\alpha_S$ incident on the specimen. The collector aperture allows electrons in an exit cone, angle $2\beta_S$, to reach the point detector B.

with angles of incidence and exit, respectively and use the sub-
scripts c and s to refer to CTEM and STEM.) More complex apertures
are possible. For example the use of hollow cone illumination in
CTEM is equivalent to an annular detector in STEM.

(3) The electron source and detector are both points (see below).

(4) Although for illustrative purposes we have drawn "geometrical op-
     tics" type ray paths in Fig. 11.1, reciprocity holds whatever the
     ray path between A and B (in practice there will of course be dif-
     fraction).

(5) An identical electron-optical system in Figs. 11.1(a) and (b) means
     that, for example, the spherical aberration coefficient of the CTEM
     objective is the same as that of the STEM probe forming lens. They
     must be identical lenses.

(6) In Fig. 11.1 we have drawn an in-focus imaging situation. Recipro-
     city also applies to out-of-focus imaging since the principle is
     quite general for elastic scattering and a point source and detec-
     tor. For example, in Fig. 11.1(a), (CTEM), consider the specimen
     to be lowered (i.e. moved towards A) so that the Gaussian plane
     (the in-focus plane) is between the specimen and the source. By
     reciprocity, in Fig. 11.1(b) the specimen must also be lowered
     (moved towards A) but the Gaussian plane is now between the speci-
     men and the detector. Identical out-of-focus signals would be ob-
     tained in both cases.

(7) In STEM the incident beam is scanned over the specimen and the time
     resolved STEM image is related to the spatially resolved CTEM image
     by reciprocity (but see later for various causes of the breakdown
     of reciprocity).

Reciprocity and the Coherence of the Source and Detector

     The detector in CTEM is usually a photographic plate or film. The
film is composed of a large number of small grains (typically <1μm
across) each of which acts as a "point detector". The film is not sensitive to phase, i.e. it detects the intensity of the incident beam
point-by-point. In STEM, the imaging lens shown in Fig. 11.1(b) is usually omitted and a finite detector is placed in this plane. A stationary STEM incident beam of angle $2\alpha_s$ forms a convergent-beam electron-
diffraction (CBED) pattern (see Chapter 15 by Steeds) below the specimen
and the finite detector collects electrons from a portion of this pattern. For STEM imaging the incident beam is swept in a raster over the
specimen and the signal from the detector modulates the intensity of the
synchronous display raster. The detector signal is the sum of the electron intensity falling upon it. Again, standard detectors are not phase
sensitive so the signal is an (incoherent) addition of intensities.

A standard heated tungsten filament has a relatively large effective source size >2μm and to a good approximation the specimen is incoherently illuminated, for normal electron-optics. Use of this finite source in STEM corresponds to adding intensities over the equivalent area of the CTEM photographic film. Use of a finite detector in STEM corresponds to adding intensities from all points of a finite incoherent source in CTEM, the source having an equivalent configuration to the STEM detector. Usually this latter condition is not realized in practical STEM and CTEM microscopes. However, in many cases we can calculate STEM images from standard CTEM theory by making use of equivalent conditions (Section 4).

A field emission source, on the other hand, has a relatively small effecitve size <50 Å and coherently illuminates the specimen, to a good approximation. Although a valid reciprocity relationship exists for the points A and B in Fig. 11.1, if the STEM imaging lens in Fig. 11.1(b) is removed and a finte detector is substituted, as is usual, reciprocity is not easily employed since the incoherent STEM detector is equivalent to an incoherent CTEM source, whereas the coherent STEM source is equivalent to a hypothetical coherent CTEM detector (see below and COWLEY, 1976).

The Inapplicability of Reciprocity for Thick Specimens

In propagating through thick specimens the incident electron beam undergoes multiple elastic and multiple inelastic scattering, and the proportion of inelastic to elastic scattering increases as the specimen thickness increases. Since reciprocity is valid only for elastic scattering (single or multiple), and a reciprocity of intensities approximately holds for only small energy loss inelastic scattering (POGANY and TURNER, 1968), it follows that the application of reciprocity progressively becomes inappropriate as the specimen thickness increases (see Section 11.7).

If the crystal contains a defect reciprocity may be inappropriate for another reason, namely that defect scattering followed by inelastic scattering is not always reversibly related to inelastic scattering followed by defect scattering since the order of the scattering is often important (HUMPHREYS and DRUMMOND, 1976a).

Qualitative Reciprocity of the Top-Bottom Effect

The top-bottom effect in both CTEM (HASHIMOTO, 1964) and STEM (see, for example, FRAZER, JONES and LORETTO, 1977) occurs in thick specimens. Although in this case reciprocity is not quantitatively valid, it is still qualitatively useful. The top-bottom effect in CTEM refers to the fact that images of defects near the bottom (exit) surface of the crystal are sharper than images of similar defects near the top (entrace) surface of the crystal. Reasons for this will be discussed in Section 11.7. Applying reciprocity to this we see from Fig. 11.1(b) that in

STEM the top-bottom effect should be reversed, with images of defects near the entrance surface being sharper than those near the exit surface. This is as observed.

Procedure if Reciprocity is not Applicable

If reciprocity is applicable we can calculate STEM images simply by reference to the corresponding CTEM images formed under reciprocity related conditions (Section 11.4). However, if reciprocity is not applicable it is necessary to calculate STEM images from first principles without invoking reciprocity. For example, the incident probe may be decomposed into plane waves the scattering of which follows the standard theory for the problem. This is discussed in Section 11.6.

11.3 IMAGE RECORDING AND SIGNAL/NOISE

STEM image recording has both advantages and disadvantages relative to CTEM which result from the fact that the STEM image is produced as a function of time whereas all points in the spatially varying CTEM image are produced at the same time. In STEM it is therefore straightforward to electronically manipulate (e.g. amplify, integrate, differentiate, etc.) the image in order to enhance the contrast. For example, contrast changes of less than 5-10% are difficult to detect on a CTEM micrograph whereas contrast changes of less than 1% may be detected in STEM images, provided the image intensity is adequate.

Electron amplification produces a bright image screen at all magnifications in STEM. This enables focusing and astigmatism correction to be carried out much more easily in STEM than in CTEM. For example, weak-beam images of defects may be easily seen and focussed at 2×10^6 magnification (BROWN, CRAVEN, JONES, GRIFFITH, STOBBS and WILSON, 1976): this is not usually possible using CTEM. For further details of image processing see Chapters 12 and 13, by Wall and Isaacson, respectively.

Current STEM image display systems are considerably inferior to the photographic film used in CTEM since the effective density of picture points on a 1000-line cathode-ray-tube is much less than that on film. This leads to a resolution attainable in STEM being about a factor of 6x less than that attainable in CTEM if the same area is viewed (BROWN, 1977). This limitation of STEM is not of course a fundamental limitation and it can be overcome by suitable improvements in the recording system: so far suggested methods have been expensive. An advantage of the STEM display over CTEM is the ability to display a line scan which can be used to produce image profiles across defects, for example.

Signal/Noise and Reciprocity

The signal-to-noise ratio is frequently a limitation in STEM imaging. For example, if reciprocity related STEM and CTEM systems are used (with $\alpha_C = \beta_S$ and $\beta_C = \alpha_S$, etc.) then one would expect to see the

same number of Fresnel fringes in the STEM as in the CTEM image. However, usually the STEM image displays considerably fewer fringes (JOY,
MAHER and CULLIS, 1976; COLLIEX, CRAVEN and WILSON, 1977). The reason
for this is not the breakdown of reciprocity but the limiting effects of
signal/noise in the STEM image (signal/noise is not usually a limitation
in CTEM in which the picture points are recorded simultaneously). We
investigate the important problem of signal-to-noise in STEM below. For
simplicity we assume no image processing.

We consider an image (on film or on a cathode-ray-tube screen,
etc.) to be composed of "picture points". A picture point is the
smallest area of the image that the eye can resolve. For example in
STEM consider the image screen to be a square of side 10 cm containing
1000 lines. The eye can resolve about 1000 points per line, hence this
image contains 10^6 picture points each of area $10^{-4} cm^2$. Let the number
of electrons contributing to a picture point be N in a given time. Now
there is a statistical distribution in the emission of electrons from an
electron gun (shot noise) so that if the mean number contributing to the
picture point is N, the statistical variation in this number is $\sqrt{N}$ (i.e.
the standard deviation of N, usually called the noise).

Consider the electron probe to move from a region of perfect crystal to a region containing a defect (or other feature). Let the image
contrast of the defect on the recording display be C, defined as

$$C = \frac{\delta N}{N} \tag{11.1}$$

where N is the mean number of electrons contributing to a picture point
due to scattering by a region of perfect crystal and $(N + \delta N)$ is the
number contributing to a different picture point due to scattering by
the defect. In order to detect visually the defect, the image contrast
of the defect must exceed the statistical contrast variations due to
noise by a certain factor k which is usually taken to be 5 (ROSE, 1948).
Thus, for the defect to be detectable visually on the image screen

$$C = \frac{\delta N}{N} \geq \frac{k\sqrt{N}}{N} = \frac{k}{\sqrt{N}} \tag{11.2}$$

In STEM, let the collector current by I_c, the area of the image
screen be A, the area of a picture point be a, and the recording time t.
Then

$$N = \frac{I_c}{e} \frac{a\,t}{A} \tag{11.3}$$

where e is the electron charge.

From Eqns. 11.2 and 11.3

$$I_c \geq \frac{eA}{at}\left(\frac{k}{C}\right)^2 \tag{11.4}$$

Equation 11.4 gives the collector current required to reveal a contrast
level C. From Fig. 11.1(b) it is clear that for amorphous specimens

$$I_c \cong \left(\frac{\beta_s}{\alpha_s}\right)^2 I_B \qquad (11.5)$$

for $\beta_s < \alpha_s$, where I_B is the incident beam current. Equation 11.5 is
not applicable to crystalline specimens owing to diffraction effects.
However, we will use it as a very rough guide. In order to maximize the
incident beam current in a small spot, α_s is usually made as large as
possible, subject to the condition that the convergent beam discs do not
overlap, i.e. $\alpha_s < \theta_B$ (the Bragg angle). Typically $\alpha_s \sim 10^{-2}$ rad for a
metallic specimen and 100 kV electrons. However, for high angular sen-
sitivity, β_S should clearly be as small as possible. By reciprocity
(see Fig. 11.1), small β_S corresponds to small α_C in CTEM, and it is
known from CTEM that α_C must be small for micrographs to exhibit many
Fresnel fringes or clearly defined bend contours, etc. (see Section
11.4).

The user thus faces a dilemma in STEM operation. To obtain micrographs similar to those in CTEM, reciprocity shows that β_S must be
small. But if β_S is small, I_C is small from Eqn. 11.4, and thus
features of low contrast cannot be detected, from Eqn. 11.3. The effect
has been well demonstrated for the case of Fresnel fringes (JOY et al.,
1976; COLLIEX et al., 1977). The optimum solution is a compromise value
of β_S. The case of crystal defects will be discussed in Section 11.4.
Finally it should be mentioned that for crystalline materials it is possible to enhance the signal/noise somewhat by using slit-shaped collector
apertures with the large dimension perpendicular to g: the small dimension then defines β_S (COWLEY, 1977).

Z-Contrast Applied to Materials

Z-contrast for single atom imaging has been very successful used in
STEM by Crewe and co-workers (see Chapter 13 by Isaacson). The usual
interpretation is more complicated when crystals are present since the
annular detector will detect Bragg reflected beams and phonon scattered
electrons (thermal diffuse scattering) (BROWN, 1977). These signals depend strongly upon diffraction conditions and temperature as well as on
Z. However, the method has recently been used (TREACY, HOWIE and
WILSON, 1979) to image crystalline catalyst particles on both crystalline and non-crystalline supports, and it seems a promising technique
for materials work of this kind.

11.4 THE OPTIMUM BEAM DIVERGENCES FOR IMAGING CRYSTAL DEFECTS

In this Section and in Section 11.5 we consider the imaging of defects in STEM using normal bright-field or dark-field diffraction contrast and show that reciprocity may be very usefully applied (subject to
the conditions outlined in Section 11.2).

Typical Values of α and β in CTEM and STEM

As shown in Section 11.2 and in Fig. 11.1, necessary conditions for CTEM and STEM to be reciprocally related are

$$\alpha_s = \beta_c \tag{11.6}$$

and

$$\beta_s = \alpha_c \tag{11.7}$$

In normal 100 kV CTEM operation the condenser is defocussed and the angle of incidence, $2\alpha_c$, is typically 5 x 10^{-4} rad. The CTEM collecting angle, $2\beta_c$, is typically about 5 x 10^{-3} rad for a medium sized objective aperture.

For STEM, in order to maximize the current in a small spot the angle of incidence, $2\alpha_c$, is usually made as large as possible subject to two conditions. The first condition arises because a stationary convergent incident beam on a crystal results in a convergent beam electron diffraction (CBED) pattern (see Chapter 15 by Steeds for further details). From simple geometry it is clear that the convergent beam discs will overlap if $2\alpha_s > 2\theta_B$, where θ_B is the Bragg angle. For imaging using diffraction contrast we require non-overlapping CBED discs, i.e. $2\alpha < 2\theta$ (overlapping discs are considered in Section 11.8). This sets an upper limit on $2\alpha_s$ of about 10^{-2} rad. Secondly, spherical aberration in the probe forming lens spreads a point source into a disc of radius $C_s \alpha_s^3$ in the focussed beam. Thus a value of $2\alpha_s$ of 10^{-2} rad may produce an unacceptably large spot size (and hence poor resolution). A typical value of $2\alpha_s$ is about 5 x 10^{-3} rad. Bragg reflection also sets an upper limit on the collecting angle $2\beta_s$. A necessary condition for the detector to collect electrons from the whole or a portion of only one diffracted beam disc is that $2\beta_s < 2\theta_B$. In order to maximize the collected signal, and hence the signal/noise (Section 11.3) β_s is usually chosen to be as large as possible, subject to the above condition. Thus typically $2\beta_s \sim 10^{-2}$ rad.

Hence for typical values of the beam divergences Eqn. 11.6 is approximately satisfied, but Eqn. 11.7 is far from satisfied, with $\beta_s \sim 20\alpha_c$.

Effects of Varying β_s on STEM Images

For the above reason typical CTEM and STEM images of the same crystalline specimen usually appear significantly different. In general the typical STEM image contains many fewer thickness fringes and bend contours than the corresponding CTEM image (see Fig. 11.2), and "dynamical" contrast effects in defect images (for example the black-white zig-zag contrast in CTEM of inclined dislocations) are very much reduced in STEM (see Fig. 11.3). The effect on the STEM image of the large collector angle normally used can be assessed by making careful use of the reciprocity principle. If the specimen is not too thick, so that inelastic scattering is small, the STEM image should be identical to the image that would be obtained using CTEM if the incident electron cone angle were increased by the appropriate factor (typically about 20) and the CTEM source is incoherent (corresponding to the incoherent STEM detector). This is essentially the problem of the modification of CTEM image quality due to finite beam divergence (WHELAN and HIRSCH, 1957) but applied to far greater beam divergences than are normally used in CTEM (BOOKER, JOY, SPENCER, VON HARRACH and THOMPSON, 1974, MAHER and JOY, 1976).

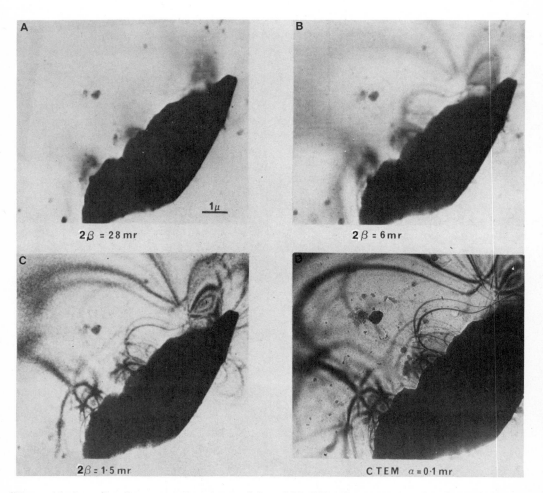

Fig. 11.2. Bend contours in gold. 100 kV electrons. A, B and C are STEM images and D is the CTEM image. In D the incident beam divergence $2\alpha_C = 2 \times 10^{-4}$ rad. In C, B and A, the detector aperture angle $2\beta_S$ is 1.5×10^{-3} 6×10^{-3} and 2.8×10^{-2} rad respectively. Note the rapid "washing out" of the bend contours in STEM as the detector collector angle $2\beta_S$ increases. Note also the poor signal/noise in the STEM image, Fig. C, which has the smallest value of $2\beta_S$. (Micrographs courtesy of Dr. D. C. Joy.)

It is important to stress again that STEM with a finite detector is equivalent to a finite <u>incoherent</u> source in CTEM, so that the CTEM condenser aperture is incoherently filled. The resultant image is calculated by first decomposing the incoherent incident beam cone into a set of beams of angles of incidence varying from $-\beta_s$ to $+\beta_s$ on either side of the mean angle of incidence. Image intensities for each incident beam direction are then calculated and the total image intensity is the sum of the individual intensities. As the STEM collector angle $2\beta_s$ is increased from a small value, corresponding to an increase in angle in the reciprocally related CTEM incident beam cone, it is clear that the image will partially blur and that fine diffraction effects will be averaged out. Such STEM images have sometimes been referred to erroneously as "kinematical images." In fact dynamical thoery must still be used to calculate the images as outlines above, and the averaged dynamical images do not correspond to kinematical images.

Two-Beam Dynamical Theory Interpretation

The two-beam approximation is often a good first approximation for calculating images if only one diffracted beam is strong and all others are weak. This approximation is particularly useful for the analysis of STEM images using reciprocity since it yields simple analytical expressions for bend contours and thickness fringes (see e.g. HIRSCH <u>et al.</u>, 1977, COWLEY, 1975, HUMPHREYS, 1979). These intensity expressions can be integrated by hand for a given incident beam cone corresponding to $2\beta_s$ and hence the effects of varying β_s on the STEM image can be readily calculated (see BOOKER <u>et al.</u>, 1974).

It is convenient to use standard CTEM notation in referring to the orientation of the incident beam with respect to the crystal. Consider Bragg reflection from lattice planes of spacing d and let the corresponding reciprocal lattice vector be $\underline{g}$. Let the crystal be deviated from the exact Bragg position so that the distance of the reciprocal lattice point $\underline{g}$ from the Ewald sphere is s_g. Let the extinction distance for the reflection be ξ_g. The dimensionless deviation parameter w is defined as

$$w = s_{\underline{g}} \, \zeta_{\underline{g}} \qquad\qquad (11.8)$$

If the incident CTEM beam has a finite divergence α_c then by simple geometry this gives rise to a range of s_g values of magnitude $\alpha_c |\underline{g}|$ and hence a range of w values, $w \pm \Delta w$, where

$$\Delta w = \alpha_c \, |\underline{g}| \, \zeta_g \qquad\qquad (11.9)$$

Applying reciprocity, a STEM detector angle $2\beta_s$ is equivalent to an incident CTEM beam from an incoherent source with a range of angles of incidence corresponding to

$$\Delta w = \beta_s \, |\underline{g}| \zeta_g \qquad\qquad (11.10)$$

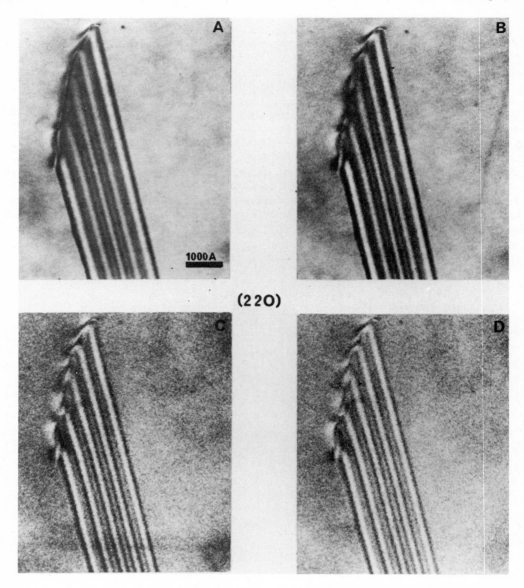

Fig. 11.3. Bright-field STEM images of a stacking fault on an inclined (111) plane in a (001) silicon crystal at the 220 Bragg position. 100 kV electrons. The incident beam divergence $2\alpha_S = 7.6 \times 10^{-3}$ rad, and the detector aperture angle $2\beta_S$ is (A) 1.5×10^{-2}, (B) 6×10^{-3}, (C) 1.5×10^{-3} and (D) 6×10^{-4} rad. Corresponding values of Δw (see text) are 2.77, 1.18, 0.3 and 0.12 respectively. In going from A to D note the increased noise, the increase contrast of the fault, particularly the center fringes, and the increase black-white contrast of the terminating dislocation (to a certain extent this is masked by the increased noise). It is also clear that in A the image has more darker-than-background features than in D (see text). Note that the dynamical defect contrast does not blur out as markedly as the bend contour contrast (Fig. 11.2) with increasing detector aperture angle (see text). (From MAHER and JOY, 1976, courtesy of North-Holland.)

For a given situation values of Δw may be calculated using Eqn. 11.10 (a table of extinction distances is given in, for example, HIRSCH et al., 1977). Experimental and theoreitcal studies (BOOKER et al., 1974, MAHER and JOY, 1976) have shown that for $\Delta w > 1$ the STEM image is often markedly different from the CTEM image owing to the averaging out of dynamical effects. The averaging is most marked for thickness fringes and bend contours, less for stacking faults and still less for dislocations (see Figs. 11.2 and 11.3).

Choice of Optimum β_S Value

In order for STEM images to closely resemble standard CTEM images it is necessary for $\Delta w \lesssim 0.3$. (For 100 kV electrons incident on Cu and using $\underline{g} = (200)$ this corresponds to $2\beta_S \lesssim 0.3 \times 10^{-3}$ rad.) However, images formed with such a small value of Δw may result in an unacceptably low signal-to-noise ratio. If $\Delta w > 1$ the STEM image will normally appear considerably simpler than the standard CTEM image since defect images will be on a less structured background, as thickness fringes and bend contours will have been largely averaged out. This may be a distinct advantage for some applications, for example counting defect densities. For applications such as these it is advantageous to use the maximum β_S value consistent with the defect retaining good contrast.

However, it should be remembered that the dynamical averaging resulting from a large β_S value always results in loss of information, particularly orientation and thickness information. It is therefore desirable for most applications to use a compromise value of β_S: a value sufficiently large for good signal/noise and sufficiently small to minimize the information loss. A suitable compromise value for many applications is $\Delta w \sim 1$ (BOOKER et al., 1974, MAHER and JOY, 1976). For low index reflections (220 and lower) this implies, from Eqn. 11.10, values of $2\beta_S \sim 10^{-2}$ rad for heavy atomic weight elements, $\sim 5 \times 10^{-3}$ rad for medium weight elements, and $\sim 3 \times 10^{-3}$ rad for light elements. For higher index reflections both g and ξ_g are greater and hence values of $2\beta_S$ must be less than those given above if CTEM-image-type information is to be retained.

11.5 THE IDENTIFICATION OF CRYSTAL DEFECTS

Well-known methods are available in CTEM for the characterization of crystal defects, for example the determination of the Burgers vector of a dislocation or the displacement vector of a stacking fault (see HIRSCH et al., 1977). Using reciprocity, it is to be expected that these same methods are available in STEM, although the large detector collection angle usually used in STEM will blur out finer points of characterization, for example oscillatory contrast on inclined dislocations. It has been confirmed experimentally (BOOKER et al., 1973, MAHER and JOY, 1976) that the usual $\underline{g} \cdot \underline{b}$ and $\underline{g} \cdot \underline{R}$ invisibility criteria for dislocations and stacking faults respectively are applicable to STEM. For completeness some basic "rules" of defect identification are given briefly in the rest of this section with comments on any modifications necessary if large STEM collection angles are used.

Properties of Dislocation Images

It has not been completely proved that different defects give rise to different and unique sets of images. However, HEAD (1969) has shown that subject to certain conditions (the validity of the column approximation, two-beam or many-beam systematic diffraction conditions, and a direction in the crystal along which the displacements are constant), the component of the displacement field in the direction of the diffraction vector $\underline{g}$ is uniquely specified by the corresponding electron micrograph. It follows that a set of three micrographs taken with noncoplanar diffracting vectors uniquely identify a defect and specify its displacement field. For the general n-beam case, three independent micrographs also serve to uniquely identify the defect displacement field, although it is no longer true that there is a direct connection between one micrograph and one component of the displacement field.

The above proof, although restricted, is very important and it underlies the "rules" given below. It should be noted that two-beam or many-beam systematic diffraction conditions yield diffraction contrast images which are easier to interpret than n-beam images. Hence both in STEM and CTEM standard diffraction contrast experiments use so-called "two-beam conditions" in which the direct beam and one diffracted beam are much stronger than all other diffracted beams.

For a dislocation to be invisible in an image the Bragg reflecting planes operating must be undistorted by the dislocations. For example for a screw dislocation in an isotropic medium planes parallel to the screw axis remain flat. Hence if the diffraction vector $\underline{g}$ is perpendicular to the Burgers vector $\underline{b}$ no image contrast is observed. The following results apply to STEM images of screw dislocations in an isotropic medium:

(a) If $\underline{g} \cdot \underline{b} = 0$ the dislocation image is invisible. Thus, the Burgers vector can be determined by finding two sets of planes, $\underline{g}_1$ and $\underline{g}_2$, for which the invisibility criterion is satisfied. $\underline{b}$ is then parallel to $\underline{g}_1 \wedge \underline{g}_2$.

(b) If $\underline{g} \cdot \underline{b} = 1$ and the dislocation lies near the middle of a reasonably thick foil oriented near the Bragg position, the image consists of a single dark peak of similar shape in both bright-field and darkfield. The dark peak is a result of anomalous absorption.

(c) If the crystal is very thin, so that absorption is negligible, bright-field and dark-field images are complementary owing to electron conservation. However, most specimens are sufficiently thick for absorption to be important.

(d) If the dislocation is inclined, running from the top to the bottom of a specimen the CTEM image appears like a zig-zag line, with black and white oscillations in contrast near the top and bottom of the specimen. This dynamical diffraction effect is blurred in STEM images, the degree of blurring corresponding to the beam divergences. A theoretical study not using the column approximation reveals that the blurring is such that the dark part of the contrast is clear, but the visibility of the bright part is substantially reduced (HUMPHREYS and DRUMMOND, 1976a). Thus, STEM images of defects often appear darker than background whereas CTEM images have more brighter-than-background features. This is observed experimentally.

(e)   The image peak has a width of about $0.2\xi_g$, i.e. typically in the range 40-200 Å depending upon the material.

(f)   For $g \cdot \underline{b} = 2$ the image of a screw dislocation near the middle of a reasonably thick foil consists of two dark peaks.

(g)   Provided the crystal is not very thin, the bright-field image of an inclined screw is symmetrical about the center, while the dark-field image is asymmetrical, being similar to the bright-field image near the entrace surface of the foil and pseudo-complementary to it near the exit surface. (Note that the entrace surface in CTEM is the exit surface in STEM and vice versa, see Fig. 11.1.) Thus, the sense of the inclination of the dislocation may be determined in STEM in a similar manner to CTEM. (Note that in both CTEM and STEM, bright-field and dark-field images are similar for the defect near the entrance surface and pseudo-complementary for the defect near the exit surface. This can be proved by a careful application of reciprocity.)

For an edge dislocation in an isotropic medium to be invisible both $g \cdot \underline{b}$ and $g \cdot \underline{b} \wedge \underline{u}$ must be zero for the diffracting planes to be undistorted by the dislocation. These conditions are fulfilled by choosing a diffracting vector which is along the line $\underline{u}$ of the dislocation. Edge dislocation images are qualitatively similar to the screw dislocation images described above except that the image width is greater than that of the screw by a factor of about two, owing to the different strain field.

For a mixed dislocation no crystal planes remain flat, neither are any crystal planes flat for pure edge or pure screw dislocations in anisotropic materials. In these cases there are no simple invisibility criteria and the Burgers vector can only be found by detailed comparison of theory and experiment using, for example, computed half-tone micrographs (HEAD, HUMBLE CLAREBROUGH, MORTON and FORWOOD, 1973). Such comparisons have only been made for CTEM images, and it is unlikely that STEM images would be as suitable owing to the dynamical averaging produced by the finite collector angle, which reduces the information content of the STEM image.

Using reciprocity it is clear that the CTEM weak-beam method of imaging dislocations (COCKAYNE, RAY and WHELAN, 1969, see also Chapter 20 by Vander Sande) may be used in STEM (BROWN et al., 1976), and the characteristics of such images will be broadly similar to the CTEM case, although the column approximation may break down more markedly in the STEM case because of the greater beam divergences (see HUMPHREYS and DRUMMOND, 1976a and 1976b and Section 11.6).

Properties of Stacking Fault Images

In STEM, as in CTEM, a stacking fault image will be invisible if
$\underline{g}^{.}\underline{R} = n$, where $\underline{R}$ is the displacement vector and n is an integer. Hence,
$\underline{R}$ can be determined by obtaining invisibility using non-orthogonal diffracting vectors. Stacking fault inclination can be determined from a
bright-field/dark-field pair taken at the Bragg position. The images
are similar at the crystal entrace surface and pseudo-complementary at
the exit surface. (This can be proved using reciprocity.) The extrinsic or intrinsic nature of the fault follows from whether the first and
last bright-field fringes are bright or dark. Further details of the
characterization of CTEM images are given in HIRSCH et al., (1977), and
thanks to reciprocity, this large amount of information accumulated over
many years can be "taken over" and applied to STEM images, provided that
reciprocity is applicable (see Sections 11.2, 11.6 and 11.7). Normally
reciprocity will be applicable at least qualitatively for diffraction
contrast images unless the specimen is very thick or very high resolution images are involved.

11.6   THE BREAKDOWN OF THE COLUMN APPROXIMATION IN STEM

We have considered so far the main features of STEM for the conventional imaging of defects. In the remainder of this chapter we discuss
some of the finer points of interpretation. However, it must be pointed
out that very basic questions, such as "what limits the penetration in
STEM?" cannot be answered without a knowledge of these finer points. As
in the previous sections of the chapter we endeavor to explain concepts
as simply as possible and to give a minimum of mathematics.

The Nature of the Column Approximation

Image contrast calculations of defects in CTEM are normally based
upon the column approximation (WHELAN and HIRSCH, 1957, HIRSCH et al.,
1960). If STEM image contrast is deduced from column-approximation calculations of CTEM image contrast by invoking reciprocity then clearly
this STEM image interpretation has the column-approximation "built in".

The column approximation divides the crystal into columns, usually
taken to lie in the z direction parallel to the reflecting planes operating. The displacement in a given column due to a defect is assumed to
be a function of the z-coordinate only and is represented by $\underline{R}(z)$. The
basic assumption of the approximation is that each column may be chosen
sufficiently narrow that the displacement within it is essentially only
a function of z, yet sufficiently wide that an electron entering the top
of a column is not scattered out of the column in its passage through
the crystal. Column widths are typically taken to be about 4 Å, but are
often greater than or less than this depending upon the image detail required.

The column approximation is very good for normal medium-resolution
electron microscopy (HOWIE and BASINSKI, 1968, JOUFFREY and TAUPIN,

1967). The column approximation also works well for reasonably high
resolution images (∼10 Å) if the incident beam is close to an exact
Bragg position or close to the symmetry position (i.e. parallel to the
crystal planes) (HUMPHREYS and DRUMMOND, 1976b). There are three main
reasons for this. First, at an exact Bragg position, or the summetry
position, the direction of the electron flux within the crystal is on
average parallel to the Bragg planes and hence parallel to the columns
(the direction is in fact normal to the dispersion surface at the wave
points excited). Second, scattering by defects (elastic diffuse scattering) is mainly through very small angles which does not appreciably
deviate the electrons within the crystal from the Bragg position, hence
the electron flux remains approximately parallel to the columns (but for
very high resolution imaging of defect cores it is the diffuse scattering through larger angles which is of interest). Third, in CTEM the
incident beam normally has a very small divergence (relative to θ_B) so
that essentially all of the beam can be near an exact Bragg position.

If the crystal is not oriented at a Bragg position or symmetry position, the electron flux is not contained within a column in the z direction, but is scattered from one column to another. For an imperfect
crystal the columns are not structurally equivalent to each other and
there is therefore a net "sideways" scattering in the direction of the
beam of interest. Hence the column approximation will hold much less
well when the crystal is not at an exact Bragg position. HUMPHREYS and
DRUMMOND (1976b) have shown that the sideways shift is linearly related
to thickness and was 70 Å in a 2000 Å thick foil, for the example they
studied. If the incident beam is parallel, however, this sideways shift
occurs without the defect image changing its shape (to within about 5
Å). In this case the column approximation could still be employed (to
an accuracy ∼ 5 Å) if the columns were redefined to lie along the direction of the diffracted beam being studied (HUMPHREYS and DRUMMOND,
1976b). If the incident beam is convergent the defect image is not only
displaced sideways, its shape changes and it is broadened (Fig. 11.4).

In STEM, the angle of incidence $2\alpha_s$ is large, typically 5 x 10^{-3}
rad. Hence even if the crystal is at an exact Bragg or symmetry position for the mean angle of incidence, a considerable fraction of the incident beam cone will be deviated from the Bragg position. Thus, for
the reasons given above the column approximation will be considerably
less valid in STEM than in CTEM for the same specimen.

High Resolution STEM Image Calculations Without the Column Approximation

For medium resolution imaging it is convenient to consider STEM images in terms of reciprocally related CTEM images for which much information exists, based on column-approximation calculations (see Sections
11.4 and 11.5). For high resolution STEM imaging in which the column
approximation is not valid it is not useful to invoke reciprocity and to
consider the equivalent CTEM case since (a) very few CTEM calculations
have been made without using the column approximation and (b) the optimum conditions for STEM imaging are very different from the usual CTEM

conditions. It is preferable to calculate the STEM image ab initio from first principles, i.e. to consider the electrons propagating through the specimen in their direction of motion in STEM.

Two main methods have been used for carrying out image calculations avoiding the column approximation. The first method, in real space, is based on the theory of HOWIE and BASINSKI (1968) in which the full three-dimensional Schroedinger equation is applied to the problem. Calculations relevant to STEM images using this method have been given by HUMPHREYS and DRUMMOND (1976a and 1976b, 1977). Numerical computations using this method involve dividing the crystal into a fine rectangular grid and propagating the electrons downwards and sideways through this grid. Using suitable numerical techniques the calculations can be carried out to any degree of accuracy and the method is fast, being only a factor of 2 or 3 slower than a normal column approximation calculation.

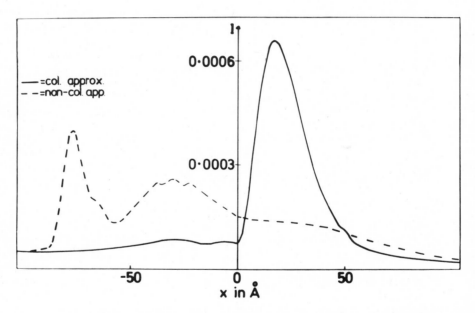

Fig. 11.4. Calculated dark-field weak-beam image of an edge dislocation near the top of a thick copper foil with a slightly divergent incident beam ($2\alpha = 5 \times 10^{-4}$ rad). The solid line calculation uses the column approximation and the dotted line calculation does not. Note the image peak shift, reduced contrast and peak broadening into two peaks when the column approximation is not used. The reduced contrast and peak broadening are due to elastic diffuse scattering by the defect from the divergent incident beam. The peak shift occurs also for a parallel incident beam. This effect gives rise to a top-bottom effect due to purely elastic scattering (see text). Calculation parameters are: $g = 2\bar{2}0$, center of incident beam fan at exact 220 Bragg position, 100 kV incident electrons, dislocation depth ξ_g in a crystal thickness $6\xi_g$, $\xi_g = 418$ Å, $\underline{u} = [11\bar{2}]$, $\underline{b} = 1/2[1\bar{1}0]$, 4-beam theory. (From HUMPHREYS and DRUMMOND, 1976b.)

The second method, in reciprocal space, regards the imperfect crystal as a complex·unit cell of a perfect crystal. In this method (GRINTON and COWLEY, 1971, COWLEY, 1975) a portion of the deformed crystal containing the defect is periodically repeated to form an "artificial" perfect crystal, having a large complex unit cell. Standard perfect crystal computing methods are then applied to this crystal, the diffracted beams being the Bragg beams of the original crystal and additional Bragg beams of the artificial crystal, these latter beams representing in discrete form the elastic diffusion scattering due to the defect. Calculations using this method may also be carried out to any degree of accuracy by increasing the size of the giant unit cell, thus decreasing the separation of the additional Bragg beams so that the diffusion scattering becomes quasi-continuous. Calculations relevant to STEM using this method have been carried out by SPENCE (1978).

A comparison of the two methods has been given by ANSTIS and COCKAYNE (1979) who conclude that the periodic continuation method is best suited to calculations in which significant image detail occurs only over a small region, whereas the Howie and Basinski Schroedinger equation formulation is best suited for calculations over more extended regions. It should also be pointed out that the real-space Schroedinger equation method calculates the electron wave-function on the exit surface of the crystal. If this is Fourier transformed we obtain the diffuse scattering distribution in the diffraction pattern. The Fourier transform procedure effectively assumes periodic continuation of the region of crystal considered, thus the real space and reciprocal space methods are linked. Apertures of any desired type can be inserted in the diffraction pattern plane and the wave-function Fourier transformed again to yield the image (HUMPHREYS and DRUMMOND, 1977).

11.7 PENETRATION IN CRYSTALS USING CTEM AND STEM

There has recently been considerable debate and confusion concerning the claims for increased penetration in STEM relative to CTEM in materials science. SELLAR and COWLEY (1973) showed that for a thick (several μm) specimen and 100 kV incident electrons STEM would have an advantage over CTEM by a factor of about 3. GROVES (1975) suggested that the factor could be as high as 10. These calculations were for amorphous specimens, and it is clear that experimentally the penetration in STEM is superior to that in CTEM for amorphous materials (e.g. biological). For crystalline materials the situation is more complex and it is necessary to understand the various factors which may contribute to limiting the penetration.

Definition of Penetration

Many definitions of penetration are possible and may be equally valid in their context. We choose the most useful definition for materials scientists to be the maximum usable specimen thickness in which defects can be characterized. Characterization is of course important: it is usually of little use to merely detect a faint "flying saucer" in the image without being able to characterize this. Others have defined

penetration in terms of the maximum specimen thickness for a given resolution. This is a useful definition for some applications, but in most
cases materials scientists studying thick crystals are satisfied with a
moderate resolution ∿ 40 Å provided the image feature can be identified.

We must also define the type of experiment used to detect and identify defects. The maximum usable specimen thickness will be very different for lattice imaging, weak-beam imaging and strong-beam imaging,
for example. Again the most useful definition for most purposes is a
conventional strong-beam diffraction contrast image. Finally it is
desirable that defects are visible throughout the foil thickness, or
else the experiment may yield misleading information.

We therefore choose a working definition of penetration to be the
maximum usable specimen thickness in which defects can be characterized
throughout the foil thickness using a conventional strong-beam diffraction contrast experiment.

Factors Limiting the Penetration in CTEM (W Filament)

The main factors which may limit the penetration in CTEM are loss
of intensity, contrast and resolution. For high atomic weight materials
(e.g. gold) experiments by HUMPHREYS, THOMAS, LALLY and FISHER (1971) at
1 MeV and by FRAZER and JONES (1975) at 100 kV have clearly shown that
loss of intensity is the dominant factor limiting the penetration. As
pointed out by FRAZER and JONES (1975) the reason is that for high
atomic weight materials inelastic scattering due to phonons has a very
high cross-section (see HUMPHREYS and HIRSCH, 1968). In addition, the
phonon-scattered electrons themselves only very weakly preserve image
contrast (see REZ, HUMPHREYS and WHELAN, 1977). It is therefore to be
expected that the penetration in heavy materials is intensity limited.

For light atomic weight materials (e.g. silicon, aluminum) the situation is more complex. In a theoretical study HIRSH and HUMPHREYS
(1968) found that chromatic aberration would limit the penetration (i.e.
result in an unacceptable resolution of much worse than 40 Å, see above)
in light atomic weight materials at 100 kV but not in heavier materials,
nor in any material at 1 MeV. This was supported by the experimental
study of FRAZER, JONES and LORETTO (1977). They found that for light
atomic weight materials at 100 kV any attempt to improve the image intensity by increasing the size of the objective aperture resulted in
chromatic aberration producing an unacceptable resolution of worse than
50 Å. The theoretical reason is that the dominant inelastic scattering
mechanisms in low atomic weight materials are plasmon and single electron excitations. Both of these mechanisms preserve contrast well in
electrons which lose energy (HOWIE, 1963, HUMPHREYS and WHELAN, 1969)
and hence both mechanisms contribute to resolution loss due to chromatic
aberration. For 1 MeV electrons the inelastic scattering cross-sections
are reduced by a factor of about 3, and experiments show that chromatic
aberration is no longer a limiting factor (HUMPHREYS et al., 1971).

For medium atomic weight materials (e.g. Cu) the factor limiting the foil thickness is loss of contrast for defects near the crystal entrance surface, defects near the exit surface remaining sharp (FRAZER et al., 1977). This top-bottom effect (HASHIMOTO, 1964) also limits the penetration in low atomic weight materials if the objective aperture is restricted so that chromatic aberration is not dominant (FRAZER et al., 1977). The mechanisms responsible for the top-bottom effect will be discussed later.

Penetration in STEM

Using a tungsten filament gun in STEM the penetration in all materials is found experimentally to be limited by lack of image intensity, which is due to the low intensity of the incident beam (FRAZER et al., 1977). The basic reason is that in order to characterize defects, e.g. resolve stacking fault fringers, a resolution of $\sim$ 40 Å is required. This requires a spot size $\sim$ 30 Å. The available current in such a small spot using a thermionic gun is extremely small (JOY, 1974) and in all cases limits the penetration in STEM to less than that in CTEM.

Using a STEM with a field-emission gun the dominant factor limiting the penetration in high atomic weight materials is lack of image intensity (FRAZER and JONES, 1975), as in CTEM and for the same reasons (see above). In light atomic weight materials chromatic aberration cannot be a limiting factor in STEM with no post-specimen lenses. The limiting factor in both light and medium atomic weight materials in STEM is found experimentally to be the top-bottom effect (FRAZER et al., 1977).

The Penetration in STEM and CTEM

The conclusions of the above results are:

(a) The penetration of any material using CTEM is better than that using STEM with a tungsten filament.

(b) A FEGSTEM has better penetration than a CTEM with a W filament in light atomic weight materials due to chromatic aberration in the CTEM image. The limitation in the FEGSTEM image is the top-bottom effect.

(c) For medium atomic weight materials both FEGSTEM and CTEM are limited by the top-bottom effect and experimentally the penetration is similar.

(d) For heavy atomic weight materials both FEGSTEM and CTEM are limited by lack of intensity. However, at present the current available in a 30 Å spot using a FEGSTEM with a field of view of 25 μm^2 is about 10 times less than that in CTEM from a tungsten filament in a 5 μm spot. Thus at present the penetration in heavy materials in CTEM is better than in STEM.

The Top-Bottom Effect

The top-bottom effect is the factor limiting the penetration of many materials in STEM and CTEM. An important question is whether this limitation can be removed by, for example, energy filtering the image. We therefore need to consider the origin of the top-bottom effect.

In CTEM, images of defects near the exit surface of a thick foil are much sharper than those near the entrance surface. In STEM images of defects near the entrance surface are sharper than those near the exit surface. Thus information is lost from the exit surface of the specimen in STEM and from the entrance surface in CTEM. Although through-focal series result in changes in image contrast the basic top-bottom effect described above remains: it is not a focussing effect (FRAZER et al., 1977).

Possible origins of the top-bottom effect are beam divergence, in-elastic scattering mechanisms that preserve contrast reasonably well, (i.e. plasmon and single electron excitations) and elastic diffuse scattering.

FRAZER et al. (1977) considered that the top-bottom effect was due to multiple contrast-preserving inelastic scattering. Phonon scattering is unlikely to be a mechanism since the contrast preservation is weak (REZ et al., 1977) and also since the scattering is mainly through larger angles so that only a small proportion of phonon-scattered electrons enter the objective aperture. Plasmon and single-electron excitations both preserve contrast well and both, especially plasmon, are through relatively small angles (those single electron excitations with large energy loss preserve contrast less well and scatter the incident electrons through large angles). Thus a mechanism for the top-bottom effect is certainly the contribution to the image of multiply inelastically scattered electrons due to plasmon and single-electron excitations. Such electrons may of course reach the image after multiple small-angle events in the same direction or in opposite directions, the important point being that they eventually enter the objective aperture in CTEM or the detector collector aperture in STEM.

In geometrical optics, beam divergence would not give rise to a top-bottom effect. However, beam divergence can give rise to a breakdown of the column approximation at quite low resolutions (Section 11.6). Figure 11.4 shows that even a small amount of beam divergence coupled with elastic diffuse scattering from a defect can lead to a top-bottom effect owing to the sideways spreading of electrons in the crystal. There is therefore a contribution to the top-bottom effect due to purely elastic scattering (HUMPHREYS and DRUMMOND, 1976b and HUMPHREYS, 1977). This contribution will be passed on to the inelastically scattered electrons (due to plasmon and single-electron excitations) by the usual contrast preservation mechanism.

In practice the situation is complex. What is required is a full dynamical calculation including multiple elastic, multiple inelastic and defect scattering for a convergent incident beam. This has not been attempted. It is not yeat clear whether the major contribution to the top-bottom effect is due to multiple elastic or multiple inelastic scattering. If the dominant mechanism is inelastic scattering then energy filtering could considerably improve the penetration. If multiple elastic scattering makes a significant contribution, then energy filtering is unlikely to be useful, which is the opinion of the author. The effect of energy filtering on the top-bottom effect is clearly an experiment which needs performing. It is of interest to note that as regards penetration using 100 kV microscopes for many materials we appear to have reached an inherent specimen limitation rather than being instrument limited.

11.8 CURRENT DEVELOPMENTS IN THE STEM IMAGING OF DEFECTS

Post Specimen Lenses

For diffraction contrast or for lattice imaging it is essential that the crystal can be precisely oriented using the diffraction pattern, and that the detector can be accurately placed. Thus variable magnification of the diffraction pattern is highly desirable, and also a parallel recording method. Hence post-specimen lenses are required (see JOY and MAHER, 1976).

On-Line Optical Image Processing

COWLEY and SPENCE (1979) have recently described an optical processor for a STEM. In this system two post-specimen lenses form a diffraction pattern of the desired size on a fluorescent screen and the optical pattern is conveyed outside the vacuum using fiber optics. The pattern is then amplified in an image intensifier and fed into an optical system in which selected diffraction discs, or portions of discs, are used to form the image.

Lattice Imaging

In order to resolve the lattice periodicity in CTEM at least two beams must pass through the objective aperture and interact coherently. Similarly in STEM two or more convergent beam discs must overlap (i.e. the angle of incident $2\alpha_S > 2\ \theta_B$) and interact coherently. The necessity for coherent interaction implies the use of a sufficiently coherent electron source, i.e. a field-emission gun. By collecting electrons from a region of overlap a high contrast lattice image is formed. The theory of STEM lattice imaging has been given by SPENCE and COWLEY (1978) and we may expect an increasing amount of experimental work in this area.

High Voltage STEM

The greatest single advantage of high voltage CTEM over 100 kV CTEM has been increased penetration. We may expect similar increases in penetration in HVSTEM relative to 100 kV STEM, although it is doubtful whether HVSTEM will offer any significant advantages in penetration over HVCTEM for crystalline materials (see Section 11.7). The greatest single advantage of HVSTEM is likely to be in high resolution. The minimum probe size of the incident beam is proportional to $C_s^{1/4} \lambda^{3/4}$, hence reducing the probe can either be accomplished by reducing C_s significantly (because of the 1/4 power relation) or by reducing λ by operating at higher voltages, which is the easier route. The gun brightness also increases at higher voltage.

In-Situ Imaging and Analysis

The advantages of STEM for analysis are well known and described elsewhere in this volume. The STEM is in principle a very powerful instrument for the complete characterization of, for example, oxidation, reduction and catalysis processes in-situ. A specific example might be the chemical analysis by XRS or EELS of small catalyst particles, the characterization of their defect structure by diffraction contrast, the characterization of their surface structure by high resolution imaging, and the study of changes in the above during an in-situ experiment using an "environmental cell". The major advantage of STEM over CTEM as regards defect imaging is the flexibility and multi-facility approach that STEM offers. The major advances in STEM are likely to lie in developing this flexibility for diffraction and imaging and in developing its use as a multi-purpose facility.

ACKNOWLEDGMENTS

The author is grateful to Professor J. M. Cowley for discussions and for providing facilities for the writing of this article, and to Dr. J. C. H. Spence for many illuminating discussions. He is also grateful to Yvette Auger for typing this very attractively!

REFERENCES

Anstis, G. R. and Cockayne, D. J. H., 1979, Acta Cryst., (in press).

Booker, G. R., Joy, D. C. Spencer, J. P., Harrach, H. Graf von, and Thompson M. N., 1974, Proc. Sevent Ann. Symp. on Scanning Microsc., ed. Om Johari, IIT Res. Inst., Chicago, p. 225.

Booker, G. R., Joy, D. C., Spencer, J. P. and Humphreys, C. J., 1973, Proc. 6th Ann. Symp. on Scanning Microsc., ed. Om Johari, IIT Research Institute, Chicago, p. 251.

Brown, L. M., 1977, in Developments in Electron Microscopy and Analysis, 1977, ed. D. L. Misell, The Inst. of Physics, Bristol and London, p. 141.

Brown, L. M. Craven, A. H., Jones, L. G. P., Griffith, A., Stobbs, W. M. and Wilson, C. J., 1976 in Scanning Electron Microscopy/1976, ed. Om Johari, IIT Research Institute, Chicago, p. 353.

Cockayne, D. J. H., Ray, I. L. F., and Whelan, M. J., 1969, Phil. Mag., 20, 1265.

Colliex, C., Craven, A. J. and Wilson, C. J., 1977, Ultramicroscopy, 2, 327.

Cowley, J. M., 1969, Appl. Phys. Lett., 15, 58.

Cowley, J. M., 1975, "Diffraction Physics", North Holland, Amsterdam.

Cowley, J. M., 1976, Ultramicroscopy, 2, 3.

Cowley, J. M., 1977, in High Voltage Electron Microscopy, 1977, eds. T. Imura and H. Hashimoto, Japanese Society Electron Microsc., Kyoto, p. 9.

Cowley, J. M. and Spence, J. C. H., 1979, ULtramicroscopy, 3, 433.

Crewe, A. V. and Wall, J., 1970, Optik, 30, 461.

Frazer, H. L. and Jones, I. P., 1975, Phil. Mag., 31, 225.

Frazer, H. L., Jones, I. P. and Loretto, M. H., 1977, Phil. Mag., 35, 159.

Grinton, G. and Cowley, J. M., 1971, Optik, 34, 221.

Groves, T., 1975, Ultramicroscopy, 1, 15.

Hashimoto, H., 1964, J. Appl. Phys., 35, 277.

Head, A. K., 1969, Aust. J. Phys., 22, 43 and 345.

Head, A. K., Humble, P., Clarebrough, C. M., Morton, A. J. and Forwood, C. J., 1973, "Computed Electron Micrographs and Defect Identification", North Holland, Amsterdam.

Hirsch, P. B., Howie, A., Nicholson, R. B., Pashley, D. W., and Whelan, M. J., 1977, "Electron Microscopy of Thin Crystals", Krieger, New York.

Hirsch, P. B., Howie, A. and Whelan, M. J., 1960, Phil. Trans. Roy. Soc., A252, 499.

Hirsch, P. B., and Humphreys, C. J., 1968, Proc. Fourth Eur. Reg. Conf. on Electron Microsc., Rome, p. 49.

Howie, A., 1963, Proc. Roy. Roc., A271, 268.

Howie, A., and Basinski, Z. S., 1968, Phil. Mag., 17, 1039.

Humphreys, C. J., 1979, Rep. Prog. Phys. (in press).

Humphreys, C. J. and Drummond, R. A., 1976a, Proc. Sixth Eur. Cong. on Electron Microsc., Jerusalem, p. 176.

Humphreys, C. J. and Drummond, R. A., 1976b, Proc. Sixth Eur. Reg. Conf. on Electron Microsc., Jerusalem, p. 142.

Humphreys, C. J. and Drummond, R. A., 1977, Developments in Electron Microscopy and Analysis, 1977, ed. D. L. Misell, The Inst. of Physics, Bristol and London, p. 241.

Humphreys, C. J. and Hirsch, P. B., 1968, Phil. Mag., 18, 115.

Humphreys, C. J., Thomas, L. E., Lally, J. S., and Fisher, R. M., 1971, Phil. Mag., 23, 87.

Humphryes, C. J. and Whelan, M. J., 1969, Phil Mag., 20, 165.

Joy, D. C., 1974, Advances in Analysis of Microstructural Features by Electron Beam Techniques (Metals Soc. London), p. 20.

Joy, D. C. and Maher, D. M., 1976, Proc. Sixth Eur. Cong. on Electron Microsc., Jerusalem, 1, 170.

Joy, D. C., Maher, D. M., and Cullis, A. G., 1976, J. Microsc., 108, 185.

Jouffrey, B. and Taupin, D., 1967, Phil. Mag., 16, 703.

Laue, M. von, 1935, Ann. Phys., 23, 705.

Maher, D. M. and Joy, D. C., 1976, Ultramicroscopy, 1, 239.

Pogany, A. P. and Turner, P. S., 1968, Acta Cryst., A24, 103.

Rez, P., Humphreys, C. J. and Whelan, M. J., 1977, Phil. Mag., 35, 81.

Rose, A., 1948, Adv. in Electronics, 1, 131.

Sellar, J. R. and Cowley, J. M., 1973, Proc. 6th Ann. Symp. on Scanning Microsc., ed. Om Johari, IIT Research Institute, Chicago, p. 243.

Spence, J. C. H., 1978, Acta Cryst., A34, 112.

Spence, J. C. H., and Cowley, J. M., 1978, Optik, 50, 129.

Treacy, M. M. J., Howie, A., and Wilson, C. J., 1979, Phil. Mag., (in press).

Whelan, M. J. and Hirsch, P. B., 1957, Phil. Mag., 2, 1307.

Zeitler, E. and Thompson, M. G. R., 1970, Optik, 31, 258, 359.

CLASSIC REFERENCES

(a) General References on Electron Microscopy

Cowley, J. M., 1975, "Diffraction Physics," North Holland, Amsterdam.

Hirsch, P. B., Howie, A., Nicholson, R. B., Pashley, D. W. and Whelan, M. J., 1977, "Electron Microscopy of Thin Crystals", Krieger, New York.

Humphreys, C. J., 1979, "The Scattering of Fast Electrons by Crystals", Rep. Prog. Phys. (in press).

The first two references above are standard books on electron microscopy, written from different viewpoints. The third reference is a review article written at a simpler level than the books.

(b) Reciprocity

Cowley, J. M., 1969, Appl. Phys. Lett., 15, 58.

Cowley, J. M., 1976, Ultramicroscopy, 2, 3.

Pogany, A. P. and Turner, P. S., 1968, Acta Cryst., A24, 103.

Pogany and Turner (1968) uses reciprocity to deduce a large amount of information concerning CTEM images. It is not an easy paper to read but it demonstrates the power of the reciprocity principle and it repays careful study. COWLEY (1969) is a short definitive letter relating STEM to CTEM by reciprocity. COWLEY (1976) examines the question of reciprocity for finite sources and detectors, coherent and incoherent.

(c) Reciprocity Related Defect Images in STEM and CTEM

Booker, G. R., Joy, D. C., Spencer, J. P. and Humphreys, C. J., 1973, Proc. 6th Ann. SEM Symposium, Chicago, p. 251.

Booker, G. R., Joy, D. C., Spencer, J. P., Harrach, H. Graf. von, and Thompson, M. N., 1974, Proc. 7th Ann. SEM Symposium, Chicago, p. 225.

Maher, D. M., and Joy, D. C., 1976, Ultramicroscopy, 1, 239.

The above papers cover most aspects of the interpretation of STEM images of defects by reference to reciprocity related CTEM images.

(d) Penetration in STEM and CTEM

Frazer, H.C., Jones, J. P. and Loretto, M. H., 1977, Phil. Mag., 35, 159.

A careful study of penetration including experimental results on the top-bottom effect.

(e) Lattice Imaging in STEM

Spence, J. C. H. and Cowley, J. M., 1978, Optik, <u>50</u>, 129.

The theory of producing lattice fringe images in STEM.

CHAPTER 12

BIOLOGICAL SCANNING TRANSMISSION ELECTRON MICROSCOPY

J. WALL

BIOLOGY DEPARTMENT, BROOKHAVEN NATIONAL LABORATORY

UPTON, NEW YORK 11973

12.1 INTRODUCTION

Since the demonstration of the visibility of single heavy atoms (CREWE et al., 1971) and unstained biological molecules (CREWE and WALL, 1970) using the scanning transmission electron microscope (STEM), biologists have been excited by the potential of the STEM for biological structure determination. Although heavy atom selective staining will probably become the most important future application of the STEM, several other applications are already well established and will be essential to development of selective staining. These applications derive from the incoherence (not to be confused with the common usage of this term) of the signal measured by the STEM annular detector, particularly if large angle scattering (40-200 mRadian) is recorded (FERTIG and ROSE, 1977). Coherence length is a measure of the distance between two scattering centers at which it is possible for the presence of the second center to influence scattering detected from the first. FERTIG and ROSE calculate that this coherence length for the conditions normally used in the STEM (.015 radian illumination half angle, .04-.20 radian annular detector acceptance, see Fig. 12.1) in the plane of the specimen is ∿0.9 Å and in the direction of the beam, ∿30 Å. Thus, it is unlikely that more than two atoms could give rise to coherent scattering, except in a crystalline specimen. This means that the STEM large angle annular detector signal is proportional to the number of atoms in the beam at any instant, the proportionality constant being the atomic scattering cross-section, $\sigma_A (Å^2)$. It can be shown that if the specimen is scanned with a focused or defocused beam to give a uniform dose $D(el/Å^2)$, the number of scattered electrons from a given atom will be $D \cdot \sigma$, independent of the focal setting. This linearity between STEM large angle detector signal and specimen mass thickness makes quantitative interpretation of STEM images straightforward, even at high resolution. The major limitation to interpretability then becomes radiation damage.

Fig. 12.1. Comparison of imaging and electron collection in the conventional and scanning transmission electron microscopes. Scattered electrons are excluded from the image in the conventional microscope and shifted from the axial bright field detector to one of the annular detectors in the STEM. In some instruments the axial detector is replaced with an energy loss spectrometer for detection of inelastically scattered electrons.

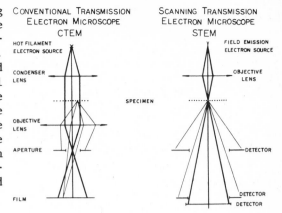

Several authors have stated that the dark field signal is proportional to the square of the specimen thickness. This may be true for a perfectly coherent dark field image (plane illumination, axial beam stop conventional microscope). However, for the incoherent dark field STEM image there is evidence that the linear approximation is valid (see below).

The linearity of the STEM annular detector signal can be employed in a number of useful ways: 1) For unstained specimens the number of carbon atoms in a given area can be determined, giving the total mass, mass per unit length, or mass per unit area. 2) For selectively stained samples, the increase in scattering power gives a measure of the number of heavy atoms bound. 3) For negatively stained samples, direct interpretation in terms of mass thickness is possible, since no phase contrast is present (image detail becomes less distinct symmetrically on both sides of focus, does not reverse contrast). Mass measurement will be described in more detail below.

From a practical point of view, the STEM annular detector arrangement provides a convenient means of detecting a large fraction of those electrons elastically scattered from the specimen (LANGMORE et al., 1973). Those electrons can be detected with quantum efficiency using either solid state detectors or scintillators and photomultipliers. In the resulting dark field image, thin samples (single heavy atoms, unstained DNA) are easily visualized above the background scattering of the carbon substrate (~10 Å thick). Image viewing is convenient since both image brightness and contrast are controlled electronically. Finally, the STEM is ideally suited for interfacing to a small computer which can record first scan data and also control the microscope. From the point of view of this chapter, this is one of the most important features of the STEM; having the quantitative data in hand one asks what can be learned from it.

In addition to the mass thickness measurements mentioned above, it is possible to measure each contribution to image signal and image noise, and to measure microscope performance. Our goal in these measurements is twofold: 1) to develop a statistical model which will allow us to predict the probability of success of an experiment given anticipated specimen geometry, probe and detector geometry, and dose to the specimen, and 2) to identify improvements which would be required to make various experiments feasible, and assess progress in making desired improvements. The ability to quantify the concept of image quality will make possible a systematic effort to improve specimen preparation, imaging, recording, and image analysis.

12.2 QUANTITATIVE MEASUREMENT WITH THE STEM

Length

Accurate length measurements may be required for comparison to structures derived from x-ray data, measurement of particle size or contour length (DNA), or measurement of changes in dimensions due to variation in specimen preparation conditions. In the conventional microscope this usually requires use of an internal standard imaged at the same magnification as the object of interest. In the scanning microscope, magnification is controlled by attenuating the scanning currents, which can easily be accomplished with less than 0.1% error. Thus, the scan calibration is important for mass measurement, since the picture element width squared is the basic unit of measurement.

Distortion of the STEM scan can result from several sources: stray magnetic or electric fields and geometric aberrations of the scan coils and lens. The presence of stray fields will be evident at high magnification and can be reduced as required by proper shielding and attention to circuit grounding. Geometrical aberrations, if present, can be corrected to arbitrary accuracy with standard pincushion correction circuitry. Image distortion has been suggested as a limitation on analysis of images of large arrays recorded with the conventional microscope.

Due to limitations on digital storage capacity, most STEM data acquistion systems employ a scan composed of 512 x 512 or 1024 x 1024 picture elements. Care must be exercised, therefore, that the specimen is sampled at sufficiently fine increments to make measurements of the desired accuracy. Most instruments are able to offset the scan raster electrically to record a montage if this is necessary.

Mass

Use of the STEM for mass measurements has been reviewed recently (LAMVIK and LANGMORE, 1977; WALL, 1979). This application arises from the direct relationship between specimen mass thickness and annular detector signal (or energy loss signal) in the STEM. The number of electrons striking the annular detector, n_S, is given in Beer's Law:

$$n_S = DA(1-e^{\frac{-N\sigma}{A}}) \cong DN\sigma \qquad (12.1)$$

where D is the dose (el/$Å^2$) incident on a picture element of area A containing N atoms of a certain type, and σ ($Å^2$) is the cross-section for an atom in A to scatter an electron from the conical illuminating beam onto the annular detector. The approximation is useful for thin specimens (Nσ/A<<1). If more than one type of atom is present, the above equation can be generalized. Values for single atom cross sections for various geometries are available in the literature (LANGMORE et al., 1973; WALL et al., 1974).

If a particle of interest is separated from its neighbors, which are all supported on a uniform substrate, the total number of atoms in the particle can be computed by summing the number of atoms from each picture element within the particle image and subtracting the number of atoms measured within the same total area of "clean" background.

Errors in mass measurement result from four sources: 1) counting noise in the finite number of scattered electrons measured, 2) random variations in substrate thickness which limit the accuracy of background subtraction, 3) radiation damage during imaging (most importantly mass loss), and 4) specimen preparation artifacts. The effects of counting noise and substrate noise (see below) on measurement accuracy are shown in Fig. 12.2. Note that at high dose substrate noise predominates, while at low dose counting noise is dominant.

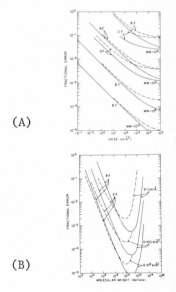

Fig. 12.2. Fractional error is mass measurement, ΔMW/MW, as a function of molecular weight (A) and dose (B) from WALL (1979). A cubic particle of uniform density, (v = 9$Å$/atom) and composed of atoms of uniform atomic number (Z = 6) is assumed. The scattering cross-section is assumed to be 0.01$Å^2$, roughly equal to that of carbon at 100 KeV. In (A) the solid curves represent the result for the bright field signal, and the dashed curves the result for the dark field signal. The substrate is assumed to be 10$Å$ thick. In (B) bright field and dark field curves are indicated (for MW = 10^8 the curves are identical for MW = 10^{10} the D. F. curve is not shown). The solid curves are calculated for a 10$Å$ thick substrate and the dashed curves for a 100$Å$ thick substrate.

(A)

(B)

From the point of view of mass measurement, mass loss is the most troublesome manifestation of radiation damage. At room temperature many biological molecules lose as much as half their mass with a characteristic dose of 10-100 el/$Å^2$. Mass measurements can be corrected for this effect if a dose-response curve for mass loss is measured (see Fig. 12.3). Correction can be done in one of two ways: low dose data can be corrected to zero dose conditions by extrapolation of the dose-response curve, or high dose data can be corrected using the final equilibrium value of mass loss. The first method is preferable for large objects where low dose images give adequate accuracy (see Fig. 12.2), since a correction factor close to unity is used. For smaller objects it may be necessary to use high dose images and apply a correction factor ~2. A better approach for small particles is to average several low dose images of different particles to obtain the desired accuracy. The rate of mass loss can be reduced significantly by cooling the specimen (RAMAMURTI et al., 1975). The rate of mass loss in Fig. 12.3 at -130°C is ~5 times less than that observed at room temperature.

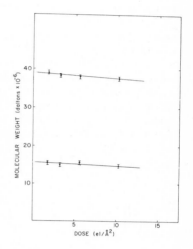

Fig. 12.3. Dose response curve for mass loss. Eight fd ciruses (lower curve) and 5 TMV viruses (See Fig. 12.5) were measured on successive scans, each imparting a dose of 1.2 el/Å². Plotted points are mean values of mass and vertical bars are ± one standard deviation. The initial portion of the exponential decay is fitted with a straight line. At larger doses, the mass becomes stable at approximately 60% of the initial value for TMV and fd.

Specimen preparation technique can result in significant errors if salt is dried close to the specimen (as in negative staining), or if the structure is denatured making parts indistinguishable from the background. Salt artifacts can be avoided by freeze drying from low ionic strength solutions (<10^{-3} molar). Denaturation can often be prevented by mild fixation with glutaraldehyde (little increase in mass). For cold specimens it is important to study the dose response curve for total mass as a function of temperature, since a significant amount of gas or water may be selectively adsorbed to the specimen.

Typical results of mass measurements are shown in Fig. 12.4 for fd phage Tobacco Mosaic Virus (TMV), and apoferritin. A typical area of an fd/TMV specimen is shown in Fig. 12.5.

This mass measurement technique provides a way to identify particles in an image if their molecular weights are known. It has become a useful technique in its own right for several reasons: 1) a small amount of sample is required, 2) particles are measured one at a time so the spread in size as well as the average can be determined accurately, 3) mass per unit length and mass per unit area can be measured directly, 4) complex systems can be studied in the active state (e.g. protein bound to DNA), 5) the mass increase upon binding heavy atoms can be used to calculate the number of heavy atoms bound.

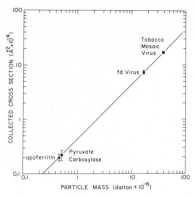

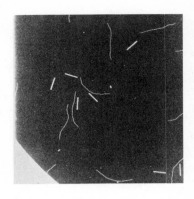

Fig. 12.4. Absolute value of total scattering cross-section measured with the STEM large angle annular detector (40-200 MRadian) vs. particle mass for several test par-. ticles. Vertical bars are ± one standard deviation of the measured distribution (20 to 50 different particles in each case).

Fig. 12.5. Dark field image of fd phage and TMV. This 512 x 512 element image was recorded at a dose of 1.2 el/$Å^2$ using the STEM large angle annular detector (40-200 MRadian, 40% collection efficiency, 100% detection efficiency). Viruses were deposited from solution onto a 20Å thick carbon substrate, freeze-dried, transferred under vacuum, and observed at -120°C. 4.15 micron full scale.

Substrate Noise

It has generally been assumed that the substrate supporting the specimen is amorphous with a statistical fluctuation in the number of substrate atoms in an area A given by $\sqrt{At/\bar{v}}$, where $At/\bar{v}$ is the average number of carbons in the area A. This parameter is of considerable interest for calculation of the expected signal to noise ratio (or statistical error, see Fig. 12.2) for high dose images ($D\sigma \gg 1$), especially heavy atom images. The noise of real substrates can be measured either by computing a power spectrum of the image, or more directly by integrating the number of carbon atoms in an area, A, of the image, and computing a distribution of the values obtained by placing this square at all possible positions within the image (as in mass measurement). The width of this distribution can be characterized by a standard deviation, S.D. Fig. 12.6 shows a typical area of clean carbon film ∿18 Å thick imaged with a probe having a diameter <3 Å (see Fig. 12.9). The bottom area was imaged with the beam defocused to determine the contribution of counting noise to the image. In Fig. 12.7 the width of the histogram distribution, S.D., is plotted as a function of the width of the measuring area. The expected square root dependence is shown for reference. For this particular substrate the noise in a large integrated area is significantly higher than expected, while the noise in a small area is significantly less than expected. This effect does not appear to result from the contrast transfer function of the STEM.

Fig. 12.6. Clean carbon film ~18A thick imaged using the STEM large angle annular detector. The lower portion of the image was recorded with the beam defocussed to ~1000Å to show the contribution of counting noise to the image. 648Å full scale.

Fig. 12.7. Carbon film noise as a function of width of measuring square. The number of carbon atoms inside a measuring square was computed using the mass measuring program described previously. The measuring square was moved over the image, and the measurement repeated. The standard deviation of these repeated measurements is plotted as (X) for the image shown in Fig. 12.6 and (·) for a similar 5188Å width image. Both sets of measurements were corrected to remove the effects of counting noise. Vertical bars represent the uncertainty in the noise measurements based on repeating the measuring procedure on other areas of film.

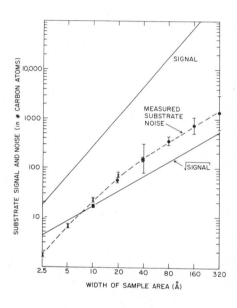

The unexpected dependence of carbon film noise on measuring area may be explained by the observation (OHTSKI et al., 1979) that a carbon substrate appears to be composed of small uniform plates arranged randomly. The data in Fig. 12.7 could be reconciled with plates of about 10 Å width, each containing ~10 atoms. Since the shape of this curve is so important in computing the visibility of both heavy atoms and biological molecules, these preliminary measurements are being repeated. They are shown here only as an illustration of the type of data which can be collected with the STEM. Noise curves for various substrates are being studied to select the substrates most suitable for various applications. The curves shown in Fig. 12.2 (Error in Mass Measurement) may be recalculated with this noise distribution, resulting in an increased error under high dose conditions. The low dose portions of Fig. 12.2 will not change, however, as this is dominated by counting noise which is accurately described by a Poisson distribution.

Heavy Atom Signal

An important parameter in the statistical model for calculating visibility of specimen features is the atomic scattering cross-section as a function of atomic number (7). The cross-section for low Z can be determined quite accurately using standard specimens as described above in the mass measurement section. Three strategies are being employed to extend these cross-section measurements to higher Z values: 1) single atom images are being analyzed using the mass measurement techniques (high dose), 2) heavy atom cluster compounds are being studied using the same programs (moderate dose), and 3) standard mass reference specimens are being modified to incorporate an accurately known number of heavy atoms (1 to 3 per subunit) and the increment in scattering between modified and unmodified specimens determined (low dose). Results of this study will be presented in a later publication.

The feasibility of the first study can be demonstrated by reference to Fig. 12.8, which shows a typical heavy atom specimen and Fig. 12.9, which shows the average intensity as a function of radius for seven atom spots. Spots were selected and centered by computer without interaction, so large numbers of atoms can be analyzed in an objective fashion. Not shown is the integrated scattering as a function of radius, which is also available from this program. Results to date are too preliminary to discuss but appear to give heavy atom scattering cross-sections lower than expected from theory, as previously found by RETSKY (1974). Methods 2 and 3 have also been shown to be feasible and give results consistent with single atom measurements. One advantage of method 3, the study of large molecules specifically labelled, is that it permits measurement of heavy atom loss starting at doses < 1 el/$\overset{\circ}{A}^2$.

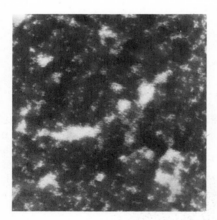

Fig. 12.8. Dark field STEM image of Uranium atoms and small crystals on a thin carbon film at 130°C obtained using the large angle annular detector. 324 $\overset{\circ}{A}$ full scale.

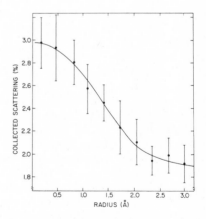

Fig. 12.9. Points represent the average radial intensity profile of 7 single atoms from Fig 12.8. Vertical bars represent ± 1 standard deviation of the measured intensities (see text).

Resolution

One by-product of the heavy atom study described above is production of heavy atom profiles as shown in Fig. 12.9. This profile represents a convolution of the beam profile with the atom profile. At > 2 Å spot size, the atom may be regarded as a pinhole over which the beam is scanned, giving essentially the beam profile. Averaging together several spots tends to reduce the effects of counting noise and substrate noise, while averaging the effects of vibration. It can be seen from Fig. 12.9 that the full width at half maximum of the atom profiles is 2.8 Å. It should be noted that this is slightly larger than the beam profile due to the difficulty in aligning individual atom spots. This provides a rapid and quantitative method for measuring "resolution."

Specimen Modification During Imaging

One of the central problems in biological electron microscopy is the attempt to maximize image information. At low dose the image is statistically poorly defined, at high dose the image may be meaningless, somewhere in between the biological microscopist must struggle to find some reliable information. The STEM is well suited to study specimen modification since dose can be controlled accurately (especially by computer) and many data channels can be used. Radiation damage measurements are described in detail elsewhere in this volume. The use of the STEM for studying mass loss was described earlier. Contamination and etching have also been described elsewhere, however, it is worth noting that in a STEM operating with a vacuum < 10 ^{8} Torr in the specimen chamber and with specimen temperature < -40°C, no contamination is observed on biological specimens, even if they have had no pretreatment (WALL et al., 1977; VOREADES, in press).

The fate of heavy atoms originally bound specifically to a biological molecule is of considerable interest in biological structure studies. COLE et al. (1977) found that osmium atoms bound to DNA were found close to the strand initially, but farther away on subsequent scans. This atom motion was sufficiently great to make it impossible to assign the starting positions of the atoms. We have been studying atom motion as a function of temperature to see if this motion could be reduced by cooling the specimen. Initial results showed significantly less motion at -90°C as compared to -40°C for uranium atoms deposited on a thin carbon film (WALL and HAINFIELD, 1978). These studies are now being extended to cover the range +20°C to -130°C. A further reduction of approximately a factor of 2 has been noted in going from -90°C to -130°C. The fraction of atoms moving between the first and second scans at -130°C is now sufficiently low that it is worthwhile to resume work on heavy atom stained biological molecules.

12.3 CONCLUSION

The analytical capabilities of the STEM have allowed us to determine many of the parameters required to develop a statistical model capable of predicting the image obtainable under a given set of conditions (dose, temperature, detector signal, etc.). Such a model will be useful in determining the experimental parameters necessary to obtain the desired information from a biological specimen, and give confidence that the observed structure is relevant. If the experiment is not feasible, the limiting factors can be identified and overcome if possible. In the course of this work it has been shown that mass measurement with the STEM is a particularly powerful technique, providing the long needed link between the solution biochemical final image. The linearity of the relationship between scattering cross-section and molecular weight is evidence of the validity of the use of incoherent theory for interpreting STEM images.

ACKNOWLEDGMENTS

J. Hainfeld, D. Voreades, and K. Ramamurti provided helpful discussions and encouragement in this work. G. Latham prepared the biological specimens and K. Thompson implemented the computer programs required for mass measurement. This work was supported by the U. S. Department of Energy and by the National Institute of Health Biotechnology Resources Branch, Grant RR00715.

REFERENCES

Cole, M. D., Wiggins, J. W., and Beer, M., 1977, J. Mol. Biol., 117, 387.

Crewe, A. V., and Wall, J., 1970, J. Mol. Biol., 48, 375.

Crewe, A. V., Wall, J., and Langmore, J., 1970, Science, 168, 1338.

Fertig, J. and Rose, H., 1977, Ultramicroscopy, 2, 269.

Lamvik, M. K. and Langmore, J. P., 1977, Scanning Electron Microscopy, (O. Johari, ed.) 401.

Langmore, J. P., Wall, J., and Isaacson, M. S., 1973, Optik, 38, 335.

Ohtsuki, M., Isaacson, M. S., et al., 1979, Scanning Electron Microscopy, (O. Jahari, ed.) in press.

Ramamurti, K., Crewe, A. V., and Issaacson, M. S., 1975, Ultramicroscopy, 1, 156.

Retsky, M., 1974, Optik, 41, 127.

Voreades, D., 1979, Manuscript in preparation. (EMSA)

Wall, J., Isaacson, M., and Langmore, J. P., Optik, 39, 359.

Wall, J. S., 1979, Scanning Electron Microscopy, (O. Johari, ed.) in press.

CHAPTER 13

ELECTRON MICROSCOPY OF INDIVIDUAL ATOMS

M. ISAACSON*†+, M. OHTSUKI†, and M. UTLAUT†+

DEPARTMENT OF PHYSICS+ AND THE ENRICO FERMI INSTITUTE†

UNIVERSITY OF CHICAGO, CHICAGO, ILLINOIS

*PRESENT ADDRESS -- CORNELL UNIVERSITY

13.1   INTRODUCTION

Thirty years ago in a classic paper, Otto Scherzer calculated the contrast expected from a single atom in an electron microscope and concluded that with the resolution about 2A, individual heavy atoms should be visible if they were supported on very thin, low atomic number substrates, [SCHERZER, (1949)]. Since that time, the attempt to visualize individual atoms has been one of the main themes of modern electron microscopy. The goal was ellusive until this decade when the use of dark field techniques enabled such visualization to be achieved in a number of different instruments, albeit not necessarily on a routine basis. [For example, see CREWE, et al., (1970); HENKELMAN and OTTENSMEYER, (1971); HASHIMOTO, et al., (1971); THON and WILLASCH, (1972); WALL, et al., (1974); RETSKY, (1974); DORIGNAC and JOUFFREY, (1976); COLE, et al., (1977). For a more comprehensive list of attempts at the visualization of individual atoms (prior to 1977) the reader is referred to the review by LANGMORE, (1978)].

The rationale for developing single atom microscopy, apart from pure intellectual satisfaction, stems from some of the limits of crystallography. The most fundamental limit is that of necessarily having the structure of interest in crystalline form. Many structures are disordered or amorphous by nature and in biology, many either cannot be crystallized or the crystals produced are too small to be studied by conventional techniques. The possibility of visualizing heavy atoms placed at strategic locations within biological molecules was explored more than a decade ago [MOUDRIANAKIS and BEER, (1965)], but it was not until this decade that the exploration neared reality.

Furthermore, the potential of being able to view atomic configurations directly and to explore the possibilities of directly studying relationships between isolated atoms and clusters brings us to the point of wanting to understand atomic microscopy techniques to see what the future may have in store. Potentials of near atomic resolution microscopy using the Conventional Transmission Electron Microscope are discussed in the paper by COWLEY and IIJIMA (1977). It is not the purpose of this chapter to elaborate on the theoretical details of atomic visibility in the electron microscope, but rather discuss some of the practical aspects of single atom microscopy as it pertains to the Scanning Transmission Electron Microscope (STEM) operated using a single dark field annular detector and to show how far the field has come in the nine years since the first attempt at single atom electron microscopy was published [CREWE, et al., (1970) (see Figure 13.1)].

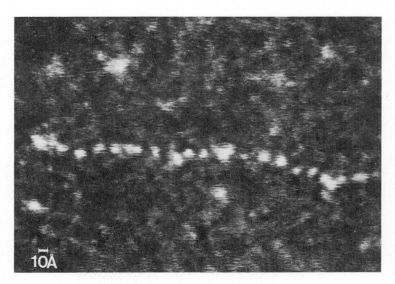

Fig. 13.1. One of the first published micrographs taken with a scanning transmission electron microscope which attempted to demonstrate the visibility of single heavy atoms. In this case the atoms were thorium and the compound was thorium-benzyl tetracarboxylic acid. The smallest bright spots along the chain were presumed to be individual thorium atoms. [CREWE, WALL, and LANGMORE (1970)]. The micrograph was taken using the annular detector signal and an incident electron energy of 37.5 keV. The probe size was approximately 5A.

13.2a BASICS - ELECTRON SCATTERING

What happens when an electron beam (of greater than 10 keV kinetic energy) is incident upon a free atom? These incident electrons either pass by without being scattered by the atom at all, they are elastically scattered, or they are inelastically scattered and lose a non-negligible amount of energy. These three classes of transmitted electrons are separated in space (to some extent) since they have different scattering angular distributions. That is, if we detect these electrons at some plane far from the atom, a plot of the intensity detected at a given angle from the initial direction will be different for the three classes.

The unscattered electrons have the same angular distribution as that of the incident beam. The inelastically scattered electrons tend to be scattered in the forward direction with an angular distribution varying as $[\theta^2+\theta_E^2]^{-1}$ with θ being the scattering angle and $\theta_E = E/pv$ being the characteristic scattering angle for an incident electron of momentum P and velocity V to lose energy E in passing the atom. Thus, for 100 keV electrons ionizing valence shell electrons, $\theta_E \sim 10^{-4}$ rads whereas for inner shell ionizations θ_E $10^{-3}-10^{-2}$ rads. Of course, this intensity distribution is convoluted with that of the incident beam so that if the convergence half angle of the incident beam, α, is much greater that θ_E, the angular distribution of the inelastically scattered electrons is approximately that of the incoming electron beam.

It is the third group of scattered electrons that are of most use for single atom microscopy. These are the electrons that have interacted with the atomic nucleus of the atom and been deflected with negligible energy loss. The angular distribution of these elastically scattered electrons varies approximately as $(\theta^2+\theta_0^2)^{-2}$ where $\theta_0 = \lambda/2\pi a$ is called the screening angle for an atom of radius a and λ is the wavelength of the incident electron beam. Since a is of the order of a fraction of an Ångstrom, θ_0 is of the order of typical Bragg angles for electron scattering in crystals and is generally larger than the convergence half angle, α, of the incident beam. Since a decreases with increasing Z, the angular distribution tends to get broader for the heavier atoms. Furthermore, since the probability of elastic scattering is approximately proportional to $Z^{3/2}$, where Z is the atomic number [e.g., LANGMORE, et al., (1973)], it means that heavier atoms scatter more electrons than lighter atoms and the fractional scattering outside a given angle increases with Z. This is the basis for single atom microscopy in the STEM. An annular detector is placed at the beam exit side of the sample and it collects more scattering due to heavy atoms as opposed to lighter ones.

We have thus far considered only individual atoms. What complications arise when we consider a real specimen consisting of many atoms? The potential complication is due to the fact that the measured intensity, I, on a detector far from the specimen, due to an assembly of N atoms is not necessarily the sum of the individual intensities, I_i. That is $I \neq \sum_{i=1}^{N} I_i$. This departure from the linear sum is due to the coherence of the incident beam and the scattering process. If we assume truly incoherent scattering (i.e., no definite phase relationship between scattering from different atoms) then the intensity is the sum of the individual intensities. If the scattering is coherent, the measured intensity can have maxima and minima depending upon the atomic arrangement.

In practice, if we illuminate the specimen with a very small beam that has a large convergence angle and detect the scattered electrons with an annular detector whose central hole subtends an angle at the specimen greater than the convergence angle of the incident beam, we can, under suitable conditions, neglect the component due to interference (coherence) effects [e.g., FERTIG and ROSE, (1976); and THOMSON, (1973)].

The measured intensity then becomes the sum of the individual intensities due to all the atoms. For the purposes of this chapter, we will assume (and give some justification) that incoherent imaging theory is valid and that the intensity measured with an annular detector depends upon the type and number of atoms illuminated by the incident electron beam.

13.2b   BASICS - OPERATION

In the STEM, the image is formed by scanning a focused beam across a specimen in a raster fashion while detecting the transmitted elastically scattered electrons with an annular detector located at the beam exit

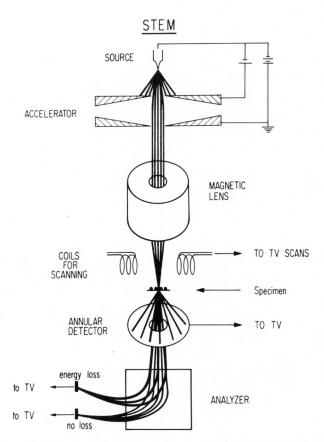

Fig. 13.2. A schematic diagram of the scanning transmission electron microscope (STEM). In the detector configuration shown, one obtains three simultaneous signals from the microscope. Note, that although the specimen is represented as being outside the lens, it is within the lens in systems capable of producing very small diameter probes.

side of the specimen (Fig. 13.2). The diameter of the beam at the specimen plane is fundamentally determined by the spherical aberration of the
final probe forming lens, diffraction from the defining aperture, and
the brightness and monochromaticity of the electron source. Practically,
it can be limited also by the electrical and mechanical stabilities of
all the optical components and by the alignment of all the lenses.

The electron current that strikes the annular detector is used to
modulate the intensity of a synchronously scanned display (usually a CRT
or a TV). In practice, the image is viewed by utilizing a scan converter and a storage device. In this way, the beam can be scanned at slow
frame rates within the microscope ($\sim$10 seconds/frame) to increase the
signal-to-noise in the image, while the image is stored in quasi-real
time and displayed continuously on a TV screen. Complete descriptions
of this kind of display are given in the literature [BECK, (1973); and
ISAACSON, et al., (in press)].

In addition to the annular detector signal, one can also obtain a
signal due to electrons that have lost energy and passed through the
hole in the annular detector, or a signal due to electrons which have
passed through the annular detector hole and lost no energy. In Figure
13.2, these three signals or any linear combination can be obtained and
used to create an "image". For the reader interested in the types of
signals and signal processing available using this basic three signal
scheme, a complete discussion can be found in the review by ISAACSON,
UTLAUT and KOPF (in press).

For imaging individual atoms, the annular detector signal tends to
be the most useful because it is mainly the elastically scattered electrons which strike that detector. Since they are due to collisions with
the nucleus, they are quite local collisions and thus allow us to accurately pinpoint the atom location. Furthermore, since the probability
of elastic scattering is approximately proportional to $Z^{3/2}$, this signal
is greatest for heavier atoms. On the other hand, the probability for
inelastic scattering is only proportional to $Z^{1/2}$ and since an electron
can (with reasonable probability) undergo an inelastic collision with an
atom while still 5-10Å away from it, the inelastic scattering cannot be
used to pinpoint atom locations [ISAACSON, et al., (1974)]. This is
shown explicitly in Figure 13.12.

13.3a PRACTICAL CONSIDERATIONS — ELECTRON OPTICAL

Probe Formation

The whole basis of single atom microscopy using the STEM rests upon
the ability to form an electron probe of diameter as close to atomic
dimension as possible (i.e., in the neighborhood of 2Å. Of course,
this is the case in the CTEM as well. However, other optical considerations must be taken into account for STEM imaging and we will attempt
to briefly discuss all the ingredients necessary to form a near-atomic
dimension probe.

In the STEM, the probe is formed by optically demagnifying the electron soucr. If we assume the source to be infinitely small, the effective aperture of the probe forming lens is then illuminated by an incident plane wave of electrons. For a perfect lens, the diameter of the probe is determined by diffraction at that aperture and is given by $0.61\lambda/\alpha$ where λ is the incident electron wavelength and α is the half angle that the probe subtends at the specimen (i.e., $\alpha = r/f$ where r is the radius of the effective aperture at the back focal plane of the objective lens and f is the focal length). However, the aberrations of the lens limit the size of the aperture angles that can be used and the effects of these aberrations must be balanced with the diffraction effects (since they have different dependences on the aperture angle).

For a lens with a spherical aberration coefficient, C_s, the optimum aperture half angle for forming the smallest diameter probe is given by [SCHERZER, (1949)]:

$$\alpha_{OPT} = (4\lambda/C_s)^{1/4} \qquad (13.1)$$

One can then show [e.g., HAINE and COSSLETT, (1961)] that the intensity distribution in a probe formed with such a lens and an aperture angle given by Eqn. 13.1 is modified Airy disc, with the full width at half maximum (FWHM) and the distance to the first zeroes being the same as for a perfect lens with the same aperture angle. Thus, the probe size is given by $0.61\lambda/\alpha_{OPT}$ and results in:

$$\delta = .43C_s^{1/4}\lambda^{3/4} \qquad (13.2)$$

being the FWHM of the probe intensity distribution. It is this value that ultimately determines the "resolution" in the STEM in the incoherent imaging mode. An image of two point scatterers placed a distance δ apart will have an intensity midway between them of 75% of the peak intensity and thus be "resolved".

Now the approximation of an infinitely small source needs to be considered. The degree to which this is valid depends upon the brightness of the electron source and the current that we require in the probe (we can demagnify any source any arbitrary amount, but could wind up with a negligible current in the probe).

We can show that the current obtainable in a probe of convergence half angle α, produced by demagnifying a source of radius δ_s and specific brightness β is given by (e.g., CREWE (1973)):

$$I_p = \beta\pi^2\alpha^2(M\delta_s)^2V_o \qquad (13.3)$$

where V_o is the total kinetic energy of the electrons and M is the total magnification of the source. Thus, if $\alpha = \alpha_{OPT}$, we get the current in this modified Airy disc to be:

$$I_p = 574\beta(M\delta_s/\delta)^2 \quad \text{(in amps)} \qquad (13.4)$$

if β is given in units of amps/$\overset{\circ}{A}^2$/steradian/eV. $M\delta_s$ is the effective size of the source projected onto the specimen plane and δ is the probe size limit due to diffraction and spherical aberration. The extent to which the actual probe size approaches the limit given in Eqn. 13.2 depends upon the demagnification of the source. That is, we want $M\delta_s/\delta$ to be small. On the other hand, if $M\delta_s/\delta$ is small, so is the probe current unless we use a very high brightness source. In order to obtain adequate probe current for obtaining high resolution pictures in 10 seconds, we require a field emission source. This can give a typical specific brightness of about $\beta \sim 1.4 \times 10^{-11}$ A/Å^2/Ster/eV and thus a probe current of $I_p \gtrsim 8 \times 10^{-11}$A if $M\delta_s = 1/10\delta$.

The total probe diameter is obtained by combining in quadrature the effects of the source size and the diffraction and aberration limits, thus:

$$\delta_{probe} = \delta\sqrt{1+(M\delta_s/\delta)^2} \tag{13.5}$$

We see that for $M\delta_s = 1/10\delta$, δ_{probe} is less than 1% larger than the limit given in Eqn. 13.2. If we want to record a 500 x 500 element picture in 10 seconds (so that focusing and photography does not become too tedious) and have 10^4 electrons incident on the sample per picture element (1% beam statistics), then we would require a beam current of about 4.0×10^{-11} Amps. The field emission source would meed this requirement. On the other hand, the specific brightness of a LaB$_6$ source is orders of magnitude less than that of a cold field emission source and so $M\delta_s$ would have to be sufficiently large such that the probe diameter would be non-negligibly larger than δ. Thus, field emission sources are essential for single atom microscopy.

We have still not considered all the contributions which determine the probe diameter in a STEM, in that we have neglected effects due to chromatic aberration, misalignment and electrical and mechanical instabilities. In order to determine exactly how chromatic aberrations and misalignments affect the probe size, we require wave optical calculations. However, reasonable estimates can be made by assuming these effects to be independent of one another (and of spherical aberration) and adding the contributions of the different aberration discs in quadrature.

The aberration disc due to chromatic aberration depends linearly on α_o and on the ratio $\Delta V/V_o$ where ΔV is the total energy spread of the incident electron beam (of kinetic energy V_o). This includes the natural energy spread of the electrons leaving the source and the stability of the high voltage supply. For cold field emission sources (i.e., those operating at room temperature) the energy spread of electrons leaving the source is about 0.25 eV. Thus, the stability of the high voltage supply need not be much better than this. If we want the chromatic aberration disc to be not more than 20% of the diffraction limited probe size δ, we get that

$$\Delta V/V \leq 0.12(C_s\lambda)^{1/2}/C_c \tag{13.6}$$

where C_c is the coefficient of chromatic aberration. For 100 keV electrons and a system where $C_s \sim C_c \sim 1$mm we get that $\Delta V/V$ must be less than 7 parts per million. Thus, a high voltage stability of at least

0.5 eV would be needed to prevent high voltage instability from increasing the spot diameter a noticeable amount.

The effect of the chromatic aberration of the final lens on the probe size enters also through the electrical stability of that lens. A current fluctuation, ΔI, in the lens results in a change in focal length that has a similar effect to a change in the high voltage. However, since $\Delta I/I \sim 1/2\ \Delta V/V$, the net result is that the lens current stability must be twice as good as the high voltage stability for the same effect. So for the above example, we would require objective lens current stability better than 3.5 ppm.

We are still not finished. It has been implicitly assumed that the aberrations of all the optical elements other than the final probe forming lens are small. In well-designed optical systems, this is indeed the case, since their effect at the specimen plane is demagnified to one extent or the other. Thus, we can ignore them here, but warn the reader to keep them in mind.

The final contributions to the ultimate probe size are practical ones due to misalignment of various of the optical components, either electrical or mechanical misalignments. The two misalignments that we need to be concerned with, are tilt misalignment (the beam comes through the lens at some angle to the optic axis so that the effective image of the source is off axis by a distance r_A) and translational misalignment (the beam comes through the lens at a distance Δr from the axis). The tilt misalignment results mainly in isotropic coma which gives an aberration spot:

$$\delta_{TILT} \sim r_A (r/f)^2 \qquad\qquad (13.7A)$$

where r is the radius of the beam at the entrance to the lens and f is the focal length. Similarly, the translational misalignment results mainly in additional spherical aberration and gives an increase in the abberation spot of:

$$\Delta\delta \sim 3C_s\ r^2/f^3 \cdot \Delta r \sim \Delta r(r^2/f^2) \qquad\qquad (13.7B)$$

Thus, the fractional increase in the spot size (due to misalignments, ΔX) over that due solely to spherical aberration is then given by

$$\delta_{ALIGN}/\delta \sim \frac{2\Delta X(r^2/f^2)}{r(r^2/f^2)} = \frac{2\Delta X}{r} \qquad\qquad (13.8)$$

This then gives us our alignment tolerance ΔX (which is a combination of translational and tilt alignment). If we require the fractional increase in the spot size to be less than 1%, we require the beam alignment to be better than 1/2% of the beam radius. For a 40 μm diameter beam (i.e., a 40 μm objective aperture) we have to align the beam to better than 1000Å from the optic axis. These are just rough estimates. In practice, for some of our single atom microscopy, we align the beam to better than 500Å, a misalignment of 1000 - 2000Å being noticeable in a deterioration of the probe size.

Insofar as the alignment of other lenses in the system are concerned, the alignment tolerances can be relaxed in proportion to the
fractional contribution that that lens makes to the spot size. Therefore, if the optical system is adjusted such that contributions due to
misalignment are less than 1% of the spot size given in Eqn. 13.2 and
the electrical stabilities of the high voltage and objective lens sup-
plies are given by Eqn. 13.6, then to a good approximation the real
probe size can be given by Eqn. 13.5 and with a reasonable demagnification of the source (and the corresponding aberrations) one can get a
probe size very near to the diffraction limit with sufficient beam
current for obtaining micrographs. This limit has been shown to be as
small as 2.4Å in diameter using 40 keV electrons and an objective lens
with C_s = 0.45 mm and 1 mm focal length (e.g., WALL, et al., (1974);
ISAACSON, et al., (1974)]. Such a small probe size then becomes suitable for single atom microscopy.

Further Stability Requirements

In order for successful single atom microscopy to proceed, there
are a few additional stability requirements which must be met; a small
diameter probe with sufficient current in it is not by itself enough.
One needs sufficient mechanical stability. In particular, since one is
projecting a demagnified image of the source on the specimen, the source
must be extremely rigid with respect to the specimen, or else the probe
will appear to move. That is, instead of the image of the probe appearing to be circular in shape as it was scanned across a point object, it
would appear to be jagged (or possibly elongated) due to the vibration
(or motion of the source with respect to the specimen. Even though the
source is demagnified by a factor of 10 to 100, 100Å vibration of the
field emission source could cause an effective vibration of the image
of 1-10Å, a non-negligible effect and one which could preclude effective
single atom microscopy. A good example of such vibration can been seen
in the first micrographs obtained with the STEM that attempted to show
single atoms (CREWE, et al., (1970), see Fig. 13.1].

In a similar vein, the specimen must be mechanically stable as well.
There can be no vibrational instability much greater than one quarter
the probe diameter or else it becomes noticeable. And although slow
mechanical drift of 1/10-1/4 or the field of view can be tolerated, it
is not sufficient for studying diffusion phenomena where one must be
able to observe the same 200-300 Å square field of view for several hours.
In that case, drifts of less than one Å per minute are necessary
(ISAACSON et al., 1976, 1977).

In addition to mechanical stability, one requires electrical
stability of the scans. For example a requirement of 1 Å/minute stability
means only a scan stability of one part in 200 when scanning a 200 Å field
of view. But since real scanning systems must scan areas from about
200 Å to 20 μm in size, this stability becomes a 5 ppm/minute absolute
stability on the scan circuits (i.e., if a 20 μm scan corresponds to
a 10 amp current in the scan coils, then a 1 Å/minute stability when
scanning a 200 Å field of view requires a stability of 50 μamps/minute).

Furthermore, there are similar requirements on the high frequency stability of the scans; we want them to be stable to less than one picture element.

Finally, we require the beam current to be reasonably stable over the period required to focus and obtain a micrograph. The term "reasonably stable" can have a different meaning to a user or to a builder. In practice, it means that fractional beam current fluctuations in the image must be less than the contrast that one expects to see on the specimen.

Typically, absolute beam current stability for cold field emission sources is about 2-6% per 30 minutes, and occasionally with non-optimal vacuum conditions in the electron gun, it can exceed that. There are several methods for reducing beam current instability (after the fact) which rely on normalization procedures. The one we have found to be most successful for use with single atom microscopy is based upon dividing the desired signal by the sum of all the detector signals, this sum being very close to the incident beam current. The rationale is that any beam current fluctuation appears in all the signals and can thus be cancelled by division. Division by the sum of all signals does not change the signal contrast and furthermore allows us to produce a signal that corresponds to the fraction of electrons scattered which strike a particular detector (e.g., ISAACSON et al., 1974; LANGMORE, 1975; CREWE et al., 1975; ISAACSON et al., in press).

13.3b   PRACTICAL CONSIDERATIONS - SPECIMEN PREPARATION

Once we have succeeded in reducing the instrumental factors to acceptable levels we then need to tackle the task of specimen preparation for single atom microscopy. This aspect of the problem is just as important as producing the small probe. The problem boils down to producing supporting films that have: 1) low scattering power; 2) little background that could interfere with visualizing bright atom spots; and 3) little trace of organic contaminants that could cause contamination buildup under the beam.

Low Noise Support Films

As SCHERZER (1949) had pointed out, one of the major impediments to the visualization of individual atoms was to produce substrates of very low mass thickness. In addition to low mass thickness, there are other "noise" problems associated with atom visualization. The background noise of the substrate has to be reduced to tolerable amounts. This noise is due to either thickness fluctuations in the support, or extraneous heavy atoms in the support. The thickness fluctuations produce noise due to the fact that as the beam is scanned across the specimen, there are different numbers of substrate atoms encompassed within the beam cylinder. This number is equal to $\rho T \pi \delta_p^2 / 4$ where ρ is the atom density of the substrate of thickness T and δ_p is the beam diameter. The thinner the substrate, the fewer atoms from which the beam can be scattered; thus, the search for very thin substrates.

However, the fluctuation in the substrate signal will go as the square root of that number for amorphous films. Should this fluctuation be equal to the signal expected from an adsorbed atom, one would have a hard time visualizing the atom with confidence. Thus, in addition to trying to produce thinner substrates, there is the search for crystalline substrates in which these fluctuations occur at regular spatial intervals.

As it turns out, nature has been relatively kind to us in our search for suitable substrates for single atom microscopy. Very thin (<25 Å) carbon films can be routinely produced provided that fenestrated plastic or metal films are used as microgrids to support the carbon films on regular electron microscope grids (see Fig. 13.3). Such films have

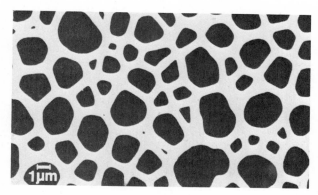

Fig. 13.3. Dark field micrograph showing a typical plastic microgrid used as a support for thin substrates, [OHTSUKI and ISAACSON, (1979)].

been produced by indirect evaporation of a carbon arc onto a mica substrate, using a standard vacuum system (e.g., WHITING and OTTENSMEYER, 1972), by direct or indirect evaporation of graphite rods (National, AGKSP) onto NaCl in an ultra-high vacuum system kept oil-free by ion pumps and cryosorption pumps (e.g., ISAACSON et al., 1974) and by direct evaporation of graphite rods onto either NaCl or mica in a modified standard bell jar system in which all internal parts were replaced with stainless steel ones (OHTSUKI and ISAACSON, 1979). The structure of all such substrate films produced for single atom microscopy seems to be roughly similar in that it does not seem to depend drastically on whether mica or NaCl is used as the substrate onto which the carbon is evaporated and the thickness fluctuations seem to be proportional to the square root of the thickness down to films 10 Å thick (on the average). Films as thin as 2 atomic layers thick (on the average) have been produced and there are indications that these layers are graphite-like (OHTSUKI and ISAACSON, 1979). In Fig. 13.4 we show dark field micrographs of typical amorphous carbon films. The annular detector signal from a 14 Å average thickness film is shown in Fig. 13.5.

Other amorphous substrates made from aluminum oxide, beryllium, and other light elements have been tried and reported to cause less background noise than amorphous carbon films. But the comparisons have not been complete and have not always been at the same thickness so that it is difficult to say whether they in fact are more useful than thin carbon films for single atom microscopy (for a discussion of these attempts see the review by LANGMORE, 1978).

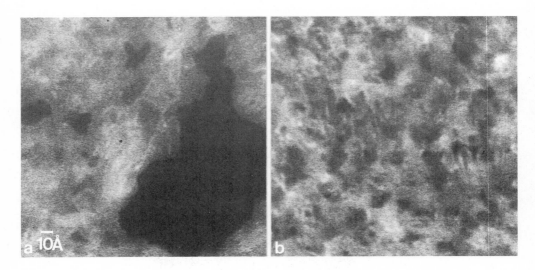

Fig. 13.4. Dark field micrographs of typical thin, clean evaporated carbon films useful for single atom microscopy. The preparation of the films is given in the paper by OHTSUKI and ISAACSON (1979). The films shown here have been evaporated from spectroscopically pure graphite rods onto NaCl at room temperature. (a) A film of 6.8 Å average thickness. The large black area is a hole in the film. (b) A film of 14 Å average thickness. Note the absence of bright spots that could be interpreted as due to heavy atoms.

Fig. 13.5. The CRT trace of the annular detector signal obtained from a 14 Å average thickness carbon film. The fluctuations in the signal (and thus the fluctuations in the number density of carbon atoms) is about 30%. This is consistent with a 2.5 Å diameter beam and a 14 Å average thickness film if one assumes an average density of 0.11 atoms/Å^3.

Attempts have been made to produce low noise crystalline substrates, but no one technique has yet to be reproduced as easily as are thin amorphous carbon films (e.g., see LANGMORE, 1978). Thus, although very thin, low noise substrates can be produced from graphite (see Fig. 13.6), the difficulty in production outweighs the reduction in noise achievable so that most single atom microscopy, at present, is performed using amorphous carbon film substrates.

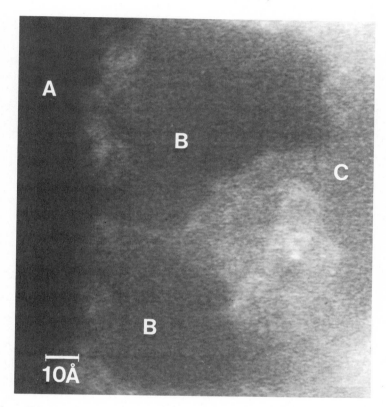

Fig. 13.6. Dark field micrograph of a thin graphite crystal. Region A is a hole; region B corresponds to one atomic layer of carbon atoms; region C corresponds to two atomic layers.

13.3c   PRACTICAL CONSIDERATIONS - CLEAN SUPPORT FILMS

Heavy Atom Contamination

It is not enough just to be able to fabricate thin light element support films in which the fluctuations in thickness are sufficiently small. One must also prevent extraneous heavy atoms from being incorporated into or onto these substrates. The contamination of the substrate by heavy atoms is a serious problem and although it is next to impossible to suggest a recipe for eliminating extraneous heavy atoms, the reader should realize that it is a problem (see Fig. 13.7).

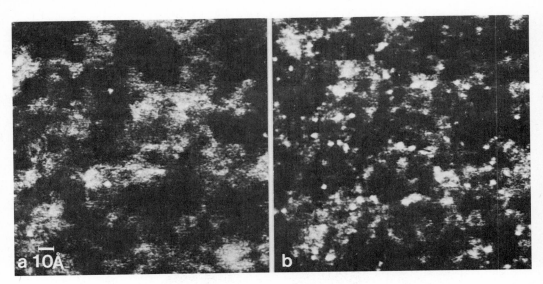

Fig. 13.7. An example of a clean carbon film and one containing heavy atom contaminants. Both micrographs are printed with identical gain and high contrast to emphasize the lack of "spots" in the clean film preparation. (a) a clean film about 15 Å thick prepared using the procedures outlined in the text. Only one heavy atom spot is evident. (b) a similar thickness film prepared using tap water and copper grids.

We have found that with proper preparative and handling techniques, we can reduce the number of extraneous heavy atoms (intense bright spots visualized in micrographs of the substrate) to about 0.5-2 per $10^4 Å^2$ on a clean carbon substrate (LANGMORE, 1975; ISAACSON et al., 1974; OHTSUKI and ISAACSON, 1979). It is found that: 1) many types of carbon rods commonly used for arc evaporation can contribute extraneous heavy atoms to the film; 2) copper EM grids cause "spot" contamination of the film presumably due to the slight solubility of copper in water (which is used to float off the films onto the EM grid); 3) the same level of heavy atom contamination is found on substrates prepared by deposition on mica or NaCl; and 4) some extraneous heavy atoms in the vacuum evaporator and the microscope can diffuse onto the support film.

As a result of such observations and much trial and error, we find that if the following procedures are used, one can routinely keep the heavy atom contamination level below the 0.5-2 per $10^4 Å^2$: use 1) titanium grids and tweezers to minimize heavy metals dissolving in the water; 2) spectroscopic grade graphite rods (National, AGKSP) with listed impurity levels less than 1 ppm; 3) polyethylene and teflon containers, reserved for use with only high Z or low Z solutions; 4) only vacuum components which have been acid etched and rinsed in deionized water as the final cleansing treatment; and 5) charcoal filtered, deionized and millipore filtered water (Millipore Super Q).

13.3d   PRACTICAL CONSIDERATIONS - ORGANIC CONTAMINANTS

Organic contaminaton is just as serious a problem as heavy atom contamination. The level at which one has to reduce organic contamination is much more than for standard electron microscopy. If one is using a substrate 10 Å thick, a fractional monolayer increase of organic contamination can noticeably increase the detected signal as well as change the character of the surface, so that measurements of the distribution or movement of atoms or the surface become meaningless. A 10 Å thick layer of contamination on a film that was 200-300 Å thick would be barely noticeable. But that thickness would double the effective thickness of a 10 Å thick substrate and in some cases this added thickness could be the difference between visualizing an atom or not.

Organic contamination can come from almost anywhere and it is this fact that makes it difficult to eliminate completely. It can come from obvious things such as using tweezers that have not been properly cleaned (i.e., acid etched), not using clean plastic gloves during specimen preparation, not working in a sheltered area, or working in the vicinity of "smokers," to not so obvious things such as having a specimen holder being contaminated by several dirty specimens such that the contamination diffuses along the surface of the holder to any later specimen. It is this surface diffusion nature of contamination that makes it difficult to control. A "clean" plastic glove is no longer "clean" if it has touched a dirty surface. Driving contaminants off the sample by heating will generally only have a positive effect if the sample holder is at a higher temperature than the surroundings. Otherwise organic junk from the surroundings can end up on the sample.

Again, as with the case for reduction of heavy atom contamination, there is no clear-cut method for eliminating contamination but we have evolved procedures that appear to work. The techniques we have found useful in reliably eliminating contamination can be found in the literature (ISAACSON et al., 1979; OHTSUKI and ISAACSON, 1979; LANGMORE, 1978).

13.4   HOW TO VISUALIZE AN ATOM

Since the cross section for elastic scattering of electrons is approximately proportional to $Z^{3/2}$ [LANGMORE et al., (1973)], the number of elastically scattered electrons will increase as the electron beam is scanned across a heavy atom, causing the image of the atom to appear as a bright spot on the video display (see Fig. 13.8). To determine whether or not the atom will be visible on the substrate (or rather, whether the bright spot will be seen over the substrate background), we have to consider the scattering due to the atom relative to the amount of scattering from the substrate. In the discussion to follow, we will assume that there is sufficient current in the beam so that the only source of noise is the thickness fluctuation of the substrate.

The only requirement on the substrate (for atom imaging) is that it be of sufficiently small mass thickness. The maximum allowable mass thickness is determined by the atomic number of the substrate, Z_s, the atomic number of the atom, Z_A, the probe diameter, δ probe, and the substrate structure. There are different criteria that can be used to determine the visibility of an atom on a substrate. We conservatively choose to assume that the atom will be visible if the peak of the elastically scattered signal from the atom, E_{EL}, is at least as large as that from the substrate beneath it, E_s. This is experimentally demonstrated in Fig. 13.8b where we have shown the elastically scattered current detected with an annular detector as a 2.4 Å diameter electron beam was scanned across a mercury atom on an approximately 20 Å thick carbon film substrate. With the above definition, we obtain the following relationship that [ISAACSON et al. 1976)]

$$\rho T \leq (Z_A/Z_S)^{3/2}/1.06 \; \delta^2_{probe} \qquad (13.9)$$

where ρ is the atom density of the substrate of thickness T.

For amorphous carbon, $\rho = 0.11$ atoms/Å³, thus Eqn. 13.9 tells us that for an 18 Å thick carbon film, and a probe diameter of 2.4 Å, all atoms would be visible on the substrate that had an atomic number greater than 30. Note that if the probe diameter is only 5 Å our atom visibility is severely reduced such that with the above criteria, only atoms with atomic number greater than 80 would be visible and this would put severe restrictions on the technique.

The line scan in Fig. 13.8b is a good example of the size of the beam. Since the atom size is much less than the beam size, the full width at half maximum of the atom spot intensity correspond to the beam diameter. In this case, it is about 2.4 Å.

At the beginning of this chapter, we mentioned that our discussion of image interpretation would be based upon incoherent imaging theory (i.e., the intensity in the image is just the sum of the scattering intensity from the individual atoms). That this assumption is reasonable in many cases of interest in single atom microscopy is shown in Fig. 13.9. Here we have a dark field micrograph (obtained with the annular detector signal) of a microcrystal containing uranium atoms. The separation between these atom "spots" is 3.3 Å, and since the probe size in this instance is 2.5 Å, we see that the intensity dip between atoms is almost 50% of the peak intensity (a larger dip than we would get if they were just separated by the beam diameter). Furthermore, in this crystal, one can see the 1,2,3 atom layers. These are indicated in Fig. 13.9b where we see the raw CRT trace of the annular detector signal as the beam is swept across the crystal. That the intensity levels of these layers are integral multiples of one another demonstrates that in this case (where the angle that the hole in the annular detector subtends at the specimen is twice the incident beam convergence angle) the image can be interpreted assuming incoherent imaging. This is to be expected based upon calculations of partially coherent scattering by FERTIG and ROSE (1977).

As an example of the different types of atom samples that have been studied, we show in Fig. 13.10, six different kinds of samples (ISAACSON

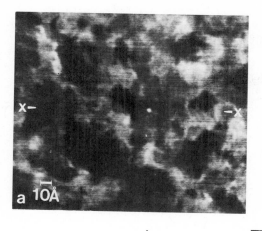

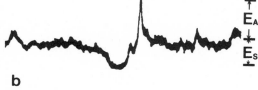

b

Fig. 13.8 A single mercury atom deposited on an approximately 20Å thick film substrate. The micrograph in (a) was recorded with the dark field annular detector signal using a 2.4Å diameter electron beam of 43.5 keV kinetic energy. In (b) is shown the CRT trace intensity as the beam is scanned across the specimen along the line indicated by X in (a). The variations in the backgound signal, E_s, are thickness variations in the substrate due to the fact that there are only about 10 atoms on the average within the beam envelope. The full width at half maximum of the signal due to the heavy atom is a measure of the beam diamater. [ISAACSON, LANGMORE, and WALL, (1974)].

Fig. 13.9. (a) A dark field micrograph of a microcrystal containing uranium atoms. The specimen was prepared by solvent evaporation of a very dilute solution of uranyl acetate [OHTSUKI and ISAACSON, (1979)]. (b) The CRT trace of the annular detector signal as the beam is scanned across one row of atoms is the microcrystal. The numbers refer to the number of atomic layers of uranium.

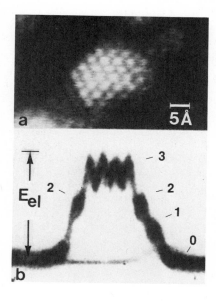

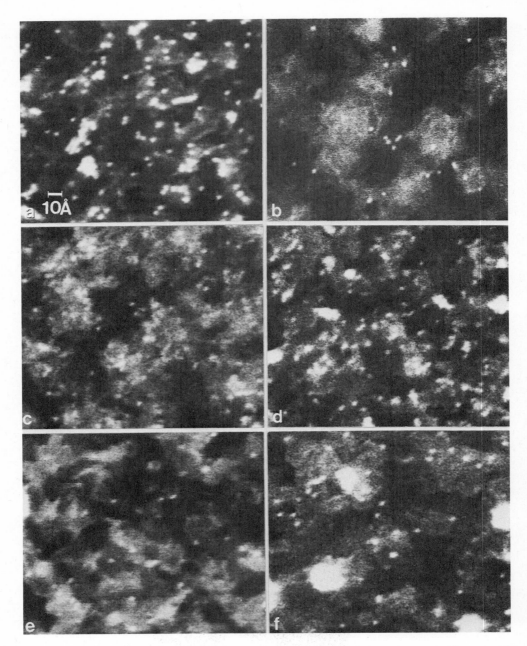

Fig. 13.10. A set of dark field micrographs showing typical examples of some of the different heavy atom specimens that we have studied. a-c have been deposited from dilute solutions of the chloride salts; d-f have been vapor deposited by resistive evaporation of high purity wire in a vacuum. (a) uranium, (b) gold, (c) cadmium, (d) indium, (e) silver, (f) 80% platinum - 20% palladium. Measurements of the scattering intensity of the bright spots in such micrographs indicate that the spots are integral multiples of single atoms (i.e., 1, 2, 3 atoms).

et al., 1977a, b). The electron optical conditions were nearly identical for all six micrographs. Measurements of the scattering intensity of the bright spots in these micrographs indicate that such spots are integral multiples of single atoms, with the lowest intensity spot corresponding to individual atoms. An indication of how far the state of the art of single atom microscopy has progressed since the first pictures were obtained can be seen by comparing Figs. 13.10-13.15 with Fig. 13.1. The main differences being evident are the smaller probe size, lower vibration and lower background noise of the carbon substrate.

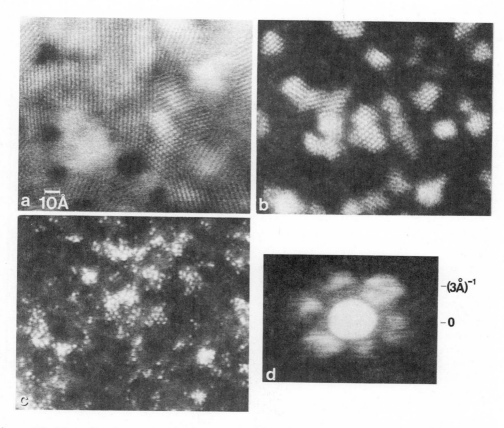

Fig. 13.11. Dark field micrographs (a-c) of a specimen prepared by solvent deposition of a dilute uranyl acetate solution. [see OHTSUKI and ISAACSON (1979)] for the details;   (a) thicker region of uranyl crystals showing a two-dimensional lattice with a spacing of about 3.1 Å   (b) a thinner region that contains much thinner, smaller microcrystals. (c) a region of very low concentration showing clusters of uranium atoms and individual uranium atoms.   All three micrographs are at the same magnification, were taken under the same electron optical conditions and the fields of view are within a few thousand Angstroms of one another. (d) shows a microdiffraction pattern obtained by stoping the probe on a small crystal like those shown in (b) and scanning the diffraction pattern across an aperture (Grigson microdiffraction).

One aspect of incoherent imaging that we have not discussed is the fact that the best contrast and resolution is obtained with a "focused" beam, i.e., the smallest beam size.   As the beam goes out of focus, the image gets fuzzier.   This is the case for atoms and atom clusters, but not necessarily the case for very thick crystal when diffraction effects dominate. However, the fact that the best image is obtained with a focused beam allows us to obtain direct lattice images and individual atom images under identical conditions. This is demonstrated in Fig. 13.11 where we show dark field micrographs of a uranium atom specimen.   Figures 13.11a-c show three different areas within a few thousand Angstroms of one another. Note that the spacings of the thicker crystals in (a) are the same as for the clusters in (c) [OHTSUKI and ISAACSON, 1979b)].

All the micrographs presented thus far have been obtained using the annular detector signal (which is due predominantly to elastically scattered electrons for specimens as weakly scatteringas these).   We had mentioned earlier the fact that the inelastically scattered electrons were of limited use for single atom microscopy due to their non-localization.   This is dramatically illustrated in Fig. 13.12 where we have shown two identical areas of a specimen prepared by evaporating a wire containing 80% platinum and 20% palladium onto a 20 Å thick carbon film. (a) shows the micrograph obtained using the annular detector signal and (b) shows the one obtained by collecting all electrons losing between 7 and 200 eV. The individual atom spots seen in (a) are not visible in (b) due to the 5-10Å range of inelastic scattering.   The general blur of the images of the clusters and the carbon film background is also evident.

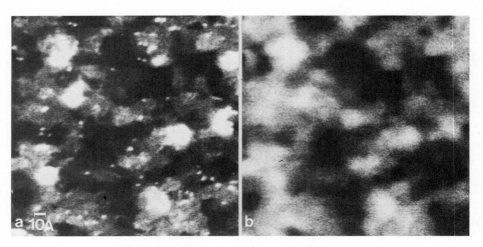

Fig. 13.12. Dark field micrographs of a sample prepared by resistive evaporation of a wire containing 80% platinum and 20% palladium (by weight). (a) annular detector signal showing the atom clusters and the individual atoms. (b) the identical field of view obtained using the spectrometer signal (collecting all electrons losing between 7 eV and 200 eV which pass through the hold in the annular detector). Notice the effect of the delocalization of the inelastically scattered electrons.

One potential application of the use of single atom microscopy is the ability to identify atomic arrangements of heavy atoms in small clusters and to observe the time evolution of these clusters (i.e., their stability). In Fig. 13.13 is shown a time sequence of an atom cluster containing seven uranium atoms. One can see the outer atoms of the cluster associate and dissociate from the cluster (a to b and b to c respectively) at what appear to be specific binding sites. This technique could prove useful in studying the stability of multiatom clusters provided beam induced effects were not dominant.

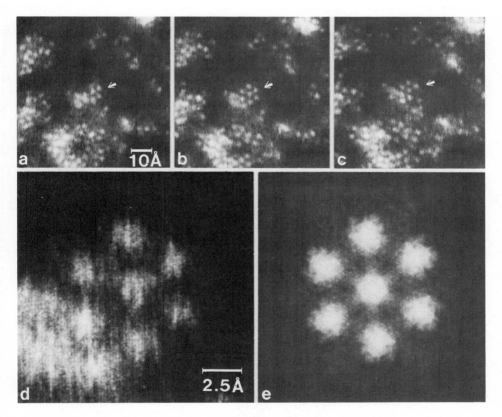

Fig. 13.13. Dark field micrographs of clusters of uranium atoms. a-c show the same field of view taken at several minute intervals. The arrow points to an interesting hexagonal cluster in which the atoms appear to be moving into and out of specific sites. The spacings between the atoms in this cluster are about 3.3A (consistent with the lattice spacings and the diffraction pattern shown in Fig. 13.11. (d) A higher magnification view of the hexagonal cluster shown in (b) indicating the stability of the beam. (e) A six-fold rotation about the center atom of the cluster shown in (d).

The ability to study atomic diffusion phenomena requires great instrumental stability (as we discussed in section 13.4). We are looking for motion on the atomic scale (several Angstroms or so) and therefore, the electrical and mechanical instabilities must be less than tens of Angstroms per hour. The short term stability is indicated in Fig. 13d !which is just a higher magnification view of (c)1, where we can see that the "vibrations" are less than .3-.5 Å. It is even more dramatically demonstrated in Fig. 13.14b which shows an individual gold atom where 15 scan lines across the atom are evident.

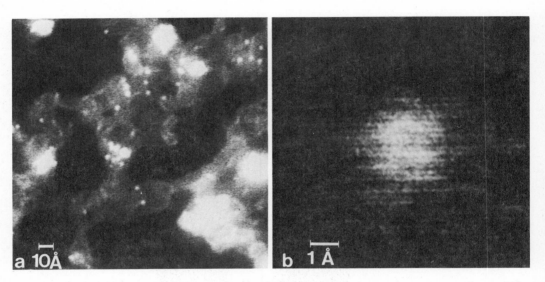

Fig. 13.14. Dark field micrographs of gold atoms on a carbon film whose average thickness is 6.8 Å [ISAACSON, et al., (1979)]. (a) shows individual gold atoms spots which appear to be situated on ledges in the carbon film. The black areas are holes in the film. Notice the angular boundaries of the carbon substrate. (b) shows a very high magnification of a single gold atom where the scan lines are evident. It is clear that although the beam has scanned across the atom about 15 times, the atom position has not changed by more than 0.5 Å indicating that there is negligible beam induced motion of the atom and that the electrical and mechanical instabilities of the instrument are small compared to the probe size of 2.5 Å.

Having demonstrated instrumental stability we are finally in a position to study atomic diffusion and an example is shown in Fig. 13.15 where we can see the diffusion of the atom spots along a ledge in the carbon film (in the center of the picture). Some aspects of our studies of atomic diffusion have appeared in the literature [ISAACSON et al., 1976; ISAACSON et al., 1977)] and will appear in the rest of these proceedings.

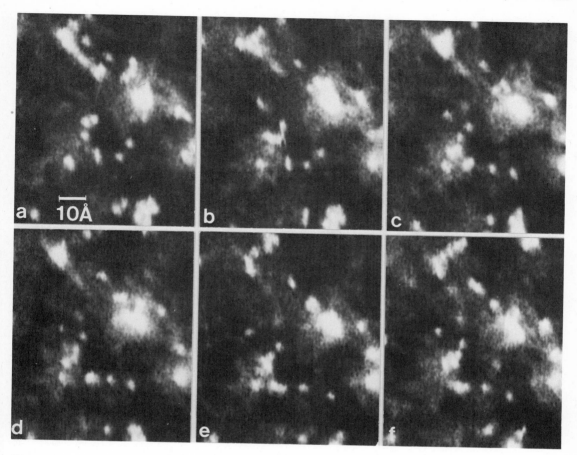

Fig. 13.15 A time sequence of dark field micrographs of uranium atoms diffusing along the surface of a thin carbon film. Each exposure is 17 seconds long and the time between micrographs is about two minutes. The sequence from a-f is in order of increasing time.

There are, of course, other techniques that can be used to study the diffusion of atoms on surfaces at near-atomic resolution, in particular, field ion microscopy. In fact, much work has been done using that technique to study the diffusion of individual adatoms on small crystal planes of some refractory metals (e.g., TSONG and COWAN, 1978). The types of substrates that can be studied using that technique are usually limited to those that can be fabricated into single crystal extremely sharp points about 1000 Å in diameter.

The STEM, can, in principle, visualize a variety of heavy atoms on substrates of arbitrary structure and composition (although we are not yet at that point). Thus, the types of materials that can be studied with the STEM should complement those studied by field ion microscopy. A more complete comparison of the STEM and other atomic surface techniques is given in the literature (e.g., ISAACSON et al., 1976).

13.6 CONCLUSION

We have tried to give an overall view of various aspects of single atom microscopy as it pertains to the STEM. The concentration has been more on the basics and some of the practical considerations rather than on the more theoretical aspects of imaging and detailed analysis of some of the microscopical studies. We hope that our casual treatment of the image interpretation has not offended the experts in the field. In many cases, we feel the simple interpretation is justified and makes life a little easier. In any event, we hope we have enlightened the readers new to the subject and given then a flavor of the problems and potentials of single atom microscopy.

ACKNOWLEDGMENTS

This work was supported by the U.S. National Science Foundation, the Department of Energy, and an NIH Biotechnology Resource Grant. We wish to thank Professor A. V. Crewe for use of his facilities.

REFERENCES

Beck, V., 1973, Rev. Sci. Inst., 44, 1064.

Cole, M. D., Wiggens, J. W., and Beer, M., 1977, J. Mol. Biol., 117, 387.

Cowley, J. M. and Iijima, S., 1977, Phys. Today, 30, 32.

Crewe, A. V., Wall, J., and Langmore, J., 1970, Science, 168, 1338.

Crewe, A. V., 1973, Progress in Optics Vol. XI (North Holland).

Crewe, A. V., Langmore, J. P., and Isaacson, M. S., 1975 in Physical Techniques of Electron Microscopy and Microbeam Analysis, eds. B. Siegel and D. Beaman (John Wiley and Sons, Inc., New York).

Dorignac, D. and Jouffrey, B., 1976, Proc. 6th Europ. Conf. Electron Micros. Vol. I, 270.

Formamek, H. and Knapek, E., 1979, Ultramicroscopy, 4, 77.

Fertig, J. and Rose, H., 1977, Ultramicroscopy, 2, 269.

Haine, M. E. and Cosslett, V. E., 1961, The Electron Microscopy: The Present State of the Art (E. and F. N. Spon, Ltd., London).

Hashimoto, H., Kumao, A., Himo, K., Yotsumoto, H., and Ono, A., 1971, Jap. Jour. Appl. Phys., 10, 1115.

Henkelman, R. and Ottensmeyer, F. P., 1971, Proc. Nat. Acad. Sci. (USA), 68, 3000.

Isaacson, M., Langmore, J. P., and Wall, J., 1974, IITRI/SEM/74, ed. O. Johari and I. Corvin (Chicago, IITRI), 19.

Isaacson, M., Langmore, J., Parker, N. W., Kopf, D., and Utlaut, M., 1976, Ultramicroscopy, 1, 359.

Isaacson, M., Kopf, D., Parker, N. W., and Utlaut, M., 1976, Proc. 34th Ann. EMSA Meeting, Miami, 586.

Isaacson, M., Ohtsuki, M., Utlaut, M., Kopf, D., and Crewe, A. V., 1979, submitted to Science.

Isaacson, M., Kopf, D., Ohtsuki, M., and Utlaut, M., 1979, Ultra-microscopy, 4, 97, 101.

Isaacson, M., Kopf, D., Utlaut, M., Parker, N. W., and Crewe, A. V., 1977, Proc. Nat. Acad. Sci. (USA), 74, 1802.

Isaacson, M., Utlaut, M., and Kopf, D., (to appear) in "Computer Processing of Electron Micrographs" (ed. P. Hawkes), Springer Topics in Current Physics.

Isaacson, M., Langmore, J. P., and Rose, H., 1974, Optik, 41, 92.

Langmore, J. P., 1975, Ph.D. Dissertation, The University of Chicago.

Langmore, J. P., 1978 in Principles and Techniques of Electron Microscopy, ed. M. A. Hayat (Van-Nostrand Reinhold Co., New York).

Langmore, J. P., Wall, J., and Isaacson, M., 1973, Optik, 38, 335.

Moudrianakis, E. and Beer, M., 1965, Proc. Nat. Acad. Sci. (USA), 53, 564.

Ohtsuki, M. and Isaacson, M., 1979, Proc. 37th Ann. EMSA Meeting, San Antonio (in press).

Ohtsuki, M. and Isaacson, M., 1979, SEM/79, ed. O. Johari (Chicago, SEM, Inc.), in press.

Retsky, M., 1974, Optik, 41, 127.

Scherzer, O., 1949, Jour. Appl. Phys., 20, 20.

Thomson, M. G. R., 1973, Optik, 39, 15.

Thon, F. and Willasch, D., 1972, Optik, 36, 55.

Tsong, T. T. and Cowan, P. L., 1978, CRC Critical Rev. in Solid State Science, 7, 289.

Wall, J. S., Langmore, J., Isaacson, M., and Crewe, A. V., 1974, Proc. Nat. Acad. Sci., 71, 1.

Wall, J. S., Isaacson, M., and Langmore, J. P., 1974, Optik, 39, 359.

Whiting, R. F. and Ottensmeyer, F. P., 1972, J. Mol. Biol., 67, 173.

CHAPTER 14

MICRODIFFRACTION*

J.B. WARREN

BROOKHAVEN NATIONAL LABORATORY

UPTON, NEW YORK 11973

14.1 INTRODUCTION

The advantage of forming diffraction patterns in a crystalline material from areas less than 500 nm in diameter, hereafter defined as microdiffraction, can be readily described. The region of interest, such as a small second-phase particle, a particular grain in a fine-grain alloy, or a portion of the strain field surrounding a crystal defect, is small in volume compared to the surrounding matrix. Any means which reduces the total diffraction-forming volume increases the contribution from the region of interest and results in a diffraction pattern with an improved signal-to-noise ratio. Before the introduction of the STEM electron microscope, the lens configuration used for diffraction was such that spherical aberration of the objective lens prevented diffraction patterns from an area less than 500 nm from being formed. This limitation, and the means for bypassing it with microdiffraction, is worth describing in detail.

Spherical aberration results from electrons passing through the outer portion of a lens being focused more strongly than those electrons passing closer to the optic axis. The electron beam cannot be brought to focus in a single focal plane, and an image of a point appears as a disc with a radius equal to

$$r = C_s \beta^3 \qquad (14.1)$$

---

*This research was supported by the U.S. Department of Energy.

where C_S = 0.4 to 2.0 cm for most electron microscopes equipped with a STEM polepiece and β is the lens aperture angle controlled by the aperture immediately below the imaging lens. In the bright-field imaging mode (Fig. 14.1(a)), the objective aperture has the dual function of preventing diffracted beams from being imaged and reducing β to an optimum value. For the selected area diffraction mode (Fig. 14.1(b)), the objective aperture must be removed and spherical aberration effects in the diffraction pattern must be controlled by other means.

Consider the formation of the selected area diffraction patterns shown in Fig. 14.1. The diffraction pattern from the entire area illuminated by the incident beam is formed in the back focal plane of the objective lens and the (000) beam is magnified to form the bright-field image. If the selected area aperture above the diffraction lens (the lens immediately below the objective) is placed such that it is coplanar with the image, tracing the rays back to the specimen plane shows that the diffraction pattern appearing on the phosphor screen below the lens column results only from the portion of the image within the selected area diffraction aperture. This is true even though the diffraction pattern from the entire illuminated area is still present in the back focal plane of the objective lens.

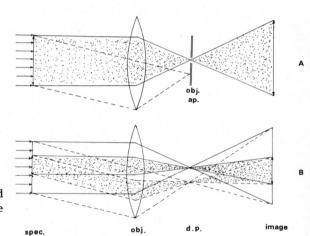

Fig. 14.1. Standard Selected Area Diffraction (after Le Poole, 1947).

Diffracted beams, however, pass through the lower objective lens inclined at an angle $2\theta_B$ from the optic axis and are subject to spherical aberration. The diffracted beam that would be focused on a flat focal plane by a perfect lens is brought to focus on a spherical surface. The practical result of this phenomenon is that the image formed by the main beam and the diffracted beam do not coincide in the image plane and this lack of coincidence increases as $2\theta_B$ increases. Thus, the diffraction spots in the selected area diffraction pattern are not formed from the same area of the specimen as the main beam. The extent of the error is described by the equation

$$x = C_s(2\theta_B)^3 \qquad\qquad (14.2)$$

where x is the distance on the image between the centers of the origin of the main beam and the diffracted beam. An additional error, resulting from defocussing the objective lens from the optimum value, increases x by an amount $2D\theta$, where D is the amount of defocus. In practice, this error is less serious than that from spherical aberration. Because of these errors, the regions forming the main and diffracted beams do not even overlap if the selected area at the specimen plane is less than 500 μm. Thus, the selected area technique cannot be used for microdiffraction.

14.2 FOCUSED PROBE MICRODIFFRACTION

Perhaps the most easily understood microdiffraction method that overcomes the spherical aberration limitation is the focused-probe technique. The lens configuration used for focussed-probe is used both to form the STEM image on the CRT, or (with a static probe) to form a microdiffraction pattern that is recorded with a camera below the lens column. Here, the two condenser lenses and the upper objective lens act as a triple condenser system to produce a highly convergent probe that illuminates only a small portion of the sample. It is now possible to use deflection coils situated above the objective lens to scan this probe across the sample. Some portion of the transmitted electrons are collected to form an image on a cathode ray tube while the diffraction pattern itself is formed in the same objective focal plane as before and can be recorded with an ordinary plate camera at the base of the lens column. Since the diffraction pattern can arise from the illuminated area alone, the spherical aberration effect resulting from a field-limiting aperture that selects only a portion of the illuminated area does not interfere, and the minimum diameter that can be examined is limited only by the spot size of the probe and scattering effects dependent on sample thickness. Ray optics for the focused probe microdiffraction mode are shown in Fig. 14.2.

In most commercially available electron microscopes with the STEM configuration, the diameter of the probe is varied by altering the first condenser lens strength, while the second condenser lens in conjunction with its aperture is used to control the convergence of the beam. The average convergence for a given probe size is now an order of magnitude greater than the TEM mode, but the spot sizes can be decreased to approx-

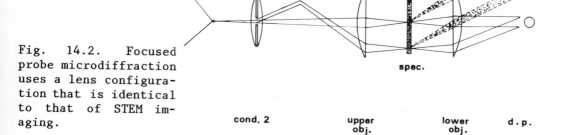

Fig. 14.2. Focused probe microdiffraction uses a lens configuration that is identical to that of STEM imaging.

cond. 2   upper   lower   d.p.
                    obj.   obj.

spec.

imately 3 nm with a standard tungsten hairpin filament or less than 0.5 nm with a field emission tip.

Once the first condenser setting and second condenser aperture size are chosen to select a particular probe size and convergence angle, the probe is rastered over the sample by using the deflection coils to twice deflect the beam below the second condenser aperture such that the beam pivots in the upper focal plane of the objective lens. As shown in Fig. 14.2, the upper objective field lens not only focuses the beam to a probe, but converts the angular beam deflection to a deflection that remains parallel to the optic axis. Trimming potentiometers used to adjust the height of this pivot point must be set carefully if the probe is to remain parallel to the optic axis over the entire scanned area. If this condition is not met, the Bragg deviation for any given diffracting plane will vary over different parts of the scanned area and the image contrast will not be comparable to that for a collimated beam used in TEM imaging.

Once the probe passes through the specimen, the diffraction pattern is formed in the back focal plane of the lower objective lens, just as it was in the standard selected area diffraction modes. The probe still remains parallel to the optic axis after passing through the specimen (solid rays in Fig. 14.2), and it always converges to the same spot in the back focal plane after being focused by the lower objective lens regardless of its position in the scan. In the same fashion, diffracted rays (dashes in Fig. 14.2) scattered through the same Bragg angle at different points of the scan are all focused to another point, forming a diffraction spot for a particular set of diffracting planes.

The most notable difference between the focused probe microdiffraction patterns and the selected area type is the former's reduced degree of angular resolution. Examples are shown in Fig. 14.3 where it is seen that each diffraction spot is now represented by a disc, whose diameter is proportional to α_i, the semi-angle between the probe and the optic axis.

Calculation of α_i from the convergent beam diffraction pattern is easily accomplished (THOMPSON, 1977). In Fig. 14.3, the convergent beam which subtends to angle of $2\alpha_i$ strikes the specimen and produces a main beam and several diffracted rays that are focused as discs in the back focal plane of the objective lens. Bragg's law shows that the distance between the main beam and any of the diffracted beams is proportional to $2\theta_B$. Also, it is evident that the width of the disc in the convergent beam diffraction pattern is proportional to the convergence angle of the incident beam. If the distances are measured as in Fig. 14.4, then by similar triangles:

$$\frac{\alpha_i}{\theta_B} = \frac{Y}{X} \tag{14.3}$$

and the degree of probe convergence can be computed directly from an experimental microdiffraction image.

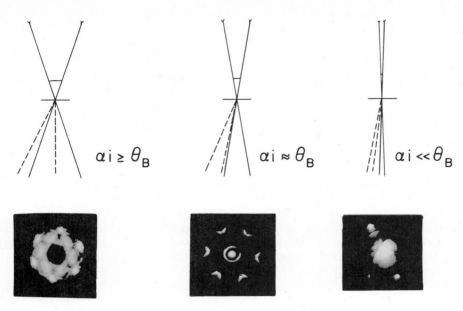

Fig. 14.3. Focused probe microdiffraction patterns for <111> direction in silicon. α_i decreases from left to right.

Fig. 14.4. Method of Measuring Angular Resolution.

14.3 FOCUSED APERTURE MICRODIFFRACTION

Focused condenser aperture microdiffraction utilizes the same objective lens used for STEM but images the specimen in the same manner as the Le Poole method of selected area diffraction. In this case, however, the strongly demagnifying upper objective lens field allows the second condenser aperture to be imaged on the specimen plane itself. For this method, only a small portion of the sample is illuminated and only this part of the sample can produce the pattern (RIECKE, 1962). The ray optics for this technique are shown in Fig. 14.5 and Fig. 14.6.

Here, the objective lens field is treated as two thin lenses, and it can be seen from Fig. 14.5 that the upper objective lens forms a demagnified image of the second condenser aperture at (a), while the lower objective lens forms a magnified image of the specimen at (b). The method of selected area diffraction involves placing an aperture at (b),

Fig. 14.5. Comparison of Focused Aperture Images from Same Diameter Aperture in Condenser and Diffraction Lens Position.

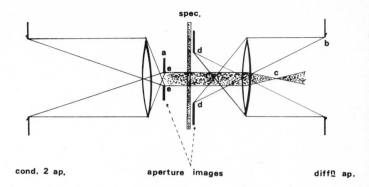

so that only the central part of the image is used to form the diffraction pattern at (c). By tracing the rays from (d) to (b), it can be seen that the selected area aperture allows only rays from the part of the specimen (e-e) to form the diffraction pattern that is recorded photographically.

However, if the specimen is raised to a non-eucentric position to coincide with the image of a small condenser aperture, a "selected area" diffraction pattern will again be formed. In this case, the condenser aperture permits only a small area of the sample to be illuminated, and it is this area alone that forms the pattern. For the Philips 301 STEM polepiece, for example, the condenser aperture is demagnified 38 times by the upper objective lens, while the selected area diffraction aperture is demagnified only 14 times at the specimen plane.

The ray diagram in Fig. 14.6 for focused condenser aperture diffraction can now be undertstood when compared to that for selected area diffraction in Fig. 14.1(b). Now, the entire illuminated area is used to form the diffraction pattern and spherical aberration no longer sets a limit on the minimum area that can be selected.

Fig. 14.6. Electron Envelope for Focused Condenser Aperture Microdiffraction

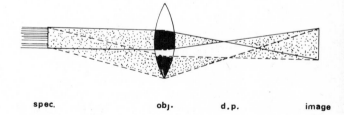

Thus, the focused aperture technique permits a quick and effective way to produce microdiffraction patterns from areas as small as 50 nm in diameter. Only the normal imaging mode is used, and the angular resolution in the diffraction pattern is equal to the older selected area method. The role of spherical aberration is now confined to a slight distortion of the diffraction pattern. According to HIRSCH (1966), the distortion increases for higher order reflections (as these rays are deflected further from the optic axis) but is only about 1% for the 3rd order diffraction spots.

14.4 ROCKING BEAM MICRODIFFRACTION

A method that circumvents the contamination and angular resolution problems common to convergent beam techniques is the rocking beam micro-diffraction method. As developed by VAN OOSTRUM (1973) and GEISS (1975), the rocking beam method makes use of the scanning coils to pivot the incident beam over the specimen surface in such a manner that the bright-field and dark-field images appear sequentially over the STEM detector at the base of the microscope. The detector, of course, senses nothing about the character of the images that momentarily appear, it just detects an increase in intensity as each dark-field image comes in view. As the pivoting of the beam is repeated in a regular fashion, the increase in electrons striking the detector result in a momentary in-crease in the signal sent to the STEM cathode ray tube, forming a dif-fraction spot each time the incident beam is tilted at the appropriate angle. In spite of its unique method of formation, the rocking beam dif-fraction pattern closely resembles the patterns produced by selected area diffraction methods, and shows both dynamic and kinematic effects.

Lens and aperture positions used for rocking beam diffraction are shown in Fig. 14.7.

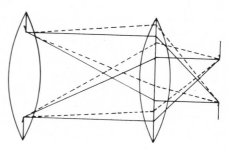

Fig. 14.7. Effect of Beam Deflection on Condenser Aperture Image.

**cond 2
aperture**

**upper
obj**

**aperture
image**

Consider the effect of the upper objective lens field on the im-aging of the second condenser aperture. As discussed in the last sec-tion, the distance of the second condenser aperture can be set such that the upper objective lens forms a strongly demagnified image of the aper-ture in its back focal plane. Using the principles of ray tracing, it is apparent that any ray, regardless of its initial direction, that emanates from the aperture plane must still strike the same point in the plane to the right of the objective lens where the aperture image is formed.

A well-collimated beam used to form a normal TEM image can then be regularly deflected with the scanning coils such that the pivot point is co-planar with the aperture. As long as this condition is satisfied, the aperture image will remain stationary in the image plane (if spher-ical aberration is zero), regardless of the angular direction of the

beam. When the specimen height is adjusted to coincide with the aperture image plane, the requirements for rocking beam diffraction are satisfied. Now, the same area of the sample remains illuminated during the
beam rocking process. GEISS was the first to recognize these requirements and produce the first true rocking beam microdiffraction pattern
(1975).

Geiss has shown that the spherical aberration causes the specimen
image to shift a linear distance

$$X = MC_s \theta^3 \tag{14.4}$$

during the rocking process. For the Philips 301-STEM, M is 1/38, C_s is
6.3 mm and the maximum rocking angle is approximately 1°. For these conditions, the maximum image shift is 9 nm, thus providing a lower limit
to the area from which meaningful diffraction patterns can be obtained.

Although the specimen must be placed in a slightly non-eucentric
position to coincide with the aperture image, the lens settings for the
rocking beam method are the same as the focused condenser aperture
method discussed before. The deflection of the beam produced by the
scanning is no different from the deflection that occurs when a dark-
field image is formed by manually tilting the incident beam to a specific position.

The angular extent of the rocking beam patterns formed on the cathode ray tube is controlled by the angular deflection of the beam, and
the deflection itself is controlled by the magnification control on the
STEM panel. Because the STEM imaging mode increases magnification by
decreasing the scanned area, it follows that adjustment of the same control when in the rocking beam mode simply reduces the angle of rock.

It is important to see that the rocking beam pattern is formed from
only the portion of the image that strikes the detector. In the Le
Poole selected area diffraction method, the portion of the image that
forms the diffraction pattern is chosen with the size of the diffraction
aperture. In the rocking beam method, it is the diameter of the solid
state detector that determines the portion of the image chosen. Thus,
by increasing the magnification an increasingly smaller image segment is
permitted to strike the detector and form the microdiffraction pattern.
The detector diameter can also be reduced by masking it with an aperture
if even small portions of the image used to form the pattern are
desired.

For the Philips 301, the diameter of the detector is 4 mm and the
maximum magnification at the detector position is 130,000X, so the minimum selected area is

$$A = \frac{D}{M} = 308 \text{ Å} \tag{14.5}$$

GEISS (1975) has reduced the diameter of the detector down to 1.0 mm and
has obtained single crystal diffraction pattern from vapor deposited
gold islands less than 10 nm in diameter. Signal strength decreases as

the active detector area is reduced or sample thickness is increased. A more practical limit for obtaining microdiffraction patterns with a reasonable signal-to-noise ratio from typical metallurgical thin foil specimens is 50 nm.

Angular resolution of the diffraction discs in the rocking beam pattern is controlled by the size of the objective aperture. A 5 mm objective aperture results in a pattern with resolution comparable to standard selected area diffraction, while removing the aperture entirely produces the rocking beam channeling patterns that will be described in the next section.

Angular resolution can be understood with Fig. 14.8, which shows that the diffraction pattern in the back focal plane of the objective lens must be swept across the objective aperture just as the image is swept over the detector. The objective aperture diameter d, limits the divergence of the rays in any diffracted beam. Thus, the maximum angular divergence is found by the ratio of the focal length of the objective lens and the objective aperture diameter:

$$\rho = d/f \tag{14.6}$$

For the Philips 301, f = 3.7 mm, so an objective aperture of 5 μm (the smallest practical size) gives a maximum angular resolution of $\rho = 1.35 \times 10^{-3}$ rad. Thus, the angular resolution for rocking beam is roughly equivalent to other microdiffraction techniques.

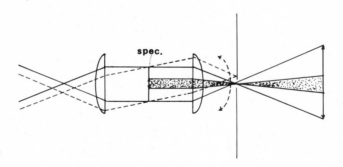

Fig. 14.8. Rocking Beam Ray Diagram.

upper obj.   lower obj.   obj. ap.   image

14.5 APPLICATIONS

By 1979, several researchers have used microdiffraction in diverse areas. Most work has been done with the convergent beam electron diffraction or focused probe method. Areas examined include accurate lattice parameter determination using upper layer diffraction effects in zone axis patterns, (RACKHAM and STEEDS, 1976) point and space group information for complex crystal structures, (STEEDS et al., 1976),

computer simulation of n-beam microdiffraction patterns near crystal defects, (COWLEY and SPENCE, 1978) short range order studies, (CHEVALIER
and   CRAVEN,   1977)   and   precipitate-matrix   orientation   problems
(CARPENTER, et al, 1977, ZALUZEC, 1976). The first two of these topics,
the extensive work of Steeds and his coworkers, is covered in the chapter on Convergent Beam Electron Diffraction and will not be discussed
further.

The other areas, however, can be divided into two major categories.
In the first, the kinematic or the dynamical theory of electron diffraction is used to interpret the details of a single microdiffraction pattern. In the second, the microdiffraction pattern is used in a more
traditional sense, as an aid to determining the orientation of an already known crystal structure with respect to the electron beam.

An example of the first class of problem is provided by COWLEY and
SPENCE (1978), who used computer simulation to predict the appearance of
a microdiffraction pattern resulting from a focused probe placed near an
end-on edge dislocation in iron. An example of their work is shown in
Fig. 14.9 where the small diagram shows the position of the probe with
respect to the dislocation's extra half plane. It was found that the
simulation was sensitive to probe position and amount of defocus. For
the example shown here, an n-beam dynamical computation using 2000 beams
and a 30 nm thick crystal were assumed.

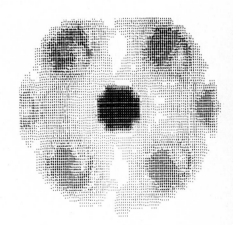

Fig. 14.9. Computer Simulation of
Microdifraction Pattern Near End-
On Dislocation of Crystals (COWLEY
and SPENCE, 1978)

Fe (100) edge ⊥
100kV STEM θ_ap=5.5 mR

WARREN and HREN (1977) used focused probe methods to form microdiffraction patterns near end-on dislocations in silicon of ~ 0.2 µm in
thickness. The sample was tilted to produce a two-beam condition and
variations in intensity of the main and the strongly excited diffracted
beam were observed when the probe was moved to different areas around
the dislocation. However, experimental difficulties such as surface relaxation and contamination resulting from the sharply focused probe make

the identification of crystal defects by the computer simulation of microdiffraction patterns a very difficult procedure.

Another class of problem is the study of the orientation of different phases such as precipitate/matrix orientations or the periodicity in short range order (CHEVALIER and CRAVEN, 1977). Here, a series of microdiffraction patterns must be compared to one another and examination of the fine detail of the pattern resulting from the dynamical theory is no longer needed. Any of the three major microdiffraction techniques can be used for this sort of analysis, but the focused probe and the focused condenser aperture methods will be described here.

To solve a problem of this sort, a series of microdiffraction patterns are obtained by manually positioning the focused probe or the shadow image of a condenser aperture over the appropriate parts of the image. The relative orientation of each part of the sample with respect to neighboring regions is then found by first determining the exact electron beam direction in the crystal coordinates of the region producing the pattern. As the specimen is usually less than 1 μm in thickness, the diffraction spots are elongated in a direction parallel to the specimen normal in reciprocal space, so simply indexing each pattern to determine the exact beam direction could result in large inaccuracies. As a result, the exact beam direction must be calculated by using Kikuchi line pairs associated with two non-parallel g vectors.

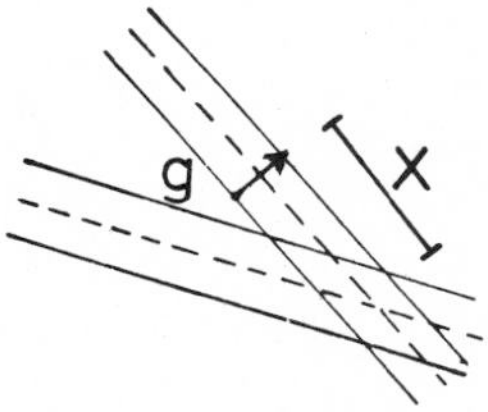

Fig. 14.10. Computation of Exact Electron Beam Direction

As shown in Fig. 14.10, the distance X in the diffraction pattern is directly proportional to the angular distance between the exact beam direction and the approximate beam direction. As long as the Kikuchi lines are distinct, the beam direction can be determined with accuracy of about $\frac{1}{2}°$. Once the beam directions are known, the orientational difference between any two cubic crystal structures (FORWOOD and HUMBLE, 1975) can be determined from:

$$BM_i = \ell_{ij} BM_j^*$$
(14.6)

where BM_i and BM_j^* are the exact beam directions for two adjacent crystals for a particular specimen orientation and ℓ_{ij} (for cubic materials) is a 3 x 3 array of directions cosines. Symmetry relations show that for the cubic case two different specimen stage-beam orientations are required to determine ℓ_{ij}. The Eqn. 14.6 can be treated as a series of simultaneous equations and solved for ℓ_{ij}. Examples of microdiffraction patterns using focused condenser apertures are shown below. The grains are approximately 2 µm in diameter and this approach made it easy to obtain distinct patterns from each one.

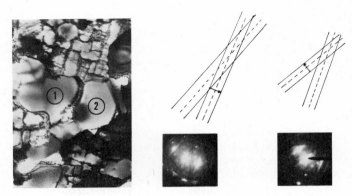

Fig. 14.11. Focused Condenser Aperture Microdiffraction Patterns from Adjacent Grains (1 and 2) in Fine-Grained Region of Nickel Alloy

It is evident that this sort of analysis is dependent upon the availability of two distinct Kikuchi line pairs in every microdiffraction pattern. In many higher atomic number elements of highly deformed metals, these lines may so diffuse that their position cannot be accurately determined. For this case, focused probe methods may again be of use.

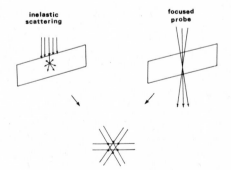

Fig. 14.12. Formation mechanism of Kikuchi lines versus transmission channeling patterns.

Compare the formation mechanism of Kikuchi lines to that for transmission channeling patterns as in Fig. 14.12. In the first case, inelastic scattering is responsible for producing an angular distribution of electrons which will encounter Bragg planes at the proper angle to form Kikuchi lines. In materials where this mechanism plays a minor role, the convergent probe can now provide primary electrons that again travel at the proper angle to undergo Bragg diffraction. Although the mechanics of formation have changed, the resulting transmission channeling pattern can supply exactly the same orientational information as the Kikuchi lines in the standard selected area diffraction pattern. An example in Fig. 14.13 shows for comparison the standard selected area and focused probe diffraction from the same area in a deformed Mo foil. The double exposure at the right shows the position of the main beam and diffraction spots with respect to the channeling pattern.

(a)   (b)

Fig. 14.13. Comparison of selected area diffraction patterns with focused probe diffraction patterns.

Rocking beam methods can be used for both classes of microdiffraction problems discussed above. The most significant advantages are low contamination rates (the specimen is illuminated by a defocused beam) and excellent control of the selected area (changes with image magnification). GEISS (1976) has developed rocking beam diffraction to its fullest extent and has used it to see if the presence of microcrystalline areas could be detected in amorphous germanium films. Examination of microdiffraction patterns taken from 2.5 nm diameter volumes showed no evidence of crystallinity.

Various methods of beam rocking have been tried, but they have essentially the same aberration limitations as the GEISS method. GRIGSON (1967), for example, rocks the beam below the objective lens. JEOL has indicated that the JEOL 1000C is capable of using both GEISS and GRIGSON methods for microdiffraction. Another rocking beam method available with the JEOL 200 microscope uses scanning coils positioned below the objective lens to scan any portion of the image or the diffraction patterns over a detector which is also situated at the bottom of the lens column. (Zalusec, private communication). In this method, the standard imaging mode is used (no scanning above the sample) and an intensity line scan is displayed on the CRT that is normally used to display the STEM image. As shown in Fig. 14.14, these intensity profiles are a useful means of quantitizing dynamical electron intensities. One novel method (KRAKOW, 1978), rotates the beam such that it traces out a conical surface and a diffraction pattern is formed similar to that expected from an annular condenser aperture. As shown in Fig. 14.15, increasing the "cone angle" to the appropriate extent (Fig. 14.15(c)) results in a situation where diffracted rays from each of the first order diffraction cones passes down the optic axis. With the objective aperture in position, all possible reflections of a particular diffraction order can be imaged at once. Such a technique can reduce dark-field images of polycrystalline materials with the assurance that all grains of a given orientation will be visible regardless of diffracting conditions pertaining to individual diffraction spots.

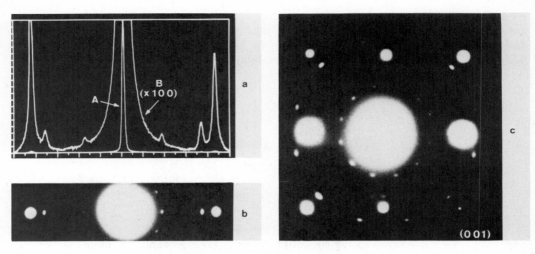

Fig. 14.14 Intensity distribution resulting from scanning an (001) diffraction pattern in a < 220 > direction over the detector.

Fig. 14.15. Appearance of diffraction patterns with increasing cone angles.

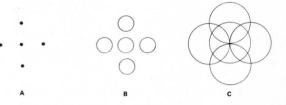

Fig. 14.16. Transmission channeling pattern produced by Geiss rocking beam technique.

Another rocking beam application uses the removal of the objective aperture from the GEISS configuration to form transmission channeling patterns. As shown in Fig. 14.16, the appearance is quite similar to the focused probe diffraction pattern in Fig. 14.13(b). In this case, a reasonably parallel beam is rocked through the necessary solid angles such that the Bragg condition for a particular diffracting plane is satisfied at opposite sides of the optic axis. In the focused probe, the same effect is realized by focusing the probe such that it encompasses the required conical volume.

Both cases produce a channel line pair at exactly the same position in the back focal plane of the objective lens. For the rocking beam case, the pattern is formed sequentially in time, and a well-collimated beam is now positioned by the scanning coils at precisely the right angle to form one segment of the channel line seen in the diffraction pattern.

Channeling patterns formed with the scanning electron microscope have been studied for several years (COATES, 1970) and can be formed with either backscattered electrons or specimen current imaging. These SEM channeling patterns are limited to solid samples. In addition, the longer focal length of the final probe forming lenses typical of commercial SEM's results in much greater spherical aberration than the corresponding STEM lens, and the minimum specimen area that can be studied is about 2 μm in diameter (VAN ESSEN and SCHULSON). Transmission channeling patterns were produced by FUJIMOTO et al. (1972) using JEOL 100B STEM, but no experimental images were shown and the minimum area that could be used to form a pattern was not described.

For transmission channeling, spherical aberration again controls the minimum diameter that can produce useful diffraction information. As discussed in the introduction, the images for the main and diffracted beam no longer coincide for an imperfect objective lens. With the objective aperture removed, both bright-field and many of the dark-field images will strike the detector and contribute to the formation of the transmission channeling pattern. Since the dark-field images are displaced, increasing the magnification to force the detector to form the pattern from a smaller region increases the displacement error in exactly the manner as using a smaller selected area diffraction aperture. Thus, the minimum usable diameter is about 1 μm for commercially available STEM's.

14.6 SUMMARY

It is clear that the successful use of any microdiffraction method depends upon a complete understanding of beam convergence and the effect of spherical aberration on the optical configuration. The convergence angle, as defined by α , directly affects:

(a) signal-to-noise ratio in the diffraction pattern (large α_i values increase signal strength),

(b) contamination (larger α_i values increase contamination rates),

(c) form of information in the different pattern (larger α_i values form channel-type patterns, while smaller ones produce discrete diffraction spots composed of small diameter discs).

Convergence angles can be calculated directly from microdiffraction patterns, so exact quantitative comparisons of the effect of α on the parameters listed above can be found and optimum α values can be repeated in subsequent experiments. It is suggested that convergence angles be listed in microdiffraction studies just as "two-beam conditions" are always recorded in selected area diffraction experiments.

Spherical aberration controls:

(1) The minimum usable diffracting volume for selected area diffraction, rocking beam diffraction, and transmission channeling patterns formed by rocking beam.

(2) The degree of distortion in the microdiffraction pattern produced by the focused probe or focused condenser aperture methods. The minimum usable diffracting volume for these two techniques is controlled by instrument stability and sample thickness.

The most efficient way to categorize microdiffraction methods is according to the electron-optic lens configuration used to form the diffraction pattern. The methods described in this chapter can be summarized as:

(a) Focused probe. The image can be viewed directly or displayed on a CRT in STEM, but the microdiffraction patterns are always recorded with the plate camera at the base of the microscope column.

(b) Focused condenser aperture. Image and diffraction pattern formed in a conventional manner, but a small (<10 μm) condenser aperture is focused on the sample and illuminates only a small area in diffraction mode.

(c) Rocking beam. Image formed in a conventional manner but the beam is rocked over the sample; the detector size and magnification setting selects a portion of the image to form the diffraction pattern on the CRT.

In addition, both the focused probe and rocking beam techniques can form transmission channeling patterns for large convergence angles.

All of the three methods have advantages and disadvantages. Focused probe methods can examine areas as small as 50 $\overset{\circ}{A}$ in diameter, but have high contamination rates and cannot be used on sensitive materials. Focused condenser aperture methods have excellent angular resolution in the diffraction pattern and intermediate contamination rates, but the minimum area that can be examined is 50 nm (for a 2 μm aperture). The rocking beam technique offers complete control of angular resolution and contamination no higher than conventional TEM, but its signal-to-noise ratio is the poorest of the three methods.

REFERENCES

Carpenter,R. W., Bentley, J., Kenik, E. A., 1977, "Scanning Electron Microscopy", V.I., Om Johari, Ed., IIT Res. Inst., p. 411.

Chevalier, J., P.A.A. and Craven, A. J., 1977, Phil. Mag. 36, p. 67.

Coates, D. G., 1967, Phil. Mag., 16, p. 1179.

Cowley, J. M., and Spence, J.C.H., 1978, Ultramicroscopy 3, p. 433.

Edington, J. W. "Practical Electron Microscopy in Materials Science", Vol. 1, p. 2.

Forwood, C. T., and Humble, P. 1975, Phil. Mag. 31, p. 1011.

Fujimoto, F., Komaki, K., and Takagi, S. 1972, Zeit. F. Naturforsch, 27a, p. 441.

Geiss, R. H. 1975, Appl. Phys. Lett. 27, p. 174.

Geiss, R. H. 1976, "Scanning Electron Microscopy", Vol. I. Om Johari, Ed., ITT Res. Inst., p. 337.

Grigson, C.W.B., and Tillett, P. I. 1967 Nature, 215, p. 617.

Hirsch, P. B., Howie, A., Nicholson, R. B., Pashley, D. W., and Whelan, M. J. 1965, "Electron Microscopy of Thin Crystals". (London: Butterworths) p. 21.

Krakow, W. 1978, Ultramicroscopy, 3, p. 291.

Le Poole, J. B. 1947, Philips Tech. Rev., 9, p. 33.

Rackham, G. M. and Steeds, J. W. 1976, "Developments in Electron Microscopy and Analysis", J. A. Venables, Ed., (Academic Press) p.457.

Riecke, W. D. 1962, Optik, 19, p. 81.

Steeds, J. W., Jones, P. M., Rackham, G. M., and Shannon, M. D. 1976, "Developments in Electron Microscopy and Analysis", J. A. Venables, Ed., (Academic Press) p. 351.

Van Oostrum, K. J., Lienhouts, A., and Jore, A. 1973, Appl. Phys. Lett. 23, p. 283.

Warren, J. B., and Hren, J. J. 1977, "Scanning Electron Microscopy", Vol. I, Om Johari, Ed., IIT Res. Inst., p. 379.

Zaluzec, N. J. and Fraser, H. L. 1976, "First Workshop on Analytical Electron Microscopy", Cornell University, p. 217.

CHAPTER 15

CONVERGENT BEAM ELECTRON DIFFRACTION

J.W. STEEDS

UNIVERSITY OF BRISTOL

BRISTOL, ENGLAND

15.1 INTRODUCTION

Development of Convergent Beam Diffraction

Convergent beam electron diffraction (CBD) is a technique with a long history of gradual development which has recently become widely available through the development of commercial TEM/STEM electron microscopes. The technique was discovered by KOSSELL and MOLLENSTEDT (1939) who obtained some quite remarkably good results when one realizes that the size of focussed probe they were working with was comparatively large. Most specimens are so irregular that there would be considerably thickness variation within such areas producing a thickness average of the information. Further, few specimens are so flat that some important angular average will not occur over such areas. Thickness and orientation are two crucial parameters of electron diffraction and it is essential to eliminate their variation within the illuminated volume if meaningful results are to be obtained.

The development of STEM has resulted in the widespread availability of microscopes capable of producing small focussed probes so that the chance of thickness or orientation variation within the illuminated volume is minimized. The way has been opened to the widespread exploitation of the powers of convergent beam diffraction.

Convergent beam patterns (CBP's) are two dimensional maps of diffraction intensity as a function of the inclination between the incident electrons and a particular crystal direction. They are normally composed of a series of discs each one corresponding to a different Bragg reflection. The intensity variation within the discs carries important information about the specimen orientation and thickness as well as other properties of the specimen.

Until quite recently work on CBD was pursued in only a few laboratories and the instruments used were either specially designed or appreciably modified. In my research group our first CBD experiments were
performed by addition of a mini-lens to a Philips EM 200 electron microscope. Somewhat surprisingly, even though many laboratories are now
well equipped to perform CBD experiments the technique does not yet seem
to be very widely used yet it is able to provide a wealth of accurate
information about crystalline specimens. In this chapter attention will
be concentrated on those aspects of the analysis of CBD patterns which
do not require a detailed knowledge of the dynamical theory of electron
diffraction for their interpretation and which are extremely useful in
practice. More general reviews can be found in the recent articles by
LEHMPFUHL (1978) and COWLEY (1978a).

The Microscope

Any modern commercial instrument designed for STEM or microanalysis
can be used for CBD experiments. It is preferable to have post-specimen
lenses so that the pattern may be formed with a fixed probe as in conventional electron microscopes and this configuration will be assumed
hereafter. A focussed probe of diameter a few tens of nanometers is
ideal for most applications. Crucial parameters for effective operation
in the convergent beam mode are the contamination rate and the angular
view in the diffraction plane. A eucentric goniometer stage is essential for application of the technique to selected small areas. In
general there are two extremes of operation which will be referred to
here as the TEM and STEM modes, as described below. In either case the
convergence angle is essentially determined by the diameter of the
second condenser aperture, and a wide range of clean apertures is called
for. One regularly uses apertures from 400 μm down to 5 μm diameter and
it would be convenient to have an aperture holder capable of carrying
ten apertures. For some purposes (see Sec. 15.6) square apertures or
circular apertures with fine wires across them are valuable. The objective aperture is only used in examining the specimen image in the conventional fashion and is removed for convergent beam diffraction. The area
selecting (intermediate) aperture is not required and is therefore with-
drawn from the beam.

The CBP is formed in the back focal plane as indicated in Fig.
15.1. A convergent cone of electrons AA is focussed on the specimen to
a probe of finite diameter BB by the pre-field of the condenser-
objective lens. Its angle of convergence is directly proportional to
the diameter of the second condenser aperture. Solid lines represent
direct ray paths, broken lines represent diffracted ray paths. The six
rays shown converging on the specimen are in fact 3 pairs of parallel
rays passing through the diametral points of the focussed probe BB. One
pair is parallel to the optic axis and therefore comes to a point focus
on the optic axis in the back focal plane of the condenser-objective
lens. The other paris are at extremes of inclination in the incident
cone and therefore cross at points CC in the back focal plane on the
circumference of a disc of illumination corresponding to all the intermediate inclinations of rays between these extremes. Diffracted rays

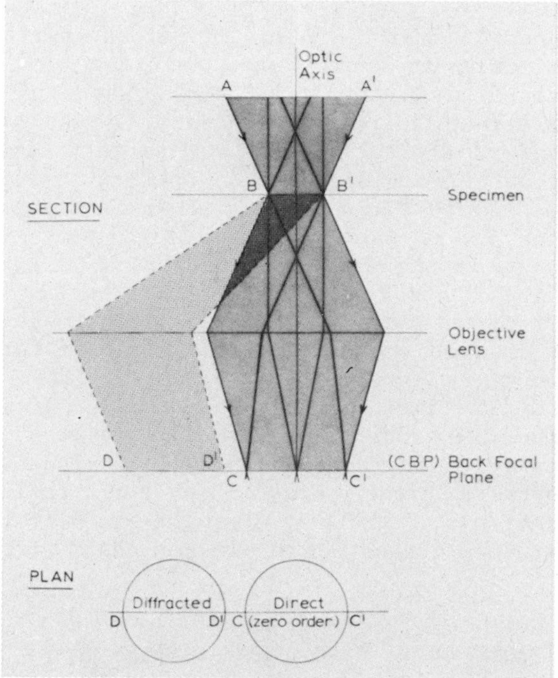

Fig. 15.1 Ray diagrams illustrating the formation of convergent beam patterns (CBP's).

fall outside the range of angles of incidence if the second condenser aperture is chosen appropriately and hence form discs outside the direct disc. In this mode of operation the illumination is extremely incoherent and it is possible to disregard any phase relationship between the different directions of propagation within the incident cone to a good approximation.

TEM Mode

This is the simplest mode of operation for performing CBD experiments if the microscope and pole pieces are designed to accommodate it (on some instruments it requires a change of pole pieces or operation of the microscope under free control of electron lenses). If, as is often the case, one wishes to work with the smallest possible probe in this mode one starts by forming an image of the specimen under standard conditions and then increases the first condenser lens to its maximum excitation while altering the second condenser lens to produce a focussed spot (adjustment of the beam alignment controls will usually be necessary to keep the spot in the area of interest). Slight adjustment of the objective lens may be required to maintain specimen focus. One can then examine this focussed spot (deflecting it beyond the edge of the specimen for convenience, if this is practical) for condenser astigmatism adjustment, for saturation and alignment of the filament and to check its stability, making the necessary adjustments before proceeding. To form a CBP it is only necessary to remove the objective aperture and to go the diffraction mode of the instrument. Once in the diffraction mode there are four adjustments to consider. The first adjustment to make is the size of the condenser aperture, selecting larger or smaller apertures as appropriate. If an aperture change is called for it is advisable to return to the image plane so that the aperture may be centered in the normal way. When the CBP contains very close orders of diffraction it is necessary to use very small second condenser apertures, down to 5 µm in diameter. One then obtains what looks like a conventional diffraction pattern with diffracted discs of only approximately 10^{-4} rad angular diameter. Longer exposures are then called for and it is rather more difficult to center the aperture and locate the spot on the area of interest. The second adjustment to make is in the fine control of the diffraction lens. This adjustment follows quite different criteria from the adjustment for conventional diffraction patterns and one has to choose one of three different conditions depending on the circumstances. There is often a caustic figure, especially when the specimen is thick, resulting from stray electrons superimposed on the CBP. One condition of adjustment minimizes the visibility of this caustic figure (Fig. 15.2). Alternatively, it may be that one wishes to observe higher order Laue zone (holz) rings with low camera length settings. In this case the diffraction lens adjustment is chosen for minimum distortion at the periphery of the pattern (Fig. 15.7).

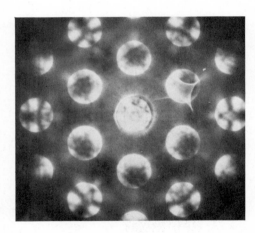 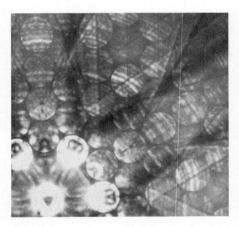

Fig. 15.2. <111> CBP from $M_{23}C_6$ precipitate from 316 stainless steel (100 KV). The caustic figure visible in the 220 reflection at 2 o'clock can be eliminated by adjustment of the diffraction lens.

Fig. 15.3 [0001] CBP from TaS (faulted crystal). Elliptical distortion of the circular discs at the center of the pattern (bottom left) can be seen to increase with distance from center.

Finally, when neither of the above considerations are important one chooses the setting which minimizes the off axis distortion of the discs from their ideal circular shape (Fig. 15.3). The final two adjustments are only necessary if there is cross-talk between the imaging lenses (as is inevitable unless electrostatic intermediate and diffraction lenses are employed). In adjusting the diffraction/intermediate lenses to form the diffraction pattern the spot may no longer be focussed on the specimen and the illuminated region may have shifted somewhat. One can re-

focus the spot by moving to an edge or some other prominent feature of
the specimen. If one is in an exact reciprocal relationship with the
image plane no spatial information will be present in the convergent
beam pattern. The presence of a shadow image of the specimen in the CBP
indicates that slight adjustment of the objective or second condenser
lens is called for. The sense of adjustment is that which increases the
magnification in the shadow image until the point is reached where the
shadow image suffers an inversion (Fig. 15.4). At this point the spatial information is lost from the CBP (all granularity vanishes) and the
probe and the back focal plane (fixed) of the objective are in reciprocal relationship to each other, i.e. the probe is focussed. To determine whether some sideways shift of the probe has occurred it is necessary to have some idea of what to expect in the CBP and, if the pattern
is obviously not coming from the desired region, to make such small adjustments as are necessary. An obvious example would be the case of a

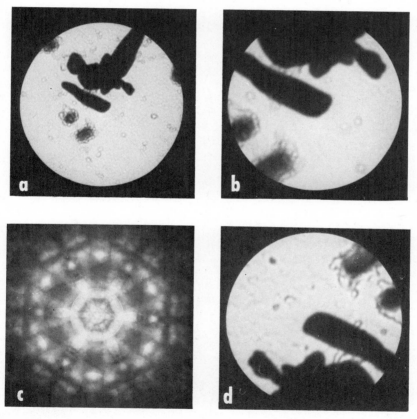

Fig. 15.4. Sequence of pictures illustrating the inversion of the shadow
image of a specimen from which CBD is required. The sequence (a) - (d)
is for increasing exitation of the second condenser lens. The CBP in
(c) is obtained from the central particle in the other micrographs
(<111> axis of $M_{23} C_6$ at 100 KeV). All micrographs show the central
disc in the diffraction mode of operation but (c) is shown at a lower
magnification so as to reveal several orders of diffraction in addition.

tiny extracted particle on a carbon replica. If, on forming the CBP, no Bragg reflected discs are present one either adjusts the beam deflectors or the specimen traverse to relocate the beam on the particle.

STEM Mode

The conventional STEM mode of operation uses the first condenser lens at its maximum excitation and an unexcited second condenser lens. The objective lens runs at a higher excitation than in the TEM mode and is generally near its maximum setting. Under these conditions it will not usually be possible to form an image of the probe, although an image of the specimen can be formed if a STEM unit is available. One works, therefore, entirely in the diffraction mode of the instrument and chooses the area of interest from a shadow image of the specimen, formed when the probe is not properly focussed on the specimen. The CBP is then obtained, in the manner described above, by adjusting the objective lens until the point of inversion of the shadow image is located. As the final position is approached the magnification of the shadow image increases rapidly and allows one to position the probe more accurately on the chosen area. To give a symmetrical probe condenser astigmatism correction can be carried out while studying a shadow image of the specimen. If the angular view in the CBP is unduly restricted, as it may be in some cases, the situation can be improved by changing the specimen height, although at the price of sacrificing eucentric tilt.

Intermediate Configurations

The TEM and STEM modes may be regarded as extreme configurations and, depending on the instrument, various intermediate configurations can be set up. The detailed behavior can vary considerably. A ray diagram for the STEM configuration is shown in Fig. 15.5(a). As the excitation of the second condenser lens is increased from zero it is evident that the objective lens current will have to be decreased to maintain a focussed probe on the specimen and the angular convergence on the specimen will decrease. In some microscopes (e.g. Philips EM 400) this situation persists up to the second condenser excitation of the TEM mode. However, for other microscopes, a new cross over may occur before the prefield of the condenser objective lens (Fig. 15.5(b)) and then further increase of the second condenser excitation produces an increase of convergence on the specimen. It is evident that in the latter case one cannot pass continuously from the STEM to the TEM mode and there will be a range of excitations over which it is impossible to perform CBD. The chief advantage of these intermediate configurations is that continuous adjustment of the convergence angle is possible. However, not all microscopes are provided with the necessary free control of the second condenser lens over all its range. In addition, the sample will not normally be focussed in the image mode of the microscope unless its height is adjusted and then one loses eucentricity of tilt.

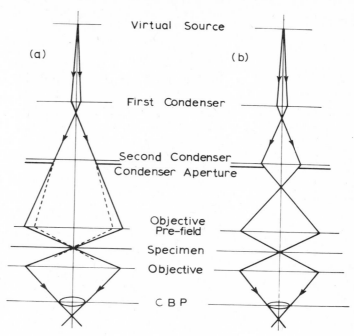

Virtual Source

(a)   (b)

First Condenser

Second Condenser
Condenser Aperture

Objective
Pre-field

Specimen

Objective

C B P

Fig. 15.5. Ray diagrams for CBD
using a condenser-objective lens
(a) STEM mode (continuous line):
(b) TEM mode (broken line).

EFFECTS CONNECTED WITH THE SPECIMEN

Beam Broadening

It often happens that the samples which one wishes to examine are
at least 100 nm in thickness. Under these circumstances a small probe
will become considerably broadened in traversing the specimen. The degree of broadening will depend on parameters such as the thickness of
the sample and its mean atomic number (KYSER and GEISS, 1979; HUTCHINS
et al., 1979), but as a rough guide one can reckon on an effect
amounting to some tens of nanometers. This is the chief factor limiting
spatial resolution in CBD and the one which makes it hard to do better
than the excellent results which one can obtain with a conventional
electron source on a microscope operated in the TEM mode.

Beam Heating

As CBD from thick specimens permits accurate determination of the
lattice parameter (Sec. 15.4) it can be used to measure the temperature
rise in the irradiated area. The sensitivity of the measurements permits
temperature rise of about 50°C to be detected for common materials and,

for example, no observable effects are obtained from bulk specimens of metals or diamond structured materials. However, experiments on LiTaO$_3$ (a very poor thermal conductor) indicate that temperature rise of greater than 300°C could be produced for rather large beam currents (LOVELUCK and STEEDS, 1977) and it is likely that a substantial increase could occur in the case of a small particle with poor thermal connection to a thin supporting foil.

Perfection of the Specimen

On the whole a crystalline specimen which is suitable for examination by transmission electron microscopy is suitable for CBD. However, there are some points of difference. CBD patterns with holz lines are very sensitive to strain fields. The origin of the strain field may visible, as in the case of dislocations and precipitates or invisible, as in the case of thin surface films in compression or tension. It is essential to avoid such regions to obtain results which can be related to the crystal symmetry. Buckled specimens can be tolerated so long as the resulting strains are not significant within the diffracting volume (see Sec. 15.5).

Samples with point disorder or planar disorder in a direction perpendicular to the plane of the foil may appear indistinguishable from more perfect specimens in transmission microscopy but the disorder, if severe, can have a devastating effect on the quality of CBP's. Point disorder can be thought of as introducing a static term into the Debye-Waller factor which destroys large angle scattering so that holz rings either vanish or at least become indistinct (SHANNON and EADES, 1978). Planar disorder in a direction perpendicular to the plane of the foil smears out the holz rings and produces characteristic patterns from thick crystals with many orders of diffraction and considerable diffuse scattering (STEEDS and FUNG, 1978).

Samples with large wedge angles are not very suitable for two reasons. In the first place the theory of CBD generally assumes that the surfaces of the specimen are approximately perpendicular to the beam direction (for the same reason strongly tilted specimens should be avoided if possible). The effects of surface inclination have not been very thoroughly investigated so far but can be quite significant on occasion (GOODMAN, 1974). The second disadvantage of samples with large wedge angles is that it is extremely difficult to avoid thickness averaging even with very small probes incident on the specimen.

Contamination

It is vital to reduce contamination effects to as low a level as possible if meaningful results are to be obtained from CBP's. It is not uncommon to be working on one small area of a specimen for more than one hour with a fine focused probe and this provides a strigent test of the cleanliness of the vacuum system and the specimen. Columns with high hydrocarbon partial pressures can be improved by carefully constructed cold traps in the vicinity of the specimen; sample holders should be de-greased and handled with gloves. If the specimen is not subject to radiation damage and can take a mild rise in temperature it is advisable to flood it over a large area with a high electron flux for several minutes prior to commencing CBD.

Specimen heating and cooling both produce a significant reductions in the contamination rate. A modest rise in temperature of little more than 100°C is sufficient to desorb most of the hydrocarbon responsible for contamination. However, the increased Debye-Waller factor may cause an appreciable reduction in large angle scattering. It is therefore preferable, for good CBP's, to cool the sample to liquid nitrogen tem-peratures when the surface diffusion of hydrocarbons is greatly reduced, producing a dramatic reduction in contamination rate. This subject is reviewed extensively in Chapter 18.

Goniometry

A eucentric double-tilt or tilt-rotate stage is essential. Various techniques exist for orienting small particles on particular zone axes, but the method which is most useful in general involves working in the shadow image mode in the diffraction plane. One then has diffraction and spatial information simultaneously displayed and this facilitates controlled goniometry. It is a good idea to have available a small pro-grammable computer with a plotter so that programs can be written to convert the goniometer readings into points on a sterographic projection so that orientation experiments can be followed with ease.

Final adjustment of the specimen onto the zone axis can be compli-cated by back-lash and is often achieved more easily by a small displace-ment of the second condenser aperture.

15.2 THREE DIMENSIONAL ELECTRON DIFFRACTION

Higher Order Laue Zones

It is helpful to discuss electron diffraction effects in terms of the Ewald sphere construction in the reciprocal lattice [see for example KITTLE (1976)]. Because of the small wavelength of high energy electrons the radius of the Ewald sphere is much greater than the spacing of re-ciprocal lattice points (typically 50 times greater for 100 kV elec-trons). When an electron beam is incident along a zone axis a recipro-cal lattice plane perpendicular to the axis (the zero layer plane) is a tangent to the nearly planar surface of the Ewald sphere (see Fig. 15.6) and may reflections are excited ("cross grating diffraction"). However, the separation between the Ewald sphere and the reciprocal lattice points in the zero layer plane increases with distance from the origin (at the point of contact) until reflections in this plane are no longer excited to any appreciable extent. Eventually, with increasing distance from the origin, the Ewald sphere will intersect the next layer of the reciprocal lattice giving rise to a circle of reflections known as the first order Laue zone (folz). Successive intersections with other layers of the reciprocal lattice give rise to different orders of holz.

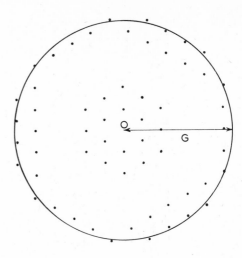

Fig. 15.6 Plan and sectional views
of the Ewald sphere construction
for high energy electron diffraction illustrating the formation
of higher order Laue zones.

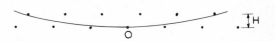

The visibility of these holz effects is dependent on a number of
factors. First, and most fundamentally, it depends on the spacing of
the layer planes along the beam direction. This in turn depends on the
particular axis chosen, the crystal structure and its dimensions. For
the dimensions given in Fig. 15.6 the radius of the folz is given approximately by

$$G = \sqrt{2KH} \qquad\qquad (15.1)$$

where $K = 1/\lambda$; λ is the electron wavelength. Second, the visibility of
the various Laue zones depends on the strength of the large angle scattering which in turn depends on the amount of thermal or static disorder
in the crystal structure. The dramatic increase in large angle scattering which results from cooling to liquid helium temperatures is illustrated in Fig. 15.7. In order to make the transition from Fig. 15.6,
which is drawn for plane wave fronts, to the CBD case it is only
necessary to draw discs centred on each reciprocal lattice point in
planes perpendicular to the mean direction of incidence. Third, the
effects which have been described can only be observed if electron
microscope being used provides an adequate angular view of the back
focal plane of the objective lens. To obtain a view of such as that
shown in Fig. 15.7 it may be necessary to change the objective pole
pieces or else a small circuit modification may be required.

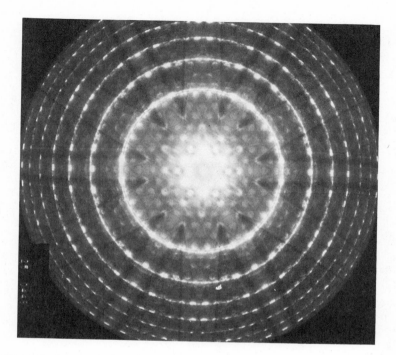

Fig. 15.7. [0001] CBP of 2H Ta Se$_2$ at 120 KV showing many holz rings. The large angle scattering has been increased by cooling the sample to approximately 80°K [Fig. 15.8(b) taken at room temperature].

Diameter of Holz Rings

There are several reasons why one might wish to determine the diameter of the holz rings for a particular zone axis of a known crystal structure. It may be that limitation to the field view in the diffraction plane is an important consideration or else that either strong or weak three dimensional diffraction is required. On the whole, the larger the diameter of the holz ring the weaker its diffraction (see below for a more accurate statement about intensities). For the <uvw> zone axis of a simple face centered cubic crystal the spacing H_{uvw} of reciprocal lattice planes along the axis is given by

$$H_{uvw} = \frac{p}{a_0(u^2+v^2+w^2)^{\frac{1}{2}}} \qquad (15.2)$$

where a_0 is the lattice constant and p takes the value 1 or 2 according to

    p = 1 if (u + v + w) is odd

    p = 2 if (u + v + w) is even.

For a simple bcc crystal the same result is valid but p = 2 if u, v and
w are all odd integers, otherwise p = 1. The diameter of the folz ring
may be obtained by substituting H_{uvw} into eq. (15.1).

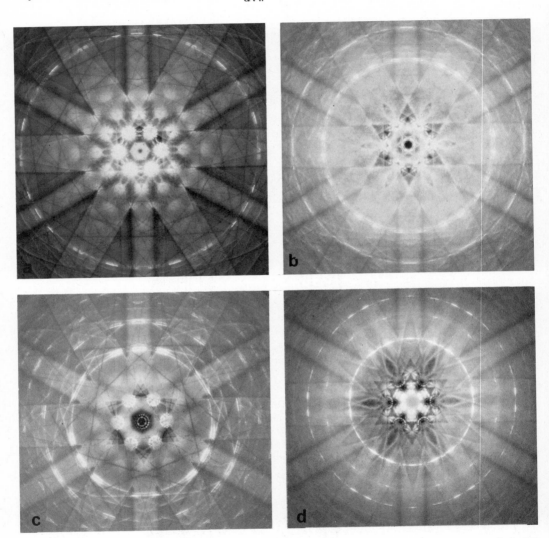

Fig. 15.8. [0001] CBP's of layer structured materials at 100 KeV (a) IT
Zr Se$_2$, (b) 2H Ta Se$_2$, (c) 3R NbS$_2$, (d) 4Hb TaS$_2$. In the one layer
repeat material (a) the holz ring corresponds to a spacing of reciprocal
lattice planes along [0001] of 1/6.14 (Å)$^{-1}$. For the two-layer repeat
material (b) the inner ring corresponds to 1/(2x6.35) (Å)$^{-1}$. For (c)
the inner ring corresponds to 1/(3x5.97) (Å)$^{-1}$ and for (d) to 1/(4x5.91)
(Å)$^{-1}$.

In working with more complex crystal structures there may be some uncertainty over the correct identification of reflections in a cross grating diffraction pattern. It is then often possible to remove the uncertainty by measuring the diameter of the holz ring. However, a word of caution should be sounded. The ring diameter is generally subject to an uncertainty of a few percent for a variety of reasons. The chief of these is the effect of lens distortions, which are apparent in low camera length patterns and depend on the fine control setting of the diffraction lens (Sec. 15.1). In addition, further inaccuracies arise from the dynamical nature of electron diffraction (see below).

One problem where a measurement of the diameter of the folz ring diameter is particularly useful is in polytype identification. As an example, consider the case of layer structured materials prepared for transmission electron microscopy by cleavage. By forming a holz pattern at the [0001] axis one can immediately determine whether there is a 1, 2, 3, 4 or 6 layer repeat along the axis (Fig. 15.8). In this respect CBD is a competitor to lattice resolution and overcomes the need for tricky specimen preparation so that the basal planes can be viewed edge-on.

INDEXING AND ORIGIN OF HOLZ LINES

Indexing

When, as in Fig. 15.9(a), one can see bright holz lines within discs in a particular zone and correlate them by orientation and position with dark holz lines in the zero order disc at the center of the pattern the lines may be indexed directly. First index the reflections in the zero layer pattern. Next, use the result quoted in the previous section for a fcc crystal to deduce that all folz reflections will have $4h + k + 1 = 2$ (because $4 + 1 + 1 = 6$, and is even so $p = 2$) and hence reflections such as 002, 020 and 111 lie in the first layer. Finally use vector addition of the reflections in the zero layer to these reflections near the origin of the first layer to index reflections in the folz ring as shown. Having indexed the discs in the folz take one of the discs whose outline can be fairly clearly distinguished and focus attention on the inner, straight line in it. In most cases there is a parallel dark line crossing the zero order disc, Fig. 15.9(b), in precisely the same position as the bright line crossing the folz disc. Stronger bright lines in the folz ring such as $2\bar{1}04$ correlate with the darker lines in the zero order disc. Repetition of this operation for each of the discs in turn permits indexing of all lines except the line marked $4\bar{8}\bar{6}/4\bar{6}\bar{8}$ in the zero order disc which has no equivalent in the holz ring. Such effects are quite common when two adjoining discs in the holz are strongly excited ($4\bar{8}\bar{6}$ and $4\bar{6}\bar{8}$) leading to hyperbola formation as illustrated later in Fig. 15.11. The comparatively weak lines at $\bar{2}\bar{2}12$ in the folz do not appear to correlate well with any of the lines in the zero order disc. As a final check on the indexing it is helpful to work with small variations in the operating voltage. Overlapping lines then separate and it can be checked that the lines move in the expected directions.

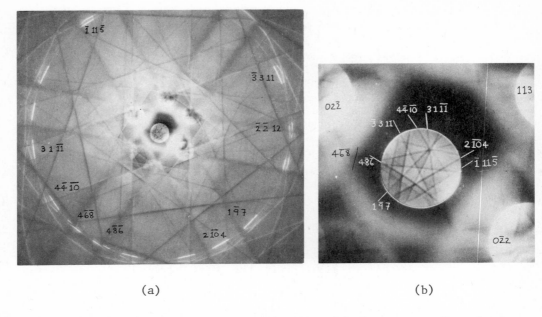

(a)   (b)

Fig. 15.9. [411] CBP of ZrO_2 at 100 KV. (a) Low camera length micrograph showing folz ring; (b) enlarged view of zero order disc visible at
the centre of (a).

When the lines are too numerous for this technique to work, possibly because the lines come from more than one holz, or else the specimen is so thick that diffuse scattering obscures the disc position,
other methods must be used. The most successful technique under such
conditions seems to be a computer simulation of the lines using a
Kikuchi (or Kossel) line program (LOH and ASHBEE, 1971) appropriate to
the chosen zone axis and the microscope operating voltage. However, we
introduce one slight subtlety into the calculation which calls for a
brief digression into the origin of the holz lines for its explanation.

Origin of Lines

Although holz lines in electron diffraction are the equivalent of
Kossel lines in x-ray diffraction there are some important differences
of detail which are a consequence of the dynamical nature of electron
diffraction. Strong dynamical interactions between the zero layer reflections lead to rather flat (orientation independent) branches of the
dispersion surface (BERRY et al., 1973). Symmetry properties may be
associated with the different branches which give rise to strong holz
effects (JONES et al., 1977). Each strongly excited symmetric branch of
the dispersion surface can give rise to a separate holz line for a given
holz reflection. As there are in general at least two such symmetric
branches for zone axis cases there is the possibility of at least two,
approximately parallel, holz lines with the same index hkl. An example
is shown in Fig. 15.10 when the pairs of lines are indicated. One line
of each pair is simple in form, the other has a rope-like form. The

slight curvature of the lines is a consequence of the slight curvature of the two associated branches of the dispersion surface. If the operating voltage was evaluated using the rope-like line of the pair in Fig. 15.10, this would result in an overestimate; if the other was used the operating voltage would be underestimated. The difference between these two voltages is typically of the order of 5% of the actual voltage which lies somewhere between the two estimates. Now it is unusual to have both these lines for a particular material at a particular zone axis and it is more often the case that just one of the pair appears. Fortunately, it is generally the same line of each pair which appears in a particular pattern so that it is possible to obtain a convincing fit between the observed line pattern and the Kikuchi line simulation as long as the operating voltage is treated as a variable parameter.

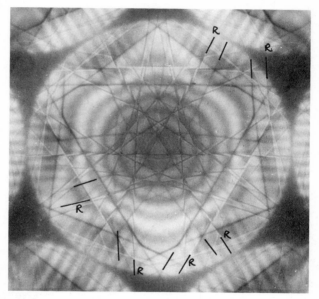

Fig.15.10. Si ⟨111⟩ zero order CBP at 100 KV, illustrating the pairing of nearly parallel holz lines.

    In the case of zone axes for which the projected potential can be thought of as arising from an approximately close packed array of simple atom strings (Sec. 15.4) it is possible to predict the sense in which the voltage must be adjusted and the approximate magnitude of the adjustment. For diffraction situations with comparatively weak string potentials such as the ⟨111⟩ axes of aluminum, diamond, stainless steel or copper the voltage must be decreased by about 1%. For stronger string potentials such as those for the ⟨111⟩ axes of germanium or gold, the voltage should be increased by about 4%. The silicon ⟨111⟩ axis is an intermediate case. A typical example of indexing in this way is shown in Fig. 15.11 for a ⟨111⟩ axis of diamond at an operating voltage of 100 kV. An artificial voltage of 99 kV was employed for the simulation.

In addition to the dynamical effects described above crossing holz
lines can interact (particularly at low angle intersections) to produce
hyperbolae in the vicinity of the intersection. One branch of the hy-
perbole is enchanced in visibility relative to either of the lines
separately, the other is weakened, producing the situation depicted in
Fig. 15.11(a), the consequence of interactions between $\overline{3}59$ and $\overline{5}39$. For
a strong interaction between lines with indices $\underline{G}_1$ and $\underline{G}_2$ the vector
$(\underline{G}_1 - \underline{G}_2)$ must be small.

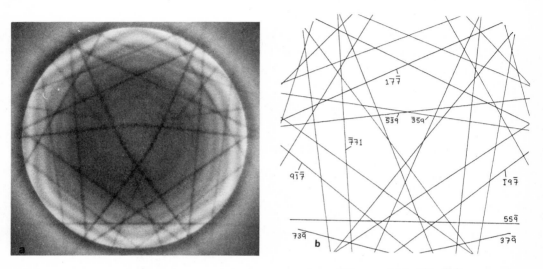

Fig. 15.11 (a). Diamond <111> zero order CBP at 100 KV (b) computer
simulation at 99 KV for a = 3.567 Å.

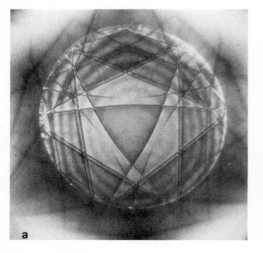
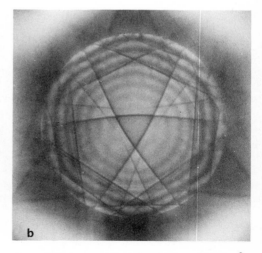

Fig. 15.12(a). Nickel 111 zero order CBP at 100 KV (a = 3.524 Å);
(b) 316 stainless steel 111 zero order CBP at 100 KV (a = 3.595 Å).
Using the result in Fig. 15.11 (a) as a standard, one deduces a = 3.516
for Ni and a = 3.589 for 316 stainless steel.

The extinction lengths associated with holz lines are long, typically in the range 200-600 nm. It is for this reason that thick crystals are necessary if sharp lines are to be obtained. The angular width
of the lines is also dependent on this extinction length (ξ_G) as in the
line visibility. One can say, as a generalization, the longer is ξ_G the
fainter and sharper are the lines. It is therefore of some importance
to have a means of calculating ξ_G and to a first approximately, making
several simplifying assumptions, one can write

$$\xi_G \sim \pi V_c \, k\cos\theta / (fF_G) \qquad (15.3)$$

Where F_G is the structure factor for the holz reflection $\underline{G}$; f may be regarded as an amplification factor and takes values in the range 2-4 for
symmetric branches of the dispersion surface; V_c is the volume of the
unit cell, and θ is the Bragg angle for the reflection $\underline{G}$.

We can, further, write an approximation expression for the intensity of the holz reflection (ignoring absorption) as

$$I_{\underline{G}} \quad G_i^2 \sin^2 (\pi t / \xi_G) \qquad (15.4)$$

where i refers to the branch of the dispersion surface which gives rise
to the line and ε_i is the excitation of this branch at orientations away
from its interaction with the reflection $\underline{G}$; t is the specimen thickness.

As ξ_G is so long the intensity given by Eqn. 15.4 increases essentially monotonically with specimen thickness for many specimens and one
recovers, as is unusual in electron diffraction, a simple relationship
between the intensity of the reflection and its structure factor. This
important result often proves useful in practice (see Sec. 15.4).

It remains to comment of the relationship between holz lines and
Kikuchi lines. That they are closely related is evident from Fig. 15.12
where one can see continuity between holz lines within the zero order
disc and Kikuchi lines in the diffuse background between the discs.
However, there are important distinctions between the lines (KOSSEL and
MÖLLENSTEDT, 1939). Holz lines shift slightly in position as a function
of the specimen thickness (JONES et al., 1977) and are therefore slightly narrower than the associated Kikuchi lines which are necessarily
thickness averaged. Further, Kikuchi lines become ever more prominent
as the specimen thickness increases but are very hard to observe in
thinner specimens. Holz lines are most visible from specimens where
Kikuchi lines are comparatively weak and become "washed out" as the
thermal diffuse scattering becomes a major component of the intensity
within a disc.

LATTICE PARAMETER DETERMINATION

It is evident that changes in the lattice parameters of a crystalline material will produce shifts in the holz lines. Geometrically
speaking, this is because the deviations between the reciprocal lattice
points and the Ewald sphere alter. Using the small angle approximation

to Braggs Law, one has

$$\Delta\theta/\theta = \Delta G/G \qquad (15.5a)$$

Hence, for a cubic material with lattice constant a,

$$\Delta\theta/\theta = -\Delta a/a \qquad (15.5b)$$

Thus, for a given change, Δa, the change in line position (determined by $\Delta\theta$) is large for large Bragg angles, i.e., holz lines should be used. As a generalization, the larger is θ, the finer are the lines and the more accurate is the measurement which can be made. However, as the scattering angle becomes greater, the lines become weaker and the associated extenction lengths become longer. It is therefore beneficial to use cooling to reduce the Debye-Waller factor and necessary to use thick specimens. The limitation to the accuracy which can be obtained is set either by the weakness of large angle scattering or else by energy losses in the specimen.

To investigate the effect of energy losses let us consider the effect of a small change (ΔE) of the operating voltage (E) of the instrument. According to the small angle approximation to Bragg's Law

$$\Delta\theta/\theta = \Delta E/(2E) \qquad (15.6)$$

That is, changes of the operating voltage change the holz line positions just like lattice parameter changes (RACKHAM et al., 1974). There are several consequences of this conclusion. First, for routine measurement of lattice parameters, it is helpful to have a modification to the microscope which permits continuous adjustment of the microscope operating voltage so that a standard reference voltage can be maintained. Small voltage shifts occur after flash-overs or other instabilities in the H.T. tank and the operating voltage also depends slightly on the Wehnelt bias in most instruments. Second, energy losses (ΔE) which occur in traversing the specimen will produce a broadening of the holz lines. The thicker the crystal the greater the losses and hence the fundamental limitation of the technique at present. For accurate measurements a specimen of 200 nm thickness might be used for which energy losses of say 40 eV occur. Hence, the limit to the accuracy with which lattice parameter changes may be measured

$$\Delta a/a = \Delta E/(2E) = 2/10,000 \qquad (15.7)$$

From this argument we deduce that higher operating voltages might yield more accurate results.

A third consequence of the effect of voltage variation is that lattice parameter changes may be directly converted to the voltage change (ΔE) which is required to return the lines to some reference position and then

$$\Delta a/a = \Delta E/2E \qquad (15.8)$$

Lattice parameter changes can be measured without even using photo-graphic material and the measurements are independent of uncertainties about the microscope camera length.

From the above arguments it would seem that accurate measurements can be made of changes of lattice parameter with respect to a known standard with the same crystal structure. The only limitation to the conclusion comes from the dynamic origin of the holz lines. If the standard has a very different strength of projected potential from the material under examination then inaccuracies could arise. For example, one would not expect accurate results by comparing gold with copper. An illustration of the use of this technique is given in Fig. 15.12. By measuring the height (p) of the "triangle" at the centers of Fig. 15.12 (a) and (b) and using Fig. 15.11 (a) as a reference it is possible to deduce the lattice parameters of nickel and 316 steel as follows. Simple geometry shows that the change in line position on the micrograph is approximately one third of the change of height of the "triangle". Hence, if the spacing between equivalent points in the central and any 220 disc is q_i, then

$$\frac{\Delta\theta}{\theta} = \frac{2}{3}\left[\frac{P_i}{q_i} - \frac{P_{ref}}{q_{ref}}\right](8/115)^{1/2} = \frac{\Delta a}{a_{ref}} \qquad (15.9)$$

where q_i $(115/8)^{\frac{1}{2}}$ is the radius of the folz ring $((115/8)^{\frac{1}{2}} = |G_{5\bar{3}9}|/|g_{220}|$) and a further factor of two is introduced because this radius is equivalent to 2θ. Note that if low camera length CBP's are available it is not necessary to index the holz lines since the radius of the holz ring can be measured directly. The resulting values for the lattice parameter are given in the caption to Fig. 15.12. They are not as good as might be expected simply because the micrographs shown in Figs. 15.11 and 15.12 were chosen for their general quality but were taken on two different microscopes (EM's 301 and 400) with a gap of at least three years between the nickel and 316 steel picture. Some volt-age differences are therefore inevitable and account for the discrep-ancies. Further, if more accurate results are desired, it would be necessary to make measurements on lines which were not affected by in-teractions between holz reflections themselves (the "triangle" chosen is evidently formed by hyperbolae) or else introduce a correction for dia-meter of the hyperbola in each case. Finally, the results presented for nickel are evidently subject to some strain as the three heights of the "triangle" differ quite considerably and this is a further source of in-accuracy.

The prospects for absolute determination of lattice parameters do not appear so promising. The best that has been achieved so far is about a factor of ten lower than the figure for relative determinations, and was only achieved with the aid of extensive computer calculations (JONES et al., 1977).

MEASUREMENT OF CHEMICAL VARIATIONS AND STRAINS

When the change of lattice parameter of an alloy or compound is directly and uambiguously related to its chemical composition, the chem-ical composition may be deduced from shifts in the holz line postions.

The effect has been demonstrated for fcc Cu Al alloys (MERTON LYN, 1977) and may be used to determine the local concentration of aluminum to an accuracy of l at %.

The spatial resolution which can be achieved depends on the probe size and its broadening by the thick specimen which is required for an accurate measurement. The diffracting volume may be approximated by a cylinder 30 nm in diameter and 200 nm long. In order to exploit the technique it is essential that there is no appreciable change of lattice parameter within the cylindrical diffracting volume. Problems with planar interfaces where the zone axis is chosen approximately parallel to the interface are ideal for the holz method. Under these circumstances a spatial resolution of about 30 nm is feasible.

Strains at planar interfaces can be measured in an exactly equivalent fashion to the chemical changes discussed above (RACKHAM and STEEDS, 1976). Strains can also be measured in the matrix surrounding large precipitates which penetrate through a thin foil, but it seems unlikely that the technique would be suitable for examining a specimen with a fine dispersion of precipitates because of the strain gradients.

15.3 CRYSTAL POINT AND SPACE GROUPS

Use of High Symmetry Zone Axes

The underlying theme of this section is the fundamental importance of using high symmetry axes in solving crystallographic problems. In return for the extra effort of tilting the crystal to special orientations one minimizes the number of zone axes to be studied. Since the specimen tilting may be performed at low levels of illumination the radiation damage or contamination of the specimen is also minimized. One thus reduces ambiguities and increases the chance of a successful symmetry determination.

It will also be assumed in this section that the specimens are 100 nm or so in thickness. For thinner crystals one has to pay careful attention to the surfaces and the effect of incomplete unit cells as has been elegantly demonstrated by Goodman in a number of articles (for example GOODMAN, 1974, GOODMAN and MOODIE, 1974) and also by CHERNS (1974) in another connection. For thicker crystals, the surfaces are less important and one is safe in assuming that the CBPs relate directly to the ideal crystal structures. A review of methods for point and space group determination was recently published by GOODMAN, (1978).

Point Group Determination

The symmetry of CBP's has recently received the attention of several authors (GOODMAN 1975; TINNAPPEL and KAMBE 1975; BUXTON el al., 1976). In the last of these papers connections were found between the 32 crystal point groups and the 31 possible symmetries (diffraction groups) of CBP's from ideal plane parallel foils with normal electron incidence. The diffraction groups are described by the symbols of the ten two dimensional point groups together with an inversion operation (designated by a subscript R) through the Bragg position of the dark field discs. It is the object of this section to illustrate the use of the tables of BUXTON, et al., (1976) by a particular example of point group determination.

 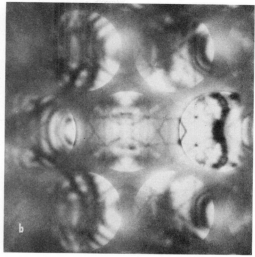

Fig.15.13.   4H b TaS$_2$ [0001] CBP at 90 KV.   (a) Zone axis orientation, (b) pattern centred on Bragg reflection condition ofr 10$\bar{1}$0 reflection.

A sample of TaS$_2$ was cleaved to produce a foil from which the CBP's illustrated in Fig. 15.13 were generated. It may be seen by inspection that the pattern in Fig. 15.13a has 6 mm symmetry both as a whole and also within the zero order (bright field) disc. According to Table 15.1 (second and third columns) this observation limits the diffraction group to one of two, 6 mm or 6mm 1$_R$ (first column). From column 5 of Table 15.1, it follows that dark field reflections in special positions (on the mirror lines or mid-way between them) would have m or 2 mm symmetry respectively for the two diffraction groups. The centre of a dark field disc is its Bragg position and so to distinguish the two possible diffraction groups it is necessary to displace the condenser aperture (§1.5) so that the centre of one of the dark field discs is visible (at the centre of Fig. 15.13b). The zero order disc is then on the right of the centre. Examination of the central disc of Fig. 15.13b reveals some left-right asymmetry which may be explained by diffuse scattering in the background. The finer details of the pattern and the holz lines clearly exhibit 2 mm symmetry; the diffraction group is determined as 6 mm 1$_R$. Finally, reference to Table 15.2 identifies the crystal point group as 6/mmm. Diffraction group 6mm 1$_R$ occupies the first row of Table 15.2. By running along this row to the cross we find the column through the same cross to be labelled 6/mmm, the point group of the crystal.

How sensitive are these properties to weak asymmetries or light atoms in the presence of heavy ones? As an attempt to examine this question experiments have been performed at <110> zone axes of Ga,As, and Ge (BUXTON et al., 1976). The asymmetric part of GaAs as compared with Ge is one atomic number in 31 yet the difference of symmetry is

Fig. 15.14. Pseudohexagonal [100] axis of (orthorombic) Mo   at 100 KV.

Table 15.1. Relation between the diffraction groups and the symmetry of convergent beam patterns. Where a dash appears in column 7, the special symmetries can be deduced from columns 5 and 6 [reproduced from Buxton et al. (1976) by courtesy of the Royal Society].

diffraction group	bright field	whole pattern	dark field		$\pm\,G$		projection diffraction group
			general	special	general	special	
1	1	1	1	none	1	none	1_R
1_R	2	1	2	none	1	none	
2	2	2	1	none	2	none	21_R
2_R	1	1	1	none	2_R	none	
21_R	2	2	2	none	21_R	none	
m_R	m	1	1	m	1	m_R	$m1_R$
m	m	m	1	m	1	m	
$m1_R$	2mm	m	2	2mm	1	$m1_R$	
$2m_Rm_R$	2mm	2	1	m	2	—	$2mm1_R$
2mm	2mm	2mm	1	m	2	.—	
2_Rmm_R	m	m	1	m	2_R	—	
$2mm1_R$	2mm	2mm	2	2mm	21_R	—	
4	4	4	1	none	2	none	41_R
4_R	4	2	1	none	2	none	
41_R	4	4	2	none	21_R	none	
$4m_Rm_R$	4mm	4	1	m	2	—	$4mm1_R$
4mm	4mm	4mm	1	m	2	—	
4_Rmm_R	4mm	2mm	1	m	2	—	
$4mm1_R$	4mm	4mm	2	2mm	21_R	—	
3	3	3	1	none	1	none	31_R
31_R	6	3	2	none	1	none	
$3m_R$	3m	3	1	m	1	m_R	$3m1_R$
3m	3m	3m	1	m	1	m	
$3m1_R$	6mm	3m	2	2mm	1	$m1_R$	
6	6	6	1	none	2	none	61_R
6_R	3	3	1	none	2_R	none	
61_R	6	6	2	none	21_R	none	
$6m_Rm_R$	6mm	6	1	m	2	—	$6mm1_R$
6mm	6mm	6mm	1	m	2	—	
6_Rmm_R	3m	3m	1	m	2_R	—	
$6mm1_R$	6mm	6mm	2	2mm	21_R	—	

Table 15.2. Relation between the diffraction groups and the crystal point groups (reproduced from Buxton et al. by courtesy of the Rocal Society).

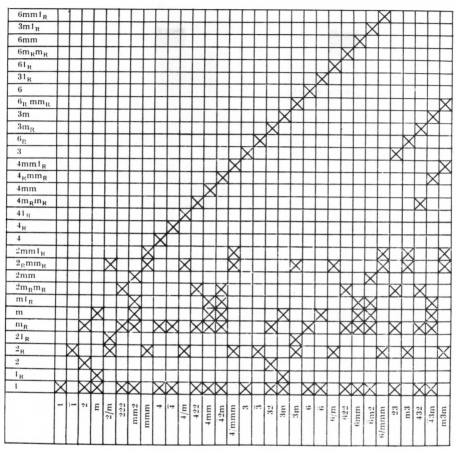

clearly visible over a wide range of specimen thicknesses. To investigate the importance of light atoms in the presence of heavier ones the pseudo-hexagonal axis of Mo_2C was studied by CBD (COOKE, unpublished). In this crystal structure the Mo atoms sit on a hexagonal lattice and it is only the carbon atoms which determine the orthorhombic symmetry. Nevertheless, the holz lines at this axis clearly reveal 2 mm rather than 6mm symmetry (Fig. 15.14).

Determination of the Reciprocal Lattice

Once a principal axis of a crystal structure has been located it is simple matter to deduce the reciprocal lattice from the three dimensional data present in a zone axis pattern (zap). An example is provided by the polytype of TaS_2 studied in the previous section. The [0001] CBP is illustrated at lower magnification in Fig. 15.15. The reflections in the holz's lie directly over the zero layer reflections and the diameter

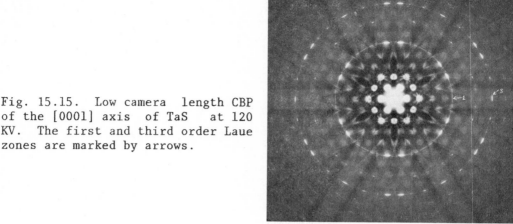

Fig. 15.15. Low camera length CBP
of the [0001] axis of TaS at 120
KV. The first and third order Laue
zones are marked by arrows.

of the folz ring indicates a four-layer repeat, i.e. a 4H polytype. It
polytypes also have the reflections of the folz directly over the zero
layer in [0001] patterns. However, in the case of 3R polytypes and [111]
axes of fcc crystals the reflections of the folz lie over the centroids
of triangles formed by the reflections of the zero layer.

Sometimes the reflections of a folz have a different spacing from
those in the zero layer, indicating the presence of forbidden reflections
and providing information about the space group of the crystal structure.
A case in point is the <001> axis (Fig. 15.16) of η carbide (M_6C),

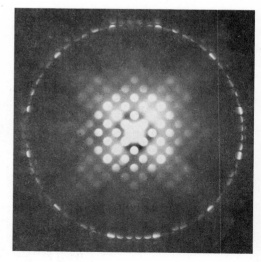

Fig. 15.16. Low camera length <100>
pattern of an η carbide precipitate
taken at 100 KV. The spacing of the
discs in the folz is closer than in
the zero layer pattern.

a fairly common precipitate in 316 stainless steel with space group Fd3m, but the closely related precipitate $M_{23}C_6$ has a space group of higher symmetry (Fm3m) and identical disc spacings in both the folz and the zero layer of <001> pattern. This difference allows the precipitates to be distinguished (COOKE, et al., 1978).

Space Group Determination

The kinematically forbidden reflections are not always as obvious as in the example illustrated above. Indeed, it will shortly be demonstrated that the space group Fd3m, cited above, can provide some surprising results in other circumstances.

Fortunately, there are clearly identifiable effects in CBP's which indicate the presence of glide planes or screw axes in crystal structures (GJØNNES and MOODIE, 1965). In the case of 2D diffraction (vanishingly weak holz effects) from a crystal aligned precisely on its zone axis it is sometimes found that patterns with 2 mm symmetry have alternate reflections with a band of negligible intensity running through them along one of the mirror lines in the pattern (GOODMAN and LEHMPFUHL, 1968). Such dynamic absences are a result of either a screw axis or a glide plane and the two effects may be clearly distinguished by obtaining patterns with 3D diffraction information in them. The symmetry of the zero order pattern is then reduced to m and if the mirror lies as m_1 in Fig. 15.17, then the zone axis is perpendicular to a screw axis in the crystal, while a mirror m_2 indicates that the zone axis is parallel to a glide plane (STEEDS et al., 1978).

Fig.15.17. Schematic diagram of CBP for axial illumination parallel to a glide plane or perpendicular to a screw axis, showing orientation of mirrors m_1 and m_2 with respect to the line of dynamic absences [reproduced from STEED et al., (1978) by courtesy of the Institute of Physics].

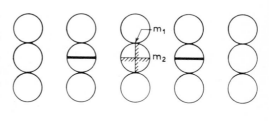

It has recently been shown that weak 200 reflections can sometimes be observed, although kinematically forbidden, in diffraction from diamond structured crystals with the space group Fd3m (TU and HOWIE 1978). Intensity reaches the reflection by diffraction out to a holz reflection and then back again. Unlike allowed reflections in the zero layer, where intensity is normally lost to a holz line, the 200 reflections are only present because of holz diffraction of intensity back into the zero layer and so appear as bright rather dark relative to the surroundings. An example is shown for a spinel structure (Fd3m) in Fig. 15.18. According to GJØNNES and MOODIE (1965), a radial line of dynamic absence is required by symmetry and this effect has recently been demonstrated by EVANS (unpublished work) using voltage variation on the <001> pattern of spinel. The bright lines in Fig. 15.18 first move together and then apart again as the voltage is varied but the radial line of missing intensity remains at all voltages.

Space group information can also be obtained by studying the intensity distributions in holz discs (STEEDS et al., 1978).

Fig. 15.18.   <001> CBP of spinel structured inclusion in 316 stainless steel examined at 100 KV.

The theory behind the dynamic absences discussed above is based on the assumption that the zone is perpendicular to the plane of the foil. If this condition is not fulfilled the cancellation will not be complete although discrepancies of 20° or so are probably not sufficient to produce an observable effect.

Handedness of a Crystal

GOODMAN and SECOMB (1977) (see also GOODMAN and JOHNSON 1977) have recently shown how to determine the handedness of quartz by two different techniques requiring reorientation of a specimen. LOVELUCK (private communication) has now found a way of obtaining the same information from a single [0001] zone axis pattern.

15.4 ATOMIC ARRANGEMENTS

A general method for determining the atomic arrangement within a unit cell remains an unsolved problem for electron diffraction (COWLEY 1978b). Lattice resolution studies have proved enormously informative in crystallogrpahic problems where a resolution of about 0.3 nm is sufficient to permit a deduction of the atomic arrangement and the crystal structure is sufficiently resistent to radiaion damage. However, it is only when the kinematic approximation holds that established routes exist for structure determination from electron diffraction patterns (COWLEY 1978a, UNWIN and HENDERSON 1975). As the kinematic approximation is only valid for thin specimens with potentials (O'KEEFE and SANDERS 1974) the condition is very restrictive and it might seem that we should leave this topic to the specialists for the present. Such a view would ignore two approaches to the problem which show considerable promise at the moment and do not call for expertise in the dynamical theory of electron diffraction or elaborate experiments. These two methods which concentrate on different aspects of the great wealth of detail which can be found in CBP's, will be discussed briefly.

Intensities of Holz Reflections

The simple relationship which often exists between the intensities of holz reflections and the appropriate structure factor (Eqs 15.2 and 15.3, Sec. 15.2) can be very useful in confirming the deduction of a particular crystal structure. Once the unit cell is determined and the space group known standard crystallographic tables will generally indicate the likely structure. A final confirmation is then very helpful. The case of 4 Hb TaS_2, examined previously, serves as an interesting illustration of the point. From the atomic arrangement of this polytype one deduces that the reflections of the first and third order Laue zones depend only on the atomic scattering factor of sulphur, but that reflections in the second Laue zone are kinematically forbidden. As can be seen in Fig. 15.15 the first and third Laue zone reflections are much stronger than those of the second, supporting the identification. The presence of the weak second ring of reflections is presumably the result of multiple diffraction.

A clearer confirmation of the relationship between structure factor and intensity for holz reflections can be found by studying 2H polytypes of WSe_2, $MoSe_2$ and MoS_2. The structure factors for the folz depend on $(f_m - \sqrt{2}f_c)$, for the second they depend on f_m and for the third on $(f_m + \sqrt{2}f_c)$ (where m stands for metal and c for chalcogen). For WSe_2 and MoS_2 the decrease in atomic scattering factor with increase of scattering angle is offset by the increase in scattering power as one moves from the first to the third Laue zone. Hence three rings of approximately equal intensity are observed. However, for $MoSe_2$ $(f_{Mo} - 2f_{se})$ is only a very small quantity and so the folz is practically invisible in spite of the comparatively small scattering angle (see Fig. 15.19).

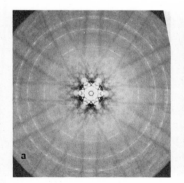 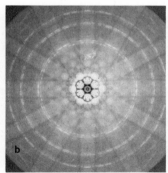 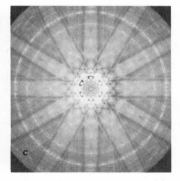

Fig. 19. Low camera length [0001] CBP's at 100 KeV. (a) 2H WSe_2, (b) 2H MoS_2, (c) 2H $MoSe_2$. Note the very weak inner holz ring for (c).

A word of caution must be sounded. The above approach relies on the assumption that the crystals are only subject to approximately equal thermal disorder. In crystals where different sub-lattices are subject

to very different degrees of thermal disorder, or one sub-lattice has
static point disorder, the conclusions could be grossly in error. The
sensitivity of holz lines to static disorder in the crystal, although a
drawback in the present context, can be very informative under other
circumstances (see Sec. 15.1).

Atomic String Approximation

The second method which can help in the structure determination depends on the eye's ability at pattern recognition to select zone axes
corresponding to particularly simple forms of projected potential (or
projected atomic arrangement). In the projection approximation, the zone
axis diffraction is regarded as essentially two dimensional, that is,
holz effects are sufficiently weak to be ignored. Under this approximation the periodicity of atoms along lines (or strings) parallel to the
zone axis is not important and the diffraction is primarily determined
by the scattering power (or strength) of the atom strings and, to a
lesser extent, by their disposition (hexagonal, square, etc.). For the
present, the string strength may be taken as $\gamma S_0 \Sigma_i Z_i / d$ where γ is the
relativistic mass factor, and there are i atoms in the repeat distance d
along the string with atomic numbers Z_i; S_0 is the area which can be associated with each string (a more accurate definition of string strength
has been given in STEEDS et al., 1977).

The simplest diffraction situation is when identical atom strings
with circular cross-section are in a close packed or nearly close-packed
array. The resulting zone axis patterns have been studied in a wide
variety of cases and have been found to pass through a series of characteristic forms as the string strength varies (SHANNON and STEEDS et al.,
1977). In order to use the results of this work the first step is to
identify simple string zaps. Two methods have been regularly used for
this purpose. The conventional diffraction pattern from a simple string
zone axis has intensities which are a monotonically decreasing function
of $|g|$ within the zero layer plane. In addition, particular points of a
zap are sensitive to small differences of string strength, if they
exist. An illustration of these differences is provided by string arrangements with hexagonal symmetry. It is helpful to indicate such differences by the larger angle intensity map provided by a bend contour
pattern (BCP), although CBD is just as useful and more widely applicable. Fig. 15.20a shows a BCP simulated by making a montage of many zero
layer CBP's for a silicon < 111 > axis (close packed array of simple
strings). On it are superimposed parts of the Brillouin zone boundaries as a reference system. For comparison, the [0001] BCP for $TiSe_2$
(close packed array of strings with slightly different strength) is
shown in Fig. 15.20b. The important region to examine is in the 4th zone
near the point K_A. Simple string axes only have two lines crossing at
this point but if there are small differences in string strength then
two additional lines can be seen emerging from this point, at an
agle of 60° to each other, with enhanced visibility in the 4th zone. As
the diffrence in string strength increases these lines become stronger
and the general character of the zap changes so that it differs in
several other respects from zaps from single string arrangements (SHANNON

and STEEDS,1977), or from simple string arrays which are not close packed
(e.g. MoS_2, TATLOCK, 1975; $LiTaO_3$, LOVELUCK and STEEDS, 1977).

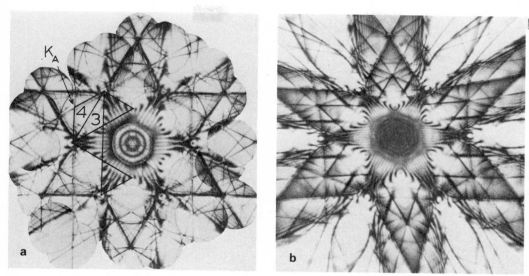

Fig. 15.20(a). <111> Si zero beam pattern at 100 KV formed by montage
from CBD pictures [reproduced from RACKHAM and EADES (1977) by courtesy
of Optik]. Part of the Brillouin zone boundary structure has been
superimposed on the micrograph. The central hexagonal first zone is
covered by six equilateral triangles constituting the second zone. The
third and fourth zones are as indicated.

Fig. 20(b). [0001] TiS_2 bend contour pattern at 100 KeV for comparison
with Fig. 20(a). The similarity between the two pictures permits identification of the point in the TiS_2 pattern which is similar to the
point marked K_A in the Si pattern, but the Brillouin zone boundaries are
in fact different.

Another form of deviation which can occur is that the cross section
of the individual strings may distort from the circular form assumed
until now. Only one case of this sort has been examined in detail so far
(BUXTON et al., 1978).

Having identified an appropriate zone axis the pattern within its
first Brillouin zone may be compared with those given in SHANNON and
STEEDS (1977) to fix a value for the string strength. The knowledge
thus obtained, basically by pattern recognition, can be of great value
in determination of the atomic arrangement of an unknown crystal structure or in testing a hypothesized structure. Note that the method does
not depend on the size of the unit cell or the number of different atoms
in it or on the resolution of the microscope. It depends on the existence of simple projections with the characteristics described.

15.5 FINGER PRINTING TECHNIQUES

The work outlined in the previous section depended on an understanding of the basic physics behind the formation of certain zone axis
patterns. It is however possible to produce a catalogue of diffraction
effects from prominent zone axes of commonly encountered materials without any attempt at understanding them. The resulting catalogue is so
effective for the purpose of identification that no case of confusion
has yet arisen among the 60 or so different materials so far studied in
this laboratory involving more than 20 space groups. There are three
different methods which are commonly used, either separately or in conjunction.

For zone axes of thin crystals with complicated projections of large
unit cells elaborate and distinctive patterns are formed of individually
featureless discs with large differences of intensity between them.
Numerous examples of this sort have been encountered and used in particle
analysis (Fig. 21a). For thicker crystals, at zone axes without appreciable 3D diffraction, the detail within individual direct and diffracted
discs provides a distinctive fingerprint (Fig. 15.21b). Finally, for
zone axes with significant 3D diffraction, elaborate meshes of lines or
complicated patterns containing a lot of fine detail serve as distinctive
identifying features (Fig. 15.4c). For reference purposes it is helpful
to arrange the correctly oriented CBP's of a particular material on the
smallest representative unit of a stereographic projection of the crystal
structure. Unfortunately, these zap maps do not produce very well but
examples have been published by LOVELUCK, et al., (1977).

One problem with such finger printing techniques is that the results may be both voltage and thickness dependent. The voltage variation may be overcome by adopting a standard reference voltage (preferably adjusted relative to a standard zone axis, such as a Si <111>

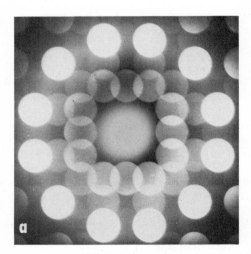 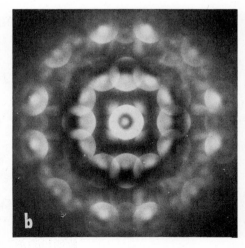

Fig. 15.21. [001] CBP's of σ phase precipitates in 316 stainless steel
taken at 100 KeV; (a) thin crystal; (b) thick crystal.

axis). At the present it would seem sensible to fix on 100 KV as the voltage most widely available to microscopists. Thickness variation is not generally under good control and it will probably be necessary to generate patterns for a series of different thicknesses, depending on the circumstances. Featureless disc patterns, although strongly thickness dependent, generally keep unique characteristics which permit unambiguous identification. 2D CBP's from thick crystals also do not change their essential character with change of thickness for approximately close packed arrangements of simple strings. 3D CBP's from thick crystals can vary greatly in some cases, but are remarkably constant in others. Each case has to be carefully investigated.

Another point which requires considreation when 3D diffraction effects are to be used as the identifying feature is the preferred choice of zone axis. There are generally may distinctive examples to choose from but two factors should be borne in mind in making a choice. The first is the frequency with which the zone axis occurs in the crystal structure, i.e. there are only 3 distinct <001>'s in a cubic crystal but 12 <114>'s (not taking account of sign). Secondly, it is not advisable to select zone axes with particularly fine lines because specimens of exceptional quality will be required to obtain results. For both these reasons <114> axes have been chosen for routine identification of rock-salt structured precipitates (COOKE et al., 1978).

One of the chief attractions of the finger printing method for identifying particles is that, unlike energy dispersive x-ray spectroscopy, it is not limited to the heavier elements. Carbides, borides, nitrides and oxides can all be identified with ease as can diamond particles which are often artefacts of the preparation of mechanically polished surfaces (COOKE et al., 1978). Another advantage is the clear identification of the particle from which the diffraction information is obtained.

15.6 CRYSTAL POTENTIAL AND THICKNESS DETERMINATION

In trying to measure either of these quantities accurately it is prudent to choose as simple a diffraction situation as possible so that analysis of the results can proceed with the minimum of effort. For this reason one dimensional or systematic diffraction situations are preferred. In the simplest case the analytical results of two-beam dynamical theory can be applied to yield foil thicknesses (KELLY et al 1975) or extinction distances (BLAKE et al., 1978). In slightly more complex diffraction situations dynamical calculations may be performed using systematic interactions to deduce structure factors (by a variety of methods) or specimen thickness (COCKAYNE et al., 1967, GOODMAN and LEHMPFUHL 1967). A good deal of effort has been put into these problems in the past decade but it is probably fair to say that standards of accuracy are not yet agreed and for this reason the Electron Diffraction Commission of the International Union of Crystallography has set up an inter-laboratory project on structure factor determinaton under the leadership of Dr. Goodman and Dr. Humphreys. GOODMAN (1978) has recently reviewed the literature on structure factor determination.

As an introduction to this work it is helpful to make a distinction similar to the simple string approximation for two dimensional diffraction in Sec. 15.4. When the diffracting planes for the one dimensional diffraction situation are equally spaced and composed of equal atomic densities of coplanar atoms the scattering potential may be called a simple-plane potential. The associated bend contours or convergent beam patterns then have a characteristic and well-known form which has been widely studied. The structure factors decrease monotonically with increase of g along the systematic row of the reciprocal lattice. When the simple-plane potential is weak (light atoms, low atomic density, small plane spacing) the two beam approximation applies and the results of KELLY et al., (1975) may be applied with confidence for specimen thickness determination, with two reservations discussed below. According to KELLY et al., (1975), thicknesses may be determined to an accuracy of 2% or better. All systematic diffraction situations of simple face centred and body centred cubic crystals are of the simple planar form. Other examples include h00 and hh0 systematics in rock-salt and diamond structured materials and hh2h2 systematics for hexagonal close packed materials. Examples of more complex diffraction situations are hhh systematics for rock-salt and diamond structured materials, hoo or hhh systematics for zinc blend structures, ooom for wurtzite structures and h0h0 systematics for hexagonal close packed materials. For strong simple-plane potentials (such as gold hhh or h00) the two beam approximation fails badly (STEEDS 1970). For intermediate cases, dynamical calculations are necessary to confirm the resluts obtained by application of the two beam approximation but no systematic investigation has yet been made of the errors which can arise. It would seem to be helpful to introduce a parameter to characterize problems of this sort which might be called the plane strength and to a first approximation would be defined as $\gamma D \Sigma_i Z_i / A$ where D is the separation between the planes, the summation is performed over the i atoms in the unit cell with any one plane and A is the area of this unit cell. A more rigorous definition has been given by BUXTON and BERRY (1976).

Two factors can degrade the accuracy of the two beam method even if the approximation holds. When using a STEM system to form convergent beam patterns the signal suffers an angular average, determined by the matching of the rocked beam to the STEM detector. Quite large inaccuracies can be introduced in this way (BROWN, private communication). In addition, the two-beam theory used by KELLY et al., (1975) assumes a parallel sided specimen perpendicular to the incident beam. Thick wedge angles and inclined or curved specimens can introduce systematic errors into the thickness determination unless a modified form of two beam theory is used. A recent investigation, taking account of the effect of foil inclination, only achieved an accuracy of $\pm$ 5% (SALDIN et al, 1978).

When two-beam theory is not adequate to interpret the results of simple-plane experiments, or when the diffracting planes have a more complex form, dynamicla calculations are required. To perform these with adequate accuracy it is necessary to have accurate values for the structure factors and, to a lesser extent, for the absorption parameters. To untangle the various parameters involved in such calculations is outside the scope of this survey but thickness determinations to an accuracy of betterh than 1% have been claimed (GOODMAN and LEHMPFUHL 1967) and structure factor determinations to better than 0.2% have apparently been achieved. What is lacking at present is any independent check of the accuracy of the claims for thickness determination by completely independent techniques although several alternative possibilities exist (NIEDRIG 1978). An investigation of this sort using plasmon loss spectroscopy (LYNCH and BROWN, private communication) has not produced very consistent results so far. One important development towards the attainment of more accurate results is the subtraction of diffuse background from the experimental results (a more important exercise for convergent contour studies). Recent investigations have used a second condenser aperture with a wire stretched across it (BLAKE et al., 1978) or square condenser apertures TANAKA (1978).

ACKNOWLEDGMENTS

It would not have been possible to write this review without the efforts insights and co-operation of the past and present members of my research group and I thank them all for the parts they have played. I should like, in particular, to thank Bernard Buxton, and Alwyn Eades for the stimulation they have provided and the influence which they have exercised on this work. Many of the results presented in this chapter were generated with a Philips EM 400 electron microscope purchased with the aid of a grant from the Science Research Council which is gratefully acknowledged.

Figs. 15.2, 4, 16 and 21 were kindly provided by Mr. N.S. Evans;
Figs. 15.3, 7, 8, 13, 15, 18 and 20(b) by Mr. K K. Fung;
Figs. 15.10, 12 and 20 (a) by Dr. G. M. Rackham:
Fig. 15.11b by Mr. D. Merton Lyn and Figs. 15.11a and 14 by
    Mr. K. E. Cooke.

REFERENCES

Berry, M.V., Buxton, B. F. and Ozorio de Almeida, A. M., 1973 Radiat. Effects, 20, 1

Blake, R. G., Jostons, A., Kelly, P. M. and Napier, J. G., 1978, Phil. Mag. A37, 1.

Buxton, B. F. and Berry, M. V. 1976 Phil. Trans. Roy. Soc. A282, 485.

Buxton, B. F., Eades, J. A., Steeds, J. W. and Rackham, G. M. 1976, Phil. Trans. Roy. Soc. A281, 171.

Buxton, B. F., Loveluck, J. E. and Steeds, J. W. 1978, Phil. Mag. 38, 259.

Cherns, D., 1974, Phil. Mag. 30, 549.

Cockayne, D. J. H., Goodman, P., Mills, J. C. and Moodie, A. F., 1967, Rev. Sci. Inst. 38, 1097

Cooke, K. E., Evans, N., Stoter, L. and Steeds, J. W., 1978 "Electron Diffraction 1927 - 1977" Institute of Physics, Bristol and London, p. 195.

Cowley, J. M. 1978a chapter entitled "Electron Microdiffraction" in Advances in Electronics and Electron Physics, 46, 1

Cowley, J. M. 1978b "Electron Diffraction 1927 - 1977" Institute of Physics, Bristol and London, p. 156.

Fung, K. K. and Steeds, J. W. 1977 "Developments in Electron Microscopy and Analysis 1977" Institute of Physics, Bristol and London, P.289.

Goodman, P. 1974, Nature 251, 698

Goodman, P. 1975, Acta Cryst, A31, 804

Goodman, P., 1978 "Electron Diffraction 1927-1977" Institute of Physics, Bristol and London, p. 116.

Goodman, P. and Johnson A.W.S. 1977, Acta Cryst A33, 997

Goodman, P. and Lehmpfuhl, G., 1967, Acta. Cryst, 22, 14

Goodman, P. and Lehmpfuhl, G., 1968, Acta. Cryst, A24, 339

Goodman, P. and Moodie, A. F., 1974 Acta. Cryst. A30, 280

Goodman, P. and Secomb, T.W. 1977 Acta. Cryst. A33, 126

Gjønnes, J. and Moodie, A. F. 1965, Acta Cryst. 19, 65

Hren, J.J. 1979, Ultramicroscopy, 3, 375

Hutchins, R., Loretto, M. H., Jones, I. P. and Smallman, R. E. 1979 Ultramicroscopy, 3, 401

Jones, P. M., Rackham, G. M. and Steeds, J. W. 1977, Proc. R. Soc. London, A354, 197.

Kelly, P. M., Jostons, A., Blake, R. G. and Napier, J. G., 1975, Phys. Stat. Sol. (a) 31, 771.

Kittel, C., 1976, "Introduction to Solid State Physics", 5th Edition, John Wiley.

Kossel, W. and Mollenstedet, G., 1939, Ann. Phys. 36, 113.

Kyser, D. F. and Geiss, R. H. 1979, Ultramicroscopy 3, 397.

Lehmpfuhl, G., 1978, "Electron Microscopy 1978" (9th International Congress on Electron Microscopy), Microscopical Society of Canada, Vol. III, p.304.

Loh, B. T. M. and Ashbee, K. H. G., 1971, Oak Ridge National Laboratory Report, ORNL-TM-3557.

Loveluck, J. E., Rackham, G. M. and Steeds, J. W., "Developments in Electron Microscopy and Analysis 1977", Institute of Physics, Bristol and London, p. 297.

Loveluck, J. E. and Steeds, J. W. 1977 "Developments in Electron Microscopy and Analysis 1977," Institute of Physics, Bristol and London, p. 293.

Merton Lyn, D. N. 1977, M.Sc.Thesis, University of Bristol.

Neidrig, H. 1978, Scanning Electron Microscopy, SEM, Inc., AMF O'Hare, Vol. 1, p. 841.

O'Keefe, M. A. and Sanders, J. V., 1974 "Diffraction Studies of Real Atoms and Real Crystals," Australian Academy of Science, p. 373.

Rackham, G. M. and Eades, J. A. 1977, Optik, 47, 227.

Rackham, G. M., Jones, P. M., and Steeds, J. W. 1974, "Electron Microscopy 1974," Proceedings of 8th International Congress on Electron Microscopy, The Australian Academy of Science, pp. 336 and 355.

Rackham, G. M. and Steeds, J. W. 1976 "Developments in Electron Microscopy and Analysis," ed. J. A. Venables, Academic Press, p. 457.

Saldin, D. K., Whelan, M. J., and Rossouw, C. J. 1978 "Electron Diffraction 1927-1977," Institute of Physics, Bristol and London, p. 50.

Shannon, M. D. and Eades, J. A. 1978 "Electron Diffraction 1927-1977," Institute of Physics, Bristol and London, p. 411.

Shannon, M. D. and Steeds, J. W. 1977, Phil. Mag. 36, 279.

Steeds, J. W. 1970, Phys. Stat. Sol. 38, 203.

Steeds, J. W. and Fung, K. K. 1978, "Electron Microscopy 1978" (9th International Congress on Electron Microscopy), Microscopical Society of Canada, Vol. I, p. 620.

Steeds, J. W., Jones, P. M., Loveluck, J. E., and Cooke, K. 1977, Phil. Mag. 36, 309.

Steeds, J. W., Rackham, G. M., and Shannon, M. D. 1978 "Electron Diffraction 1927-1977," Institute of Physics, Bristol and London, p. 135.

Tanaka, M. 1978, JEOL News 16E, No. 3.

Tatlock, G. J. 1975, Phil. Mag. $\underline{32}$, 1159.

Tinnappel, A. and Kambe, K. 1975, Acta Cryst. $\underline{A31}$, 6.

Tu, K. N. and Howie, A. 1978, Phil. Mag. $\underline{B37}$, 73.

Unwin, P. N. T. and Henderson, R. 1975, J. Mol. Biol. $\underline{94}$, 425.

CHAPTER 16

RADIATION DAMAGE WITH BIOLOGICAL SPECIMENS AND ORGANIC MATERIALS

ROBERT M. GLAESER

DIVISION OF MEDICAL PHYSICS AND DONNER LABORATORY

UNIVERSITY OF CALIFORNIA, BERKELEY, CALIFORNIA 94720

16.1 INTRODUCTION

In this chapter we will use the expression "radiation damage" to refer to any changes in the physical structure or the chemical makeup of the specimen which occur as a result of exposure to the electron beam. The different types of such changes that can occur are indeed quite diverse. Furthermore, the electron exposure that results in major structural or chemical changes varies according to the type of specimen being observed and according to the type of "damage endpoint" that is being studied. For further details, beyond what is described here, the reader is referred to a number of recent reviews on the subject of radiation damage in biological and organic materials (COSSLETT, 1978; ISAACSON, 1977; SEVERAL AUTHORS, 1975).

Radiation damage occurs in organic materials predominantly as a consequence of inelastic scattering of the incident electrons, which is a process that frequently results in molecular ionization. Molecular ionization leads in turn to chemical bond dissociation (radiolysis). Primary radiolysis products, such as molecular ions and free radicals, are highly reactive, and they are therefore frequently involved in secondary chemical reactions. These secondary reactions depend in detail upon the specimen composition, and they are too numerous and too complicated to be described by any simplified picture.

Other types of "radiation damage" can also occur, however, and under certain conditions their effects cannot be neglected. The phenomenon of "specimen etching" represents the synergistic action of electron irradiation and the presence of certain residual gasses, such as oxygen or water. Damage which is associated with etching is noticeable only at electron exposures that are much greater than those which cause extensive damage due to direct ionization events in most organic materials. A second phenomenon, which also occurs only at electron current

densities that are large on the scale of ionization damage, is specimen
heating. Finally, it should be mentioned that atomic displacement collisions, which are the primary cause of radiation damage in metals and
some other inorganic materials, can also occur in organic materials.
However, atomic displacement collisions are not usually considered as
being a significant factor in radiation damage of organic materials, because most of the damage that can occur due to ionization is already completed at such low electron exposures that atomic displacement collisions can still be ignored.

The type of radiation damage that is important for studies involving high resolution structural details (e.g. structural features in
the range of 5 to 10 Å) is pretty much completed by the time that the
electron exposure applied to the specimen has reached the level of 1 to
10 electrons/$Å^2$. There is, of course, some variation in the sensitivity
of different organic materials, and there is some variation in sensitivity of any one material as a function of specimen temperature. Finally,
the damage produced depends also upon the velocity of the incident electron. The value of 1 to 10 electrons/ $Å^2$ refers to 100 keV electrons,
and the damage rate scales with electron velocity approximately as $1/v^2$.
If a more precise figure of the "critical dose" for damage is needed in
any particular case, then it is necessary to carry out a direct study of
the matter.

Rather gross structural damage, for example the loss of as much as
half of the mass of the specimen, can also occur after electron exposures of approximately 1 to 10 electrons/$Å^2$. Specimen etching can be
observed to occur on a time scale of seconds (when it is not obscured by
the inverse problem of specimen contamination) at a current density of
approximately one ampere/cm^2. Thus we can estimate that specimen
etching is only significant for electron exposures in the range of thousands of electrons/$Å^2$, and only under conditions of sufficiently high
pressure of the appropriate residual gas. Specimen heating depends upon
a large number of parameters, of which the thermal conductivity of the
specimen material, the illumination spot size, and the electron current
density are of critical importance. For an illumination spot size of
one micrometer and a current density of one ampere/cm^2, a temperature
rise of tens and perhaps hundreds of degrees can be expected in materials of poor thermal conductivity. However, with much smaller illumination spot sizes, the temperature rise decreases to such an extent that
it can normally be neglected.

The types of radiation damage effects that one can expect to see
vary depending upon the size scale at which one examines the specimen.
For electron exposures in the range from 1 to 10 electrons/$Å^2$, one might
expect to see the complete loss of crystalline structure in all but the
more resistant organic materials such as the aromatic hydrocarbons. Proteins can be expected to lose up to half of their mass, with a rather
greater percentage loss of oxygen and nitrogen than of carbon and hydrogen. The high resolution features of the molecular structure are accordingly destroyed completely by electron exposures in this range. Aromatic residues are somewhat more resistant to ionization damage, and accordingly their characteristic energy loss spectrum, which generally lies in

the region from 3 to 10 eV, is seen to persist up to electron exposures varying from approximately 10 to 1000 electrons/$\overset{\circ}{A}{}^2$.

In the case of biological materials that are stained with heavy metal salts, a certain amount of stain migration and crystallization of the staining material is observed for electron exposures in the region of 10 to 10^3 electrons/$\overset{\circ}{A}{}^2$. Structural detail down to the level of 25 to 35 $\overset{\circ}{A}$ is readily lost in stained materials after such electron exposures. This observation should alert us to the possibility that the "physiological" distribution of salts might also easily be relocated and redistributed during microanalysis, on a size scale of 10 to 100 $\overset{\circ}{A}$, due to the processes of ionization damage and mass loss. Gross morphological changes can also occur in biological materials during electron bombardment, particularly under conditions of very high current density, and particularly in hydrated specimens. The formation of bubbles, voids, and the general disorganization of fine structure is a phenomenon that is not entirely out of the question, particularly in frozen hydrated materials.

16.2 PRIMARY EVENTS IN RADIATION PHYSICS AND RADIATION CHEMISTRY

Inelastic scattering is the primary cause of radiation damage that is of interest for work with biological specimens. It is therefore especially useful to refer to the characteristic features of the energy loss spectrum for organic materials. An example for the nucleotide base is shown in Chapter 8 by Johnson, and examples of energy loss spectra for inorganic materials are also shown in the contributions by Cowley and Joy. For biological specimens, the region of energy losses from approximately 1 to 10 electron volts may show some detailed structure, due to electronic transitions to various bound, excited states. Molecular ionization (excitation into the continuum) can begin in the region of about 5 eV and continue upwards in energy.

In the region above 10 eV the interesting phenomenon of "collective excitation" also occurs (HUBBARD, 1955; Chapter 10 by Silcox). The process of collective excitation is referred to as "plasmon excitation" in metals, since the collective excitation process is easily interpreted in that case in terms of the "free electron" model (RITCHIE, 1957). Such a picture is of lesser significance or usefulness in organic materials and other non-conductors. The collective excitation state is known to decay very rapidly in organic materials by processes that can frequently result in a number of ionized molecules being produced in close proximity to one another. The combined processes of ionization and collective excitation lead to a peak in the energy loss spectrum of organic materials at approximately 30 eV, and the energy loss spectrum declines monotonically thereafter. At rather higher energies, fine structure is again seen in the spectrum, corresponding to the excitation of K-shell electrons. The spectroscopic and structural significance of this fine structure is dealt with by Isaacson.

A microscopic picture describing the physical structure of primary inelastic events has been developed by Magee and collaborators. In Magee's picture it is imagined that there are intermittent, randomly dis-

tributed "spurs," "blobs" and "short tracks" decorating the track of the
primary incident electron as it travels through the specimen material
(MOZUMDER, 1969). The "spurs" represent the very short branching
tracks, typically 20 Å or so in length, that are produced by the low
energy secondary electrons which result from molecular ionization. The
term "blobs" refers to the larger and often more concentrated distributions of ionization of the specimen material, which result from inelastic scattering events that produce secondary electrons with an energy of
roughly 100 eV to 500 eV. The term "short tracks" refers to the path of
those infrequent "secondary electrons" which have kinetic energies in
the range of 500 eV to 5 keV. These energetic secondary electrons are
also frequently referred to as "delta rays" in the radiation physics and
radiation chemistry literature. According to the "equipartition"
theorem of the distribution of energy in inelastic scattering events,
approximately half of the energy lost by the 100 keV primary electron
goes into the formation of spurs and blobs and is thus confined to a distance of ~50 Å or less from the primary track, and the other 50% of the
energy goes into much "harder" ionizations leading to the formation of
short tracks (MOZUMDER, 1969; CHATTERJEE et al., 1973). In thick specimens, of course, the energy carried off by the energetic secondary electrons further decays in a series of spurs and blobs, but for thin specimens the energetic secondary electrons may in fact escape the specimen
and thereby cause little further damage.

A not insignificant consideration to be included among the primary
processes of radiation damage is the ultimate capture of secondary electrons, which must occur at the end of the electron track. Capture by
neutral molecules leads to the formation of negatively charged molecular
ions, which in turn are chemically very unstable and decompose to give a
variety of reactive molecular species. In solid materials, however, the
secondary electron frequently does not escape the coulomb potential of
the parent ion, and capture most likely occurs by one of the positively
charged ion-products of the radiolysis step.

The primary step in radiation chemistry is chemical bond scission
or radiolysis. Chemical dissociation or other reactions involving intact molecules in an excited electronic state are normally of lesser importance. Bond scission occurs with high probability following ionization in aliphatic or saturated-bond materials. The ionization process
itself occurs in a time that is comparable to the passage of the electron through the molecule, and which has been estimated to be in the
range of approximately 10^{-18} to 10^{-17} seconds. Bond scission, in turn,
may be expected to occur on a time scale comparable to the vibrational
period of the bond, since the potential energy for nuclear motion in an
ionized molecular bond is generally a purely replusive potential. The
vibrational period of chemical bonds is generally in the region of 10^{-14}
seconds. Further separation of the "loose ends" formed by bond scission
is very likely to occur in complex molecular structures on a time scale
that is once again comparable to a few vibrational periods, that is,
again on a time scale in the region of 10^{-14} seconds. Some of the first chemical reactions of the primary radiolysis products can also occur on this
same time scale. Thus, once a chemical bond is broken, it is exceedingly rare for the bond to reform of "heal" even if the thermalized secondary electron is recaptured by the primary molecular ion fragment.

Following bond scission, the molecular ion fragments and free radicals that are produced are likely to become involved in processes of secondary chemical reactions. These reactions are too numerous and too complex to summarize here, particularly since no conclusions of a general character can be applied to all biological materials. Suffice it to say that common products of radiolysis, and of the subsequent steps of radiation chemistry, are species of low molecular weight such as CO_2, NH_3, CH_4, and H_2. The preferential loss of oxygen and nitrogen leads to an enrichment of the specimen in carbon, and the loss of hydrogen leads to a certain degree of unsaturation in the residual carbon skeleton. A certain amount of cross linking between adjacent molecules also occurs. At the same time, other portions of the molecules are split off into small, low molecular weight fragments.

In addition to the microscopic description of radiation physics and radiation chemistry given above, it is useful to equip one's self with the insights provided by the macroscopic, "stopping power" theory of energy loss and with the macroscopic concept of the "radio-chemical yield" of the different products of radiolysis. The stopping power theory gives a statistical average of the amount of energy lost by the primary electron per unit of path length within the specimen material. Detailed tables of stopping power values are given for different primary electron energies in a variety of different materials (BERGER and SELTZER, 1964 and 1966). For 100 kV electrons and biological materials, the stopping power is approximately 4×10^{-2} eV/Å. The stopping power at different energies is approximately related to the stopping power at 100 kV by the ratio $(v_{100}/v)^2$. The total energy deposited per unit volume in the specimen can be obtained from the stopping power after multiplying by the net electron exposure:

$$\text{Energy deposited/vol} = \frac{dE}{dx} \cdot N \qquad (16.1)$$

where N is the exposure in units of electrons/area. This estimate of the energy deposited in the specimen supposes, of course, that no energy is lost from the specimen due to the escape of the energetic delta rays.

An estimate of the energy deposited in the specimen is very valuable, since one can then consult the radiation chemical literature in which empirical measurements have been made of the amount of molecular damage caused per 100 eV of energy deposited in the specimen. Thus, the number of product molecules generated by radiation chemical processes, per 100 eV deposited in the specimen, is referred to as the G value, or "radiochemical yield." In the radiation chemistry of some crystalline amino acids, for example, the yield of H_2 varies from 0.2 to 0.5; the yield of NH_3 varies from 4.1 to 5.2; the yield of CO_2 varies from 0.3 to 1.7, and numerous other products are observed (GEJVALL and LOFROTH, 1975). Whenever possible, it is certainly enlightening and valuable to use stopping power theory to calculate the energy deposited in the specimen during a specified electron exposure, and to then consult the radiation chemical literature to estimate the number of molecular fragments of different type that will be produced from the parent specimen material, for that dose of deposited ionizing energy.

16.3 EMPIRICAL STUDIES OF RADIATION DAMAGE EFFECTS MEASURED UNDER CONDITIONS
      USED IN THE ELECTRON MICROSCOPE

One of the easiest radiation damage effects to observe is the phenomenon of mass loss, which has already been described above. Mass loss
can manifest itself by an increased transparency to the electron beam,
that is, by an apparent thinning of the specimen, as is seen in bright
field conditions of image formation. The decrease in mass, of course,
gives rise to a decrease in electron scattering, and as a result, mass
loss is also easily observed as a decrease in the dark field image intensity. In a wide variety of biological materials the mass loss occurs
exponentially with the accumulated electron exposure, down to a final,
residual mass value. The exponential decay constant is typically in the
range of 1 to 10 electrons/$Å^2$ for most biological materials at room
temperature. The mass loss decay constant may be larger for aromatic
compounds and other "radiation resistant" organic materials. The absolute amount of mass that is lost varies considerably from one type of
chemical substance to another. For polyethylene, and presumably for
other aliphatic hydrocarbon materials, as little as 5% of the mass is
lost, while for proteins as much as 20% to 40% of the mass is lost
(STENN and BAHR, 1970). The percent mass loss in carbohydrates is apparently not accurately known, but common experience with materials such as
cellulose-nitrate support-films would indicate that the mass loss is
likely to be between 25% and 50% in carbohydrate materials.

Only a limited number of studies have been reported on the prevention of mass loss by the use of very low specimen temperatures. One
might suppose that mass loss could be almost completely eliminated by
using a specimen temperature close to that of liquid helium. This idea
has been confirmed in a number of experiments, while at the same time it
has been observed that a rather significant reduction in mass loss is
already obtained at liquid nitrogen temperatures (RAMAMURTI et al.,
1975; HALL and GUPTA, 1974). Further details of the way in which the
mass loss phenomenon can affect measurements of elemental composition or
measurements of the molecular weight of individual biological molecules
by dark field microscopy, are contained in the individual contributions
by Hall, Johnson and Wall.

The processes of mass loss and of mass redistribution can give rise
to changes in the gross morphology (fine structure) and in the high resolution morphology (ultrastructure) of the specimen. In the typical
fixed, embedded and sectioned specimens and in the typical negatively
stained or shadowed specimens, the fine structure is normally not affected by the processes of radiation damage. However, in unstained specimens, particularly those which are kept in a hydrated state, electron
exposures of 1 to 10 electrons/$Å^2$ or more can easily result in the complete obliteration of the native fine structural features and in the introduction of a number of gross distortions and obvious artifacts such
as bubbles, voids and residual debris. At the ultrastructural level,
that is to say, at resolutions exceeding approximately 25 $Å$, electron
exposures of 1 to 10 electrons/$Å^2$ or more are assured to lead to alterations in the native structure, if not the complete loss of the original
structural features.

The loss of such high resolution structure is most directly demonstrated by examples in which the native structure is originally crystalline. When the structure within one unit cell of a crystalline specimen
is identical in structure in every other unit cell, then the electron
diffraction spots will extend out to high resolution. The processes of
radiation damage progressively change the structure in one unit cell in
a way that has no correlation to the changes occurring in any other unit
cell, and this leads to a progressive loss of high resolution diffraction spots. In stained specimens, the diffraction pattern often fades
down to a final resolution of only 25 or 35 Å, while in unstained specimens all orders of diffraction are normally lost, even on a size scale
of hundreds of Å. The fact that the diffraction patterns of stained,
crystalline specimens are partially destroyed suggests that there can be
a diffusion and redistribution of heavy metal salts on a size scale of
up to 25 Å or more. This point is further confirmed by the fact that
the diffraction intensities, in the spots that do remain, continue to
change for a long period of electron exposure, even after the complete
loss of the higher resolution diffraction spots has ceased. A change in
the lower resolution diffraction intensities, of course, indicates both
that matter has continued to be rearranged within each unit cell of the
crystalline specimen and that the rearrangement is very similar within
each unit cell. The typical electron exposures needed to cause these
effects are ∿100 electrons/Å.

Aromatic molecules, or aromatic residues within biological molecules, give rise to characteristic spectroscopic features in the energy
loss spectrum. These characteristic energy losses generally fall in the
range from about 3 to 10 eV, and they are potentially very useful for
analytic purposes at moderate resolution. Aromatic residues have the
double advantage of providing characteristic and relatively sharp energy
loss peaks, and at the same time being considerably more resistant to
radiation damage than are the majority of biological materials. Nevertheless, even the structure of aromatic residues is destroyed, and consequently their "signature" energy loss spectra are destroyed, by electron exposures in the range from 10 to 1000 electrons/Å^2. In a terminal
state of radiation damage, aromatic materials continue to produce a
marked energy loss spectrum in the 5 to 10 eV range, but the spectrum
that is produced is no longer characteristic of the starting material.
Indeed, even non-aromatic materials generate a substantial absorption
band at about 7 eV in the terminal state of radiation damage, although
no band in that energy region is observed in the native, undamaged material (DITCHFIELD et al., 1973).

16.4 SIGNAL-TO-NOISE CONSIDERATIONS AT "SAFE" ELECTRON EXPOSURES

The effects of radiation damage can be clearly avoided in electron
microscopy, and in microanalysis, by taking care to record the image
with such a small exposure to the specimen that the effects of damage
are negligible. However, this approach is often of no practical use,
since the images obtained with a "safe" exposure may turn out to have
very poor statistical definition. Because of the statistically noisy
character of the image, it may not be possible to visually discern the
object features that one hopes to see.

The limitations in the visual detectability of small objects, in a statistically noisy image, were put on a quantitative basis by Rose in his psychophysical studies of noisy images of the type that might be obtained in poor quality television transmission (ROSE, 1948; 1973). Rose determined in quantitative terms the degree of statistical definition that is needed in order to see an object, as a function of its size and the contrast relative to its immediate surround. The object size and the contrast must satisfy the "Rose equation," which for our purposes can be expressed as

$$dC \geq \frac{5}{\sqrt{f\ N}} \tag{16.2}$$

where d is the characteristic object size, C is the contrast relative to the immediate surround, f is the efficiency of "electron utilization" and N is the number of incident electrons per unit area. The area is expressed in the same units as the resolution. The factor of 5 applies, strictly speaking, to the case of a circular disc of uniform intensity surrounded by an adjacent clear area of different intensity; this factor was obtained from practical experience with human observers. From these empirical studies we conclude that the contrast difference must be at least 5 standard deviations above the statistically expected fluctuations in intensity, in order for a human observer to say with confidence that an object feature can be "seen." However, when the object features have some regular pattern, the psychophysical visibility factor can be reduced from a value of 5 to values as low as 1 or 2.

Application of the Rose equation is exceedingly simple to carry out. For example, one might wish to know whether the electron exposure needed to obtain a statistically defined image, for an object feature of known size and contrast, is greater than the electron exposure that will, in fact, destroy that object feature. For example, an unstained protein molecule of molecular weight 40,000 will have a globular diameter of approximately 37 Å. The bright field image contrast for such an object will depend, of course, upon the support film thickness, but may be typically in the region of 0.1 or less. The electron exposure necessary to see this object is then given by the expression

$$fN \geq \frac{25}{(37\mathring{A} \times 0.1)^2} \tag{16.3}$$

which calculates to be 1.8 electrons/$\mathring{A}^2$. We see that the electron exposure needed to obtain a "statistically defined image" of an entire protein molecule is approximately the same as the exponential decay constant for mass loss. Thus the residue of the molecule that is left, by the time that a statistically defined image is obtained, will represent perhaps only 60% to 80% of the original mass of the parent molecule.

Alternatively, one might be interested in knowing what are the smallest object features that can be observed, assuming that we know something about the contrast to be expected in a certain size range, and assuming that the safe electron exposure is as great as 10 electrons/$\mathring{A}^2$. As an example, we may take the contrast to be 0.1, and we might assume that f = 1.0, the most favorable case possible. From the Rose equation we then find

$$d \geq \frac{5}{0.1\sqrt{10}}$$

(16.4)

which leads to the conclusion that individual objects of dimensions smaller than 15 Å. cannot give rise to statistically defined images if the electron exposure is kept below a value that would otherwise lead to a total state of radiation damage.

Similar calculations can be carried out for problems in microanalysis, where the image-formation signal might be either a fluorescent x-ray or an inelastically scattered electron which has suffered a specific energy loss. In these cases the utilization factor, f, is likely to be very small (e.g., $<10^{-3}$), while the contrast may be about 1 in favorable cases. The safe exposure must be determined for each type of application and for each type of signal. Such a representative calculation is given in Chapter 8 by Johnson.

The use of crystalline specimens provides a special advantage in overcoming the problem of obtaining a statistically defined image, while still keeping the electron exposure to such a low level that radiation damage causes negligible adverse effects. The advantage inherent in the use of a crystalline specimen is, of course, related to the spatial redundancy of the object. Thus, assuming that it is possible to record useful image data at exceedingly low electron exposure levels, then one can imagine a process by which the image of one unit cell is superimposed upon the image of another, and the process is continued throughout the entire lattice so as to build up a statistically defined image of one unit cell. In practice, perhaps the best way in which this can be achieved is by digitizing the image intensity obtained under conditions of low electron exposure, followed by calculation of the Fourier transform of the image intensity. The Fourier transform will show distinct peaks, corresponding to the regular lattice, even though the image itself represents a statistically noisy version of the lattice. An inverse Fourier transform is then calculated, using only the amplitudes and phases of the crystalline diffraction spots and ignoring the "noise values" lying in between the diffraction spots. This "Fourier transform" approach is completely equivalent to a real space superposition, such as that introduced by Markham for the photographic processing of periodic objects.

Spatial averaging of images obtained at a "safe" electron exposure was used by UNWIN and HENDERSON (1975) in their work on the structure of unstained, hydrated membranes from the microorganism, Halobacterium halobium. The protein in these membranes is known as bacteriorhodopsin, and it is known to function as a photon-driven proton-pump. This membrane is one of a large number of examples in which a specialized area of the membrane is found to be crystalline in the native, functional state. By using low exposures imaging conditions and the method of spatial averaging as described above, image data can presently be obtained to a resolution of 7 Å. In the spatially averaged images (Fig. 16.1),

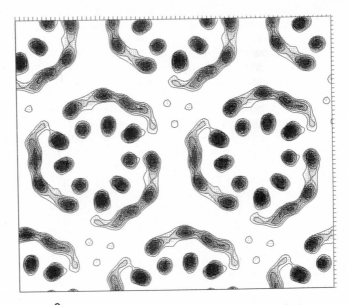

Fig. 16.1. A 7.1 Å resolution, spatially averaged image obtained from "safe-exposure" micrographs of crystalline arrays of bacteriorhodopsin. The regions of dense contours correspond to the α-helixes of the protein, which span the full membrane thickness. For further details of the experimental method, refer to HAYWARD et al., (1978). The bacteriorhodopsin molecules are packed together as trimers, which in turn are packed into a hexagonal lattice with a center-to-center distance of ~ 63 Å.

the main structural features seen in the image of the bacteriorhodopsin molecule are the individual alpha helixes, which span the membrane from inside to outside. It may be expected that the spatial averaging technique of overcoming radiation damage will find very wide application in many areas of molecular structure analysis of biological materials in the years ahead.

16.5 ADDITIONAL PROCESSES OF "RADIATION DAMAGE" THAT OCCUR AT VERY HIGH ELECTRON EXPOSURES

Structural damage can occur in all materials due to the atomic displacements that arise from so-called "knock-on" collisions (see Chapter 17 by Hobbs). These atomic displacement collisions occur as a result of the elastic scattering of the incident electron, when the change in momentum imparted by the scattered electron is sufficiently great to actually displace the scattering atom from its original, chemically bonded position. The frequency with which such collisions occur depends in part on the binding energy that holds the atom into its original position and also in part upon the initial kinetic energy of the incident electron. In fact, there is a threshold energy for the incident electrons, below which atomic displacement collisions do not occur. For biological materials, the threshold energy for displacement of hydrogen atoms is only a few keV, while the threshold energy for displacement of carbon atoms is estimated to be about 25 to 30 keV (COSSLETT, 1970). The cross-section for atomic displacement collisions is very small, however. An experimental study of mass removal in an amorphous carbon film, under conditions where etching was not likely to be a factor, has demonstrated a reasonably good agreement between the atomic displacement cross-section and the actual observed removal of carbon atoms (DIETRICH et al., 1978). In that study, it was observed that 50 Å of carbon were removed by an electron exposure of $\sim 10^7$ electrons/Å^2, at an incident energy of 230 keV. From this we may conclude that atomic displacement damage is not normally a matter worth considering for the usual conditions of microscopy and analysis in biological materials. However, it can obviously become a problem for special cases in which the atomic binding energy is considerably less than 5 or 10 eV, or in cases where the electron exposure of the specimen must exceed $\sim 10^5$ electrons/Å^2.

The amount of temperature rise produced in biological specimens due to the electron beam has always been difficult to estimate, either by theoretical calculations or by direct experimental measurements. Theoretical formulas for thin specimens are based primarily on a model of a uniform disc of illumination, of radius R_1, and a uniform heat sink of radius R_2. The temperature rise in thin specimens is then given by the formula (REIMER, 1967)

$$\Delta T = J \frac{dE}{dX} \frac{R_1^2}{2k} \ln \left(\frac{R_2}{R_1} \right) \tag{16.5}$$

where J is the current density expressed in electrons/sec-area, dE/dX is the stopping power, and k is the thermal conductivity.

Applications of this theoretical expression face the difficulty of deciding on a value of thermal conductivity that is appropriate to use for biological material. The thermal conductivity of evaporated carbon film is about 10^{-2} W/cm-degree (CHRISTENHUSZ and REIMER, 1968), but that of biological materials is presumably ten to a hundred times less. The thermal conductivity of ice at -100°C is $\sim 5 \times 10^{-2}$ W/cm-degree, and in some cases this may be a reasonable number to assume for frozen hydrated specimens. The problem is still not defined, however, until one specifies the illumination spot size and the size of the constant temperature boundary (i.e., the heat sink).

In the case of ice at -100°C, the temperature rise is given by

$$\Delta T = 6.5 \times 10^{-12} (\text{sec} - °k) J \, R_1^2 \ln \left(\frac{R_2}{R_1} \right) \tag{16.6}$$

where J is the current density (in electrons/sec-Å^2) within the illumination spot. Then, as an example, we may take $R_1 = 1$ μm, $R_2 = 50$ μm, and $J = 600$ electrons/sec-Å^2 (i.e., 1 A/cm^2). For these values of the parameters, we find $\Delta T \simeq 1.5$ °k. We can thus see that specimen heating will not be a problem, as a general rule, except when the current density appreciably exceeds 1 A/cm^2 or when the thermal conductivity of the specimen is appreciably lower than that of ice. Recently the question of the amount of temperature rise that is expected to occur in frozen hydrated materials under various conditions of microprobe analysis has been studied by Talmon and Thomas (TALMON and THOMAS, 1977a; 1977b; 1978). The interested reader is referred to the original references for further information.

Specimen heating by the electron beam is generally a matter of secondary concern, since heating can only occur under conditions where primary processes of radiation damage are essentially instantaneous. Putting this another way, if one is careful to carry out experiments under conditions where primary radiation damage processes do not occur, then a significant amount of specimen heating is impossible. Thus, for example, in high resolution electron microscopy of frozen hydrated specimens, a rise in specimen temperature sufficient to produce sublimation of ice is not normally observed. However, increasing the current density of the specimen by approximately four orders of magnitude above that which can be safely used from the point of view of radiation damage does result in an observable rate of sublimation of the ice, which is induced by the electron beam. In the case of microanalysis, the small probe size that is needed for "image formation" automatically limits the amount of heating that can occur. For example, a probe size of 200 Å and a current density of 10 amperes/cm^2 would produce a temperature rise of only $\sim 1.2 \times 10^{-2}$ degrees centigrade in ice.

In closing, we note once again that beam-induced etching can be another mechanism by which radiation damage occurs in organic materials. As described in the introduction, there is rather little quantitative information about the etching phenomenon, except that it only occurs at very high electron exposures and in the presence of certain residual gasses. A further discussion of etching is given in Chapter 18 by Hren.

ACKNOWLEDGMENTS

I wish to thank Dr. David A. Grano for providing the combination contour / grey-scale plot of the projection of the structure of purple membrane (Fig. 16.1), using data collected by Dr. Steven B. Hayward. This work has been supported by NIH grant GM 23325 and by DOE contract W-7405-eng-48.

REFERENCES

Berger, M. J. and Seltzer, S. M., 1964, in National Academy of Sciences National Research Council Publication 1133, Washington, D.C., 205.

Berger, M. J. and Seltzer, S. M., 1966, Supplement to NAS-NRC Publication 1133, titled NASA SP-3036, Clearing House for Federal Scientific and Technical Information, Springfield, Virginia.

Chatterjee, A., Maccabee, H. D., and Tobias, C. A., 1973, Radiat. Res., 54, 479.

Christenhusz, R. and Reimer, L., 1968, Naturwissenschaften, 55, 439.

Cosslett, V. E., 1970, Ber. Bunsenges. Phys. Chem., 74, 1171.

Cosslett, V. E., 1978, J. Micros., 113, 113.

Dietrich, I., Fox, F., Heide, H. G., Knapek, E., and Weyl, R., 1978, Ultramicroscopy, 3, 185.

Ditchfield, R. W., Grubb, D. T., and Whelan, M. J., 1973, Phil. Mag., 27, 1267.

Gejvall, T. and Lofroth, G., 1975, Radiation Eff., 25, 187.

Hayward, S.B., Grano, D.A., Glaeser, R.M. and Fisher, K.A., 1978, Proc. Natl. Acad. Sci. USA 75, 4320.

Hubbard, J., 1955, Proc. Phys. Soc. (Lond.), 68, 976.

Isaacson, M.S., 1977, In: Principles and Techniques of Electron Microscopy (Ed. by M.A. Hayat), Vol. 7, 1, Van Nostrand Reinhold Co., New York.

Hall, T. A. and Gupta, B. L., 1974, J. Micros., 100, 177.

Hubbard, J., 1955, Proc. Phys. Soc. (Lond.), 68, 976.

Mozumder, A., 1969, In Adv. in Radiation Chemistry (Ed. by M. Burton and J. C. Magee), Vol. 1, Wiley-Interscience, New York, 1.

Ramamurti, K., Crewe, A. V., and Isaacson, M. S., 1975, Ultramicroscopy, 1, 156.

Reimer, L., 1976, Elektronenmikroskopische Unter. u. Prep.-Methoden, Springer, Berlin, 226.

Ritchie, R. H., 1957, Phys. Rev., 106, 874.

Rose, A., 1948, Advan. Electronics, 1, 131.

Rose, A., 1973, Vision: Human and Electronic, Plenum Press, New York.

Several Authors, 1975, In Physical Aspects of Electron Microscopy and Microbeam Analysis (Ed. by B. Siegel and D.R. Beaman), John Wiley & Sons, New York.

Stenn, K. and Bahr, G.F., 1970a, J. Ultrastruct. Res., 31, 526.

Stenn, K.S and Bahr, G.F., 1970b, J. Histochem. Cytochem., 18, 574.

Talmon, Y. and Thomas, E. L., 1977a, Scanning Electron Microscopy (IITRI), Vol. I, 265.

Talmon, Y. and Thomas, E. L., 1977b, J. Micros., 111, 151.

Talmon, Y. and Thomas, E. L., 1978, J. Micros., 113, 69.

Unwin, P. N. T. and Henderson, R., 1975, J. Mol. Biol., 94, 425.

CHAPTER 17

RADIATION EFFECTS IN ANALYSIS OF INORGANIC SPECIMENS BY TEM

L.W. HOBBS

DEPARTMENT OF METALLURGY AND MATERIALS SCIENCE
CASE WESTERN RESERVE UNIVERSITY
CLEVELAND , OHIO 44106

17.1 INTRODUCTION

In the practice of analytical electron microscopy, particularly at near-atomic resolution, one is forced by the exigencies of statistics and the briefness of the encounter between fast electrons and the specimen, to use upwards of 10^4 electrons per $\overset{\circ}{A}{}^2$ in order to acquire information about the identity and position of a single atom (see Sec. 17.6). It is not therefore surprising that, in addition to the "elastic" interaction which provides positional information and the "inelastic" interaction which provides chemical identification, there should be the further prospect of a significant perturbation of the atomic structure under analysis. Such perturbations may be generically labelled radiation effects and, where they lead to permanent alterations in atomic structure, can be sensibly termed radiation damage.

Radiation effects in organic solids are of fundamental concern for TEM of biological materials and have been extensively reviewed (GLAESER, 1974, 1975, this volume; REIMER, 1975; ISAACSON, 1977, 1979). Most of the interest in TEM irradiation effects in inorganic solids has been historically in metals (MAKIN, 1971; NORRIS, 1975; PETERSON and HARKNESS, 1976) where they are seldom problematical for TEM investigations, but there is now substantial understanding of irradiation effects in covalent (CORBETT, 1966; CORBETT and WATKINS, 1971; CORBETT and BOURGOIN, 1975) and ionic (SONDER and SIBLEY, 1972; POOLEY, 1975; HOBBS, 1975) solids, some of which are exceptionally electron-sensitive in TEM (HOBBS, 1975, 1979; HOBBS, HOWITT and MITCHELL, 1978). There will be no attempt here to duplicate the bibliographical efforts of these reviews, and the reader is referred to them for further details, examples and literature.

Radiation Damage in Compact Lattices

Radiation damage in inorganic materials differs from that in organic solids in at least one aspect, in that the atomic structure concerned is in general considerably more compact and the coordination
about each atom invariably much higher. The consequence is that there
are fewer options for accommodation of atomic rearrangements. The inorganic solid, even a predominantly covalent noncrystalline one, tends
to rigorously preserve well-defined atomic positions, so that the
notions of atom vacancy and atom interstitial are correspondingly well-
understood and their possible configurations distinctly limited. Thus
the fundamental unit of damage in inorganic solids is more likely to be
an atom displaced to an interstitial site and its attendant vacancy,
known collectively as a Frenkel defect, rather than the molecular radical of organic materials. Molecule-like coordination polyhedron rotations and rebonding are also possible in some covalent inorganic solids,
but probably only in the presence of a high concentration of Frenkel
pairs.

Frenkel defects effect gross structural alterations, in general,
only at high density (typically at concentrations $c_D > 10^{-4}$) because it
is only at this stage that the longer range identity of inorganic lattices is seriously perturbed. This situation is in contradistinction to
organic solids, where molecules may exist in long-range ordered arrangements but are only weakly cognizant of each other, and for which alterations within a single molecule immediately alter the local chemical identity of the solid. At defect concentrations $c_D \cong 10^{-4}$, defects in compact lattices are on average only a few atom sites apart and, even without thermally-induced mobility, sufficient elastic and electrostatic interaction exists between defects to induce grosser atomic rearrangements. The catalogue of such structural modifications includes dislocation loops, faults and other planar condensates; stacking-fault tetrahedra, voids, bubbles, colloids and other three-dimensional aggregates;
altered order; precipitation; polytypic, polymorphic and composition-
ally-driven phase transformations (some of these are discussed in Sect.-
17.5). The atom motion attendant upon atom displacement itself represents a sort of diffusion otherwise present only at high temperatures,
so transport properties are altered. In addition, the energy stored in
displacements, as well as subtler entropic considerations, can appreciably alter the thermodynamics of phase equilibria in the presence of a
radiation flux.

Fast electrons (with primary kinetic energies $U_p > 10$ keV), being
charged particles, interact strongly with both atomic nuclei and atomic
electrons in a Coulombic manner. Both sorts of interaction can lead to
permanent atom displacements, and a distinction is conveniently made between them for the purpose of discussing radiation damage. Atom displacements resulting from fast electron-nucleus interactions are called
knock-on damage because they involve direct transfer of momentum to the
displaced atoms in elastic collisions. Atom displacements arising from
fast electron-atomic electron interactions are less direct and require a
mechanism to convert the potential energy of electron excitations into
nuclear momentum. Such damage is often (though somewhat inaccurately)

called ionization damage, while the operation of such a conversion mechanism is termed radiolysis.

Electron-Atom "Inelastic" Interaction

The Coulombic interaction of a fast electron with an atom nucleus is essentially "elastic" from the standpoint of the two-particle system, but "inelastic" from the standpoint of the electron alone, since nontrivial amounts of kinetic energy can be transferred to the nucleus. The overall energy lost to such interactions is, however, rather small, the cross-section for which (per atom) is given by the McKinley-Feshbach (1948) approximation to the Mott expression (see MOTT and MASSEY, 1965 and LEHMANN, 1977)

$$\sigma_n = Z^2 \, 4\pi a_o^2 \, U_R^2 \, \frac{\{1 - \beta^2\}}{m^2 c^4 \beta^4} \{(T'_{max}/T'_{min}) + 2\pi\alpha\beta(T'_{max}/T'_{min})^{\frac{1}{2}}$$

$$- (\beta^2 + \pi\alpha\beta) \ln (T'_{max}/T'_{min}) - (1 - 2\pi\alpha\beta)\} \qquad (17.1)$$

Here a_0 is the Bohr radius (53 pm), U_R the Rydberg energy (13.6 eV), Z the atomic number, $\alpha = Z/137$ and $\beta = v/c$. T'_{max} is the maximum energy transferable to a nucleus of mass M by an electron of mass m, velocity v and kinetic energy U_p, which is just

$$T'_{max} = 2 \, U_p \, (U_p + 2mc^2)/Mc^2. \qquad (17.2)$$

T'_{min} is the minimum transferable energy, essentially a lattice vibrational quantum (phonon) $\sim 10^{-2}$ eV. Only a fraction of the total energy transferred is available for atomic displacements, however. The McKinley-Feshbach approximation (17.1) is valid only for light elements (SEITZ and KOEHLER, 1956); for heavier elements, the full Mott expression must be used and has been evaluated numerically (OEN, 1973).

The energy lost by fast electrons to interactions instead with atomic electrons is considerably greater due to the larger number of atomic electrons and the equivalence in masses. A corresponding cross-section for interaction with Z free electrons at rest can be obtained by integrating the Rutherford differential cross-section to give

$$\sigma_e = Z \, (8\pi a_o^2 \, U_R/mv^2) \, (U_R/T'_{min}), \qquad (17.3)$$

where T'_{min} here represents the minimum excitation energy for bound or quasibound atomic electrons. The bound-electron structure of the atom substantially modifies the interaction, and more rigorous approaches (INOKUTI, 1971, 1979; INOKUTI, ITIKAWA and TURNER, 1978) show that the Z dependence is irregular. Taking $T'_{min} \cong 2$ eV approximates these results for the present purposes. Most of the energy transferred in electron-electron interactions is available for radiolysis, provided a radiolytic mechanism exists (see Sect. 17.3).

A comparison of electronic and nuclear cross-sections simplified to the nonrelativistic case is instructive:

$$\sigma_e \cong Z \ (4\pi_o^2/U_p) \ (U_R^2/T'^{e}_{min})$$

$$\sigma_n \cong (Z^2 m/M) \ (4\pi a_o^2/U_p) \ (U_R^2/T'^{n}_{min}).$$

The ratio of these two cross-sections

$$\frac{\sigma_e}{\sigma_n} \cong \frac{M}{mZ} \frac{T'^{n}_{min}}{T'^{e}_{min}} \cong 4000 \qquad \frac{T'^{n}_{min}}{T'^{e}_{min}} \cong 40 \qquad (17.4)$$

shows that most of the energy lost by fast electrons in interacting with the specimen is lost to atomic electron excitations. Therefore, the average _rate_ of energy loss can be calculated from σ_e alone to obtain the Bethe-Bloch expression (BETHE, 1933)

$$-\frac{dU}{dz} = \frac{8\pi a_o^2 \ U_R^2}{mc^2 \ \beta^2} \ \{ \ln \frac{mc^2 \ \beta^2 \ U}{2(T'^{e}_{min})^2 \ (1 - \beta^2)}$$

$$-(2\sqrt{1 - \beta^2} - 1 + \beta^2) \ \ln 2 + (1 - \beta^2) + \frac{1}{8} (1 - \sqrt{1 - \beta^2})^2\} \qquad (17.5)$$

which is plotted for a typical low atomic number solid in Fig. 17.1 and varies from $\sim 10^{10}$ eV m^{-1} for U = 10 keV to $\sim 10^8$ eV m^{-1} for U = 1 MeV. Above 1 MeV, Bremsstrahlung and electron-positron pair production again increase the energy loss rate. The β^{-2} scaling law has been verified for radiolytic damage in organic polymers (THOMAS _et al._, 1970), and use of high-voltage electron microscopy (U_p > 0.5 MeV) can significantly reduce energy deposition rates provided primary electron beam current density is not altered. Since specimen thicknesses of interest to TEM are < 1 μm (the "inelastic" energy loss in fact sets the maximum usable specimen thickness due to chromatic aberration in CTEM), relatively little energy is lost from the investigating beam, and the effective value of -dU/dz is essentially that for the initial electron energy, -dU$_p$/dz, which can be calculated from Eqn. (17.5) or deduced from more convenient tabulations (e.g. SPENCER, 1959).

The electron current densities $j_p \cong 10^2$ - 10^7 A m^{-2} in TEM probes are such that these small losses nevertheless represent a phenomenal energy density, typically ~ 1 kW - 1 MW mm^{-3} or 1 G Gy s^{-1} to 1 T Gy s^{-1} (1 Gy = 1 J kg^{-1}). The energy _stored_ in atomic displacements (if any exist) is far less than the energy expended in producing them, so the major portion of the energy loss eventually ends up as heat and attendant elevation of the specimen temperature. Specimen temperature is important, often critically important, to the chemical stability of specimens undergoing irradiation.

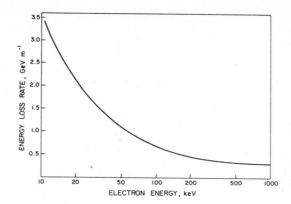

Fig. 17.1. Energy deposition rate for fast electron irradiation calculated from Eqn. (5) for a low atomic number material.

Electron-Beam Heating

The rise in specimen temperature may be calculated (GALE and HALE, 1961; FISHER, 1970; HOBBS, 1975) by solving the radial form of the differential equation for heat conduction

$$\frac{1}{r} \frac{\partial}{\partial r} \left(r \frac{\partial T}{\partial r}\right) + \frac{1}{\kappa} j_p \ (dU_p/dz)/e = 0 \qquad (17.6)$$

where $j_p \ (dU_p/dz)/e$ is the time rate of heat input per unit volume at radius r and κ is the specimen thermal conductivity. Two specimen geometries are of interest. For a flat film of uniform thickness, thermally anchored to a good conducting medium (say, a copper grid) at a distance $r = s$ from the electron probe, the maximum temperature rise in the center of the irradiated area is

$$\Delta T_{max} = i_p \frac{(dU_p/dz)/e}{2\pi\kappa} \ (\tfrac{1}{2} + \ln s/b). \qquad (17.7)$$

The radius of the irradiated area b and j_p are both averaged over the scan for STEM. For a dish-polished specimen, such as results from chemical jet-polishing or ion-beam thinning, having radius of curvature ρ, local thickness z and anchored at its periphery to a thermal sink, the temperature rise neglecting radiative heat loss (HOBBS, 1975) is

$$\Delta T_{max} = i_p \frac{(dU_p/dz)/e}{2\pi\kappa} \ (\tfrac{1}{2} + \ln \frac{\sqrt{\rho z}}{b}). \qquad (17.8)$$

The temperature rise is notably proportional to the total primary beam current $i_p = \pi b^2 j_p$ and not to the current density alone. Approximate temperature rises for various combinations of i_p and κ, calculated from Eqn. (17.7) but appropriate to either geometry, are plotted in Fig. 17.2

and are, for the most part, modest. However, if contact to the thermal
sink is poor, as often happens with specimens simply lying on grids,
temperature rises can be excessive. It is, in fact, possible to melt
refractory ceramics! Caution should also be exercised in avoiding irradia
tion of grid bars, since the <u>entire</u> electron beam is stopped by the bar
thickness and can cause local heating. Good thermal contact is best provided by cementing foils (one point on the periphery is sufficient) with
a thermally-conducting cement such as a suspension of colloidal silver.
The organic binder in such cements can prove a source of contaminants in
some experiments, however.

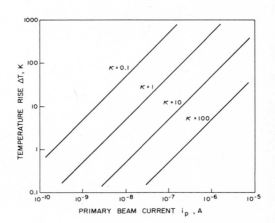

Fig. 17.2. Specimen temperature
increase arising from electron-
beam heating for primary beam current i_p and specimen thermal conductivity κ in J-m^{-1} K^{-1} calculated from (Eqn. 17.7).

Charge Acquisition by Insulating Specimens

A special additional caution should be mentioned in respect to insulating solids. Many inorganic materials have low electrical conduc-
tivity, typically $\sigma < 10^{-15}$ Ω^{-1} m^{-1}, and can acquire a static charge under
electron bombardment. The ranges of electrons utilized in TEM (U_p = 20
keV to 3 MeV) lie between 50 μm and 1 mm, which is considerably greater
than the foil thicknesses used (<1 μm), so that electrons from the primary beam do not accumulate in the foil, contrary to what is sometimes
supposed. A large number of secondary electrons is, however, generated,
many of which are sufficiently energetic to escape the specimen foil.
Their departure leaves insulating specimens with a local positive charge
which, if uncompensated, gives rise to a positive potential difference
+V with respect to the surrounding conducting media. This potential difference initiates flow of charge by ionic or electronic conduction or
attracts secondaries emitted from other parts of the microscope (for example the nearby objective aperture) until some potential V* is reached
at which the electron conduction or arrival current i_c equals the secondary emission current i_s. The magnitude of V* may be such that the
accompanying internal field in the specimen is large enough to induce
electrical breakdown. This phenomenon has been analyzed in some detail
by the author (HOBBS, 1975), an abbreviated treatment of which follows.

The positive foil potential V* prevents electrons of energy U < U *
= V*e from leaving the foil. The secondary emission current consequent-

ly consists of all secondaries with energies $U_s > U_s *$. Thus $V*$ is itself a function of the energy distribution of emitted secondaries. This distribution may be obtained from differential cross-sections for electron-electron scattering of the sort used to derive Eqn. (17.3). The total cross-section for production of secondary electrons of energy $U_s > U_s *$ is therefore from Eqn. (17.3) nonrelativistically

$$\sigma(U_s > U_s*) \cong N_e \; 4\pi a_o^2 \; U_R^2/U_s* \; U_p = N_e \; 4\pi a_o^2 \; U_R^2/eV* \; U_p \qquad (17.9)$$

for $\sim 10 \; V \ll V* \ll U_p/e$, where the electron density N_e is restricted to the least tightly-bound valence electrons. The range of the more probable secondaries is of the order of the specimen thickness z, so the secondary current is

$$i_s \cong i_p \; z \; \sigma(U_s > U_s*). \qquad (17.10)$$

The corresponding (radial) electric field strength is maximum at the periphery of the irradiated area

$$E_{r \; max} = V*/b \qquad (17.11)$$

or $\sim 10^6 \; V* \; V \; m^{-1}$ for $b \cong 1 \; \mu m$ in CTEM and $\sim 10^9 \; V* \; V \; m^{-1}$ for $b \cong 1 \; nm$ in fine STEM probes. Conduction becomes non-ohmic for most insulators at high field (typically for $E \cong 10^6 \; V \; m^{-1}$), so even a small foil potential $V*$ will result in field-dependent conductivity of the form

$$j_c = j_o \; exp \; A \; E^{\frac{1}{2}} \qquad (17.12)$$

where j_0, the high-field conduction current density extrapolated to zero field, and A are material constants (typical values $j_0 \cong 10^{-13} \; A \; m^{-2}$, $A \cong 10^{-3} \; (V/m)^{\frac{1}{2}}$). The conduction current i_c conducted through the specimen depends on the specimen cross-sectional area $2\pi bz$ at the periphery of the irradiated area; thus

$$i_c = 2\pi \; b \; z \; j_o \; exp \; A(V*/b)^{\frac{1}{2}} \qquad (17.13)$$

At equilibrium, setting $i_c = i_s$,

$$V* \; b \; exp \; A(V*/b)^{\frac{1}{2}} = \frac{2 \; N_e \; a_o^2 \; U_R^2}{j_o e} \; \frac{i_p}{U_p}$$

$$(17.14)$$

$$\frac{V*}{b} \; exp \; A(V*/b)^{\frac{1}{2}} = \frac{2\pi \; N_e a_o^2 \; U_R^2}{j_o e} \; \frac{j_p}{U_p}$$

Equations (17.14) show that decreasing the radius of the irradiated area gives an approximately proportionate decrease in $V*$, but $E*_{r \; max}$

from Eqn. (17.11) remains the same, so there should be little difference between CTEM and STEM for the same probe current. For $i_p \cong 1$ nA, typical values of V^* and $E^*_{r\,max}$ are several hundred volts and 10^8 V m^{-1}. The latter is of the order of dielectric strengths, and there is the possibility that electrical breakdown can occur. In addition, the charged foil region produces an electrostatic lens, the electron beam generating its own electrostatic deflection field emanating from the irradiated area and extending irregularly to neighboring <u>external</u> surfaces (holder, objective aperture, etc.) at ground potential. (The <u>internal</u> electric field within the specimen does not act over sufficient distance to noticeably deflect the electron beam.) Inevitable asymmetries in the external field can deflect the investigating beam through annoyingly large angles (>1°).

The simplest remedy to both difficulties involves coating the entire specimen (one side is sufficient because of radiation-induced conductivity within the irradiated area) with a 10-20 nm film of evaporated carbon or aluminum which makes contact with contiguous metallic components and reduces V^* to zero. Films of this thickness remain in the single-scattering regime for fast electrons and reduce image intensity by <20% (LENZ, 1954). Loss of resolution due to chromatic aberration could prove a problem, however, for highest resolution TEM or convergent-beam electron diffraction studies. Evaporation on the electron-exit surface is preferable since this alters only diffracted intensities and not Bloch-wave excitations. Like contamination, such (albeit intentional) surface films may additionally complicate or impede analytical TEM (see Hren, Chapter 18). Alternative remedies involve flooding the irradiated area with low energy electrons or restricting study to small flat flakes on thin carbon substrates.

17.2 KNOCK-ON DISPLACEMENT

The cross-section for knock-on displacement of atoms of atomic mass A is provided by Eqn. (17.1) with T'_{max} (in eV) for primary electrons of energy U_p (in MeV) given by

$$T'_{max} = 2147.7 \; U_p \; (U_p + 1.022)/A \qquad (17.15)$$

and T'_{min} given by the kinetic energy T_d necessary to displace at atom from its normal position through a saddle point to a second stable (or metastable) position. Atom displacements therefore occur whenever T' > T_d, which can occur for primary electron energy $U_p > U_{th}$, where U_{th} is the threshold electron energy corresponding to $T'_{max} = T_d$ in Eqn. (17.15). The displaced atom must necessarily occupy an interstitial position in the undisplaced solid and consequently leaves behind an atom vacancy; the identity, number and distribution of these Frenkel pairs constitute the primary defect spectrum. The notion of Frenkel pairs is most often applied to crystalline solids, but the concept has at least limited validity in amorphous solids as well (though the immediate surroundings of the vacancy and the interstitial will not necessarily be identical for every Frenkel pair), particularly if an average environment can be well-defined.

Displacement Energy

Displacement energies T_d are clearly related to atomic bonding energies, since several bonds must be broken and others seriously perturbed in order to displace an atom to a normally unoccupied position. Bond strengths vary according to the bond character of the solid (metallic, covalent, ionic, Van der Waals). Since most bonding energies vary from ~1 eV to ~200 eV per atom (Table 17.1), displacement energies will be of this order or larger.

TABLE 17.1

Bonding energies per atom (after HUGHES and POOLEY, 1975)

Bond	Bond Principle	Bonding Energy
Van der Waals	permanent or induced dipole-dipole interaction	0.1 - 1 eV
metallic	conduction electron delocalization	1 - 3 eV
covalent	partial, directed electron transfer	4 - 8 eV
ionic	electrostatic	8 - 200 eV

The displacement process is usually considered highly adiabatic, with considerably more than the energy of broken bonds dissipated in the displacement event.

The small bonding energies in metallic solids, arising almost solely from the cohesion provided by delocalization of conduction electrons, suggests that atom displacements should be relatively easy to produce, but these solids are invariably close-packed, and to remove an ion core through several saddle point configurations to a stable interstitial position where there is little room involves seriously perturbing the bonding of upwards of ten neighboring atoms, so that the minimum expected T_d is ~10-20 eV, which is observed. Saddle-point configurations differ for different directions of knock-on atom momentum, and T_d is orientation dependent up to a factor of about 2-3. Dynamic effects and the possibility that the interstitial is mobile after production introduce temperature dependence into T_d. _Average_ displacement energies for metals are found to be in the 20-40 eV range (Table 17.2).

The situation for ionic crystals is similar in that nondirected bonds are involved , but the ionic bonding strengths are much larger and given (per ion) by the lattice Coulomb energy

$$U_\ell = \alpha(r) \ (ze)^2/4\pi \ \varepsilon_o \ r \cong 8 \ z^2 \tag{17.16}$$

TABLE 17.2

Displacement energies T_d for several elemental and diatomic solids near room temperature			
Material	T_d (eV)	Material	T_d (eV)
Al	16–19	GaAs	9 (Ga) 9.4 (As)
Cu	19–22	CdTe	5.6 (Cd) 7.9 (Te)
Au	33–36	ZnS	9.9 (Zn) 15 (S)
Si	11–22	ZnO	57 (Zn) 57 (O)
C (graphite)	28–31	MgO	64 (Mg) 60 (O)
C (diamond)	80	Al_2O_3	18 (Al) 72 (O)

Here $\alpha(r)$ is the Madelung constant, (ze) the lowest common ion charge, ε_0 the vacuum permittivity and r the characteristic distance (usually nearest neighbor) used for calculating $\alpha(r)$. Ionic crystals, like metals, are relatively close-packed, and displacement energies are likely to be several times the bonding energy. The z^2 term in the bonding energy results in large displacement energies for ions with $z > 2$, typically $T_d \cong 60$ eV in ionic oxides. The presence of two or more ion species presents the possibility that T_d may differ for each ion (Table 17.2). Directional effects exist, but are probably averaged out more effectively than in metals by the long-range Coulomb interactions. Using Eqn. (17.16) amounts to assuming that it is an <u>ion</u> which is displaced and the resulting vacancy and interstitial both charged. Their Coulomb interaction may require larger separations and higher displacement energies for stable Frenkel pairs than for metals. However, if an <u>anion</u> loses an electron in being displaced, requiring energy U_A (the anion electron affinity), the displacement energy may actually be <u>reduced</u> if the energy gained in localizing the electron in the resulting vacancy is $>U_A$ (see Section 17.3). The departing interstitial is then smaller in size and additionally may bind to a lattice ion, further lowering the displacement energy.

Covalent solids differ from both metals and ionic solids in that, while the bonding is stronger than metals and comparable to univalent ionic solids (Table 17.2), the directed nature of the bonding means that covalent solids are less likely to be close-packed, and displacement energies are closer to bonding energies. Displacement energies for III-V and II-VI semiconductor compounds are especially small and again can be different for the two atom types. It has been assumed here that what is displaced is an <u>atom</u>, but there is evidence that the interstitial atom may be ionized as it departs, resulting in an additional Coulomb interaction; the latter will be large for materials with small high-frequency dielectric constants (high-frequency because atom displacement takes place in ~10 fs) and may account for the large displacement energy ($T_d \cong 80$ eV) in diamond ($\varepsilon_\infty = 5\varepsilon_0$) compared to that ($T_d \cong 14$ eV) in silicon ($\varepsilon_\infty = 12\ \varepsilon_0$) where the Coulomb interaction is diminished by polarization.

There is considerable evidence that displacement thresholds increase with temperature, due to loss of close Frenkel pairs otherwise stabilized at low temperature, but also that at a still higher temperature they may fall again (Fig. 17.3). These phenomena are not well understood; the latter may arise from the nearby presence of defect sinks forming readily at higher temperatures perhaps by different mechanisms (see Section 17.5). Surfaces are good sinks, and atoms at or near surfaces may be displaced more easily because there are fewer bonds to break and fewer other atoms to disrupt. Thus sputtering energies should be lower than displacement energies (see Hren, Chapter 18). Indeed, an early estimate (SEITZ and KOEHLER, 1956) put the displacement energy at roughly four times the sublimation energy.

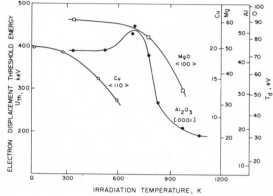

Fig. 17.3. Variation of threshold energy for aggregate damage in MgO, Al$_2$O$_3$ and Cu with irradiation temperature. (Data for MgO from BARNARD, 1977; Al$_2$O$_3$ from PELLS and PHILLIPS, 1978; and Cu from DROSD, KOSEL and WASHBURN, 1978).

Momentum Transfer

Electron displacement threshold energies U_{th} corresponding to $T'_{max} = T_d$ for an atom of atomic weight A can be calculated from Eqn. (17.15) as

$$U_{th} = (\sqrt{104.4 + 0.186\ A\ T_d} - 10.22)/20$$

$$\cong (\sqrt{100 + A\ T_d/5} - 10)/20 \qquad (17.17)$$

for U_{th} in MeV and T_d in eV.

Small values of U_{th} therefore arise for light elements with small displacement energies. Several of these are plotted in Fig. 17.4. For typical displacement energies in the region 10-40 eV, electron displacement threshold energies for light elements range from 0.1 to 0.5 MeV. It is therefore natural that the HVEM has been used to measure displacement thresholds and study knock-on displacement damage. By the same token, it is unlikely that knock-on displacement damage is a concern at or below 100 kV except for elemental hydrogen, lithium, beryllium, perhaps nitrogen and oxygen, and some of their compounds (viz hydrides, hydrocarbons, ice, azides, LiF, BeF$_2$). In many of these materials electronic excitations may generate displacements much more efficiently through radiolysis mechanisms (see Sect. 17.3).

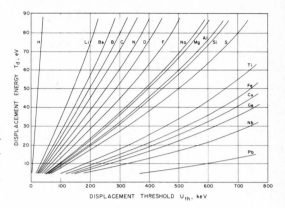

Fig. 17.4. Electron displacement threshold energies U_{th} for a representative collection of elements as a function of displacement energy T_d.

Displacement cross-sections σ_d for $T' > T_d$ can be calculated from Eqn. (17.1) with T'_{max} given by Eqn. (17.15) and T'_{min} by T_d. A typical result for $A = 6$ is plotted in Fig. 17.5 for three displacement energies ($T_d \geqslant 10$ eV). Even at 1 MV, T'_{max} is at most only a few times T_d for most solids, so TEM irradiation is unlikely to produce much more than <u>single</u> displacement events. For $2T_d > T'_{max} > T_d$, the concentration of displacements is just

$$c_d = \sigma_d \, (j_p \, t/e) \qquad\qquad (17.18a)$$

where $(j_p \, t/e)$ is the total electron fluence after time t. For $T' \geq 2\ T_d$, each primary knock-on atom of energy T' produces $\sim T'/2\ T_d$ defects; integrating over all T' between $2\ T_d$ and T'_{max} yields the displacement concentration given approximately (KINCHIN and PEASE, 1955) by

$$c_d \cong \sigma_d \, (j_p \, t/e) \, (1 + \ln \frac{T'_{max}}{2\ T_d}). \qquad\qquad (17.18b)$$

For $U_p \gg mc^2$ (0.511 MeV), Eqn. 17.1 and 17.2 reduce to an asymptotic displacement cross section

$$\sigma_d = Z^2 \, \frac{8\pi \, a_o^2 \, U_R^2}{Mc^2 \, T_d} = 140.5 \, \frac{Z^2}{T_d(eV) \, A} \cong 70 \, \frac{Z}{T_d(eV)} \text{ barns.} \quad (17.19)$$

Since σ_d rarely exceeds 100 barns (1 barn = 10^{-28} m^2) for <u>any</u> material even at 1 MV, the displacement rate c_d is unlikely to exceed 10^{-5} displacements per atom (dpa) s^{-1} in conventional medium-resolution TEM ($j_p \cong 100$ A m^{-2}). For high-resolution imaging or microanalytical TEM, however, electron fluences can exceed 1 to 10 MC m^{-2} (16 C m^{-2} = 1 electron/Å^2), and displacement densities ~ 1 dpa could be sustained for $U_p > U_{th}$. The nature of damage sustained at these displacement levels is characterized in Section 17.5.

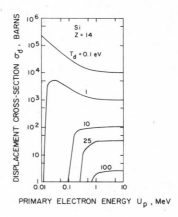

Fig. 17.5. Variation of electron displacement cross-section σ_d with electron energy for several displacement energies T_d in silicon. Accepted value of T_d for displacement of silicon in a perfect crystal lies between 10 and 25 eV. (Data from CORBETT and BOURGOIN, 1975; and OEN, 1973.)

Calculating σ_d from Eqn. (17.1) assumes a sharp displacement threshold T_d; this assumption proves adequate for $T' \gg T_d$. The value of U_{th} and thus T_d using this assumption may be obtained by plotting some damage parameter proportional to σ_d against U_p and extrapolating to zero damage. However, T_d is unlikely to be a step function (Fig. 17.6 due to orientation dependence, channelling, focussing and the probabilistic nature of close-pair recombination; the uncertainty introduces curvature into the tail of σ_d (Fig. 17.5) and U_{th} loses its rigid meaning. The requirement for widely-separated Frenkel pairs in ionic solids means that the probability of such pairs remaining separated for any given separation (which is roughly proportional to T') is less than that for a metal in equivalent circumstances; thus the displacement threshold is even "softer" for ionic materials. In addition, some displacements are apparently produced at constant rate well below U_{th}; these are subthreshold events and may owe their presence to existing defects (lowered T_d at defect sites such as dislocations, displacement of anions into existing cation vacancies, etc.) or inefficient operation of radiolysis mechanisms. Also, light atoms, particularly hydrogen, incorporated (often as an impurity) into the solid can act as intermediaries, acquiring sufficient energy from collision with an incident electron to in turn displace other nuclei whose electron displacement thresholds may be higher.

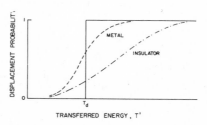

Fig. 17.6. A step function and two more realistic forms of the displacement probability function for a metal and an insulator. (After HUGHES and POOLEY, 1975.)

An important feature of knock-on displacement damage in multiatomic solids (HOBBS and HUGHES, 1975) is that different displacement thresholds may exist for the several atom species present. Even if the T_d are similar, the different atom masses can yield very different U_{th} and σ_d, and the resulting primary defect spectrum is unlikely to remain stoichiometric. This is particularly true for TEM irradiation ($U_p < 1$ MeV) where T' is never much larger than T_d and displacement cross-sections are very sensitive to the position of the threshold energy. Even for $T' \gg T_d$ when σ_d loses its primary energy dependence, σ_d still depends on Z and T_d from (17.19). In extreme cases, it may prove impossible to displace one or more of several atom species altogether, but it is wrong to suppose this necessarily precludes accumulation of significant damage. Either possibility has significant consequences, however, for defect stabilization and aggregate defect structures (see Sect. 17.5).

Even if it is not possible to displace an atom in the perfect solid, it may be possible to transfer sufficient momentum in the vicinity of existing defects to enhance defect or atom diffusion. As examples, an atom may be displaced more easily (typically with $T' \cong 1$-2 eV) into a neighboring vacancy, resulting in both vacancy diffusion and atom diffusion by a vacancy mechanism; similarly, the energy required to propagate an existing interstitial atom (typically $T' \cong 0.1$ eV) is considerably less than that required to create a Frenkel pair. In the limit, we may set T_{min} approximately equal to the defect migration enthalpy, and this results (Fig. 17.5) in high cross-sections for radiation-enhanced diffusion for U_p well below 100 keV. More importantly, the large number of defects created during irradiation (generally much larger than equilibrium thermal defect numbers) removes the defect formation enthalpy contribution to the diffusion enthalpy, so that thermally-activated diffusion may take place at lower temperatures. Radiation-enhanced diffusion phenomena are well-documented (see ADDA, BEYELER and BREBEC, 1975; LAM and ROTHMAN, 1976), and the effect of the continuing radiation field in TEM cannot be ignored in gauging subsequent defect behavior.

17.3 RADIOLYSIS

The necessary conditions for fast electron-atomic electron interactions to generate atomic displacements are localization of an electronic excitation with sufficient energy and for times long enough (longer than the time for an atomic vibration, or ~ 1 ps) that mechanical relaxation of the surrounding ion cores leads to a bonding instability, in many cases involving displacement of an ion core through a saddle point. In this way the potential energy inherent in electronic excitations can be converted to momentum of a departing atom nucleus. Metals fail these criteria in several respects: electron excitations are rapidly delocalized via the conductions band in times ~ 1 fs and any bonding instability is likely to be screened before there can be a mechanical response. Semiconductors are marginal in that, while sufficiently long-lived excitations exist (and even give rise to phosphorescence on a 1 s time scale), the excitations are seldom sufficiently localized or of sufficient energy. It is possible that radiolysis does occur with low efficiency in semiconductors, and there is evidence (e.g. DEARNALEY, 1975) which suggests synergism with knock-on displacement mechanisms. This leaves insulating solids, for a large number of which (Table 17.3) there exist efficient and documented (though not always well-characterized) radiolysis mechanisms.

TABLE III

Inorganic solids in which radiolysis is known to occur

---

alkali halides (LiF, LiCl, LiBr, LiI, NaF, NaCl, NaBr, NaI, KF, KCl, KBr, KI, RbF, RbCl, RbBr, RbI, CsF, CsCl, CsBr, CsI)

alkaline earth halides (CaF_2, SrF_2, BaF_2, MgF_2)

silver halides (AgCl, AgBr, AgI)

cadmium halides (CdI_2)

lanthanum halides (LaF_3)

lead halides (PbI_2)

perovskite halides ($NaMgF_3$, $KMgF_3$)

silicas (quartz, cristobalite, fused silica)

silicates (alkali feldspars, some amphiboles, mica)

ice (H_2O)

alkali hydrides (LiH)

alkali azides (LiN_3, NaN_3, KN_3)

sulfides (MoS_2)

carbonates ($CaCO_3$)

alkali perchlorates (NH_4ClO_3, $NaClO_3$, $KClO_3$)

alkali bromates ($NaBrO_3$)

The conditions to be met for radiolysis to occur with any efficiency are that electronic excitations must have energies at least as large as the energy stored in the resulting displacements, and that an efficient mechanism operates to convert this energy into substantial momentum of displaced atom cores.

Electronic Excitations

Fast electron irradiation of insulators results in four distinguishable sorts of atomic-electron excitations: ionization of core electrons (the process eventually responsible for characteristic x-ray generation,

T' > 100 eV); ionization of valence electrons and covalent bond breakage
(T' ≅ 10-20 eV); elevation of valence electrons to locally-bound electron-hole pair states (excitons, the highest of which are degenerate
with the conduction band, T' ≅ 5-10 eV) and collective excitation of
weakly-bound valence electrons into long-wavelength semilocalized (∿10
nm) oscillations (Bohm-Pines plasmons, T' ≅ 25 eV). These various excitations are reflected in the electron energy-loss spectrum (EELS) shown
in Fig. 17.7 for two halides. Inner-shell electron excitations admittedly
involve large energies, but these excitations are among the least frequent and most of their energy is lost in internal atomic-electron rearrangements, resulting in characteristic x-ray emission or ejection of
Auger electrons, too quickly (≅1 fs) to effect motion of atom nuclei.
Plasmon excitations are among the most probable and can involve appreciable energies, but they are much too delocalized to be of use. Most,
in fact, of these higher energy excitations eventually decay to one or
more low-energy localized single-electron excitations with exciton character (see, for example, the treatment of plasmon decay in semiconductors and insulators by ROTHWARF, 1973). Even complete ionization which
results in electrons excited into the conduction bands eventually results also in holes appearing in the valence band; the latter have large
effective masses in wide-gap insulators because of an accompanying local
configurational lattice distortion (small polaron). In the absence of
alternative electron-trapping sites, any mobile electrons in the conduction band must inevitably recombine with the holes in the valence band
via intermediate exciton configurations. It is these directly-excited
or secondarily-formed exciton states which are responsible for radiolysis. What is important then to radiolysis is, broadly, the total energy
deposited in electron-electron interactions given by the Bethe-Bloch expression, Eqn. (17.5).

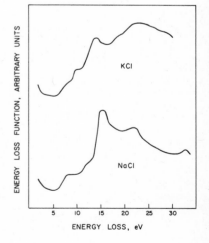

Fig. 17.7. Electron energy loss function for two halides irradiated with
100-kV electrons, revealing sharp exciton peaks and broad plasmon reson-
ances. (After HOBBS, 1979.)

The energy to displace an atom from its lattice site in an ionic
crystal is at least equal to the lattice binding energy U_ℓ given in Eqn.
(17.16). The energy available from exciton states depends on the
electron-hole pair energy which is roughly (POOLEY, 1975)

$$U_{e'h\cdot} = U_\ell + U_A - U_I \qquad (17.20)$$

where U_A is again the electron affinity of the anion and U_I the ionization potential of the cation. The necessary condition for radiolysis to
occur, $U_{e'h\cdot} > U_\ell$, thus reduces to

$$U_A - U_I > 0. \qquad (17.21)$$

Table 17.4 lists these parameters for NaCl and MgO.

TABLE 17.4

Lattice and exciton energies (in eV)
for two ionic solids

Solid	U_A	U_I	U_ℓ	$U_{e'h\cdot}$	T_d
NaCl	3.8	5.1	7.9	6.6	6.6
MgO	-9.5	15	47	22	27

It is clear from Table 17.4 that radiolysis is a possibility for NaCl
but unlikely for MgO. Such a simple criterion is less useful for more
covalent solids, such as SiO_2 and silicates, and instead a specific comparison must be made between covalent bond energy and available exciton
energy as for organic solids (PARKINSON et al., 1977). For example, in
SiO_2 exciton states occur with energies between 8 and 11 eV, while the
energy of the Si-O bond is in the region of 5 eV. One imagines that
several bonds would have to be broken simultaneously to provide the
requisite freedom for distortions and rotations of coordination polyhedra to ensure permanent structural alterations. However, an alternative Frenkel defect mechanism (discussed below) could provide a way of
accumulating the necessary spatial freedom.

Energy-To-Momentum Conversion

The best-documented example of an intrinsic energy-to-momentum conversion mechanism occurs in alkali halides (ITOH, 1976; KABLER and
WILLIAMS, 1978) where one or more exciton states (T' $\cong$ 7 eV) are con-
figurationally unstable. In such states (Fig. 17.8) the hole is tightly
localized in an X_2^- homonuclear molecular bond between two neighboring
X^- halogen ions, while the excited electron resides in hydrogen-like
orbitals in the vicinity of the surrounding alkali cations. The distinct separation of charge, together with the bonding instability introduced, permit diabatic mechanical dissociation of the exciton involving
departure, in times $\cong$10 ps, of the X_2^- molecule away from the position
of the initial excitation. The X_2^- species moves away by a combination

of hole tunnelling and interstitialcy propagation and comes to rest at a _single_ anion site where it resembles a crowdion interstitial (an H center). It therefore leaves behind a halogen vacancy with a trapped electron (an F center). The F and H center together constitute the fundamental Frenkel pair. The mechanism also illustrates a frequent feature of radiolysis, namely that it need involve only a single sublattice.

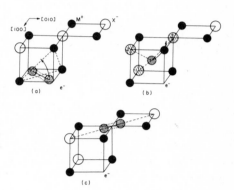

Fig. 17.8. Radiolysis sequence in NaCl-structure alkali halides involving migration of an X_2^- crowdion interstitial away from the site of the initial exciton, forming an H center and an F center. (After KABLER and WILLIAMS, 1978.)

Since, in radiolysis of halides, the interstitial anion is removed essentially as an _atom_ instead of an ion (because of the trapped hole), the radiolysis criterion, Eqn. (17.21), must be modified. An electron is removed from an anion (requires energy U_A) and localized in the F-center vacancy (where it is bound with energy U_F), while the corresponding interstitial atom is bound to the lattice with energy U_H. The overall displacement energy is thus

$$T_d \cong U_\ell + U_A - U_F - U_H \qquad (17.22)$$

which is also shown in Table 17.4 for NaCl and MgO and is seen to be close to $U_{e'h}$ for NaCl but still not for MgO.

Radiolytic displacement in alkali halides as a mode of exciton decay occurs with a high quantum efficiency between 0.15 and 0.8 depending upon temperature ($\cong 0.3$ near room temperature), or one displacement per 10-50 eV absorbed. Alternative exciton decay modes are radiative (exciton luminescence) or involve phonons (generate heat) and are anticorrelated with radiolytic displacement. Such efficient radiolysis has enormous consequences for radiation damage in TEM. Setting $T'_{min}{}^e$ and $T'_{min}{}^n$ both equal to T_d in Eqn. (17.4) reveals that, for materials like halides in which efficient radiolysis mechanisms exist, damage arising from radiolysis will always be far more important than damage from direct knock-on displacement.

Related mechanisms operate in alkaline earth fluorides (CaF_2, MgF_2) and perovskite halides, though radiolysis of silver halides may proceed by a somewhat different mechanism (SLIFKIN, 1969, 1975; TOWNSEND, 1975). Details of radiolysis in other broad classes of materials, like silicates, remain obscure, even though in some cases one or more end-products are known. For example, the E_1' center in SiO_2, an oxygen vacancy accompanied by asymmetric relaxations of neighboring silicon atoms, is the analogue of the F center (Fig. 17.9, YIP and FOWLER, 1975) and is found in both crystalline and amorphous irradiated silica. The corresponding interstitial component of the Frenkel pair has not been identified, but a peroxy O_2^- or $O_2^=$ molecular species would be a good candidate and has been detected recently (FRIEBELE et al.[*],1979; STAPLEBROEK et al., 1979). Radiolysis in SiO_2 could, in fact, proceed as follows. Initial breakage of the predominantly covalent Si-O bond, followed by relaxation of neighboring tetrahedra, produces a nonbridging oxygen which forms an O_2^- peroxy linkage with a neighboring oxygen and leaves behind an E_1' oxygen vacancy center. The stability of the peroxy linkage (or of the O_2^- peroxy radical to which it is easily converted) is very high and, together with the Si-Si bonding across the oxygen vacancy, may stabilize what is essentially a close Frenkel pair. While many more ionic oxides like MgO do not undergo radiolysis for the reasons outlined in Table IV, it is significant to note that O_2^- or $O_2^=$ interstitial species have been detected by EPR in both neutron-irradiated MgO and CaO, suggesting that the molecular form of anion interstitials may be a general feature of radiation damage in ionic solids, even in the absence of radiolysis.

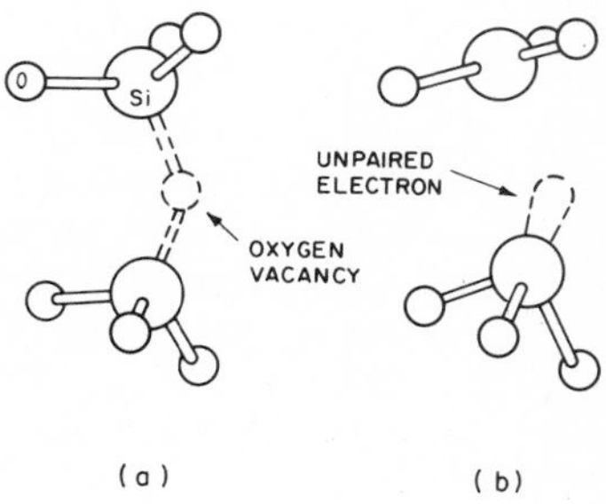

Fig. 17.9. The E_1' center in SiO_2 (a) before and (b) after asymmetric tetrahedron relaxations. (After YIP and FOWLER, 1976.)

(a)   (b)

Influence of Temperature, Impurity and Radiation Flux

The efficiency of radiolysis mechanisms can be temperature-dependent if an activation barrier exists somewhere in the forward radiolysis sequence, for example in the transition from exciton to nearest-neighbor Frenkel pairs in halides. Such a barrier ($\cong 0.1$ eV) is found in KI, NaCl, NaBr and probably many other irradiation-sensitive solids, and accounts for a dramatic decrease in radiolytic efficiency at low temperatures (Fig. 17.10). Little information is presently available concerning the temperature dependence of other radiolytic mechanisms. An additional influence of temperature occurs in correlated recombination of close, only metastable Frenkel pairs (see Section 17.4).

Many radiolytic systems are sensitized by the presence of impurities because these provide preferred sites for localized excitations and locally-decreased lattice binding energies. For example, the incorporation of water into SiO_2 and other silicates leads to substitution of hydrolyzed Si-OH:HO-Si bonds in place of the much stronger Si-O bonds of the perfect lattice, and these weaker bonds may be readily·excited and severed. There is in fact experimental evidence for considerable variation in radiolysis efficiency with water content in silicates. Hydrogen impurities are responsible for at least three other radiolysis

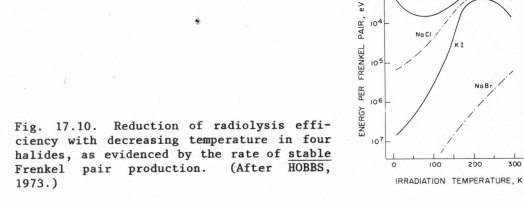

Fig. 17.10. Reduction of radiolysis efficiency with decreasing temperature in four
halides, as evidenced by the rate of stable
Frenkel pair production. (After HOBBS,
1973.)

mechanisms. In alkali halides, hydrogen substituted for halogen may be
radiolytically ejected from anion sites, resulting in formation of interstitial hydrogen (U_2 centers) and halogen vacancies (F centers). In
MgO, incorporation of hydrogen at Mg sites (bound to a neighboring oxygen) results, upon ionization, in interstitial hydrogen (forming bru-
cite, $Mg(OH)_2$) and cation vacancies. Hydrogen in germanium, too, apparently can be relocated by ionization. These impurity pathways are,
of course, limited to the impurity level and will thus lead to observable degradation in TEM only for impurity concentrations $\gtrsim 10^3$; also,
they are unlikely to function repeatedly as would radiolysis mechanisms
involving only the perfect lattice.

Impurities conversely can impede radiolysis by providing alternative pathways for exciton collapse, leading to electron-hole recombination instead of radiolysis. For example, at temperatures where excitons
or self-trapped holes are mobile, electron-trapping impurity ions (such
as aliovalent cations) can provide alternative sites to the perfect crystal for exciton localization and may additionally provide favored radiative decay modes for the resulting localized excitations, effectively
short-circuiting the radiolysis process (POOLEY, 1966). Since diffusion
of excitons or self-trapped holes to impurity sites is required for this
process to operate, depression of the radiolysis rate is ionization
dose-rate, as well as impurity-level, dependent. Rather substantial impurity concentrations are in fact necessary ($\cong 1\%$), in view of the relatively low hole or exciton mobilities in wide-gap insulators and the
enormous ionization rates encountered in electron microscopy, to effect
substantial changes in radiolytic efficiency (HOBBS, 1975).

An important additional influence of the ionizing radiation field,
whether on the products of radiolysis or those of knock-on displacement,
is its effect on the electronic structure of defects and their associated mobilities. An F center, which is an anion vacancy which has
trapped one or more electrons to become neutral with respect to the lattice, is more mobile in its ionized state (F+) when one or more of its

electrons have been removed, and still more mobile in its excited state (F*). Similarly, excited states of the H-center molecular bond may result in its enhanced mobility. BOURGOIN and CORBETT (1975) have discussed a number of analogous thermal and athermal ionization-enhanced diffusion mechanisms in semiconductors. The mobility of defects and their consequent aggregation are enormously important to overall damage kinetics, particularly in the advanced stages of damage often experienced with TEM of electron-sensitive materials (see Sect. 17.4). In addition, the identity of defects may be altered altogether. H centers in alkali halides (essentially crowdion atoms) can be converted by continued irradiation at low temperature into true ion interstitials; similarly, F center-H center close pairs in alkaline earth fluorides can be reconverted into extended excitons.

The ionization density of the electron beam of the TEM (even more so in STEM than in CTEM) is prodigious (1 kW – 1 MW mm^{-3}) and is probably our most intense energy deposition source. This enormous power density implies a 10 eV excitation at every atom site up to a thousand times every second. For radiolysis mechanisms as efficient as that in alkali halides, such ionization density forbodes displacement rates of 1-1000 dpa s^{-1}, far in excess of the rates arising from knock-on displacement. That such radiation-sensitive solids can be observed at all is a tribute to the long-range ordering forces in these crystals. Indeed, lattice-fringe imaging of alkali halides has been surprisingly successful (YADA and HIBI, 1969), but it is clear that what is being recorded is a dynamic average over stable lattice sites but far from stable lattice atoms. The necessarily dynamic character of the environment has a measurable influence on defect recombination rates (Sect. 17.4) which can alter expected degradation kinetics. In addition, the resulting defects produced in such intense ionization fields may differ substantially in kind from those observed in irradiation fields orders of magnitude lower in ionization density such as x-rays, heavy ions or neutrons. These facts should be borne in mind when attempting extrapolations from more conventional radiation studies.

17.4 DEGRADATION KINETICS

The rate at which permanent atomic alterations occur, either by radiolysis or knock-on displacement, depends on the frequency of subsequent restorative reactions, as well as on the efficiency of the forward displacement mechanism. Both factors depend in turn critically on temperature, bond character and the impurity state of the crystal. Restorative processes can occur as both correlated and uncorrelated events. Correlated restorations involve undoing what the forward process has just effected, e.g. recombination of Frenkel pairs or bond fragments insufficiently separated to be stable against collapse to the perfect lattice configurations. It is clear that additional cooperative lattice relaxations are normally required to stabilize such close Frenkel pairs or severed bonds. Correlated recombinations, if they occur quickly enough, simply adjust the forward rate and to first order do not depend on defect concentration. There is then little dose-rate dependence of damage fraction and little dose dependence of the damage rate.

Correlated recombinations in alkali halides result typically in loss of >90% of all initial Frenkel pairs within microseconds (which is not quite fast enough to avoid dose-rate effects in TEM). Thus, the efficiency of stable defect production in alkali halides is ~250 eV per Frenkel pair. Recent evidence (KABLER and WILLIAMS, 1978) suggests that the quantum efficiency of radiolytic exciton decay in fluorite- (CaF_2), rutile- (MgF_2) and perovskite- ($NaMgF_3$) structure halides is similar to that in rock-salt structure halides (NaCl), but that the much lower stable defect production rates (~10 keV per Frenkel pair) are a consequence of lattice geometries which favor more efficient close-pair recombination. In CaF_2, for example, although F-H pairs are the primary stable products of radiolysis, the separations of F centers and nearest-neighbor H centers are small enough that such pairs can be considered as extended excitons with a radiative electron-hole recombination mode.

Radiolysis in crystalline silicates, which appears to involve Si-O bond breakage instead of or as the means to close Frenkel pair production, leads eventually to reduction of the crystal to an amorphous state (metamictization, see Sect. 17.5). For such broken bonds to contribute to degradation, they must be stabilized against correlated bond reformation by, for example, [SiO_4] tetrahedron relaxations and peroxy linkages, and the overall radiolytic damage efficiency can be estimated by dividing the overall metamict dose (~10 GGy) by the bond density to yield ~10 keV per bond, close to that for stable defect production in CaF_2. In metals and semiconductors, much less is known about correlated back reactions because many of the defect spectroscopy tools employed in wide-gap insulators are not available; however, it is known from various kinetic and annealing studies that a similar fraction (>0.9) of initial displacements do not survive correlated recombination.

Uncorrelated restorative processes, where restoration takes place at a site different from that of the initial excitation, have an important and very different effect on damage kinetics, because the rate of back reactions then depends on the fraction of damage material. For example, interstitials have a larger probability of recombining with un-correlated (other than their original) vacancies as the overall concentration of Frenkel pairs increases. Such recombinations can occur entirely athermally when an interstitial is created within some volume υ surrounding an existing vacancy, called the athermal recombination volume, within which recombination takes place spontaneously; similarly, a vacancy can be created within the recombination volume of a nearby interstitial. Recombination volumes are typically of order 1000 lattice sites, which implies spontaneous recombination with a vacancy-interstitial separation of 5-10 lattice sites. With so many atoms involved and recombination times similar to those for correlated recombination, recombination volumes inevitably become dose-rate dependent and are found to diminish significantly under the enormous displacement rates associated with TEM when efficient radiolysis mechanisms exist.

Uncorrelated athermal recombination has been shown (HUGHES and POOLEY, 1971) to lead to a continually decreasing overall damage rate and alteration of damage kinetics from linear to logarithmic. For Frenkel defect production or its equivalent, actual damage kinetics depend on the state of defect aggregation (or existence of irreversible defect sinks) for each complementary defect species, because aggregation of defects of even one kind decreases the probability of recombination with complementary defects compared to that for defects in isolated distributions. If complementary defects with similar recombination volumes υ are kept immobilized and isolated, at low temperature (temperature T_1, Fig. 17.11) or by impurity pinning, the stable defect concentration c_D at time t eventually follows kinetics of the form

$$c_D \cong (1/4\ \upsilon)\ \ln\ (4\ \upsilon\ \dot{c}_D\ t).\qquad\qquad (17.23)$$

Here $\dot{c}_D$ is the defect production rate surviving correlated recombination. Kinetics, Eqn. (17.22), lead to a slowly saturating defect concentration.

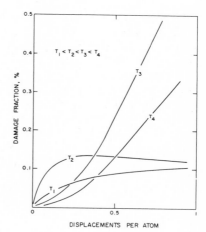

Fig. 17.11. Damage kinetics for four characteristic irradiation temperatures. (After HOBBS, 1979.)

Where at least one defect species is mobile (temperature T_2) and aggregates or is otherwise removed from consideration (for example at foil surfaces), the athermal recombination volume for that species becomes vanishingly small; for aggregating defects, this occurs because their recombination volumes overlap substantially. Damage kinetics for stable (or stabilized) defects then follow the form

$$c_D \cong (1/\upsilon)\ \ln\ \upsilon\ \dot{c}_D\ t\qquad\qquad (17.24)$$

which represents considerably faster accumulation (and saturation) of stabilized defects than does Eqn. (17.22). The alteration in logarithmic damage kinetics from Eqn. (17.23) to Eqn. (17.22) confers an enormous advantage to low-temperature TEM of sensitive specimens; even the most sensitive alkali halides, for example, can be examined briefly at temperatures (<30 K) where the more mobile halogen interstitial products of radiolysis are restricted to athermal aggregation. Such statistical aggregation of defects incidentally leads to eventual acceleration of damage kinetics from Eqn. (17.22) to Eqn. (17.23), effectively bracketing the permissible low-temperature dose. Trapping of interstitials at existing defects can produce the same advantage as lowering temperature. Smaller substitutional ions (e.g. Na in KI) and charge-compensating vacancies on another sublattice (e.g. cation vacancies, formed by incorporation of Ca into NaCl, which trap anion interstitials) have been shown to act in this fashion but must exist in concentrations similar to saturation defect concentrations to remain effective.

The existence of aggregate sinks for <u>all</u> defect species (which occurs at temperature T_3, not too far above room temperature for many inorganic solids) leads to an alarming linear relation between damage and electron dose

$$c_D \cong \dot{c}_D \, t \qquad\qquad (17.25)$$

because the recombination probability again no longer depends on overall damage fraction (but only on the much lower steady-state concentrations of defects on their way to sinks). Without substantial recombination, damage increases to encompass all or a significant portion of the lattice, and eventual saturation is due largely to exhaustion of unaffected material. The latter principle underlies the "target" or survival approaches popularly applied to damage in organic solids (see GLAESER, Chapter 16) and involves similar kinetics. Lowering radiation temperature can be advantageous in avoiding this regime if any of the nonreversible (sink) processes is temperature-dependent, for example if they rely on the thermally-activated diffusion of defects. Increasing radiation temperature (T_4, Fig. 17.11) may cause one or more defect sink to lose its stability, whereupon recombination becomes again increasingly probable. Usually, however, the high-temperature regime is unattractive owing to its deleterious effect on existing specimen features intended for examination.

Foil surfaces are an interesting special case because, there, defects are not only lost to irreversible sinks but also removed altogether. This produces <u>denuded</u> regions (typically a few tens of nm) closest to the foil surfaces. The thinnest specimen regions are therefore freed of high point-defect concentrations and aggregate defect structures, but are being continually reconstructed (a point sometimes overlooked).

17.5 RADIATION-INDUCED STRUCTURAL CHANGES DURING ANALYSIS

The fact that point defects will be produced in large numbers if they are produced at all in TEM has a significant bearing on the stability of the material under investigation. This is because point defects interact strongly at high densities and are seldom found in isolated distributions; their interaction can readily alter material structure in a number of ways. Such defect-induced alterations can be classified into three characteristic forms of attendant degradation (HOBBS, HOWITT and MITCHELL, 1978).

i) identifiable point defects accumulate at well-defined localized sinks; thus a portion of the solid may decompose or transform, but the major portion retains its structure intact. Examples are nucleation of dislocation loops, voids and precipitates formed by point defect accumulation.

ii) The entire solid rearranges, retaining its short-range chemical integrity (near-neighbor coordination), and thus its stoichiometry, but altering its long-range order relationships. Examples are polymorphic phase changes, order-disorder reactions and crystalline-to-amorphous transformations.

iii) The solid alters its short-range chemical identity by local loss or gain of constituents, altering its local composition and eventually its long-range order relationships as well. The composition fluctuations may be periodic and the recomposition or decomposition spinodal, or precipitation may occur by nucleation and growth. Analogous loss of constituents in organic solids leads to molecular reactivity and extensive restructuring (see GLAESER, Chapter 16), but inorganic solids are less likely to lose atomic species altogether (except near surfaces) and are more likely to retain some form of long-range order.

Frenkel Defect Condensation

Defect aggregates are generally two- or three-dimensional, though more or less one-dimensional aggregates (lines of dilatation, decorated dislocations, narrow dislocation dipoles) do occur but will not be discussed here. Point defects aggregate because their aggregation lowers the overall energy of the lattice by either the elastic interaction energy (though consolidation of defect strain fields) or electrostatic or electronic interaction energies (Coulomb attraction, charge delocalization). Vacancies, as well as interstitials, can be accompanied by appreciable relaxation of the surrounding lattice; outward strains are observed from vacancies in some systems, particularly in insulating or semiconducting solids where the vacancy may be charged or contained trapped electrons or holes, contrary to what might be naively supposed. Charged defects can additionally interact electrostatically, and even neutral defects may interact covalently, as do two F centers.

i Planar Aggregates

Atom vacancies aggregating in two dimensions generate a planar void which will, unless otherwise stabilized, collapse to restore continuous lattice bounded by a planar ring of dislocation core. Such dislocation loops may be unfaulted or faulted, depending on whether the proper stacking sequence across the loop plane is preserved or altered. Analogous planar condensation of interstitial atoms generates new lattice planes bounded by a dislocation loop core (of opposite sense) into which the accommodation strain is similarly localized (Fig. 17.12). The elastic energy of a round loop of radius r is

$$U_\ell \cong \frac{G\,b^2\,r_\ell}{2(1-\nu)}\,\ln\{\frac{r_\ell}{b} + \frac{5}{3}\} + \pi\,r_\ell{}^2\,\gamma_f \qquad (17.26)$$

where b is the loop Burgers vector, G the shear modulus, ν Poisson's ratio and γ the stacking-fault energy if the loop is faulted. The latter restricts faulted loops to the fault plane but also drives large faulted loops to unfault (if possible) by propagation of a (partial) lattice shear across the loop, after which the resulting unfaulted loops are free to rotate within the confines of their glide cylinders. Since the number of defects contained in the loop is of order $N = \pi r^2/b^2$, the elastic strain energy per defect for unfaulted loops,

$$\frac{U_\ell}{N} \cong \frac{G b^4}{r_\ell} \qquad (17.27)$$

decreases monotonically as the loop radius increases. Large loops and dislocation segments with large radius of curvature are thus particularly good sinks for interstitials and vacancies. The elastic interaction energy is slightly biased in favor of interstitials, and the majority of large loops encountered in irradiation situations is interstitial in character, since vacancy loops tend to shrink by preferential absorption of interstitials.

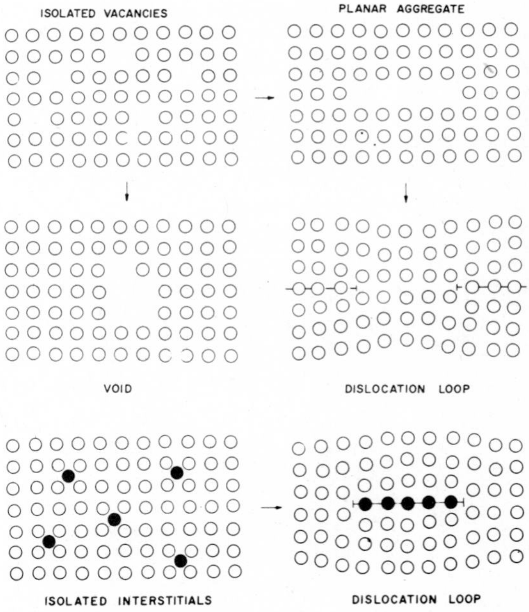

Fig. 17.12. Stylized formation of dislocation loops and voids from aggregation of vacancies and interstitials in a monatomic crystal.

Dislocation loops usually nucleate heterogeneously on impurities or other existing point defects, so their initial density is often a function of impurity content and distribution and radiation dose rate. Their density at a later stage of development (and therefore size for a given radiation dose) depends on the ease of coalescence, a process principally controlled by glide parallel to Burgers vector b on a glide cylinder and translation normal to b by self (i.e. conservative) climb

(Fig. 17.13). Elongated or faceted loops frequently develop [Fig. 17.14(a)], particularly at high growth rates, because loop growth is a climb process proceeding by a jog mechanism and depends on jog geometry. At high density, dislocation loops intersect frequently and reduce to

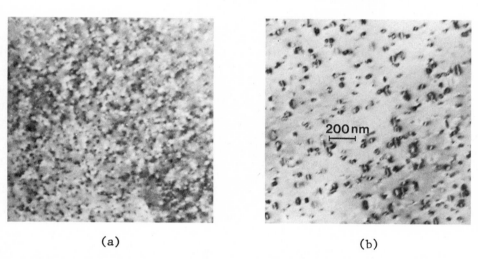

(a)   (b)

Fig. 17.13. Interstitial dislocation loops in NaCl electron-irradiated at room temperature to (a) 10 C m^{-2} (2 MGy or $\sim$ 0.1 dpa) and (b) 100 C m^{-2} (20 MGy or $\sim$ 1 dpa) showing decrease in loop density due to coalescence.

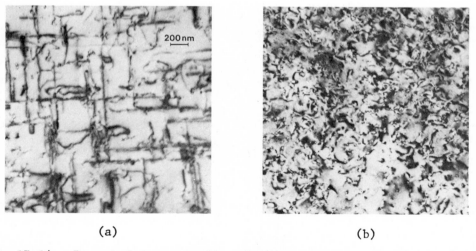

(a)   (b)

Fig. 17.14. Repeated intersection of interstitial dislocation loops in NaCl electron-irradiated at 423 K to (a) 200 C m^{-2} (40 MGy or $\sim$ 2 dpa) and (b) 2 kC m^{-2} (0.4 GGy or $\sim$ 20 dpa) leads to formation of dense dislocation networks.

dense dislocation networks [Fig. 17.14(b)], the details of which depend on the geometry of crystallographically equivalent Burgers vectors and on loop morphology.

Dislocation loops in multiatomic solids involve atoms from all sublattices in stoichiometric ratio, although they may be _formed_ by condensation of sublattice defects in nonstoichiometric ratios. A good example is halides, for which the primary products of radiolysis are displacements on the halogen sublattice only and in which stoichiometric interstitial dislocation loops (Fig. 17.13) form by aggregation of halogen interstitials alone. In this case, lattice energetics favor formation of stoichiometric dislocation loops (with increasingly smaller energy per defect condensed) involving all sublattices at the expense of generating additional secondary defects, here cation interstitials. In essence, the halogen interstitials dimerize and form neutral halogen molecules which displace anions and cations into stoichiometric interstitial loops and themselves occupy the resulting vacancy pairs as substitutional halogen molecules (HOBBS, HUGHES and POOLEY, 1973; CATLOW et al., 1979). These substitutional halogen molecules may condense as well into planar aggregates [Fig. 17.15(a)].

ii Volume Inclusions

Large three-dimensional aggregates of interstitial atoms are unlikely because the strain associated with such aggregates becomes rapidly prohibitive. Volume inclusions therefore arise only from condensation of vacancy or substitutional defects. Planar inclusions can result from condensation of interstitial or substitutional atom species or from vacancy aggregates under certain condtions. The energy associated with an unstrained inclusion is essentially the interfacial energy. For a two-dimensional unstrained inclusion of radius r_I this energy is $U_I = 2\pi r_I^2 \gamma$, where γ is the interfacial energy per unit area of interface. The number of defects, each occupying volume Ω, contained in such an inclusion is $N = \pi r_I^2/\Omega^{2/3}$, so the energy per defect

$$\frac{U_I}{N} = 2 \gamma \Omega^{2/3} \tag{17.28}$$

remains constant. For a three-dimensional unstrained inclusion, $U_I = 4\pi r_I^2$ and $N = 4\pi r_I^3/3\Omega$, so the energy per defect

$$\frac{U_I}{N} = 3 \Omega \gamma/r_I \tag{17.29}$$

decreases as the inclusion radius increases at the same rate as for a dislocation loop. Planar inclusions, like the halogen precipitates in Fig. 17.15(a), therefore usually consolidate into three-dimensional inclusions [Fig. 17.15(b)] unless there are strong epitaxial forces operating. Strain-energy considerations in the case of strained inclusions additionally complicate these simple surface energy criteria.

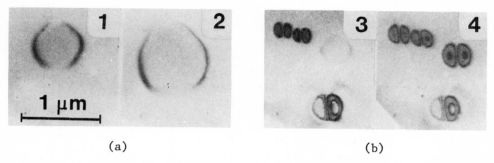

(a)   (b)

Fig. 17.15. Halogen bubbles in a thin foil of KCl TEM-irradiated somewhat above room temperature in the form of (a) planar aggregates collapsing to (b) several volume inclusions. (From HIBI and YADA, 1965).

For the case of stoichiometric vacancy aggregates, planar condensation results in collapse to form dislocation loops, while condensation in three dimensions forms voids (which are unstrained inclusions). For the same number of vacancies N condensed

$$\frac{U_I}{U_\ell} \cong \{ \frac{\gamma}{5Gb^3} \} \, N^{3/4} \qquad\qquad (17.30)$$

and voids are ultimately preferred to loops; voids must however be stabilized against collapse in the early stages, e.g. by internal pressure of trapped impurity gas atoms or displaced atoms with a volatile but insoluble phase. Aggregation of vacancies on a single sublattice into three-dimensional inclusions in a multiatomic solid is stabilized by the remaining sublattice(s). A classic example is condensation of F-center halogen vacancies in halides which creates local regions containing only cations and free electrons that soon revert to metallic inclusions (Fig. 17.16). Condensation of substitutional species (like halogen molecules) is of course autostabilizing, resulting in this case in a fluid inclusion under considerable pressure, in other cases in more conventional precipitation.

Notice that point-defect condensation has provided the means in this example for decomposition of a solid into separate phases of its constituent elements. Such decomposition will occur rapidly for irradiation in a temperature regime in which both interstitial and vacancy defects are mobile and their respective sinks (halogen inclusions, metal precipitates in halides) are stable against redissolution. In metals, this is just the regime of the void-swelling peak. Many nonmetallic systems which undergo radiolysis exhibit a similar peak (Fig. 17.17) at or just above room temperature. Here damage kinetics approach linearity with dose, as discussed in Sec. 17.4, and this regime should be avoided if gross damage is to be prevented.

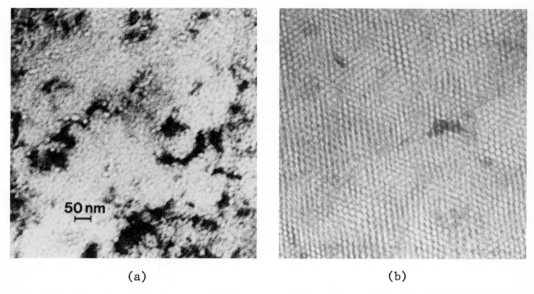

(a)   (b)

Fig. 17.16. Inclusions, probably metallic, formed from condensation of F centers (a) in NaCl electron-irradiated to 2 kC m^{-2} (0.4 GGy or $\sim$ 20 dpa) and (b) in CaF$_2$ TEM-irradiated to 50 kC m^{-2} (10 GGy or $\sim$ 100 dpa). [(a) reproduced from HOBBS, 1976; (b) from CHADDERTON, JOHNSON and WOHLENBERG, 1976.]

Fig. 17.17. Damage peaks analogous to void-swelling peaks in metals for four halides electron-irradiated near room temperature. (After HOBBS, 1976.)

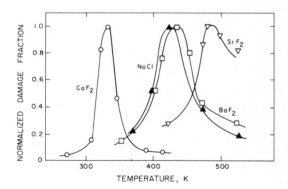

Ordering and Disordering

Alteration of the ordered arrangement of different atom species among a fixed set of atom sites by atom displacements is an anticipated result of electron irradiation above threshold. In strictly-coordinated ionic solids (which are perfectly ordered in the above sense) it is unlikely that an anion could be stably displaced to a cation site, and ionic lattices accommodate disorder by precipitating defects in ordered stoichiometric units. In metal alloy systems, ordering forces are not so strong, and disorder in the distribution of two or more atom species among available atom sites is more easily effected.

The state of order is expressed by the conventional order parameter S ($0 \leq S \leq 1$), and the rate of disordering by irradiation is expected to depend on both the displacement rate and the existing state of order

$$- \frac{dS}{dt} = \bar{\sigma} \, (j_p/e) \, S \tag{17.31}$$

where $\bar{\sigma}$ is an effective displacement cross-section which is a suitable weighted average of the displacement cross-sections for the different atomic species. In a binary alloy, it is generally assumed that both species must be displaced to effect disorder, but this criterion may not be strictly necessary since it is relatively easy for an atom of one species to be displaced to an adjacent site vacated by a displaced atom of the other species. Integrating Eqn. (17.31),

$$- \ln S = \bar{\sigma} \, (j_p/e) \, t \tag{17.32}$$

and ln S should decrease linearly with dose ($j_p \, t/e$). Since atom diffusion will tend to restore order, disordering to completion will only occur at temperatures below which atom diffusion (usually by a vacancy mechanism) is negligible. Several simply ordered binary alloys (Fig. 17.18) are found to obey Eqn. (17.32) reasonably well at temperatures where vacancies are immobile. The dose range in which these measurements are made corresponds exactly to the 1-10 MC m^{-2} dose required for atomic scale investigations in TEM. Significant disordering may therefore accompany high resolution analysis of ordered structures for electron energies above threshold for atom displacement.

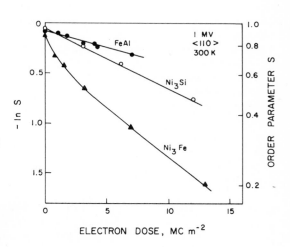

Fig. 17.18. Kinetics of displacement disordering for TEM irradiation at 1 MV at room temperature (vacancies not mobile). (Data for FeAl from KINOSHITA, MUKAI and KITAJIMA, 1977; for Ni$_3$Si and Ni$_3$Fe from BUTLER and SWANN, 1977.)

At higher temperatures, radiation-enhanced diffusion results in a competitive reordering; in fact, defect mobility is the main factor determining whether or not a system disorders or orders under irradiation. Since these are competing processes, it might be expected that a dynamically stable compromise value of S is reached for which $dS/dt = 0$

and where the rate of disordering due to displacement balances the rate of ordering due to irradiation-enhanced diffusion. Under steady-state irradiation conditions dominated by interstitial-vacancy recombination of the sort discussed in section 17.4, this balance is expressed analytically (BUTLER and SWANN, 1977) as

$$\frac{dS}{dt} = (\bar{\sigma} \, j_p/e)^{\frac{1}{2}} \exp \, (-U_m/kT) \, \{\sinh(T_c/T)/S - S \cosh(T_c/T)/S\} \qquad (17.33)$$

where U_m is an activation energy for radiation-enhanced diffusion and T_c is the critical ordering temperature in the absence of irradiation. For a given dose rate (j_p/e), Eqn. (17.33) predicts a limiting value of S, which is observed experimentally (Fig. 17.19).

Fig. 17.19. Order/disorder kinetics for TEM irradiation at 1 MV at a temperature where radiation-enhanced defect diffusion is evident, showing steady-state value of order parameter S where dS/dt = 0. Circles are data points, solid line is theory from Eqn. 17.33. (After BUTLER and SWANN, 1977.)

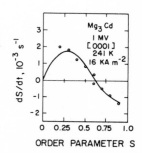

Strictly coordinated covalent solids are subject instead to spatial disordering of intact coordination polyhedra. The loss of long-range order relationships between [SiO$_4$] polyhedra in irradiated SiO$_2$ polymorphs has already been cited in section 17.3. Crystalline α-quartz undergoes a 14% decrease in density and transforms to an amorphous (metamict) state under electron irradiation. There is evidence from irradiation of other polymorphic forms that transformations en route to intermediate modifications occur, and also that a similar end state can be achieved by irradiation-induced compaction of still less dense conventional silica glass. Although these transformations are largely a consequence of particular connectivity and network topology, there are sufficient materials with these features (viz silicates) for example to prove illustrative of at least one broad class of irradiation-induced polymorphic phase transformations. Amorphization of other ceramic solids under intense ion irradiation (NAGUIB and KELLY, 1975) probably proceeds by other mechanisms.

In quartz, a critical density of point defects appears initially necessary to provide the requisite freedom for reorientation of coordination polyhedra. The crystalline-to-amorphous transformation proceeds in two stages. Local nuclei with some disordered structure form first and grow [Fig. 17.20(a)]. These initially strain the crystalline matrix (presumably due to density differences), but long-range order (and presumably the density difference) is eventually lost as well in the surrounding lattice by a more homogeneous transformation [Fig. 17.20(b)]. The identity of the nucleating agent is not known, but could be weak Si-OH:HO-Si hydrolyzed bonds associated with water impurity. The absence of nuclei adjacent to boundaries and surfaces suggests that point defect or exciton migration is required for their continued growth. On the other hand, rotation and rebonding of coordination polyhedra represents an irreversible sink process in the language of section 17.4, and the kinetics of the matrix transformation should resemble those for damage in organic solids (see Glaeser, Chapter 16). The efficiency of this process is such that crystalline quartz becomes completely amorphous at electron doses (~10 kC m^{-2}) considerably less than those required for atomic resolution. Similar difficulties beset TEM study of many silicate minerals. Analogous processes occurring during examination of silica glasses will not be so evident but have serious implications for high-resolution studies of glassy structures, since the structure of the radiation glass is not that of the thermal glass.

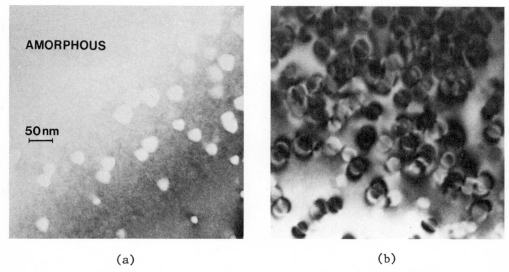

(a)   (b)

Fig. 17.20. Inclusions in α-quartz TEM irradiated at 125 kV (a) at an early stage as strained inclusions in a crystalline matrix and (b) as amorphous inclusions in a partially-transformed matrix.

Segregation and Precipitation

Several examples of local change of chemical composition due to radiolysis have been discussed in the context of point defect condensation in ionic crystals **(see above)** and were shown to lead to wholesale precipitation of anion or cation species as separate phases. The effects of electron irradiation on phase stability have been more extensively studied in metal alloy systems, where displacements are due solely to direct knock-on. Here, TEM irradiation above threshold has been shown to selectively enhance diffusion of one or more atomic species, leading to redistribution of alloying elements, enrichment or depletion of solute in the vicinity of extended defects (dislocations, surfaces, boundaries, interfaces and other point defect sinks), dissolution, growth or coarsening of precipitates and even homogeneous precipitation of phases not normally expected.

The origin of radiation-induced segregation is found in the coupling between point defect fluxes and solute fluxes (WIEDERSICH, OKAMOTO and LAM, 1977). Any preferential association of defects with solute atoms or preferential participation of solute atoms in defect diffusion (for example because the primary defect spectrum is biased towards or away from the solute) will provide the necessary coupling. The fate of solute atoms, and thus the diffusion coefficient for radiation-enhanced diffusion of solute, therefore depends on the principal fate of the Frenkel defects responsible. In the simplest case, two possible fates exist for Frenkel defects: they annihilate at fixed sinks (e.g. precipitates), or they annihilate by mutual recombination (LAM and ROTHMAN, 1976). At high temperatures, in the presence of a high concentration C_s of fixed sinks, annihilation at fixed sinks predominates and yields a radiation-enhanced diffusion coefficient

$$D_{rad} = (\sigma_d \, j_p/e) \, \Omega/2\pi \, r_s \, C_s \qquad (17.34)$$

where $(\sigma_d \, j_p/e)$ is the displacement rate, Ω an atomic volume and r_s the capture radius for vacancies or interstitials at fixed sinks. In this regime, D_{rad} is independent of temperature and scales linearly with displacement production rate. For lower irradiation temperatures more typical of TEM analysis, vacancies are likely to be much less mobile than interstitials and thus likely to act as the principal sink for interstitials, so that athermal recombination as outlined in section 17.4 predominates and yields

$$D_{rad} = 2 \{(\sigma_d \, j_p/e) \, \Omega \, D_v \, D_i/4\pi \, r_{iv} \, (D_i + D_v) \}^{\frac{1}{2}}$$
$$= \{(\sigma_d \, j_p/e) \, \Omega \, D_v/\pi \, r_{iv}\}^{\frac{1}{2}} \qquad (17.35)$$

where D_v and D_i are the vacancy and interstitial diffusion coefficients and $r_{iv} = (3\upsilon/4\pi)^{1/3}$ is the radius of the athermal recombination volume. In this regime, radiation-enhanced diffusion is temperature dependent through D_v and again scales with displacement rate.

RUSSELL (1977) has pointed out that equilibrium phase diagrams can no longer be expected to apply under irradiation conditions. In addition, the displacement products, including vacancies, act as chemical components to stabilize certain phases at the expense of others (MAYDET and RUSSELL, 1977). TEM irradiation above threshold has indeed been shown to induce precipitation from undersaturated solid solutions as well as from supersaturated solutions. Examples are precipitation of Ni_3Si from dilute Ni-Si alloys (Fig. 17.21), Si from Al-Si alloys and Mg_3Cd from Mg-Cd alloys. Extended defect aggregate structures (dislocation loops, etc.) often provide nuclei for precipitation, but more homogeneous precipitation can occur, possibly by spinodal decomposition. There is recent evidence that modulated structures may arise from irradiation-induced disordering of ordered metallic and nonmetallic phases, for example. Apart from obvious dose and temperature parameters, BARBU and MARTIN (1977) have shown that significant dose-rate effects exist for precipitation from undersaturated solid solutions during TEM irradiation (Fig. 17.22).

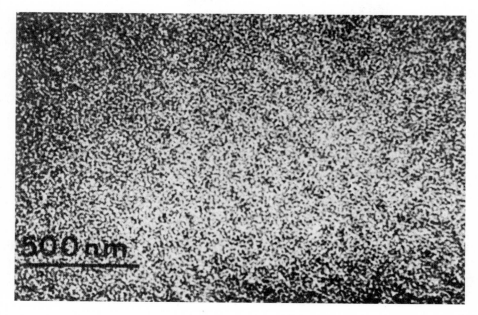

Fig. 17.21. Ni$_3$Si γ' precipitation in an undersaturated Ni-6 at.% Si alloy TEM-irradiated at 1 MV at 468 K to a dose of 0.5 dpa at a dose rate of 6 × 10^{-4} dpa/s. Dark-field image using 021 superlattice reflection. (Reproduced from BARBU and MARTIN, 1977.)

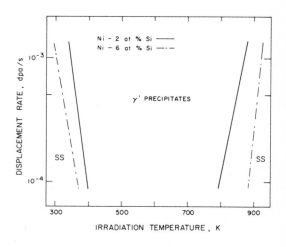

Fig. 17.22. Phase stability in dilute Ni-2 at.% Si (solid line) and Ni-6 at.% Si (broken line) for TEM irradiation at 1 MV as a function of dose rate and irradiation temperature. (After BARBU and MARTIN, (1977).

17.6   MINIMIZING THE EFFECTS OF RADIATION

Several avenues are available for minimizing the uninvited effects of electron irradiation in TEM when it is not possible to avoid them altogether. Some of these (e.g. reducing the electron dose) aim at reducing the primary effects, but are subject to easily calculable statistical limitations; others (e.g. lowering the temperature) attempt to mitigate secondary manifestations. Both approaches require knowledge about the specimen and details of the electron-specimen interaction, for which there is no substitute in the intelligent practice of TEM analysis. Some aspects of their implementation are discussed here.

Reducing the Electron Dose

For specimens subject solely to knock-on displacement damage above some electron displacement threshold energy U_{th}, it is a simple matter to reduce the electron energy U_p below U_{th}. However, this is not always practicable, for example if a given accelerating voltage has been chosen for optimum resolution or penetration, nor efficacious if other specimen deterioration mechanisms (radiolysis, specimen heating) are additionally troublesome. In these cases, one obvious remedy is to reduce the electron current, electron current density and total electron dose all to a minimum. The first reduces heating effects, and next dose-rate effects, the last overall radiation damage. Limitations exist in all three cases.

Electron emission and signal collection in electron-optical instruments are essentially incoherent in time and space and governed by Poisson statistics. For imaging with elastically-scattered electrons, the minimum electron fluence through the object plane for a given object resolution d at contrast level Γ is given by the Rose criterion (ROSE, 1948; HOBBS, 1975; GLASER, 1975; this volume)

$$(j_p\, t/e) > K^2/d^2\, \Gamma\, \eta\; ; \qquad\qquad (17.36)$$

Here K is a physiologically-determined signal-to-noise ratio, typically 5-30, required for recognition of detail and η is a factor <1 reflecting the efficiency of electron collection and the mode of imaging. The electron density at the object plane used for imaging will be considerably smaller than the total electron dose incident on the specimen because objective apertures in CTEM or collection apertures in STEM subtend only a portion of the solid angle through which electrons are scattered. Image resolution on an atomic scale (d = 0.1 nm) requires from Eqn. (17.36) a fluence $(j_p\, t/e) > 1$ MC m^{-2} (see Sinclair, Chapter 19); this dose amounts to $>10^4$ e/Å^2. A similar dose is required for single-atom imaging (see Isaacson, this volume). The requirements of microanalysis parallel those of imaging (ISAACSON, 1979). The number of counts C detected for some particular inelastic process (electron energy loss, x-ray emission) with cross-section σ from N atoms is

$$C = (j_p\, t/e)\, N\, \sigma\, \eta\; ;$$

thus the electron density required to detect a minumum of N atoms is

$$(j_p\ t/e) > \frac{C}{N\ \eta\ \sigma} \tag{17.37}$$

By Poisson statistics, $C = 100$ counts are needed for 10% counting statistics. For typical η and σ, Eqn. (17.37) yields doses $(j\ t/e) > 10\ e/\overset{\circ}{A}{}^2$ as well to (theoretically) detect single atoms. Note that these criteria are fundamental limitations and cannot be improved upon by any instrumental arrangement.

At the very least, detector efficiency can be made as high as possible. Photographic emulsions, for example, are widely used for parallel recording in CTEM, but surprisingly seldom used at maximum efficiency. Such emulsions are virtually noiseless detectors of high-energy electrons, and their response to fast electrons (unlike light) is approximately linear with dose (VALENTINE, 1965; COSSLETT, JONES and CAMPS, 1974) up to useful developed densities (D < 2). Maximum contrast for all emulsions occurs for D > 3 which is opaque, so the criterion for selecting the most efficient photographic emulsion becomes the simple one of ensuring that for a maximum tolerable electron dose to the specimen the emulsion is as fully exposed as practicable for printing.

Sensitivity data for emulsions useful in TEM applications have been collated by HOBBS (1975), and suitable emulsions are available for recording down to the resolution limit imposed by damage in the most sensitive specimen. X-ray and nuclear-track emulsions are enormously useful under conditions of marginal illumination, but are considerably under-utilized, perhaps out of ignorance. It is not widely understood, for example, that the "graininess" in these faster emulsions when developed is essentially due to the reduced electron statistics, not to the emulsion itself. Since the exposure process involves radiolysis of silver halides, emulsions exhibit reduced sensitivity to higher energy electrons as predicted from Fig. 17.1. Cosslett et al. have investigated this effect and have shown that increasing the thickness of the emulsion to recover sensitivity does not reduce definition proportionally, so that the advantage of using higher voltages to reduce ionization damage in radiolytically-sensitive specimens is not altogether lost in the recording process. Electronic detection systems for serial or parallel recording are generally better optimized but suffer from lower collection and quantum efficiencies. For redundantly periodic objects, image enhancement techniques are possible; these are discussed by GLAESER (1975).

Instrumental limitations (fixed probe size in STEM, finite focusing and manipulation times in both STEM and CTEM, as well as deleterious dose-rate effects) in many cases require that electron current density be reduced in addition to the recording electron dose in order to reduce the overall electron dose suffered by the specimen. It is important, for example, that the time of recording occupy the largest portion of the total irradiation time, with the fewest number of electrons wasted in microscope manipulation, area selection and focusing. With STEM, one eventually runs into detector noise at low electron density levels and into stage stability for long recording times; in CTEM, one encounters stage stability and the response of the phosphors used for observation and focusing.

Comparatively little attention has been paid to optimizing the response of phosphors to fast electrons and the response of the human eye to phosphor light emission. Phosphor response was investigated during the early period of instrumental development (HAINE, 1961) and again recently (IWANAGA et al., 1968; COSSLETT et al., 1974) when its limitations become apparent for high resolution, weak illumination, or high voltage microscopy. Phosphor screens exhibit an ultimate resolution of only ∿100 μm, but more importantly the visual acuity of the dark-adapted human eye degrades beyond this at low illumination levels, falling rapidly with decreasing phosphor brightness. The consequence is that very often the microscopist is working in a very inefficient portion of the phosphor-eye response curve. The current density <u>on the phosphor screen</u> is, however, a function of electron-optical magnification, and a simple anlaysis (HOBBS, 1975) using the Rose equation reveals that an <u>optimum</u> magnification M* exists for observation of object detail for a given contrast level and available current density at the object plane. Figure 17.23 indicates perceived object resolution as a function of magnification for 100 kV electrons at two (modest) electron current densities and two contrast levels. There is no sense in increasing magnification beyond M*, though this limitation is not generally appreciated.

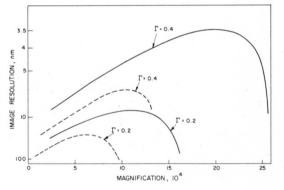

Fig. 17.23. Object resolution perceived on a phosphor screen as a function of electron-optical magnification M for two image electron current densities (referred back to the object plane) of 500 Am^{-2} (solid line) and 150 Am^{-2} (broken line) for two contrast levels,

Image intensifiers for CTEM cannot improve upon electron statistics; in fact, they introduce additional noise and are less efficient quantum detectors than the dark-adapted human eye (ENGLISH, GRIFFITHS and VENABLES, 1973; GLAESER, 1975). Their chief utility is in dynamic recording, but they are of some aid in focusing weak images because they collect information over the entire 2π sr of the image plane. The eye is at a disadvantage because it subtends only about 1 msr, but HAINE (1961) some time ago pointed out that, by using an additional optical magnifying lens of magnification m and of such aperture as to just fill the aperture of the eye, it is possible to collect m times more   light from the phosphor screen. This contributes a roughly m-fold increase in visual acuity.

An effective implementation of this little-heeded suggestion is to effect part of the final image magnification <u>optically</u>; the size of the illuminated area on the phosphor screen is reduced by reducing projector lens excitation so as just to fill the field of view of a binocular telescope (m < 10). There is, of course, no point in increasing m beyond the point where either the ultimate resolution of the phosphor screen or limiting electron statistics over the small persistence time of the phosphor begin to confuse image definition.

Reducing the Temperature

It was indicated in section 17.4 that reducing specimen temperature during examination at least circumvents linear damage kinetics and possibly provides further advantage by utilizing the probability of increased athermal recombination. Use of cryogenic TEM provides close to a hundred-fold increase in permissible dose in halides and should provide similar advantage for any system in which Frenkel pair production by knock-on or radiolysis is the principle damage mode. Where radiolytic bond-breaking followed by atom rearrangement is the dominant mode, low temperature may still provide useful advantage, as it does for some organic molecular solids (see JONES <u>et al</u>., 1977).

There are three principle concerns in the practice of low-temperature microscopy: Temperature stability, thermomechanical rigidity, and contamination (HOBBS, 1973, 1975). Additional specimen concerns (differential cooling stresses, mounting techniques) have been discussed by HOBBS (1972, 1975). Temperature stability is important because even small changes in cold-stage temperature may lead to unacceptable thermal drift. Thermal expansion coefficients for most materials (including the specimen) are of order 10^{-6} K^{-1}, which for 1 nm image stability implies the need for temperature stability to better than 0.1 K throughout the stage. Gas-cooling is more desirable than use of cryogenic liquids owing to more uniform heat transfer coefficients. Unstable two-phase flow and gaseous transfer films are characteristics of cooling surfaces at temperatures above the boiling point of the liquid coolant. Thermal eddy currents are, in fact, the largest problem and should be minimized by stage design.

Fused quartz is the single most useful stage support material because of its negligible thermal expansion coefficient and small thermal conductivity (VENABLES, BALL and THOMAS, 1968). Differential contraction, due to use of different stage materials or to inevitable temperature gradients, will always be present, so critical parts (such as stage supports) should be spring-loaded to ensure mechanical stability. On the other hand, the parts of the stage exclusive of the stage supports need be no more mechanically rigid than in a high-resolution room-temperature stage, and indeed conventional stages suitably supported and cooled make perfectly acceptable cryogenic stages.

Contamination is clearly a problem in cryogenic microscopy, since sticking coefficients are close to unity even for species not cracked by the electron beam. Water is the chief culprit, and below 30 K, nitrogen as well. It is desirable to shield the specimen from radiation from nearby warm surfaces anyway, and complete enclosure in a cryogenic box (with apertures) can reduce contamination to less than the levels encountered in more conventional analytical microscopy. Contamination layers can, of course, be at least partially removed by periodic warming up. For insulator specimens, evaporated carbon coatings to reduce charge acquisition are ineffectual at cryogenic temperatures, and aluminum is better used instead.

17.7 CONCLUSIONS

A few inorganic solids of interest to solid-state science suffer efficient radiolysis under electron irradiation in TEM. All solids suffer knock-on displacement damage for electron energies above threshold for atom displacement, and irradiation above threshold is likely to lead to defect generation rates unacceptably large under the highest resolution conditions. Radiolytic damage will always prove more serious, provided a radiolysis mechanism operates, while knock-on displacement damage will be absent or negligible for 100 kV electron irradiation in all but a few solids. However, the increased use of microscopes with accelerating voltages in the range 0.2-1 MV for highresolution and analytical studies will bring inevitable problems of specimen damage to the interpretation of TEM information. It is important in such experiments to understand fundamentally how and why the specimen could have altered during irradiation. In some favorable cases, the material under examination may prove dynamically stable, so that the information obtained is averaged over a dynamic but invariant structure. In many other cases, the structure will have changed appreciably due to defect aggregation, radiation-enhanced diffusion or loss of long-range order stability. In such cases, a few avenues are open to reduce the rate of damage production, to reduce the grosser effects of unavoidable damage and to maximize the information obtained from each electron.

REFERENCES

Adda, Y., Beyeler, M., and Brébec, G., 1975, Thin Solid Films 25, 107-56.

Barbu, A. and Martin, G., 1977, Scripta Met. 11, 771-5.

Barnard, R. S., 1977, Ph.D. Thesis, Case Western Reserve University.

Bethe, H. A., 1933, in Handbuch der Physik, ed. H. Geiger and K. Scheel (Springer Verlag, Berlin), Vol. 24, p. 273.

Butler, E. P. and Swann, P. R., 1977, in High Voltage Electron Microscopy 1977, J. Electron Microsc. Suppl. 26, 551-4.

Catlow, C. R. A., Diller, K. M., Hobbs, L. W., and Norgett, M. J., 1979, "Irradiation-Induced Defects in Alkali Halide Crystals," Phil. Mag. (in press).

Chadderton, L. T., Johnson, E., and Wohlenberg, T., 1975, in Developments in Electron Microscopy and Analysis, ed. J. A. Venables (Academic Press, London), pp. 299-302.

Corbett, J. W., 1966, Electron Irradiation Damage in Semiconductors and Metals (Academic Press, New York). Corbett, J. W. and Watkins, G. D., 1971, ed. Radiation Effects in Semiconductors (Gordon and Breach, New York).

Corbett, J. W. and Bourgoin, J. C., 1975, in Point Defects in Solids, ed. J. H. Crawford, Jr., and L. M. Slifkin (Plenum Press, New York), Vol. 2, pp. 1-161.

Cosslett, V. E., Jones, G. L., and Camps, R. A., 1974, in High Voltage Electron Microscopy, ed. P. R. Swann, C. J. Humphreys, and M. J. Goringe (Academic Press, London), pp. 147-154.

Dearnaley, G., 1975, Appl. Phys. Lett. 26, 499-501.

Drosd, R., Kosel, T., and Washburn, J., 1978, J. Nucl. Mat. 69 & 70, 804-6.

English, C. A., Griffiths, B. W., and Venables, J. A., 1973, Acta Electronica 16, 43.

Fisher, S. B., 1970, Radiation Effects 5, 239.

Friebele, E. J., Griscom, D. L., and Stapelbroek, M., 1979, "Fundamental Defect Centers in Glass: The Peroxy Radical in Irradiated High-Purity Silica," submitted to Phys. Rev. Letters.

Gale, B. and Hale, K. F., 1961, Brit. J. Appl. Phys. 12, 115.

Glaeser, R. M., 1974, in High Voltage Electron Microscopy, ed. P. R. Swann, C. J. Humphreys, and M. J. Goringe (Academic Press, London), pp. 370-8.

Glaeser, R. M., 1975, in Physical Aspects of Electron Microscopy and Microbeam Analysis, ed. B. M. Siegel and D. R. Beaman (Wiley, New York), pp. 205-29.

Haine, M., 1961, The Electron Microscope (Spon, London).

Hobbs, L. W., 1972, D. Phil. Thesis, Oxford University.

Hobbs, L. W., 1973, J. Physique 34, C9, 227-41.

Hobbs, L. W., Hughes, A. E., and Pooley, D., 1973, Proc. Roy. Soc. A332, 167-85.

Hobbs, L. W., 1975, in Surface and Defect Properties of Solids, ed. M.
    W. Roberts and J. M. Thomas (The Chemical Society, London), Vol. 4,
    pp. 152-250.

Hobbs, L. W. and Hughes, A. E., 1975, "Radiation Damage in Diatomic
    Solids at High Doses," UKAEA Research Report AERE R-8092.

Hobbs, L. W., 1976, J. Physique 37, C7, 1-26.

Hobbs, L. W., Howitt, D. G., and Mitchell, T. E., 1978, in Electron
    Diffraction 1927-1977, ed. P. J. Dobson, J. B. Pendry, and C. J.
    Humphreys (Institute of Physics, London), Conf. Ser. No. 41, pp.
    402-10.

Hobbs, L. W., 1979, Ultramicroscopy 3, 381-6.

Hughes, A. E. and Pooley, D., 1971, J. Phys C 4, 1963-76.

Hughes, A. E. and Pooley, D., 1975, Real Solids and Radiation (Wykeham,
    London).

Inokuti, M., 1971, Rev. Mod. Phys. 43, 297-347.

Inokuti, M., Itikawa, Y., and Turner, J. E., 1978, Rev. Mod. Phys. 50,
    23.

Inokuti, M., 1979, Ultramicroscopy 3, 423-7. Isaacson, M., 1977, in
    Principles and Techniques of Electron Microscopy, Biological
    Applications, ed. M. A. Hayat (Van Nostrand Reinhold, New York),
    Vol. 7, pp. 1-78.

Isaacson, M., 1979, Proc. Specialist Workshop on Analytical Electron
    Microscopy, Cornell University, 25-28 July 1978, ed. P. L. Fejes,
    pp. 73-87.

Itoh, N., 1976, J. Physique 37, C7, 27-37.

Iwanaga, M., Ueyanagi, H., Hosoi, K., Iwasa, N., Oba, K., and
    Shiratsuki, K., 1968, J. Electron Microsc. Chiba Cy 17, 203-14.

Kabler, M. N. and Williams, R. T., 1978, Phys. Rev. B 18, 1948-60.

Kawamata, Y. and Hibi, T., 1965, J. Phys. Soc. Japan 20, 242-50.

Kinchin, G. H. and Pease, R. S., 1955, Rept. Prog. Phys. 18, 1.

Kinoshita, C., Mukai, T., and Kitajima, S., 1977, in High Voltage
    Electron Microscopy 1977, J. Electron Microsc. Suppl. 26, 551-4.

Lam, N. Q. and Rothman, S. J., 1976, in Radiation Damage in Metals, ed.
    N. L. Peterson and S. M. Harkness (ASM, Metals Park, Ohio), pp.
    125-56.

Lehmann, Chr., 1977, Interaction of Radiation with Solids (North-Holland, Amsterdam).

Lenz, F., 1954, Z. Naturforsch. 9a, 185.

Makin, M. J., 1971, in Electron Microscopy in Material Science, ed. U. Valdre (Academic Press, London), pp. 388-461.

Maydet, S. I. and Russell, K. C., 1977, J. Nucl. Mat. 64, 101-14.

McKinley, W. A. and Feshbach, H., 1948, Phys. Rev. 74, 1759-63.

Mott, N. F. and Massey, H. S. W., 1965, The Theory of Atomic Collisions (Clarendon Press, Oxford).

Naguib, H. M. and Kelly, R., 1975, Radiation Effects 25, 1-12.

Norris, D. I. R., 1975, in Electron Microscopy in Materials Science, ed. E. Ruedl and U. Valdre (Commission of the European Communities, Luxembourg), EUR 5515e, Vol. III, pp. 1099-1144.

Oen, O. S., 1973, "Cross Sections for Atomic Displacements in Solids by Fast Electrons," Oak Ridge National Laboratory Report ORNL-4897.

Parkinson, G. M., Goringe, M. J., Jones, W., Rees, W., Thomas, J. M., and Williams, J. O., 1976, in Developments in Electron Microscopy and Analysis, ed. J. A. Venables (Academic Press, London), pp. 315-18.

Peterson, N. L. and Harkness, S. M., 1976, ed. Radiation Damage in Metals (ASM, Metals Park, Ohio).

Pells, G. P. and Phillips, D. C., 1978, "The Temperature Dependence of the Displacement Threshold Energy of α-Al$_2$O$_3$," UKAEA Research Report AERE-R9138.

Pooley, D., 1966, Proc. Phys. Soc., 89, 723-33.

Pooley, D., 1975, in Radiation Damage Processes in Materials, ed. C. H. S. Dupuy (Noordhof, Leyden), pp. 309-23.

Reimer, L., 1975, in Physical Aspects of Electron Microscopy and Microbeam Analysis, ed. B. M. Siegel and D. R. Beaman (Wiley, New York), pp. 231-45.

Rose, A., 1948, Adv. Electron. 1, 131.

Rothwarf, A., 1973, J. Appl. Phys. 44, 752-6.

Russell, K. C., 1977, in Radiation Effects in Breeder Reactor Structural Materials, ed. M. L. Bleiberg and J. W. Bennett (AIME, New York), pp. 821-39.

Saidoh, M. and Townsend, P. D., 1975, Radiation Effects 27, 1-12.

Seitz, F. and Koehler, J. S., 1956, Solid State Physics 2, 305.

Slifkin, L. M., 1969, in Solid State Dosimetry, ed. S. Amelinckx (Gordon and Breach, New York), pp. 241-60.

Slifkin, L. M., 1975, in Radiation Damage Processes in Materials, ed. C. H. S. Dupuy (Noordhof, Leyden), pp. 405-34.

Sonder, E. and Sibley, W. A., 1972, in Point Defects in Solids, ed. J. H. Crawford, Jr., and L. M. Slifkin (Plenum Press, New York), Vol. 1, pp. 201-90.

Spencer, L. V., 1959, "Energy Dissipation by Fast Electrons," National Bureau of Standards, Washington, D.C., Monograph 1.

Thomas, L. E., Humphreys, C. J., Duff, W. R., and Grubb, D. T., 1970, Radiation Effects 3, 89.

Valentine, R. C., 1965, Adv. Opt. Electron Microsc. 1, 180-203.

Venables, J. A., Ball, D. J., and Thomas, G. J., 1968, J. Phys. E 1, 121-6.

Yada, K. and Hibi, T., 1969, Bull. Res. Inst. Tohoku Univ. 17, 87-100.

Yip, K. L. and Fowler, W. B., 1975, Phys. Rev. B 11, 2327-38.

Wiedersich, H., Okamoto, P. R., and Lam, N. Q., 1977, in Radiation Effects in Breeder Reactor Structural Materials, ed. M. L. Bleiberg and J. W. Bennett (AIME, New York), pp. 801-19.

CHAPTER 18

BARRIERS TO AEM: CONTAMINATION AND ETCHING

J.J. HREN

DEPARTMENT OF MATERIALS SCIENCE AND ENGINEERING
UNIVERSITY OF FLORIDA
GAINESVILLE, FLORIDA 32611

18.1 INTRODUCTION

Contamination and etching are terms used quite negatively by the analytical electron microscopist. The mental "images" evoked are of a loss of resolution, a decrease in signal/noise ratio, destruction of the microstructure, etc. Yet more propitious results may be achieved by using our empirical understanding of the phenomena to manufacture microcircuits, measure foil thickness, or to preferentially etch microstructures. For the most part, however, the effects of rapid contamination buildup or the local loss in specimen mass are viewed as worse than annoying. They are, in fact, significant barriers to continued advances in true microchemical and high resolution microstructural analyses. The present chapter attempts to summarize our understanding of the physical processes contributing to each phenomenon, to illustrate their effects on AEM, to describe some working cures (and even a few applications), and to project some avenues for further improvement.

The description of the present state of our understanding will follow a largely chronological course. Although this is the general approach to a review, we will not try to be comprehensive but rather place our emphasis here on circumscribing our present bounds of understanding. That is, some of the intermediate stages of understanding will be minimized and those results which are still debatable will be left inconclusive. On the otherhand, working solutions which have proven successful will be described and still others which appear useful, but remain largely untested, will be suggested.

18.2 SOME DEFINITIONS

There are some uncertainties in the precise meaning of the terms contamination and etching and other terms such as mass loss, radiation damage, bulk and surface impurities, etc., all commonly used in the literature of electron microscopy. Some confusion may be avoided by specifying a working definition. Simply put, by contamination, we mean the unintentional act of adding mass to the surface(s) of a thin specimen during observation or analysis by an electron beam. This definition is well illustrated to the modern analytical microscopist by Fig. 18.1. Note that the mounds of contaminant were produced by a stationary electron beam of approximately 10 nm diameter.

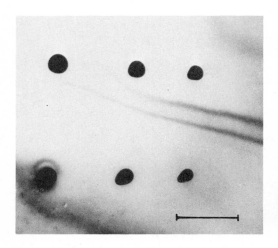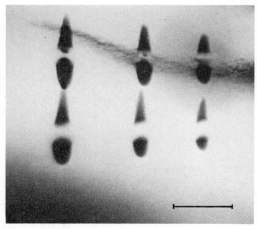

Fig. 18.1. A series of contamination deposits on an Al alloy specimen produced by a stationary electron probe of ∿ 10nm diameter at 80 keV (upper) and 100 keV (lower). Time increases from right to left: 1/2, 1 and 2 minutes. (a) Imaged in the same orientation as deposited. (b) Tilted 45° about horizontal axis before imaging.

In our definition, etching is just the reverse process of contamination, qualified only to draw a reasonable line between etching and radiation damage. We, therefore, restrict our working meaning to that given by Glaeser (Chapter 16): the removal of material by "....the synergistic action of electron irradiation and" "....certain residual gases, such as oxygen or water vapor...." Of course, to the practicing microscopist, local mass loss is local mass loss, whatever the mechanism, and a careful reading of both chapters 16 and 17 is thus mandatory. One result of this somewhat arbitrary division is that the present discussion of etching will be largely restricted to organic, carbonaceous, and polymeric mterials. However, contamination effects, as here defined, extend to all known materials.

18.3 EARLY OBSERVATIONS OF CONTAMINATION

Historically, all of the early observations of contamination and etching were observed with large electron beams (>1 μm diameter) varying in energy from a few hundred volts to one hundred kilovolts. As we shall see, both effects become significantly more severe as the beam size decreases (e.g. Fig. 18.1). Etching was, in effect, "discovered" by the attempts to control the buildup of contaminating layers.

The earliest reference to the subject of contamination deposits induced by electron beams is probably untraceable. However, STEWART (1934) is a suitable starting point. He points out: "In an evacuated tube in which the slightest traces of organic vapors may occur,.... insulating layers are formed on surfaces subject to electron.... bombardment. These layers may be attributed to carbon compounds and their formation is related to the polymerization of organic vapors...." Stewart's analysis, as far as it went, remains unimpeachable today and has been re-deduced many times since.

Stewart was not an electron microscopist and his measurements were made at very low electron energies (~200 eV). Contamination effects were soon encountered and studied by a number of practicing electron microscopists. WATSON (1947) was the first to report a significant change in the mean particle size and shape of carbon black while under examination in the electron microscope. He concluded that the contaminants were condensed organic vapors polymerized by the electron beam. Watson's observations were incomplete and even somewhat contradictory. He recognized the need for further studies and called for an exchange of experience. About the same time, KINDER (1947) noted that a transparent skin of contamination was left around crystals after prolonged observation. The crystals themselves had often been completely removed under the influence of the beam, presumably by evaporation. Specimen degradation in a number of substances, mainly ionic crystals, was soon observed by BURTON et al. (1947), who reported both increased electron transparency and debris tracks as a consequence of electron irradiation. Although they had no conclusive explanation of the phenomena, BURTON et al. thought the debris to be the same as that observed by STEWART (1934) and WATSON (1947). Sublimation and redeposition due to beam heating was ruled out as an explanation. COSSLETT (1947) responded to Watson's call for further observation of contamination. His observation on zinc oxide and magnesium oxide crystals complicated the picture still further. Cosslett's observation tended to minimize the influence of organic vapors and emphasize the importance of ejected particles from the grid or specimen itself. He suggested that the contaminants observed by Watson came from the carbon black.

18.4 THE NATURE OF THE CONTAMINANT

HILLIER (1948) and KÖNIG (1948) also responded to the now obvious need for more controlled studies. Some notable observations of the nature of the contaminant itself were made by Hillier. He observed that: a) the scattering power was slightly less than carbon and that the deposit was amorphous as deduced from diffraction; b) only carbon was found when analyzed with his electron microanalyzer (a very early version of an electron energy loss spectrometer); c) surface migration of the contaminant occurred; d) shielding the specimen (at room temperature) reduced the rate of deposition only slightly. Hillier concluded that the contaminant product was formed by the polymerization of hydrocarbon molecules as originally proposed by Stewart and Watson and that material in the gas phase and from the instrument surfaces were both important sources. The responsible vapors were not just from the diffusion pumps. Hillier proposed several solutions, one based upon a low

energy electron spray of the chamber (DAVIDSON and HILLIER, 1947), a
method since adopted by others. KÖNIG (1948) confirmed the hypothesis
that the contaminants were primarily hydrocarbons and were vacuum system
dependent. He pointed out that improvements in the general vacuum of an
electron microscope could and should be made and that these were well
within the state of the art at that time. Subsequently, KÖNIG (1951)
showed by electron diffraction that organic material such as collodian
could indeed be converted into amorphous carbon by intensive electron
bombardment. KÖNIG and HELWIG (1951) also showed that a thin coating of
polymerized hydrocarbon was formed on top of underlying thin films in
the electron microscope. Two other researchers were investigating the
contamination phenomenon at about the same time and soon reported further results.

ELLIS (1951) suggested that the principle source of organic molecules was condensate on the specimen supports which then migrated to the
irradiated area of the specimen. Extensive studies by ENNOS (1953,
1954) concluded that surface migration was not an important supply mechanism, but that organic molecules were continuously absorbed onto the
irradiated surfaces from the vapor phase. The relative importance of
various vacuum based contaminant sources in descending order was found
by Ennos to be: diffusion pump oil, vacuum grease, and rubber gasket
materials, followed by other vacuum system components. Uncleaned metal
surfaces were also recognized to be substantial sources of the contaminating vapor. In short, the application of good high vacuum practices
was strongly urged by Ennos. Two expedients were demonstrated that considerably reduced contamination: specimen heating (to ~200°C, if possible) and cryoshielding (i.e. a cold finger). Rates of buildup were
monitored by measuring the step height of the contaminating layer using
optical interference methods.

18.5 RELATIONSHIP BETWEEN CONTAMINATION AND ETCHING

Following the publications of Ennos, cryoshielding was commonly employed to decrease the rate of contamination buildup. However, in
cooling the area around the specimen, the specimen itself was also often
cooled.* For example, in using a Siemens Elmiskop I, LEISEGANG (1954)
noticed that below about -80°C (for both the cryoshield and the specimen), not only was carbon contamination stopped, but that the structure
of the organic specimens under observation was destroyed. At about the
same time, a number of other investigators noted that the introduction
of various gases (e.g. air, hydrogen, nitrogen, and even inert gases)
resulted in the reduction of carbonaceous deposits. For example,
CASTAING and DESCAMPS (1954) discovered that the introduction of air
would prevent the buildup of a contaminating layer during specimen analysis in their early electron microprobe. That is, contamination buildup could be made to slow to a stop and even occasionally to reverse. In

*There are other reasons why specimen cooling may be useful to prevent
radiation damage [Glaeser (Chapter 16) and Hobbs (Chapter 17)], but we
concentrate here only on the effects of specimen contamination and
etching as defined in the introduction.

fact, similar observations were originally reported by RUSKA (1942) in an electron microscope, but never investigated further.

The combined effects of low specimen and cryoshield temperatures and the introduction of certain gases were studied in detail by HEIDE (1962, 1963) in an attempt to find the optimum conditions to prevent specimen contamination and damage. Although his work was concerned mainly with carbon removal (C-Abbau), it was also the forerunner of subsequent studies on specimen etching. The link between specimen contamination and specimen etching was now forged. (Carbon removal, after all, is not always desirable if it is an important constituent of the specimen under study.)

Heide's view of the contamination process was in the following mechanistic sequence: condensation of hydrocarbon molecules onto the surface - ionization - polymerization and reduction to an immobile carbon deposit. Since in all previous attempts to prevent contamination both the specimen and its surroundings were cooled (with or without the introduction of gases into the specimen chamber), these almost invariably led to specimen damage by carbon removal, particularly if the specimens were biological in nature. Naturally, one preferred to observe a specimen with neither contamination buildup nor carbon removal (unless only the contaminant was removed). Heide thus built an independently regulated cryoshield in an electron microscope. That is, he separately controlled the temperature of the shield surrounding the specimen and the specimen itself. The results were dramatic and are best illustrated by Figs. 18.2 (a) and (b). Heida found that keeping the specimen at room temperature while cooling the surrounding cryoshield below about -130°C seemed to achieve the simultaneous goals of contamination prevention without carbon removal. Heide's interpretation of these data, involved a mechanism that required the presence of vapors other than those of hydrocarbons. He assumed that under normal operating conditions (in the Elmiskop I) the partial pressures of various gases were as follows: $H_2O-2x10^5$ Torr, $\Sigma CH-5x10^6$ Torr, Co, N_2, CO_2, and H_2 - $1x10^6$ Torr each, O_2-1x10^7 Torr. He then concluded that the temperature was lowered the partial pressure of the hydrocarbons (ΣCH) also lowered such that at temperatures below -60°C their partial pressures were negligible. No further specimen contamination due to the deposition of hydrocarbons thus took place. The partial pressures of the other gases, however, were very little affected until the temperature dropped to between -100°C + -130°C. At these temperatures the partial pressure of water vapor dropped dramatically from $\sim 1x10^5$ Torr to $\sim 1x10^8$ Torr. The process of carbon removal was then postulated to take place through reactions of the type:

$$H_2O \rightarrow H_2O^+ + e \qquad (18.1)$$

and

$$H_2O^+ + C \rightarrow H + H^+ + CO \qquad (18.2)$$

under the action of the electron beam. If the specimen was cooled along with the cryoshield, water molecules would absorb onto its surface thereby

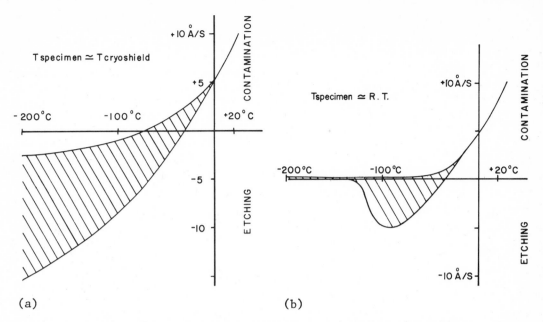

Fig. 18.2. Rate of contamination buildup or carbon removal as a function of the temperature of the specimen and cryoshield. Beam diameter ∿
2μm, microscope vacuum ∿ 5x10^5 Torr. (a) With specimen and cryoshield
cooled together, (b) Cryoshield cooled and specimen at room temperature.
Shaded areas indicate data scatter (After HEIDE, 1962).

providing a plentiful supply of reactants to remove carbon. If the specimen was not cooled, but the cryoshield was, the water vapor would preferentially absorb onto the cryoshield and protect the specimen.

It seemed that with the conclusion of Heide's work the way was open
to high resolution electron microscopy and to a considerable extent this
was true. Unfortunately as progress to better and better resolution continued further difficulties appeared. In particular, the use of smaller
and smaller electron probes for both imaging and local elemental analysis aggravated the processes of contamination and etching in a surprisingly intense manner.

18.6 SURFACE DIFFUSION AND BEAM SIZE EFFECTS

Vacuum technology has improved substantially in parallel with improvements in electron microscopes. For example, microscope column
pressures of less than 10^7 Torr in a modified electron microscope was
achieved by HARTMAN and HARTMAN (1965) using ion pumps. Differential
pumping of the specimen chamber was demonstrated by several investigators (GRIGSON et al., 1966; HART et al., 1966; VALDRÉ, 1966) to reduce
pressure to the region of 10^{-8} Torr. As a result of these improvements,
some more subtle effects of contamination and etching were detected and
the earlier explanations were found to be wanting. HART et al. (1970),
for example, concluded that surface diffusion of the contaminant to the
irradiated area was required to account for the observed growth rate in
contrast to the earlier conclusions. Differential pumping, mass spectro-

metric analysis, and some regulation of the partial pressure of the background gases was also a feature of this work. This more extensive control of the environment led to attempts to measure contamination rates quantitatively.

Confirmation of the importance of surface diffusion came from a somewhat different direction. It was found that the controlled local growth of contamination could be used to write patterns of incredibly small dimensions on a surface (BROERS, 1965; OATLEY et al., 1965; THORNHILL and MACINTOSH, 1965). Literally, the Bible could be put on the head of a pin! The carbonaceous deposits formed by the electron beam resisted subsequent removal by chemical etching or ion bombardment in the same way that the photoresists used in solid state microfabrication did and the size scale of the patterns could thus be made in the range of several hundred Angstroms rather than a few microns as limited by the wavelength of light. The control of such "electron beam writings" required at least an empirical knowledge of contamination rates, since these rates determined the total writing times for the circuits. A search for such quantitative control led MÜLLER (1971a, b) to an extensive study of the dependence of the contamination rate on the electron beam diameter and current density. He used a thin carbon film as the "recording" medium. Halftone pictures 5 μm x 5 μm could be produced in this way by varying the electron beam scan rate! Müller's derived contamination growth rates varied with the beam diameter, but not with the current density. His model also required surface migration of the precursive material from the immediate surroundings of the illuminated region to account for the observed rates. Experimental results, fitting his theory, were reported for probe diameters of 15 nm to 3 μm.

18.7 RECENT STUDIES OF CONTAMINATION AND ETCHING

Substantial differences in the kinetics of contamination associated with beam size effects were encountered in electron microscope applications at about the same time as the early beam writing studies. For example, the possibilities of STEM imaging using electron probes of nearly atomic dimensions were just beginning (e.g. CREWE and WALL, 1970). Convergent beam diffraction, using stationary probes of 30 nm, was also more and more often employed (e.g. GOODMAN and LEHMPFUHL, 1965; MILLS and MOODIE, 1968; RIECKE, 1969). Extreme measures had to be taken to avoid contamination (e.g. through the use of ultrahigh vacuums, by specimen heating, etc.). Even so, many of the cures found were impractical for the vast majority of practicing microscopists. Still further investigations of the detailed mechanisms of contamination and etching were obviously required.

The work of ISAACSON et al. (1974) was particularly informative, since they could combine very high image resolution with an analysis of the electron energy loss signal (ELS). Their specimens were placed on very thin (∼2nm thick) carbon support films and investigated under ultrahigh vacuum (UHV) conditions. Many contaminant molecules still remained absorbed to the films from prior handling unless extraordinary means to clean the specimens were not taken beforehand. Isaacson et al. could

observe the effects of absorbed contaminant molecules by changes in the
electron scattering power. It was clear to them that the adsorbates
were responsible for both etching and contamination in the beam.
Etching was observed to proceed rapidly at first and was highly specimen
dependent, being especially rapid when organic salts were placed on the
film. The measured characteristics of the contaminants were indistinguishable from those of the film and the rate of buildup depended highly
on the type of specimen and the gases to which the film was exposed.
All of the data of Isaacson et al. was consistent with the hypothesis
that surface diffusion was the principle source of the molecules which
formed their contaminating layers. They found that contamination could
be virtually eliminated by gently heating the specimen to ∿50°C in the
ultrahigh vacuum before observation.

Further semiquantitative studies of contamination were conducted by
EGERTON and ROSSOUW (1976) using a conventional TEM equipped with an
electron energy loss analyzer. The relative thickness, t, of the contaminating film could be deduced directly from the heights of the carbon
peaks for zero energy loss, h_o(elastically scattered), and the plasmon
loss maximum, h_i. The measurement could be made approximately quantitative by determining a proportionality constant, C, calibrated with a carbon specimen of known thickness. The relationship then became simply:

$$t = C \; \frac{h_i}{h_o} \tag{18.3}$$

Egerton and Rossouw proceeded to monitor specimen thickness as a function of irradiation time for a series of specimen temperatures. Their
results are summarized in Fig. 18.3. The thinning effect (etching) was
found to be confined to the area illuminated by the electron beam. The
authors evaluated a number of possible mechanisms: 1) ionization of the
molecules in the vacuum system in the gas phase which could then react
with the specimen upon being adsorbed, 2) ionization of gas absorbed on
the specimen surface (e.g. H_2O) which could then react with carbonaceous
material, 3) ionization of the atoms of the specimen surface itself or
any carbonaceous deposits which could subsequently react with the absorbed gases, 4) direct sputtering of the surface atoms by the highly
energetic incident electrons.

After careful consideration, none of these mechanisms could be
ruled out completely. Mechanisms 1) and 2) were those proposed originally by HEIDE (1963), with the latter likely to predominate. However,
mechanism 3) could also reasonably account for the observed thinning
rates (∿5 Å/min) if one assumed a surface coverage of about 0.5% absorbed gas (say H_2O) that provided reactants for the ionized carbon
atoms. In addition, Egerton and Rossouw noted that areas that had been
thinned by the electron beam appeared rougher than their surroundings.
They concluded that impurity metal atoms might be acting as localized
catalysts for the oxidation of carbon in agreement (for example) with
the observations of THOMAS and WALKER (1965) on the etching of graphite.

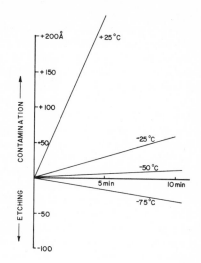

Figure 18.3. Increase or removal of mass as measured by relative peaks heights of elastically scattered and plasmon loss maximum for carbon. Film thickness ~20 nm. (After EGERTON and ROSSOUW, 1976).

A series of papers by HARTMAN et al. (1968), CALBRICK and HARTMAN (1969), HARTMAN et al (1969), and HARTMAN and HARTMAN (1971) was specifically directed towards defining the mechanisms of residual gas reactions during irradiation under controlled conditions in the electron microscope. These authors concluded that inelastic scattering events can result in the transfer of sufficient energy to the surface atoms of the specimen that the activated molecule which results can decay by a large number of possible mechanisms. In many cases, the most probable decay mechanism is simply for the activated molecule to return to its original state with no permanent damage. However, quite often damaging etching reactions (for organic specimens) such as the following could occur:

$$C_n H_{2n} + 2 H_2O \overset{e}{\leftrightarrows} C_{n-1} H_{2n} + CO + 2H_2 \qquad (18.4)$$

Hartman et al. showed that such reactions could indeed have a negative temperature dependence in agreement with the findings of HEIDE (1962).

In more recent work, etching reactions in biological molecules were studied using the loss of ^{14}C due to electron irradiation on labeled T4 bacteriophages and E. coli bacteria (DUBOCHET, 1975). He used a variety of commerical electron microscopes with both fixed and scanned beams. The irradiation doses were those which approximated practical observational conditions. Radiographic methods were employed to determine the ^{14}C distribution using high resolution photographic recording. Dubochet concluded that: 1) the sensitivity to carbon loss decreased with exposure, 2) surface migration of molecular fragments and absorbed molecules is involved in the mechanism of beam damage, 3) there is no perceptible carbon loss when irradiation takes place at liquid helium specimen temperatures.

In an extensive recent series of studies, FOURIE (1975, 1976a, 1976b, 1978a, 1978b, 1979) has attempted to develop a quantitative

theory capable of explaining the drastic difference in the kinetics of
contamination between large and small electron beam diameters and to
develop a satisfactory cure. Fourie states flatly that surface diffusion alone is the source of contaminant provided that the specimen is
suitably cryoshielded or that a UHV microscope is employed. He also
argues that the influence of an induced electric field on the specimen
surface (arising from the ejected secondary electrons) dramatically en-
hances surface diffusion as the beam size decreases. The general form
of the contamination buildup shifts with decreasing beam size as indicated in Fig. 18.4. Surface charging effects become dominant, according
to Fourie, as the beam size decreases to well below the lateral range of
the "high energy secondaries". He shows, for example, that charging effects are most dramatic with dielectric samples which are irradiated dis-
tant from conducting pathways (such as grid bars, coatings, substrates,
etc.). A further critical observation by Fourie is that <u>all</u> pretreatments of the specimen before it is in position for analysis are
critical. Pretreatment, includes specimen preparation (including
handling), cleaning or rinsing before insertion, prepumping, exposure to
hydrocarbon vapors within the microscope prior to cryoshielding or
viewing, etc. Only by carefully controlling specimen pretreatments
(e.g. prepump times) could Fourie obtain reproducible kinetic results of
the growth of contaminant deposits. Apparently, many of the inconsis-
tencies in earlier attempts to develop a quantitative theory can be
traced to a lack of control of the specimen beforehand.

Fig. 18.4. Schematic illustration
of the change in appearance of the
contamination deposit with different
beam sizes. Note the change in
scale in going from (a) → (c).

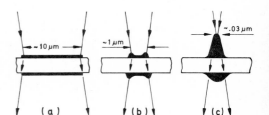

18.8 SUMMARY OF PHENOMENOLOGICAL OBSERVATIONS

This chronological review of the major literature on contamination
and etching to date has given us a reasonably complete description of
the qualitative conditions leading to specimen contamination and a somewhat less complete idea of the conditions leading to specimen etching.
Although there have been numerous attempts to provide quantitative contamination rates, the variety of experimental parameters encountered
makes such attempts necessarily incomplete or only narrowly applicable.
In any case, a full quantitative description is probably only needed in
very special cases (perhaps, for electron beam writing). The practicing
microscopist really only wants to know how to avoid the deleterious effects during analysis or how to buy sufficient time to obtain the results of interest.

As a first step to evolving practical solutions we simply summarize the most important observations (Table 18.1). The properties selected, those left out, and the brief descriptions of each are purely subjective judgements by the author. Several conclusions are evident immediately. Our qualitative understanding of contamination is significantly better than of that of etching. On the other hand, there is still much which is unclear or unstudied for both. Of course, many of the control parameters are specimen related and cannot be arbitrarily altered. Contrary-wise, there is much that we can control and yet often do not. If we strip the empirical observations down further, we are left with the following obvious parameters: the operator, the specimen, and the microscope! Of course, this should have been clear to all <u>ab initio</u>. The interaction of these three and their principal effects are described in Fig. 18.5 for emphasis. Although the operator cannot affect the nature of the specimen, he (she) can affect its pretreatment and control the mode of operation of the microscope. He (she) can also use (or add) accessories and this very act affects future microscope designs by the manufacturers.

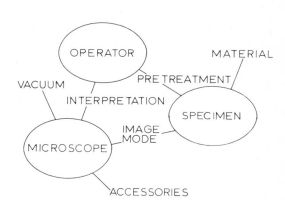

Figure 18.5. Interaction of principal factors which result in contamination or etching.

18.9 THE MECHANISMS OF CONTAMINATION

As we have shown, contamination and etching effects are often related; however, it will be less confusing to consider their mechanisms separately. The plural of mechanism has been used with forethought. To speak of a singular mechanism is a gross oversimplification in either case. We already know that working cures such as cold fingers (or cryo-shields) appear to be perfectly satisfactory for some specimens under specific working consditions. However, changing just one variable for example, decreasing the beam size or changing the mode of operation (to stationary), is known to drastically alter the rate of contamination buildup.

TABLE 18.1

Parameter or Condition	Contamination	Etching
Residual Gases	H_xC_y, other large molecules	$p(H_2O), p(O_2), p(H_2)$ unclear
Initial Surface Composition	Critical for small beam diameters	Effect unclear, but probably more severe with small beam size
Electrical Conductivity	Important for insulators	Effect unclear, probably only indirectly involved
Thermal Conductivity	Probably of some importance, but highly variable	Effect unclear, may be secondarily important
Beam Size and Current Density	Critical for small beam sizes	Effect unclear, probably more severe with smaller beam size
Secondary and Back Scattered Electrons	May be important, especially for insulators	Effect unclear, probably only indirect
Specimen Temperature	Important, but can be kept low	Important, determines reaction rate; can be kept low

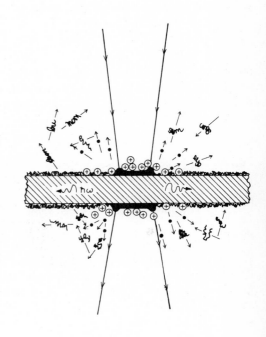

Figure 18.6. Illustrating the multitude of processes occurring during the growth of a contamination deposit (in black). Hydrocarbons are polymerized in the electron irradiated area and replenished by surface diffusion and deposition from the vapor phase. Secondary electron emission (small black dots) results in a local charge buildup near the deposit.

What then are the fundamental processes whose relative importance affects the rate kinetics and the form of the buildup so dramatically? Figure 18.6 illustrates schematically all those mechanisms that are now known. We consider these processes now in order.

Physisorbtion of Hydrocarbon Molecules

Any low vapor pressure molecule deposited before or during the observational or analytical step will probably contribute to the contamination process. The specimen itself could be such a source, but the evidence so far points overwhelmingly towards other external sources. The key property seems to be that the absorptive bonds be weak (physisorbed not chemisorbed). The large adsorbed molecules come from sources in either the microscope vacuum (or prepump chamber) or from an earlier preparation step. The incoming molecular flux f is controlled by a physical law of the following form:

$$f \propto \overline{\sqrt{m/T}} \, P \tag{18.5}$$

where m is the mass of the hydrocarbon molecules, P is their partial pressure and T is the absolute temperature.

Surface Diffusion of Hydrocarbon Molecules

Because adsorbed hydrocarbons are not tightly bound to the surface, they may diffuse rapidly under the influence of any suitable gradient (chemical, thermal, or electrical). Such gradients may be created by several mechanisms which vary strongly with the nature of the specimen and the imaging conditions. For example, the diffusivity D is determined by an equation of the kind

$$D_s \propto \exp \left[-Qs/kT \right] \tag{18.6}$$

where Q is the activation energy for surface diffusion (and thereby strongly related to the strength of the molecule surface bond), k is Boltzmann's constant, and T is the absolute temperature.

Polymerization and Fragmentation of Hydrocarbon Molecules in the Electron Beam

The specimen surfaces (top and bottom) are coated with hydrocarbons. Passing an intense, energetic electron beam through the specimen is known to polymerize and fragment the hydrocarbon molecules. In turn, polymerization immobilizes them and produces a carbonaceous layer (blackened in Fig. 18.6) which in turn creates a surface concentration (chemical) gradient. Hydrocarbons from outside the irradiated area fill the void by random walk surface diffusive processes. The local concentration changes with time in a gradient $\partial C/\partial$ according to Fick's second law:

$$\frac{\partial C}{\partial t} = Ds \frac{\partial^2 C}{\partial x^2} \tag{18.7}$$

where D_s is the surface diffusivity as defined above.

Beam Induced Thermal Gradients

No controlled study of the effects of thermal gradients on contamination in thin films has as yet been reported, probably because temperature measurement and control at the spatial levels required is very difficult. However, many theoretical calculations of beam heating have been reported (e.g. see Chapters 16 and 17). In addition, heat flow is governed by (a different form, but) the same Fick's laws as those for molecular diffusion and therefore boundary conditions such as the cross sectional area of the specimen, the distance to a thermal sink, the thermal conductivity of the specimen, etc., are all important variables. For example, even a thin gold foil can be melted under suitable conditions in an electron microscope (the author can attest to this). On the other hand, under the right conditions polymers and biological thin sections can be observed for hours without obvious image deterioration. In short, although local temperature (or a temperature gradient) can clearly affect the rate of surface diffusion and adsorption drastically (see for example Eqns. 18.5, 18.6 and 18.7), its quantitative importance is still subject to extreme variation.

Electrical Gradients in the Surface

As illustrated in Fig. 18.6 (by the escaping secondary electrons and their residual holes) a positive potential is being constantly created under and near the electron irradiated area. The electric field so created may be quickly neutralized (e.g. in a metallic specimen with little or no oxide on its surfaces) or it may build up until local dielectric breakdown takes place (e.g. in a thin, uncoated insulating film). In very large electrical gradients and small beam sizes the physisorbed hydrocarbons may become polarized by the field and drain the area surrounding that being irradiated at a surprisingly rapid rate (FOURIE, 1979). For large beams (>1 μm) and insulating specimens, contamination may actually be suppressed by the electric field gradient (FOURIE, 1977).

Although the precise electrical field and its effects on contamination are as widely variable as the thermal gradients cited above, their origins seems well-defined. Secondary electrons must be emitted by the incident beam (and its elastically and inelastically scattered successors) at a sufficient rate that charge neutralization does not take place instantaneously. Such charging effects are commonly observed in SEM imaging and may be overcome by coating with a conducting layer or intermittent flooding by an electron source.

The five processes just described seem to be at the heart of the mechanisms creating contamination as we encounter it in analytical electron microscopy. If this qualitative understanding is correct, we should be able to regulate, eliminate or at least minimize the undesirable effects of contamination. We now seem to be approaching that condition and will describe below which methods seem to be most successful to date and what further improvements we can hope for.

18.10 THE MECHANISMS OF ETCHING

We will try to follow the same format in explaining etching mechanisms as we did for contamination. Unfortunately, we find that there have been fewer explicit studies of etching than of contamination. Further, the etching phenomenon is a form of beam induced radiation damage and hence overlaps the subject matter of Chapters 16 and 17. By limiting ourselves to Glaeser's definition, however (see Introduction and Chapter 16), we can present a somewhat limited picture of our current understanding.

In place of the mechanisms depicted for Fig. 18.6, we present a somewhat different model, Fig. 18.7. Although water vapor is not the only possible reacting gas (O_2, H_2, etc., are possible), it is a proven culprit. Furthermore, water vapor is a common residual gas in all vacuum systems, certainly in most electron microscopes. It follows that a certain fraction of water molecules will always be adsorbed on the specimen surface along with the hydrocarbons responsible for contamination. In Fig. 18.7, we have described only the hypothetical situation where carbon containing molecules (comprising the specimen) react with activated water molecules adsorbed to the specimen surface. We could also have added a contaminant layer in this model with similar results. Furthermore, the mechanisms shown need not be only hydrocarbons. What generalization can we make with regard to possible mechanisms?

Physisorbtion of a Potentially Reactive Gas

Water vapor (or other gaseous molecules) can be made to react with either the contaminant or the specimen directly under the conditions of electron irradiation. Most observations of etching to date have been on organic specimens or on carbon bearing contaminants themselves. Other substrate materials can certainly be susceptible to etching. However, in order that there be a sufficient physisorbed gas, the specimen temperature must be lower than ambient (generally quite a lot lower) or the partial pressure of the reactive gases must be high. Of course, a specimen at low temperatures will act as a cryopumping surface and thereby increase its surface concentration of physisorbed gas.

Activation of the Reactive Gas by Electrons

It is difficult to separate this step from that of physisorbtion. What is required is that either sufficient number of gaseous molecules

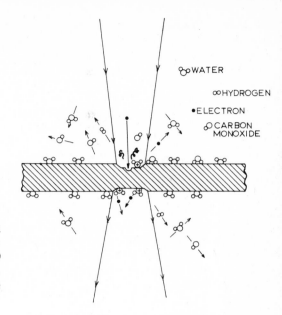

Fig. 18.7. Schematic illustration of the etching process for reactions of the types that produce CO by interaction with H_2O ions. The specimen is assumed to be carbon-containing.

be available or that their activation cross sections be high enough such that a measurable reaction rate between the gas and substrate be attained. The electron energies available to create such reactive ions are obviously more than large enough (see Chapter 16).

Specimen or Contaminant Molecules That will React With the Excited Physisorbed Gas

Mass is removed by a chemical reaction between the excited gas molecule and ionized fragments of either the specimen or the contaminant layer. The prototype reactions of HARTMAN et al. (1978) are typical (Eqn. 18.4). Gaseous products are released (e.g. H_2 and CO_2) and not readsorbed. The reacting species may be provided by fragmentation of larger molecules, but because the lifetime of such activated species is short, the supply of physisorbed gas must be large enough to provide a high reaction probability.

The Reactant Molecules must be Volatile

Mass can only be removed from the specimen if at least one of its atomic species forms a volatile gas upon reaction with adsorbates and is not readsorbed upon reacting (e.g. Eqns. 18.1 and 18.2). Naturally, this requirement is linked to the other thermodynamic variables P and T so that a limited class of potential etching conditions appears possible given the usual constraints of electron microscopy. For example, metallic specimens and most ceramics seem unlikely to be susceptible to etching under most AEM conditions, but their contaminating layer will.

Further (hypothetical) steps could be added. The effects of local heating and charging could alter conditions sufficiently to affect the reaction rate as well. There is little point in speculating further without reasonably secure verifications from experiment and these are still lacking.

It can be safely concluded that mass can be removed by beam induced chemical reactions. The most certain combination leading to etching involves mass removal of cooled organic specimens where there is also a sufficiently high partial pressure of water vapor. Since such conditions are often met, this is an important practical case. Does this imply that specimen cooling will not help prevent its destruction during observation? Fortunately, it does not, because: a) the partial pressure of water vapor and other culprit gases can be reduced below harmful levels with sufficient care and b) specimen temperatures that are low enough can also slow surface diffusion and chemical reaction rates to a snail's pace. On the other hand, etching may still be useful as an in situ contaminant removal process. This still remains to be reliably demonstrated.

18.11 WORKING SOLUTIONS: PROVEN AND POTENTIAL

Contamination and etching have been with us since the infancy of microscopy and so have many cures. We review here some of these proven solutions to reduce contamination and suggest why they are or are not adequate.

Shielding or cryoshielding the specimen has been used in numerous forms (e.g. ENNOS, 1954; HEIDE, 1963; RACKHAM, 1975 HREN et al., 1977; TOMITA et al., 1979). The philosophy behind its use is simple and emminently practical. Cooling the region surrounding the specimen to as low a temperature as possible achieves several desirable objectives. First, contaminant (or etchant) molecules are shielded by line of sight from the specimen. Second, the undesirable molecules are captured (adsorbed) and held by the cryoshield. Third, the cryoshield will act as a local pump toward the specimen, grid, and holder and clean them up with time. Fourth, a cryoshield can also help eliminate system generated background x-rays and stray electrons. For all of these reasons a cryoshield is a highly desirable addition for nearly any analytical electron microscope. In its idealized configuration it should completely surround the specimen. Since this presents some practical difficulties, the openings through it should be made as small as possible (∿1 mm openings top and bottom). The benefits of cryoshielding in AEM are obvious from Fig. 18.8. Clearly, an unworkable condition can be made marginally useful by good cryoshielding design. For example, the volume of the contaminant formed by a ten minute irradiation using a cryoshield at liquid nitrogen temperature is about the same as that for a one minute irradiation with the cryoshield at room temperature. Cryoshielding is, therefore, a first line cure; and it is useful in nearly all cases.

Better vacuum systems are, of course, nearly a cure all, but there are two remaining barriers to analysis: the specimen and the operator (see especially SUTFIN, 1977). Experience with ultrahigh vacuum

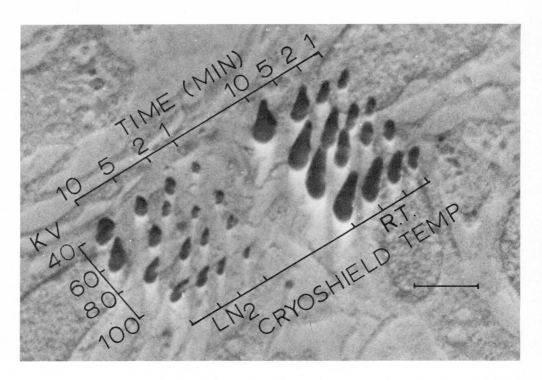

Fig. 18.8. Illustrating the substantial benefits of good cryoshielding when using a small, stationary beam for AEM on a biological thin section. Specimen at room temperature for all measurements. Beam size approximately 10 nm. Tilted at 30° to beam. Beam current held constant.

conditions is now very extensive. It stands to reason that hydrocarbon based contaminants can be eliminated by simple exclusion from the microscope (provided that they are also excluded from the specimen beforehand). Furthermore, good UHV techniques should result in substantial decreases of residual gases such as H_2O, O_2, etc., as well and therefore result in a decreased etching rate. Although a better vacuum is an obvious cure, it is placed second among the solutions described here because it will take many years for the current generation of microscopes to be upgraded to UHV capabilities. However, ceratin measures can be taken now that will help the average operator. Vacuum fluids and sealants (greases) can be substantially improved by replacement with new low vapor pressure fluids and materials, e.g. AMBROSE et al., (1972); HOLLAND et al., (1973); ELSEY, (1975a, b). Electron microscope manufacturers can help immensely by taking advantage of well-developed vacuum procedures commonly employed in other instruments (e.g. mass spectrographs, surface analytical instruments, etc.). In fact, this upgrading process is now being started.

The other major area for immediate improvement is in specimen preparation and pretreatment. Every handling and preparation step is important up until the specimen is actually in place and being analyzed.

This includes the step of insertion into the analyzing chamber. Residues from electropolishing, hydrocarbons deposited during freeze drying, fingerprints, etc., are all sources of contamination. It shouldn't be surprising that the insertion of a dirty specimen into a sophisticated STEM (even one with an ultrahigh vacuum pumping system) will still yield a cone of contaminant when examined with a fine electron beam. On the other hand, even with a dirty specimen, a UHV system will drastically slow if not eliminate etching effects. This will be true even for low specimen temperatures and with no cryoshield (e.g. WALL et al., 1977, and Chapter 12).

A number of other workable solutions exist as well. Most are restricted to specific analytical techniques or to certain classes of specimens. One widely used approach involves flooding an area much larger than that to be analyzed with electrons or ultraviolet radiation and thus polymerizing (immobilizing) the physisorbed supply of hydrocarbons. The diffusion length for a fresh supply of contaminant now becomes approximately the radius of the irradiated area. An analysis using Eqns. 18.6 and 18.7 is consistent with the experimental observation that useful working times up to ten minutes in the stationary spot mode may thus be obtained (see RACKHAM, 1975, for a review). Microdiffraction information from areas of diameter as small as 3 nm were achieved by GEISS (1975) while continuously flooding a much larger area ($\sim$1 μm) with electrons. He thus prevented contamination buildup while collecting the analytical equivalent of stationary microbeam data. The method is described in Chapter 14 by Warren. Long diffusion times, and thus long working times, may be obtained by specimen cooling provided that specimen etching effects are avoided by using suitable vacuum pumping, cryoshielding, or both. Desorption of the physisorbed contaminants before analysis provides another approach. One way to achieve desorption is by heating the specimen in a prepump chamber free of contaminating molecules (ISAACSON et al., 1974) or in situ. A simple, but time consuming, alternative is to pump on the specimen in the analyzing position for a long period (hours) prior to the actual analysis. In either case, the specimen must be free from further contaminating adsorbants when in the microscope. That is, either the vacuum system must be intrinsically contaminant free or a good cryoshield must be in place and cooled during the entire preanalytical period. Note that the cryoshield acts as a pump towards the specimen and, in effect, attracts contaminants to it and from the specimen. Etching effects may also be reduced by such procedures. Alternative methods such as glow discharge have been proposed to achieve this same end (BAUER and SPEIDEL, 1977). Still another avenue is to substitute a high vapor pressure contaminant (e.g. a volatile solvent such as ethyl alcohol which is commonly used for a final rinse) for the low vapor pressure hydrocarbons (e.g. HARADA et al., 1979). The specimen is then inserted while still wet with the solvent and the volatile molecules are quickly desorbed. Provided that further contamination in the microscope is avoided, the area analyzed will contaminante at a much lower rate than before. This procedure is not really good high vacuum practice, since the solvent may eventually cause difficulties in other parts of the vacuum system (e.g. affecting seals, greases, etc.).

FOURIE (1979) suggests that the rate of buildup may be largely suppressed by irradiation in situ with an auxiliary source of low energy
electrons. He proposes that this procedure prevents charging effects
thereby drastically cutting the kinetics of the buildup of a cone for
small spot sizes. Such procedures have been suggested before (HILLIER,
1947; LEPOOLE, 1976). These same low energy electrons will simultaneously polymerize a large area surrounding the beam and help in this way as
well (STEWART, 1934). Finally, intermittant analysis (e.g. by beam
pulsing) appears to be an attractive means to achieve a number of desirable results in AEM without requiring extensive vacuum modifications
(SMITH and CLEAVER, 1977; HREN, 1978). If the purpose of a spot analysis is to gain spectroscopic (compositional) data, then suitably controlled pulsing intervals will actually increase the useful information
per unit real time (e.g. STATHAM et al., 1974). A reduced irradiation
time also reduces the contamination and etching rates and any charging
effects in at least linear proportion to the off time. In all probability the gains will be considerably better than linear, since the kinetic
processes involved are higher order or exponential.

18.12 SOME EFFECTS OF CONTAMINATION AND ETCHING ON AEM

This section could have been placed at the beginning of this chapter as a means to convince the reader that the subject is of importance
to AEM. Such emphasis was considered redundant, since earlier authors
in this book have already achieved this result (see Chapters 3, 4, 7, 8,
9, 12, 14, 15, 16 and 17) and nearly every reference cited describes
such effects. The proceedings of a recent AEM workshop (SILCOX, 1978)
also provide an extensive summary of many of the experimental limitations resulting from contamination and etching effects, especially the
former. It seems fair to assume that those contemplating research using
the AEM are forewarned. The present section will therefore present only
a few selected encounters with these barriers to AEM.

One of the most serious and directly adverse consequences of contamination is a loss of spatial resolution. Historically, even the very
first observations (e.g. WATSON, 1947) were most concerned with a loss
of image detail (e.g. edges lost, contrast reduced, etc.). The effective loss of resolution can be made more quantitative, especially in the
extreme case of a stationary probe. Chapters 3 and 5 provide means to
calculate the extent of spreading of the electron beam as it passes
through a contaminating film. Another way is from the information contained in an electron diffraction pattern of a film of amorphous carbon
(or any light element or combination thereof). An easily measurable intensity maximum is recorded in the photographic film of such patterns at
1 Å (e.g. HEIDENREICH, 1964). From a simple Bragg's law calculation
this corresponds to $\theta \sim 0.02$ rad. Thus for a point electron source impinging at the top of a 100 nm thick film of carbonaceous contaminant,
one would expect a circle of at least 4 nm diameter striking the surface
of the specimen. Since typical probe sizes are $\leq$ 10 nm and contaminant
buildup for stationary probes are typically $\geq$ 100 nm, the accompanying
loss in resolution is quite substantial (e.g. FRASER, 1978). There is
also some evidence that a stationary electron probe may be deflected by
charging effects inherent to the poorly conducting contaminant (ZALUZEC

and FRASER, 1977; FOURIE, 1976a; TONOMURA, 1974). Obviously, this effect will seriously degrade the expected resolution even more. An indirect effect, during convergent beam diffraction experiments, is the local specimen bending induced in crystals during their examination (KAMBE and LEHMPFUHL, 1975). The lack of local stability not only affects the orientation determination, but interferes directly with observations of atomic sized step heights on the surfaces. The effects of etching on resolution are considerably more subtle and complex. Examples clearly fall into the regime of radiation damage and the reader is referred especially to Chapter 16.

The measurement of local composition provides what may be the strongest driving force towards the minimization of contamination and etching. If spires of contaminant are built up on the entrance surface to the specimen, then there are clearly limitations to the minimum volume that can be analyzed (see above). This limitation affects both EDS and ELS, particularly the latter since it is often used for its sensitivity to light elements. Etching, on the other hand, not only decreases the volume analyzed, but it may significantly alter the apparent (and even real) local composition (e.g. SAUBERMAN, 1977).

Contamination deposits will also affect the apparent local concentration measurement in a number of other ways. In the case of ELS measurements P/B ratios will be altered drastically for specimens containing light elements. Even in EDS systems characteristic peaks may be built up if the contaminating molecules contain elements $\geq$ 10. For example, Fig. 18.9 shows an expanded region of an EDS spectrum recorded from a thin sample of homogeneous β-NiAl (ZALUZEC and FRASER, 1977).

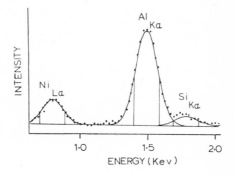

Fig. 18.9. Expanded energy spectrum from contamination deposit on β Ni Al substrate. Characteristic Si K from vacuum pumping fluid. After ZALUZEC (1978).

Three peaks are resolved: Ni L_α, Al K_α and Si K_α. The silicon came from the fluids and greases in the vacuum components. Even if such characteristic peaks from the contaminant are not obtained, the effective mass sensitivity for x-ray analysis can be substantially reduced by the presence of contaminants, because of the increased bremsstrahlung generated (ZALUZEC and FRASER, 1977; SHUMAN et al., 1976). Absorption effects from the contaminant affect not only ELS, but can substantially alter characteristic x-ray intensity ratios (ZALUZEC and FRASER, 1977). For example, Fig. 18.10 shows the intensity ratio of Ni K_α to Al K_α plotted as function of measurement time (hence increasing contaminant thickness).

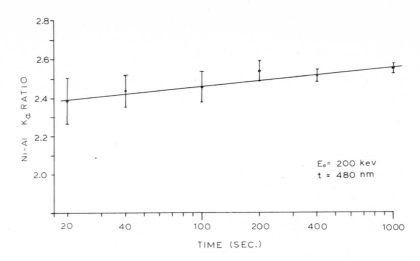

Figure 18.10. Variation in Ni K /Al K with accumulation of contaminant on surface (increasing time). After ZALUZEC (1978).

There is no need to belabor the point of this section. Clearly, contamination and etching effects have numerous deleterious influences to AEM. After all, the rates of contamination buildup have been measured by observing image or diffraction pattern degradation. Etching effects can be seen (i.e. imaged) or measured as a mass loss. These obvious results are sufficient of themselves to indicate the seriousness and magnitude of the problem. Yet we can close on a less negative note by remembering that etching can possibly be used to remove contamination (two negatives would indeed produce a positive!) and that contaminants can be used to mark areas, to measure specimen thickness and variations in thickness, and to detect drift (or its lack). These can be useful phenomena occasionally but we'd like to be able to turn them on and off at will. Perhaps we shall be able to achieve just that goal, soon.

REFERENCES

Ambrose, B. K., L. Holland and L. Laurenson, J. Microscopy 96 (1972) 389.

Bauer, B. and R. Speidel, Optik, 48 (1977) 237.

Broers, A. N., Microele. and Reliability, 4 (1963) 103.

Burton, E. F., R. S. Sennett and S. G. Ellis, Nature, 160 (1947) 565.

Calbick, C. J. and R. E. Hartman, Proc. Twenty Seventh Annual EMSA Meeting (1969) 82.

Castaing, R. and J. Descamps, C. R. Acad. Sci.: Paris, 238 (1954) 1506.

Cosslett, V. E., J. Appl. Phys., 18 (1947) 844.

Crewe, A. V. and J. Watt, J. Mol. Biol., 48 (1970) 375.

Dubochet, J., J. Ultrastr. Res., 52 (1975) 276.

Egerton, R. F. and C. J. Roussouw, J. Phys. D.: Appl. Phys., 9 (1976) 659.

Ellis, S. G., Paper read to the American Electron Microscopy Society, Washington, D. C., Nov. (1951).

Elsey, R. J., Vacuum, 25 (1975a) 299.

Elsey, R. J., Vacuum, 25 (1975b) 347.

Ennos, A. E., Brit. J. Appl. Phys., 4 (1953) 101.

Ennos, A. E., Brit. J. Appl. Phys., 5 (1954) 27.

Fourie, J. T., Optik, 44 (1975) 111.

Fourie, J. T., Proc. Ninth Annual SEM Symposium, IITRI: Chicago (1976a) 53.

Fourie, J. T., Proc. Sixth European Congress on Elec. Micros., Jerusalem, 1 (1976b) 396.

Fourie, J. T., Proc. Ninth Intl. Congress on Elec. Micros., Vol. 1 (1978a) 116.

Fourie, J. T., Optik, 52 (1978b) 91.

Fourie, J. T., Proc. Twelfth Annual SEM Symposium, Washington, D.C. (1979) in press.

Fraser, H., Scanning Electron Microscopy, I (1978) 627.

Geiss, R. H., Proc. Thirty Third EMSA Annual Meeting (1975) 218.

Geiss, R. H. and T. C. Huang, J. Vac. Sci. and Technol., 12 (1975) 140.

Goodman, P. and G. Lempfuhl, Z. Naturforsch, 20a (1965) 110.

Grigson, C. W. B., W. C. Nixon and F. Tothill, Proc. Sixth Intl. Conf. Elec. Micros.: Kyoto, 1 (1966) 157.

Hart, R. K., T. F. Kassner and J. K. Maurin, Proc. Sixth Intl. Conf. Elec. Micros.: Kyoto, 1 (1966) 161.

Valdré, U., Proc. Sixth Intl. Conf. Elec. Micros.: Kyoto, 1 (1966) 155.

Hartman, R. E. and R. S. Hartman, Lab. Invest., 14 (1965) 409.

Hartman, R. E., R. S. Hartman and P. L. Ramos, Proc. Twenty Sixth Annual EMSA Meeting (1968) 292.

Hartman, R. E., H. Akahori, C. Garrett, R. S. Hartman, and P. L. Ramos, Proc. Twenth Seventh Annual EMSA Meeting (1969) 82.

Hartman, R. E. and R. S. Hartman, Proc. Twenty Ninth Annual EMSA Meeting (1971) 74.

Heide, H. G., Proc. Fifth Intl. Conf. Elec. Micros.: Philadelphia (1962) A-4.

Heide, H. G., Zeit für Angew. Phys., XV (1963) 117.

Hillier, J. and N. Davidson, J. Appl. Phys., 18 (1947) 499.

Hillier, J., J. Appl. Phys., 19 (1948) 226.

Holland, L., L. Laurenson and M. J. Fulkner, Japan J. Appl. Phys., 12 (1973) 1468.

Hren, J. J., E. J. Jenkins and E. Aigeltinger, Proc. Thirty Fifth Annual Meeting EMSA (1977) 66.

Hren, J., Proc. Workshop on AEM: Cornell (1978) 62; Ultramicroscopy, 3 (1979) 375.

Isaacson, M., J. Langmore and J. Wall, Scanning Elec. Micros., IITRI: Chicago, IL (1974) 19.

Kambe, K. and G. Lehmpfuhl, Optik, 42 (1975) 187.

Kinder, E., Naturwiss., 34 (1947) 23.

König, H., Naturwiss., 35 (1948) 261.

König, H., Zeit. für Phys, 129 (1951) 483.

König, H. and G. Helwig, Zeit. für Phys, 129 (1951) 491.

Eisegang, S., Proc. Intl. Conf. Elec. Micros.: London (1954) 184.

Le Poole, J. B., Developments in Electron Microscopy and Analysis, ed. J. A. Venables, Academic Press (1976) 79.

Mills, J. C. and A. F. Moodie, Rev. Sci. Instru., 39 (1968) 962.

Müller, K. H., Optik, 33 (1971a) 296.

Müller, K. H., Optik, 33 (1971b) 331.

Oatley, C. W., W. C. Nixon and R. F. N. Pease, Adv. Elect. Electron Phys., 21 (1965) 181.

Rackham, G. M., Ph.D. Dissertation, University of Bristol (1975) 39.

Riecke, W. D., Z. Angew. Phys., 27 (1969) 155.

Ruska, E., Kolloid. Z., 100 (1942) 212.

Sauberman, A. J., Proc. Thirty Fifth Annual EMSA Meeting (1977) 366.

Silcox, J., Analytical Electron Microscopy: Report of a Specialist
    Workshop, July 1978, Cornell University, pp. 62-129.

Smith, K. C. A. and J. R. A. Cleaver, Developments in Electron Micro-
    scopy and Analysis, ed. J. A. Venables, Academic Press (1976) 75.

Statham, P. J., J. V. P. Long, G. Waite, and K. Kandiah, X-ray Spectro-
    metry, 3 (1974) 153.

Stewart, R. L., Phys. Rev., 45 (1934) 488.

Sutfin, L. V., Proc. IXCOM, Boston (1977) 66A.

Thomas, J. M. and P. L. Walker, JR., Carbon, 2 (1965) 434.

Thornhill, J. W. and I. M. Mackintosh, Microelec. and Reliability, 4
    (1965) 97.

Tomita, T., Y. Harada, H. Watanabe and T. Etoh, Ninth Intl. Congress
    on EM, Toronto (1978) 1, 114.

Tonomura, A., Optik, 39 (1974) 386.

Valdré, U., Proc. Sixth Intl. Conf. Elec. Micros.: Kyoto, 1 (1966)
    157.

Wall, J., J. Bittner and J. Hainfeld, Proc. Thirty Fifth Annual EMSA
    Meeting (1977) 558.

Watson, J. H. L., J. Appl. Phys., 18 (1947) 153.

Zaluzec, N., Ph.D. Thesis, Univ. of Illinois (1978); Oak Ridge National
    Laboratory Report ORNL/TM-6705.

CHAPTER 19

MICROANALYSIS BY LATTICE IMAGING

ROBERT SINCLAIR

DEPARTMENT OF MATERIALS SCIENCE and ENGINEERING
STANFORD UNIVERSITY
STANFORD, CALIFORNIA 94305

19.1 INTRODUCTION

The resolution of closely spaced lattice fringes ($\sim 2 \overset{\circ}{A}$) is increasingly being used for studying fine-scale phenomena in materials. It is natural, therefore, that lattice imaging should also be employed to obtain information about chemical composition, especially in those circumstances where the method is most powerful, viz. on a highly localised level. In this article the application of fringe imaging is considered for the determination of chemical composition, emphasising procedures for obtaining and interpreting the images in a reliable way and illustrating the situations where such images are beneficial.

Conventional bright and dark field transmission electron microscope (TEM) images rarely contain information about the distribution of elements in a specimen, and certainly do not yield quantitative chemical composition. Hence chemical analysis in a TEM has only become common with the availability of commerical TEM/STEM instruments which possess a spectroscopic capability such as the identification of characteristic x-rays created within the sample. Lattice fringe imaging, although a direct imaging technique, does provide an alternative. The fringes recorded in the micrograph are associated with the diffracting lattice planes of the specimen. The periodicities of fringe image and corresponding crystal lattice planes are identical under appropriate experimental conditions, and since the latter also depend on composition, chemical data are tractable from fringe spacing measurements. There are limitations to the general applicability of this approach both from an experimental and materials viewpoint. But, when feasible, it provides an important and sometimes complementary means of studying composition and should be included as a technique for consideration by the analytical electron microscopist.

This article deals predominantly with fringe images. However it is instructive to explore multi-beam imaging in addition, as this mode should become more widely used as instruments achieve higher resolutions.

19.2 THEORETICAL CONSIDERATIONS

Fringe Imaging

Lattice images are formed in the TEM by the interaction of two or more electron waves. When the beams are in the same systematic row a fringe pattern is produced. In order for the images to yield useful information about the structure of the object the fringes should be related in a straightforward manner to the disposition of lattice plans. A number of factors influence the image and these must be considered to establish optimum imaging conditions. Thus the normal specimen parameters such as thickness and orientation determine the amplitudes of the participating beams. Their relative phase is affected by the transfer of the electron waves from the object plane (bottom surface of the crystal) to image plane, since a real microscope objective lens introduces phase shifts principally by the action of spherical aberration and defect of focus.

Specimen-Related Parameters

Using a two-beam theory of dynamical electron diffraction, HASHIMOTO, MANNAMI and NAIKI, (1961) have treated in detail the effects of specimen parameters on two-beam fringe images. The predictions of this theory have generally been confirmed. In order to appreciate the importance of various parameters, their calculations may be expressed in the following form (AMELINCKX,1964) for the intensity of the resultant electron wave function:

$$I = A + B \sin\{2\pi gx + \phi\} \tag{19.1}$$

where the first term represents the background signal and the second term a periodically varying intensity in the $\underline{x}$ direction due to interference between the two beams, transmitted and diffracted ($\underline{g}$). The factors A, B and ϕ depend on specimen thickness (t) and deviation from the Bragg condition ($\underline{s}$) in a non-simple manner. The image of a perfect crystal has the following general properties:

(i) the background intensity does not vary periodically but is a function of t and $\underline{s}$;

(ii) superimposed on the background is a series of sinusoidal fringes whose amplitude and periodicity both depend on t and $\underline{s}$;

(iii) the origin of the fringes is displaced from the centrosymmetric origin of the lattice planes, again in a way determined by thickness and exact orientation.

In the specific circumstances that ϕ is constant, the fringe period-icity is exactly that of the diffracting planes. This relationship is no longer true when ϕ varies with $\underline{x}$. Thus it can be appreciated that for information concerning lattice periodicity there are both favorable and unfavorable experimental situations. More refined calculations in-volving multi-beam scattering theory do not appear to significantly al-ter the results of HASHIMOTO et al., (1961) and so we shall consider the implications of their work in more detail.

Chemical composition influences both the amplitude and periodicity of the fringe pattern. However the dependence of amplitude on composi-tion is complex and it would be difficult to extract data from the image that way. The fringe spacings are more readily interpretable, optimum conditions for which may identified from the above equations. The im-portant parameter is ϕ, which is given in conventional terminology by

$$\tan \phi = \zeta_g \, s \, \sin\beta \, \tan \frac{\pi t}{\zeta_g} \sin\beta \qquad (19.2)$$

where ζ_g is the extinction distance and $\cot\beta = s\zeta_g$. The instantaneous fringe periodicity (d) may be computed from the following expression (CARLSON 1972):

$$\frac{1}{d} = \frac{1}{2\pi} \frac{d}{dx} \ (2\pi g x + \phi) \qquad (19.3)$$

$$\text{i.e.} \quad d = [g + \frac{1}{2\pi} \frac{d\phi}{dx}]^{-1} \qquad (19.4)$$

The problem may be conveniently discussed in two categories:

(i) $d\phi/dx = 0$ (i.e. ϕ is not a variable of x). The fringe periodicity is identical to that of the crystal lattice planes. In terms of the specimen orientation and thickness this can occur for a number of experi-mental situations:

(a) s = 0 (i.e. $\phi = 0$). The crystal is exactly at the Bragg condition. The fringe visibility is maximum at specific specimen thickness (t = $\xi_g/4$, $3\xi_g/4$, $5\xi_g/4$ etc.) decreasing to zero at intermediate thicknesses. The origin of the fringe pattern is translated by half a fringe spacing every half extinction distance, due to a change in the sign of B (AMELINCKX, 1964), an effect which is easily recognized.

(b) s, t constant (i.e. ϕ constant). In this case the crystal is parallel-sided and deviated to a constant extent from the Bragg condi-tion. This would be rare for a self-supporting specimen made by conven-tional thinning methods but may arise for foils prepared by vapor depo-sition or by cleavage.

(c) s,t variable, but not in $\underline{x}$-direction (i.e. $\phi(x)$ constant). Any variations in thickness or orientation only occur in a direction paral-lel to the diffracting planes. HASHIMOTO et al., (1961) use the example of imaging planes perpendicular to the edge of a wedge-shaped specimen at constant $\underline{s}$. The fringe pattern varies in a more complex way than in (a) but the lattice periodicity is still maintained. This situation may possibly be realized by appropriate tilting experiments but would not normally occur.

(ii) $d\phi/dx \neq 0$ (i.e. ϕ is a function of $\underline{x}$). The fringe periodicity is no longer exactly equal to the lattice spacing. This is the general case which would be encountered most often with typical thin foils prepared from bulk material: i.e. orientation and thickness vary in the direction of $\underline{g}$ due to the bending and wedge-like shape of normal TEM specimens. At first one would therefore be very wary of using fringe spacing measurements, but a detailed consideration of the degree of deviation shows that for the experimental conditions used for lattice imaging, the magnitude of this electron-optical effect is too small to be significant. This is an important point since it also shows the limit of reliability to which fringes can be measured.

Schematic examples of the relationship between the positions of fringes and those of lattice plane have been given by a number of authors (e.g. HASHIMOTO et al. 1961, AMELINCKX 1964, HIRSCH, HOWIE, NICHOLSON, PASHLEY and WHELAN 1965). HASHIMOTO et al., (1961) also showed lattice image pictures taken of copper phthalocyanine ($\sim$12Å spacing) by MENTER (1956) in support of their calculations. There is no question that electron-optical effects do occur, but it is worthwhile considering a contemporary example to assess the extent of the danger.

Calculations have recently been made (HIBBS, PRIOUZ and SINCLAIR, 1979) on the (200) fringe spacings in aluminum (d = 2.02 Å) for a foil varying in thickness from $\xi_g/8$ to $3\xi_g/8$ over a distance L and orthogonally in orientation from $s_g=0$ to $s_{3g}=0$, also over L. These are easily identifiable experimental variations in the vicinity of the maximum fringe visibility condition (i.e. s = 0 at $\xi_g/4$). Three cases have been considered so far, with L = 100 Å, 1000 Å and 10000 Å. The first case (L=100Å) naturally has the largest deviations, because ϕ is varying most rapidly with $\underline{x}$. The maximum amount is -2% from the true fringe spacing occurring near t = $3\xi_g/8$, with most deviations being smaller. An interesting aspect which emerges from these calculations is that $d\phi/dx$ is maximum at $\underline{s} = 0$, so that the maximum deviations actually occur near the Bragg condition. Thus although at first sight it would appear from case (i) above that $\underline{s} = 0$ is the safest orientation, in fact for a typical, bent, wedge-shaped foil, this is not true. The magnitude of the deviation correspondingly decreases as L increases, and is only -0.02% for L = 10,000 Å, again with most deviations lower than this.

A 2% deviation from the true lattice spacing is large, of the same order as the changes associated with composition variation. However when it is realized that a value for L of 100 Å corresponds to a specimen wedge angle of about 60 Å, a bend radius of 0.5μ and associated elastic strain at $\xi_g/4$ of 1.5% one appreciates that such a specimen area would never be used for lattice imaging anyway. Much more realistic is the circumstance when L = 10,000 Å i.e. a wedge angle of 1° and a bend radius of 50μ. The practicing microscopist would recognize this as a good quality foil, which would normally be used for high resolution work. The figure for maximum spacing deviation of 0.02% is now quite acceptable. Measurements on 2 Å fringe images of materials known to be homogeneous (e.g. pure gold) rarely yield detectable fringe spacing variations, consistent with these considerations. Furthermore it can be speculated that the wedge angle in the specimens imaged by MENTER (1956) was large, especially since several half-spacing fringe shift (which occur every half-extinction distance) are evident in some of the published micrographs.

The implications of such calculations are important for the reliability of the lattice image analysis and will be published in full elsewhere. It is recommended that foils used for careful spacing measurements should be fully characterised for exact thickness and orientation variations, a task which should be more easily accomplished in the future when convergent beam and microdiffraction capabilities are available with high resolution objective lenses. Nevertheless the extent of the deviation does not appear to be significant for high quality foils, indicating that confidence may be placed in the interpretation of fringe spacing data. This is confirmed by various experimental examples given in section 19.6.

Microscope Parameters

An ideal objective lens at perfect (Gaussian) focus would faithfully transfer the electron waves leaving the exit surface of the specimen to the image plane. However the spherical aberration of a real lens alters the phases of beams travelling at an angle to the lens optical axis, an effect which may be compensated by a defect of focus of the lens (SCHERER 1949). The phase shift (χ) relative to an axial ray is given by:

$$\chi = \frac{2\pi'}{\lambda} \quad \frac{C_s\alpha^4 + \Delta f\alpha^2}{4 \qquad\quad 2} \tag{19.5}$$

where C_s is the coefficient of spherical aberration, α is the angle of the beam from the optical axis and Δf is the degree of defocus (taken here as positive for overfocus or strengthening of the objective lens). For images produced by the interaction of a number of electron waves, the relative phase differences so introduced are extremely important to the resultant intensity distribution in the final image.

The two-beam image case is quite straightforward: the objective lens may be suitably defocused to bring the waves in phase with one another and give the fringe pattern described previously. This occurs at periodic values of Δf for which $\chi = 0$ (every $\sim$ 1200 Å for 2 Å fringes with C_s = 2.5 mm and 100kV electrons). The situation may be simplified further (DOWELL 1962, 1963) by arranging for the beams to travel through the lens at equal, but opposite, angles to the optic axis, which may be achieved by suitably tilting the incident illumination. It may be noted that now α is the Bragg angle and the diffracting planes are geometrically parallel to the optic axis. The two waves are always in phase, independent of objective focus, and lattice fringes may be obtained over a range of focus settings particularly with more coherent electron sources such as field-emission and LaB$_6$ guns. Nevertheless, the exact focus setting is still very important for obtaining useful images. The image that is produced should represent, as closely as possible, the object structure. Therefore it is necessary to combine in the image plane waves from the same point in the object plane. Both spherical aberration and defocus shift the image points (Fig. 19.1) by an amount, y, referred to object space, given by (HIRSCH et al., 1965).

$$y = C_s \alpha^3 + \Delta f \alpha \qquad (19.6)$$

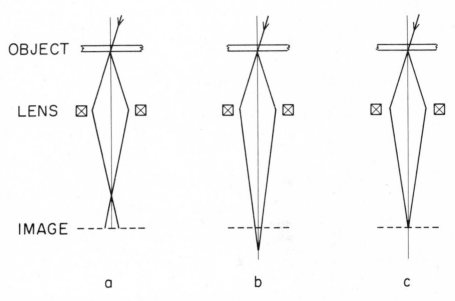

Fig. 19.1. Schematic diagrams showing the influence of spherical aberrations and defocus on the ray paths of transmitted and diffracted beams using a thin-lens approximation and tilted illumination: (a) spherical aberration alone; (b) underfocus of the objective lens alone; (c) combined spherical aberration and defocus at the coincidence focus setting (equivalent to the perfect lens case).

Perfect coincidence of the two beams occurs at

$$\Delta f_c = C_s \alpha^2 \qquad (19.7)$$

(about 2000 Å underfocus for the example considered above). A deviation from this condition brings together waves from different points on the specimen and hence degrades the information contained in the image. The image shifts produced by defocusing are easily seen during experimental imaging. It is also more useful in two-beam imaging to consider the defect in focus from the coincidence position rather than absolute defocus from Gaussian focus.

Multi-beam fringe images, produced by a number of reflections in a systematic row, are more useful than two-beam images when the lattice periodicities are large. The fringe profile shows a more complex variation with specimen thickness and defocus, because of the relative amplitudes and phases of the imaging beams. It is generally necessary to calculate the images using many-beam dynamical theory to find both the optimum imaging conditions and how to interpret the images in terms of the specimen crystal structure (e.g. ALLPRESS and SANDERS 1972, 1973; SINCLAIR, SCHNEIDER and THOMAS, 1975). The major lattice periodicity may be used in the same way as in the two beam case for determination of composition. In addition the character of the fringe profile may yield information on the local atomic arrangements within the unit cell (e.g. VAN LANDUYT and AMELINCKX, 1975). This is an important aspect of fringe imaging, but is beyond the scope of the present article.

Multi-Beam Imaging

The information contained in fringe images is limited since only one set of lattice planes is being examined. The best image would be produced by a perfect lens using all the beams coming from the specimen in the image formation process. However, in the real case, the spherical aberration phase shifts described above must be taken into account. Further complications arise from the beam divergence, chromatic energy spread of the electrons, phase differences across each beam due to spherical aberration, astigmatism, mechanical vibrations etc. The spherical aberration effect limits the resolution to about 3.5Å at 100kV (COWLEY and IIJIMA, 1977). Thus although there is more information in the image, the resolution is relatively poor compared to the tilted illumination, two-beam case. Multi-beam structure images can resolve heavy atoms so long as they are separated by more than 3.5Å, but the atomic spacings of many important materials are less than this (e.g. metals, semiconductors, commercial ceramics) and one must resort to fringe imaging to study their lattice at all.

Several approaches are currently being pursued. Firstly, for those materials whose spacings are sufficiently large, the multi-beam image can produce a projected charge density picture of the specimen. (COWLEY and IIJIMA 1972; O'KEEFE 1973). The specimen must be thin (<100Å for 100kV electrons) and oriented close (within 10^{-3} radians) to a zone-axis orientation (IIJIMA 1971, 1973). The objective lens defocus should be chosen to transfer the majority of the interacting beams with similar amplitude and phase (the so-called Scherzer defocus), typically about -900Å from Gaussian focus (IIJIMA, 1973). This method has been used with considerable success for studies of many oxides and minerals, not only for information about perfect crystal structure (e.g. IIJIMA and ALLPRESS, 1974) but also about structural imperfections, including point defects, the local atomic arrangements in various states of order and deviations from stoichiometric composition (e.g. IIJIMA, KIMURA and GOTO 1973, 1974 ; IIJIMA and COWLEY 1977; BUSECK and IIJIMA 1974, 1975; MCCONNELL, HUTCHISON and ANDERSON 1974; ANDERSON 1977; IIJIMA and BUSECK, 1978). It is a powerful technique for microanalytical microscopy, since the arrangement of atoms can be devised directly from the micrographs, and should be applied increasingly as higher resolution instruments become available.

For some materials with smaller lattice repeats, multibeam images
taken with a limited number of reflections have recently been shown to
be quite useful (KRIVANEK, ISODA and KOBAYASHI 1977; REZ and KRIVANEK,
1978). Thus for silicon and related materials with open structures it
is possible to image the channels in the structure and so obtain structural information, such as the nature of a stacking fault (Fig. 19.2)
and unusual atomic configurations near a grain boundary (KRIVANEK
et al., 1977). Such images obviate the necessity to take a series of
fringe pictures of various sets of planes for complete crystallographic
information about defects etc., although they are more experimentally
difficult to record. It must be borne in mind however that each fringe
set is displaced with respect to the others by the appropriate spherical aberration image shift and that defocus should again be carefully
chosen to optimize resolution and contrast.

t =   0   100 Å   200 Å   300 Å   400 Å

Fig. 19.2. A multibeam lattice image of the (110) section of silicon,
showing the channels through the structure. The stacking fault can be
identified directly from the image as extrinsic in this case (courtesy
of O. L. Krivanek).

Finally it has also recently been demonstrated that micrographs may
be taken at higher spatial resoltuion by operating not at the Scherzer
focus condition but rather at a defocus value which brings additional
higher order reflections into the image plane in phase (HASHIMOTO,
ENDOH, TANJI, ONO and WATANABE, 1977). At this "aberration free focus"
(AFF) condition (HASHIMITO, SUGIMOTO, TAKAI and ENDOH, 1978) resolution
on the order of 1Å has been claimed in images of gold (HASHIMOTO et al.,
1977) and silicon (IZUI, FURUNO and OTSU, 1977), an example of which is
given Fig. 19.3. Calculations have also demonstrated that different
elements possess different fine structure in such images (HASHIMOTO,
KUMAO and ENDOH, 1978) and that it is feasible to distinguish a row of
aluminum atoms in a gold crystal. While the interpretation of AFF images is still at an early stage they may offer the possibility of atomic
imaging with 100kV microscopes. A current disadvantage which may in
time be overcome is that AFF imaging requires a low C_s lens (e.g. C_s =
0.7 mms) which is now achieved with a low focal length objective: the
tilting capacity is commensurately limited. In addition, while it is
possible to arrange for the electron beams from a perfect crystal to
arrive in phase, this may not be true for imperfect crystals which contain a range of lattice spacings, as especially occurs for materials
with compositional variations.

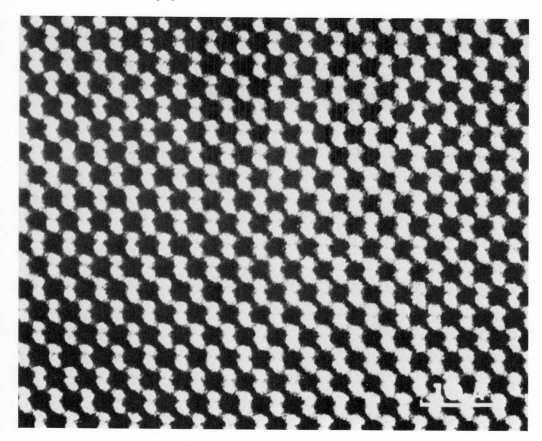

Fig. 19.3. An aberration free focus (AFF) image of the (110) section of silicon (courtesy of K. Izui).

19.3 EXPERIMENTAL PROCEDURES

A certain degree of care and experimental expertise is necessary to produce lattice images. Naturally the microscope should be operating in optimum condition, with no lens, stage or filament instability and with clean (preferably self-cleaning, thin film) objective apertures. Such practical factors are extremely important but are assumed to be under control and need not enter into the scientific considerations of this article. The following parameters significantly influence the nature of the image and need to be taken into account by the microscopist. The discussion is limited to microscopy of stable (e.g. inorganic) materials with ∿100kV electrons.

(i) Imaging reflections: It is usual to allow the transmitted and one strongly diffracted beam through the objective aperture for fringe imaging. The reflection chosen should be appropriate to study the phenomenon in question (e.g. using (200) planes for <100> composition modulations in FCC spinodal alloys). This is the simplest experimental arrangement but yields information about only one set of lattice planes at a time. It is rarely beneficial to include additional reflections for

planes with spacings ~2Å, but for larger spacing materials there are
some advantages. Thus the major periodicity becomes more sharply defined for multi-beam imaging (i.e. high order reflections contribute to
the image) and additional information can be given about the atomic
planes within the unit cell. However, various experimental factors such
as thickness and defocus must be optimized more carefully in the multi-
beam case.

At the other extreme, the most representative structure images are
obtained using as may reflections as possible. The objective aperture
size should be chosen to exclude reflections far from the optic axis
which would not arrive in-phase at the image plane. For the Scherzer
defocus this corresponds to reflections from spacings smaller than
about 3.5Å, a figure which may be reduced at the AFF condition. For
interpretation of the image in terms of projected charge density more
than twenty beams are required (LYNCH, MOODIE and O'KEEFE, 1975).

(ii) Specimen thickness: It is well-known that the amplitudes of transmitted and diffracted electron waves are critically dependent on the
distance traveled through the crystal. The maximum two-beam fringe visibility occurs when the beams have equal intensity i.e. at $\xi_g/4$, $3\xi_g/4$,
etc. for $\underline{s}$ = o. Because of the effects of absorption at increasing
thicknesses, specimens should be as thin as possible and of the appropriate thickness. This may be achieved by vapor deposition, but good
quality foils prepared from bulk material do not normally present any
problem. The approximate thickness can be found experimentally by reference to the thickness fringes. However fringe visibility does not decrease so rapidly that thickness becomes a critical parameter and so
long as there is a reasonably matched intensity in the transmitted and
diffracted beam images, successful fringe imaging should result. When
using fringe spacings, the d-spacing calculations of the previous section should also be considered: thus the variation of thickness in the
region of interest should be low to ensure fringe-lattice spacing correlation.

For the multi-beam fringe case, knowledge of the exact thickness is
more important for image interpretation. In the most reliable work
thickness should be determined experimentally by a suitable procedure
(e.g. from a stereo pair of images or by the convergent beam method
(AMELINCKX, 1964), especially for comparison with computed images. Calculations should generally be performed to identify optimum conditions
for obtaining the desired structural information. With structure imaging, the projected change density approach may be applied only for
very thin crystals i.e. <50 - 100 Å (O'KEEFE, 1973; LYNCH et al., 1975).

(iii) Specimen orientation: The maximum two-beam fringe visibility occurs exactly at the Bragg condition ($\underline{s}$=0) for the participating diffracted beam. Deviation from $\underline{s}$=0 reduces visibility and can influence
fringe periodicity. As with thickness, it is necessary to choose an
area where small variations in $\underline{s}$ occur to ensure straightforward interpretation of the spacing measurements. Greatest confidence can be
placed on the data when both thickness and orientation variations are
determined experimentally and their influence calculated by the expression for d given in section 2.

For thin specimens, Kikuchi lines are not present in the diffraction pattern to assist orientation of the crystal. Instead it is easiest to tilt the specimen until the center of the bend contour in the dark field image coincides with the area of interest. During imaging, the orientation should be periodically checked (e.g. using the diffraction pattern) as there is a tendency for the exact orientation of thin foils to drift with time. Significant deviations can easily be identified from the relative diffraction spot intensities (e.g. when $s_{2g} = 0$, a small deviation the which can easily be corrected, the second order spot has noticeably higher intensity than before).

Local changes in lattice orientation occur near a defect or when d-spacing variations are present, and these may also influence the fringe-lattice spacing correspondence. So long as the deviations are not large, which is true in most circumstances, the true local lattice periodicity is recorded, according to the considerations of section 2. This has been demonstrated for the case of a dislocation in germanium (BOURRET, DESSAUX and RENAULT 1977) in which fringe and lattice coincidence were found apart from the very core region of the dislocation (to within 10Å) where the present approximation, and that of elasticity theory, breaks down.

In structure imaging it is very critical to be at the exact zone axis, to better than about 10^{-3} radians, otherwise the image is no longer a good representation of projected charge density (IIJIMA, 1973). This has also been shown to apply for AFF imaging (IZUI, NISHIDA, FURUNO, OTSU and KUWABARA, 1978).

(iv) Angle of incident illumination: In order to minimize the influence of various aberrations on the resolution of finely spaced lattice fringes, it is necessary to operate with tilted, rather than axial, illumination (DOWELL 1962), thus decreasing the angle at which the outermost reflections travel with respect to the microscope optical axis. The condition most commonly employed is when the transmitted (i.e. incident) and diffracted beams are equally inclined to the optic axis, eliminating altogether some aberration effects (e.g. the spherical aberration phase shift is equal for both beams). For structure imaging many reflections, symmetrically positioned about the transmitted beam, are used and there is no advantage in tilting the incident beam (i.e. the illumination is axial). Multi-beam fringe images can be taken in either mode, depending on the circumstances, but for small lattice spacings (e.g. ∿2 Å) tilted illumination is usually required.

(v) Objective lens focus: Although the simple theory outlined in section 2 indicates no variation in fringe image with defocus, the exact focus setting is extremely important for interpretative purposes. It is necessary for beams from the same point in the object to be brought together in the image, which occurs at the "coincidence focus" (Fig. 19.1).

With a sharply defined feature, this may be achieved to a certainty approaching 100 Å defocus, even without observing the fringes, by making transmitted and diffracted images coincide. Planar defects or

the specimen edge are particularly useful reference objects. In practice, however, the fringes lose visibility quite rapidly as one departs
from the coincidence position (within a few hundred Ångstroms for 2 Å
fringes and a therminoic tungsten filament). This probably arises from
a combination of coherence, and spherical and chromatic aberration
phase spreading in a beam of finite divergence.

With multi-beam fringe imaging some setting away from coincidence
may be required for the information of interest (e.g. the superlattice
and fundamental lattice of ordered alloys appear at focal settings differing by about 150 Å (SINCLAIR and THOMAS, 1977)). When structure imaging the Scherzer focus setting is preferable, or for finer lattice
spacings the aberration free focus conditon (see section 2). In all
cases computed and experimetnal images should be compared for appropriate image interpretation and to establish the optimum value of defocus.

(vi) Astigmatism: The correction of objective lens astigmatism is as
important as focusing in obtaining successful lattice images. The asymmetry of the magnetic field in the objective lens introduces further
phase shifts between the beams, or alternatively displaces the images of
different beams. The extent of the effect may be expressed in terms of
the difference in focus (Δf_a) necessary to bring in turn the images to
Gaussian focus. If the relative image displacement is larger than the
resolution required then the fringes will be destroyed. An estimate may
be made as follows:

$$\Delta f_a \; \alpha \lesssim d \tag{19.8}$$

yielding a value of about 200 Å for 2Å fringes. This is similar to the
range of defocus over which highly visible fringes are obtained with a
tungsten thermionic filament, indicating the approximately equal importance of the two for successful imaging.

The astigmatism correction may be made approximately by reference
to the Fresnel fringe at the edge of the specimen: the fringe should
disappear at the same focus setting (to within Δf_a) around the perimeter
of a semicircular projection. A finer level of correction is achieved
by minimizing and making symmetrical the phase contrast background present in all TEM images, and also by maximizing the visbility of the lattice fringes.

(vii) Beam divergence and illumination conditions: Since lattice images
must be taken at high magnifications it is necessary to operate the
microscope at maximum brightness, normally with a fully focused second
condenser lens (C_2). Under these circumstance the beam divergence is
large ($\sim 10^{-3}$ radians) and is determined by the size of the second condenser aperture. This imposes a further limitation on imaging: the
range of directions in each beam is subject to the spherical aberration
phase spread and hence limits resolution. Both two-beam fringe images
(DOWELL, 1963) and n-beam structure images (O'KEEFE and SANDERS, 1975) are
affected, being significant at the 1.5 Å level for the former and the
3.5 Å level for the latter. Clearly, decreasing the beam divergence improves the situation and if a compromise can be made by defocussing C_2

and extending the exposure time, image quality is improved. The high brightness electron sources possible in higher vacuum instruments provide a favorable solution to the problem.

(viii) Magnification, photography and image drift: To record lattice images with clarity the magnification should be high enough for details to be resolved on the photographic film, especially so that fringe positions may be determined with precision. 2 Å lattice planes produce 100μ fringes when magnified 500,000 times. This is really a lower working limit for conventional electron negative emulsions the grain size of which is about 5-10μ. Image clarity certainly appears to decrease as the fringe spacings drop below 100μ, and at the 50μ level is extremely difficult to manipulate (both on the negative and by photographic processing). The microscope magnification should be chosen accordingly. Conversely, with higher magnifications, exposure times are longer and mechanical instability or image drift may limit the recording of the image. This factor becomes important for times longer than about 10-20 seconds. The situation is the same for any high resolution imaging work and a compromise must generally be made for the practical operation of the microscope.

(ix) Chromatic aberation effects: Chromatic aberration is introduced by a variation either of electron energy or of objective lens focal length. If severe, resolution may be greatly impaired as occurs during gun instability or in the imaging of thick sections ($\sim$1000 Å). There is little the operator can do to nullify its influence, apart from realising that it should be minimized whenever possible.

A chromatic spread causes a blurring of the image since different electron energies are focused to different points, or alternatively there is a range of phases of the interacting beams. The effect increases linearly with angle and so does not produce such a dramatic reduction in resolution for typical energy spreads ($\Delta E\sim$1eV) as does spherical aberration. The phase spread within each beam may be treated in a similar way to the combination of divergence and spherical aberration, but unfortunately cannot be reduced by manipulation of the microscope. In this respect it may provide ultimate resolution limit which is improved by reducing the coefficient of chromatic aberration of the objective lens, using a more monoenergetic source or by operating at higher voltages (through the term $\Delta E/E$).

For fringe imaging chromatic effects, like all effects which increase with angle, are reduced by the titled illumination condition. Since sub-1 Å fringes have been resolved there seems to be no real practical limitation. In structure images, the higher frequency beams (smaller d) are effectively eliminated from contributing to the image (i.e. chromatic aberration acts as an envelope for the contrast transfer function (HANSZEN and LEPTE 1971)). This resolution limit is also about 3.5 Å currently. In some experiments (e.g. KRIVANEK, TSUI, SHENG, and KAMGAR 1978) this effect has been used to advantage to form multi-beam images without an objective aperture: the spatial resolution in the image is determined by the chromatic cut-off rather than by a physical apodisation.

From the above consideration, a suitable experimental procedure can be adopted, taking into account both the specimen and microscope parameters. The diversity of materials now studied by the lattice imaging technique indicates that no inordinate experimental difficulties currently exist.

19.4 ANALYSIS OF FRINGE IMAGES

The determination of a lattice fringe periodicity requires a straightforward measurement on the negative or on a photographic print. However some consideration needs to be given to the competing requirements of precision of the result and the spatial resolution it represents. A full discussion of this and other related points has been given by SINCLAIR and THOMAS,(1978).

A number of workers have commented on the accuracy of making a fringe spacing determination. There appears to be little or no improvement gained by sophisticated gadgetry over the combination of a photographic enlargement, an accurate rule and the eye. With the highest quality images the accuracy of a single-fringe measurement is about ±5%, decreasing to about ±10% for lower contrast pictures, and is limited by the difficulty in locating the same position in each fringe. Averaging over N fringes reduced the uncertainty in the <u>periodicity</u> to ±(5/N)%. The determination can thus be very precise for large N (e.g. ±0.001 Å for only one hundred 2 Å fringes). However by increasing N, we are also increasing the dimension which the average periodicity represents, thus compromising spatial resolution. The purpose of a microanalysis experiment is to establish chemical composition in a small region. This relationship between precision and resolution is therefore extremely important. Basically, gross fringe spacing differences can be found in highly localized regions whilst more subtle differences require averaging over a large spatial scale. Typical working figures for 2 Å fringes are ±0.5% with 20 Å resolution, or ±0.01% with 1000 Å resolution (SINCLAIR and THOMAS,1978).

Laser optical diffraction from the lattice image provides an alternative method for determining the average fringe spacing. It turns out that direct and diffraction measurements are often complementary, because of the inverse relationship between real and reciprocal space. Thus direct measurement is best for large spacings, diffraction for small. High quality fringes are best treated in real space but low visibility, poor quality images often give remarkably sharp optical diffraction spots and correspondingly reliable data (SINCLAIR and THOMAS 1978). The area of selection can be chosen by position of a suitable aperture over the fringe image and can be as small or as large as desirable. With small apertures the diffraction spot is broadened and hence the measurement is less precise. Fraunhofer diffraction from the aperture itself, overlapping the fringe diffraction pattern, sets the useful lower limit to about 10 Å equivalent for 2 Å lattice spacings (SINCLAIR, GRONSKY and THOMAS,1976).

The precision of the optical technique, in our experience, is never as good as direct measurement for high quality images and nor does it improve significantly with increasingly large apertures (since the sharpness of the diffraction spot does not improve linearly with N). The best accuracy is about ±0.5% with one hundred fringes and above. The optical approach is therefore best employed with lower quality images of smaller lattice spacings. It is also extremely useful when a precise sequence of measurements is required along a specific direction or when mapping an area on the image: the negative can be moved about in the laser beam in controlled steps using a suitable, standard holder for the optical bench.

A further point concerns the absolute value of the interplanar spacing which is to be quoted. There is sufficient day-to-day and specimen-to-specimen variation of the microscope magnification (typically several percent) that the exact absolute magnification is unknown for any given image. The best procedure is to determine the lattice spacing accurately by alternative means such as x-ray diffraction and equate this value with the average spacing determined over many fringes. This provides a reliable calibration for the image magnification which can be used for subsequent localized measurements on that image.

19.5 COMPOSITION DETERMINATION

The variation of lattice constant (a) with composition (c) is a well-known phenomenon. Extensive data exist in the literature for almost all binary combinations of elements and materials (e.g. PEARSON 1958) and for many ternary and quaternary systems. Thus there is a reliable foundation on which we can base our conversion of fringe spacings into composition. However the analysis requires a reasonable value of da/dc for accurate composition determination and is limited somewhat in this respect. If da/dc is small it is necessary to average the spacing measurement over large distances to detect small changes hence sacrificing resolution. It is worthwhile considering this aspect in more detail to identify favorable circumstances for the application of the lattice image approach and to compare it with alternative techniques.

Figure 19.4a shows the idealized variation of lattice parameter with composition of a binary alloy. If a can be established to within ±Δa, then the corresponding uncertainty in composition is Δc. For larger da/dc, Δc is smaller value. Figure 19.4b shows the corresponding number of fringes required to achieve this degree of precision. The appropriate expressions in algebraic form are

$$a \pm \Delta a = \frac{Na \pm \delta a}{N} \tag{19.9}$$

where δa is the absolute error in determining the lattice constant from measuring a single fringe spacing. In terms of relative errors

$$\frac{\Delta a}{a} = \frac{1}{N} \frac{\delta a}{a} \tag{19.10}$$

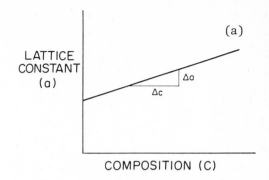

LATTICE
CONSTANT
(a)

(a)

Δa

Δc

COMPOSITION (C)

Fig. 19.4. (a) Schematic variation of lattice parameter (a) with composition (c) in a binary system, showing the uncertainty in composition (Δc) for an experimental uncertainty in lattice spacing (Δa). (b) Schematic representation of the number of fringes (N) which must be measured to achieve a given uncertainty in lattice spacing (Δa).

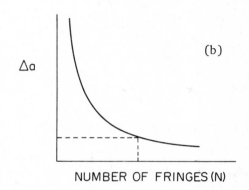

Δa

(b)

NUMBER OF FRINGES (N)

There are two extremes for our experiment. Either it is desirable to detect composition variations over a specific distance or spatial resolution in material, or it is necessary to identify whether there are specific composition differences occurring within the specimen. In the former case the resolution required determines N, hence Δa (fig. 4b) and consequently identifies the degree of segregation which may be detected. Conversely, in the latter case, the Δc to be found sets Δa, which in turn implies the distance (number of fringes, N) over which the measurement must be made. For example in a system with $d/dc(\Delta a/a) = 0.15$ (which is large) a 10% composition change requires a precision $\Delta a/a$ of 1.5%, which can be measured over four fringes in a good image ($\lesssim 10$ Å). A 1% composition change can be established with 100 Å resolution.

In systems with rather smaller d/dc ($\Delta a/a$), such as 0.015, the corresponding figures are 1000 Å to detect 1% composition changes, 100 Å resolution for 10% difference in C. A master plot showing the interrelationships of composition and resolution for various d/dc ($\Delta a/a$) is given in Fig. 19.5 (using $\delta a/a = 5\%$).

The electron optical calculations of section 2 allow a cut-off to be made when the precision of the measurement approaches the deviation from true lattice spacing in a bent foil. If electron-optical effects of about 0.01% occur in the specimen, then it is dangerous to measure over 500 fringes, or 1000 Å, in the above example. The useful range of the lattice image method appears then to be 10-1000 Å, with detectability down to 0.1% in composition for the best circumstances.

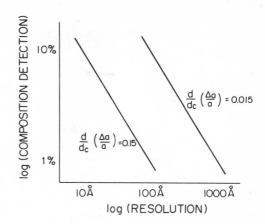

Fig. 19.5. Relationship between precision of composition determination and spatial resolution, for alloy systems with either large (0.15) or small (0.015) variations of lattice constant with composition.

It is clear from Fig. 19.5 that the present approach is very sensitive for materials with large variation of lattice constant with composition (e.g. $d/dc(\Delta a/a) \sim 0.15$). Reasonable composition differences can be detected at extremely high resolution, more subtle differences at respectable resolutions especially in comparison to current spectroscopic techniques. Such materials are not uncommon, and indeed some are technologically quite important (e.g. Fe-C, Aℓ-Cu, Cu-Be). Most interstitial solid solutions have a large $d/dc(\Delta a/a)$. It is interesting to note that the analysis of the common interstitial elements, carbon, nitrogen, and oxygen, is not possible with current x-ray energy dispersive systems but represents one of the most advantageous situations for applying the lattice image method. For systems with smaller d/dc ($\Delta a/a$) (e.g. 0.015) the analysis is not so sensitive but is still similar to that of current alternative spectroscopy (e.g. detection of several percent change in composition with a few hundred Å resolution.). One further advantage accrues because we are dealing with an _image_ of the specimen: thus composition variations may be established in the vicinity of a prominent microstructural feature (e.g. precipitate, grain boundary) to exact spatial precision.

On the other hand many materials of commercial relevance are neither simple binaries nor well-characterized multicomponent systems. Some specimens of interest may be non-crystalline, for which lattice imaging is just not applicable. Furthermore the experimental requirements for lattice imaging are quite severe, requiring careful control of a number of sensitive parameters. Thus the elegance of being able to microanalyze any region, in any material for any desired element is not possible. In summary the technique may not be applied in general, but in favorable circumstances it is both more sensitive and has higher resolution than conventional methods.

Finally, it has been assumed that the local lattice spacing is identical to that of bulk material with the same composition, which is probably valid for compositions slowly varying with distance. However when large lattice constant differences occur over short distances high coherency strains result. The mutual constraint of the two regions must locally alter their interplanar spacing. The measured difference Δa will be smaller than that of unconstrained lattices, yielding values for composition differences which are smaller than are actually present. This becomes a more important problem for segregations large in degree but small in extent. No rigorous analysis has yet been performed for such situations but its influence should always be borne in mind when interpreting fringe spacings in terms of composition.

19.6 EXPERIMENTAL EXAMPLES

A number of investigations have appeared recently in the literature which have employed fringe images either to establish local composition or to demonstrate the presence of segregation. These results illustrate the application of lattice imaging for microanalysis.

The ability to specify low atomic number, interstitial content in alloys is demonstrated by the work of KOO and THOMAS (1977). By suitable heat treatment it is possible to produce "duplex" steels which have a mixture of high carbon martensite interspersed in a low carbon ferrite. The intimate combination of hard and soft phases confers desirable mechanical properties to the alloy. From knowledge of the iron-carbon phase diagram and the temperature of treatment, the degree of carbon segregation can be predicted. Analysis of the lattice fringe spacings in the two phases is in exact agreement with the prediction. For example in Fig. 19.6, (110) fringe spacings of 2.03 Å and 2.05 Å are found in ferrite and martensite respectively corresponding to 0.0% and 0.3% carbon content by weight. This study shows the importance of lattice imaging for analysis of interstitials and that small compositon differences can be detected when $d/dc(\Delta a/a)$ is large (~0.2 in this case). In addition the excellent agreement between experiment and theory, from a knowledge of the material, gives confidence to this type of analysis especially in light of possible anomalous effects discussed in previous sections.

Spinodal decomposition produces a very fine-scale composition modulation, the wavelength of which is often too small for investigation by spectroscopic methods (e.g. $\lambda \lesssim 100$ Å). The concomitant modulation of interplanar spacing is readily detectable by lattice imaging (SINCLAIR et al. 1976). Fig. 19.7 shows examples for the situations of both large (Fig. 19.7a) and small (Fig. 19.7b) variations of lattice constant with composition. In the case of Au-Ni [$d/dc(\Delta a/a) \sim 0.15$] quite high precision can be achieved by averaging over only five fringe spacings. This is sufficiently short to show the nature of the wavelike modulation and to reveal clearly the large differences in interplanar spacing from one region to another (Fig. 19.7a). Alternatively in Cu-Ni-Cr [$d/dc (\Delta a/a) \sim 0.001$] it is necessary to average over larger distances (fifteen fringes) to reveal similar behavior (Fig. 19.7a) but since the wavelengths are generally larger in this system, the spinodal waves are still detected (WU, SINCLAIR and THOMAS 1978). It is also instructive to note the different approach to these measurements: direct print measurements for Au-Ni and optical diffraction for Cu-Ni-Cr. Once again a

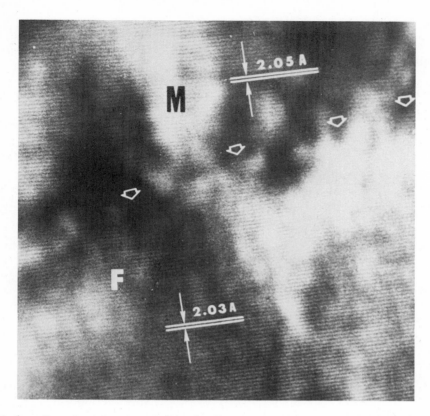

Fig. 19.6. Lattice image of "duplex" martensite-ferrite steel. The lattice constants in the two phases establish their carbon content (courtesy of J. Y. Koo and G. Thomas).

high degree of confidence can be placed in these results since the modulation wavelengths determined from the fringe images correspond exactly with those found by conventional experiments, including bright field microscopy and selected area electron diffraction.

Grain boundary segregation and precipitation are widely recognized for their influence on the failure of polycrystalline metals, and this is an important area where fringe imaging can contribute. GRONSKY (1976) has demonstrated solute segregation in the vicinity of grain boundary particles in the Aℓ-Zn system, which has small $d/dc(\Delta a/a)$. Two situations have been recognized: one with no segregation (Fig. 19.8a) and one where there are composition variations near the precipitate-matrix interface in both phases (Fig. 19.8b). These relate to the mechanism of precipitate growth and hence are observations necessary to understand the fundamentals of this phenomenon (GRONSKY and THOMAS 1977).

Composition differences are also important to the properties of ceramics, particularly to their feasibility for high temperature structural applications. Various analytical techniques have been used recent-

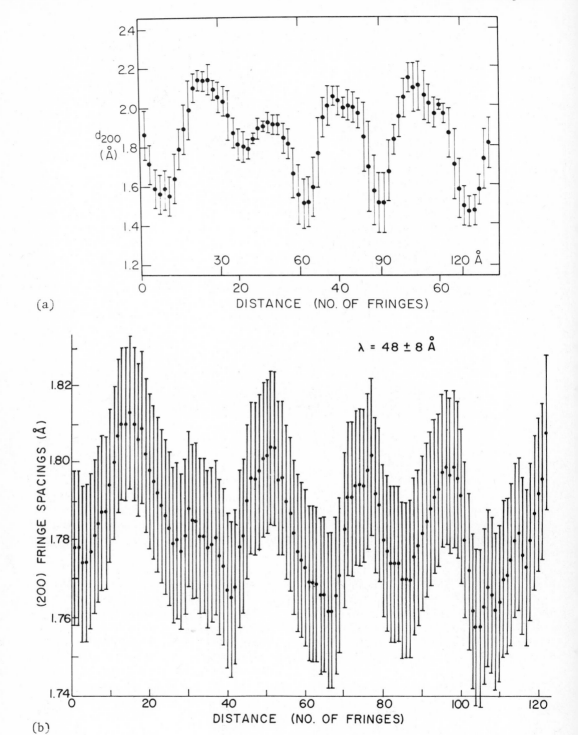

Fig. 19.7. Variations of (200) d-spacing in spinodal alloys: (a) for Au-Ni, with large lattice parameter-composition variation; (b) for Cu-Ni-Cr, with small lattice parameter-composition variation (courtesy of R. Gronsky and C.K. Wu respectively).

ly to study this problem, and an interesting combination of fringe-spacing and energy-dispersive data has been made by CLARKE, (1978). Figure 19.9a shows the lattice image near a grain boundary in a Mg-Si-Aℓ-O-N alloy. An abrupt change of fringe periodicity can be seen, corresponding to a change in the local crystal structure from a 12H polytype to 15R. The anion and cation ratios must show commensurate changes, the latter being established from x-ray energy dispersive analysis (Figs. 19.9b, 19.9c). By combining the two pieces of information the exact local composition can be determined (CLARKE, 1978). In this example the boundary layer is $Mg_{1.75}Si_{1.45}Aℓ_{1.8}O_3N_3$ while the overall composition is $Mg_{1.86}Si_{1.67}Aℓ_{2.47}O_{3.19}N_{3.81}$.

Individual point defects can not be detected in the fringe image. The ultimate capability of the technique then is the identification of crystal imperfections whose width is only a single plane spacing. Such defects may be associated with large local changes in composition or may deviate considerably from the normal atomic arrangements of the crystal lattice. The fringe method is able to detect them either by a physical translation of the fringes (e.g. at a translational antiphase boundary (SINCLAIR and DUTKIEWICZ, 1977)) or by a local change in fringe spacing for the specific case of fringes parallel to the fault. The former gives infomation about the fault displacement, the latter about the local atomic arrangements, and when combined they can be used to extensively characterize the defect. Examples previously reported include G. P. zones in Aℓ-Cu (PHILLIPS, 1973), thought to be single atomic planes enriched in Cu in the Aℓ lattice, and non-conservative antiphase boundaries in ordered Ni_4Mo (SINCLAIR and THOMAS 1978), with a local change of composition from Ni_4Mo to Ni_6Mo. Figure 19.10 shows a fault occurring at a twin boundary in MiTi martensite and the local atomic arrangements thought to be associated with it. The change in fringe periodicity is much larger than the limitation of precision for single fringe measurements. The imperfection is thus readily detected and leads to the proposed model consistent with the materensite reaction and internal twinning in this alloy.

19.7 FUTURE DIRECTIONS

Lattice fringe imaging has probably reached its ultimate sensitivity and further refinement of the experimental production of images appears unnecessary. Thus images can now be taken routinely of the whole spectrum of materials at about the 2 Å level and in any desired orientation with the availability of suitable goniometer stages. The introduction of objective lenses with the combined capability of lattice resolution and energy dispersive spectroscopy is an exciting prospect since it will then be possible to complement information from the two techniques, the role of lattice imaging being for high resolution analysis and the estimation of interstitial solute content. Some steps need to be taken to obtain images under well-known experimental conditions of exact orientation and thickness variation especially for comparision with computed images. Convergent beam and micro-diffraction methods are the most accurate for local thickness and orientation determination respectively, both of which are possible with the dual-purpose

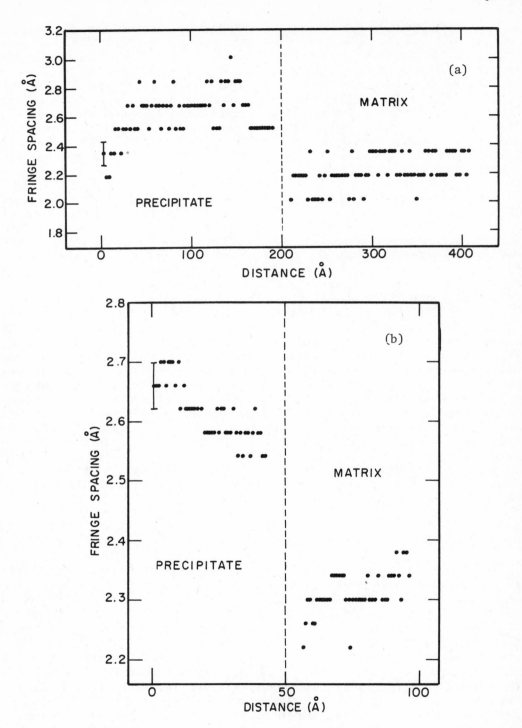

Fig. 19.8. Variation of (111) lattice spacing in the vicinity of grain boundary precipitates of Aℓ-Zn: (a) with no solute segregation; (b) with solute segregation in the vicinity of the precipitate-matrix interface (courtesy of R. Gronsky).

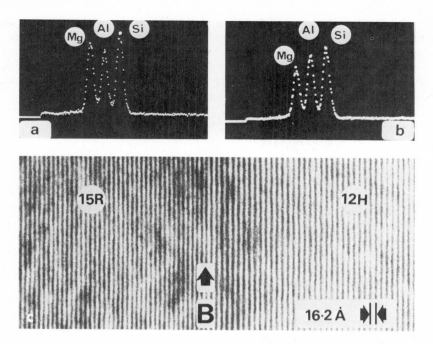

Fig. 19.9. (a) Lattice image and x-ray energy dispersive analysis of a
Mg-Si-Aℓ-O-N alloy. X-ray energy dispersive spectra are shown for the
15R boundary phase (a) and in the 12H matrix phase (b). The smaller
lattice spacing of the 15R phase, adjacent to a grain boundary, is clearly seen in the lattice image (c). Combination of the data allows identification of local alloy composition (courtesy of D. R. Clarke).

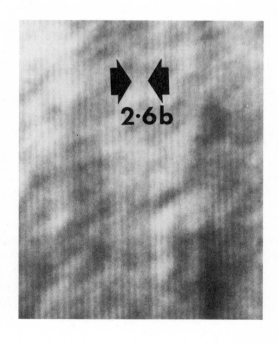

Fig. 19.10. A lattice image of
NiTi showing an anomalous lattice
spacing at a twin boundary. A
model for the local atomic arrangements can be proposed from
the image.

lens cited above. However for specimens normally used for lattice imaging (high quality foils, not severely bent) the electron optical deviations from true lattice spacing do not appear to be significant. The role of elastic constraint in the vicinity of large lattice spacing variations needs further consideration, although the results discussed in the previous section are most encouraging.

The next step forward appears to be through the direct resolution and imaging of atomic arrangements, similar to the current structure imaging of oxides and minerals. This requires an increase in resolution, probably best achieved by higher voltage (smaller electron wavelength) instruments, which would also have several other advantages for high resolution experiments and their interpretation (THOMAS, COWLEY, GLAESER and SINCLAIR, 1976). Progress through lower spherical aberration lenses on 100kV microscopes is a possible lower cost alternative for most TEM laboratories and it will be interesting to follow developments with this approach, especially with AFF imaging.

19.8 SUMMARY

The lattice fringe imaging method provides an alternative approach for obtaining information about local composition through the dependence of interplanar spacing on composition. When the images are taken under controlled experimental conditions in good quality foils, the fringe spacing is virtually identical to the true crystal lattice spacing. In favorable circumstances rather small composition variations can be detected at higher resolutions than with current spectroscopic techniques (e.g. 10% composition difference with 10 $\overset{\circ}{A}$ resolution, 1% difference with 100 $\overset{\circ}{A}$ resolution), and this includes the analysis of interstitial elements in alloys (e.g. C,N,O) which are notoriously difficult to treat spectroscopically. The technique is not universally applicable, being limited to crystalline systems with reasonably well documented variation of lattice constant with composition. Highly complex commercial materials may for instance be intractable. A number of experimental examples has been presented which supports the confidence of the present article and illustrates the major strength of the method: for highly localized composition variations.

ACKNOWLEDGMENTS

Financial support for this research is provided by the NSF-MRL program through the Center for Materials Research, Stanford University. Experimental facilities are funded by the above grant and by NSF grant #DMR 77-21629. The author would like to thank the workers cited in the figure captions for kindly providing pictures for illustrative purposes, and also members of his research group (especially M.K. Hibbs) for helpful discussions.

REFERENCES

Allpress, J.G., and Sanders, J.V., 1972, <u>Electron Microscopy and Structure of Materials</u> (Berkeley: University of California Press) P. 134.

Allpress, J. G., and Sanders, J.V., 1973, J. Appl. Crystallogr., <u>6</u>, 165.

Anderson, J. S., 1977, J. Physique, <u>38</u>, C7-17.

Amelinckx, S., 1964, <u>The Direct Observation of Dislocations</u> (New York: Academic) p. 194, p. 406.

Bourret, A., DESSEAUX, J., and RENAULT, A. 1977, J. Microsc. Spectrosc. Electron., <u>2</u>, 467.

Buseck, P.R., and IIJIMA, S., 1974, Am. Mineral., <u>59</u>, 1.

Buseck, P. R., and IIJIMA, S. 1975, Am. Mineral., <u>60</u>, 771.

Carlson, A. B., 1968, <u>Communication Systems</u> (New York: McGraw-Hill) p. 223.

Clarke, D. R., 1978, <u>Scanning Electron Microscopy</u>/1978/I, edited by O. Johari (SEM Inc.) p. 77.

Cowley, J.M. and Iijima, S., 1972, Z. Naturforsch., <u>27a</u>, 445.

Cowley, J.M. and Iijima, S., 1977, Physics Today, <u>30</u>(3), 32.

Dowell, W. C. T., 1962, J. Phys. Soc. Japan, <u>17</u> Suppl. BII, 175.

Dowell, W. C. T., 1963, Optik, <u>20</u>, 535.

Gronsky, R. G., 1976, Ph.D. Thesis (Berkeley: University of California, LBL 5784).

Gronsky, R. G., and Thomas, G., 1977, 35th Ann. Proc. Electron Microsc. Soc. Amer. (Baton Rouge, Claitor's) p. 116.

Hanszen, K-J., and Trepte, L., 1971, Optik, <u>32</u>, 519.

Hashimoto, H., Endoh, H., Tanji, T., Ono, A., and Watanabe, E., 1977, J. Phys. Soc. Japan, <u>42</u>, 1073.

Hashimote, H., Kumao, A., and Endoh, H., 1978, 9th Int'l. Congr. on Electron Microsc. (Toronto: Microscopical Society of Canada) Vol. 3, p. 244.

Hashimoto, H., Mannami, M., and Naiki, T., 1961, Phil. Trans. Roy. Soc., <u>253</u>, 459.

Hashimoto, H., Sugimoto, Y., Takai, Y., and Endoh, H., 1978, 9th Int'l
    Congr. on Electron Microsc. (Toronto: Microscopical Society of
    Canada) Vol. 1, p. 284.

Hibbs, M.K., Pirout, P., and Sinclair, R., 1979, in preparation.

Hirsch, P. B. Howie, A., Nicholson, R.B., Paschley, D.W., and Whelan,
    M.J., 1965 <u>Electron Microscopy of Thin Crystals</u> (London: Butter-
    worths) p. 21, p. 161.

Iijima, S., 1971, J. Appl. Phys., <u>42</u>, 5891.

Iijima, S., 1973, Acta Cryst., <u>A29</u>, 18.

Iijima, S., and Allpress, J.G., 1974, Acta Cryst., 22, 29.

Iijima, S., and Buseck, P.R., 1978, Acta Cryst. <u>A34</u>, 709.

Iijima, S., and Cowley, J.M., 1977, J. Physique, <u>38</u>, C7-135.

Iijima, S., Kimure, S., and Goto, M., 1973 Acta Cryst., <u>A29</u>, 632.

Iijima, S., Kimura, S., and Goto, M., 1974, Acta Cryst., <u>A30</u>, 251.

Izui, K., Furuno, S., and Otsu, H., 1977, J. Electron Microsc., <u>26</u>, 129.

Izui, K., Nishida, T., Furuno, S., Otsu, H., and Kuwabara, S., 1978, 9th
    Int'l. Congr. on Electron Microsc. (Toronto: Microscopical Society
    of Canada) Vol. 1, p. 292.

Koo, J.Y., and Thomas, G., 1977, 35th Ann. Proc. Electron Microsc. Soc.
    Amer. (Baton Rouge: Claitor's) p. 118.

Krivanek, O. L., Isoda, S., and Kobayashi, K., 1977, Phil. Mag., <u>36</u>,
    931.

Krivanek, O. L., Tsui, D.C., Sheng, T.T., and Kamgar, A., 1978, <u>Physics
    of SiO$_2$ and its Interfaces</u> (Yorktown Heights: IBM).

Lynch, D. F., Moodie, A. F., and O'Keefe, M. A., 1973, Acta Cryst., <u>A31</u>
    300.

McConnell, J. D. M., Hutchison, J. L., and Anderson, J. S., 1974, Proc.
    Roy. Soc., <u>A339</u>, 1.

Menter, J. W., 1956, Proc. Roy. Soc. <u>A236</u>, 119.

O'Keefe, M. A., 1973, Acta Cryst., <u>A29</u>, 389.

O'Keefe, M. A. and Sanders, J. V., 1975, Acta Cryst., <u>A31</u>, 307.

Pearson, W. B., 1958, <u>Handbook of Lattice Spacings of Metals and
    Alloys</u> (London: Pergamon).

Phillips, V. A., 1973, Acta Met., $\underline{21}$, 219.

Rez, P. and Krivanek, O. L., 1978, 9th Int'l. Congr. on Electron
    Microsc. (Toronto: Microscopical Society of Canada), Vol. 1,
    p. 288.

Scherzer, O., 1949, J. Appl. Phys., $\underline{20}$, 20.

Sinclair, R. and Dutkiewicz, J., 1977, Acta Met., $\underline{25}$, 235.

Sinclair, R., Gronsky, R., and Thomas, G., 1976, Acta Met., $\underline{24}$, 789.

Sinclair, R., Schneider, K., and Thomas, G., 1975, Acta Met., $\underline{23}$, 873.

Sinclair, R. and Thomas, G., 1977, J. Physique, $\underline{38}$, C7-165.

Sinclair, R. and Thomas, G., 1978, Met. Trans., $\underline{9A}$, 373.

Thomas, G., Glaeser, R. M., Cowley, J. M., and Sinclair, R., 1976
    Report of Workshop on High Resolution Electron Microscopy
    (Berkeley: Lawrence Berlekey Laboratory #106).

Van Lunduyt, J. and Amelinckx, S., 1975, Am. Mineral, $\underline{60}$, 351.

Wu, C. K., Sinclair, R., and Thomas, G., 1978, Met. Trans., $\underline{9A}$, 381.

NOTE ON KEY REFERENCES

The above references list represents only a cross-section of the
lattice imaging literature and is not exhaustive. The following
articles are thought to be important for the present treatment of lattice imaging for compositional analysis.

Scherzer (1949) for the influence of spherical aberration and defocus on
    the relative phase of electron imaging waves.

Menter (1956): The first experimental demonstration of lattice imaging.

Hashimoto et al. (1961) for considerations of electron optical effects
    on fringe periodicity.

Dowell (1963) for identification of the benefits of tilted illumination,
    two-beam fringe imaging.

Iijima (1971), Cowley and Iijima (1972, 1977) for introduction of the
    structure imaging method.

Allpress and Sanders (1973): a major review of lattice imaging, from
    experimental and theoretical points-of-view.

Phillips (1973) for the application and structural interpretation of 2 $\overset{\circ}{A}$
      fringe images with respect to a classic metallurgical problem.

Sinclair and Thomas (1978) for considerations of lattice imaging for com-
      positional microanalysis in general.

CHAPTER 20

WEAK-BEAM MICROSCOPY

JOHN B. VANDER SANDE

DEPARTMENT OF MATERIALS SCIENCE AND ENGINEERING
MASSACHUSETTS INSTITUTE OF TECHNOLOGY
CAMBRIDGE, MASSACHUSETTS 02139

20.1 INTRODUCTION

Transmission electron microscopy of thin, crystalline samples has been profitably performed for the past two decades. A significant fraction of the transmission electron microscopy performed has focused on the observation of strain-related defects, such as point defect clusters, dislocations, etc. associated with the microstructure of the crystalline thin sample. Even though these strain-related defects are often considered with reference to an angstrom-unit-sized displacement, for instance a Burgers vector in the case of a dislocation, the transmission electron microscope image of such a defect, observed in bright-field or strong-beam dark-field microscope, can be hundreds of angstroms in size. It is well known that a dislocation has a bright-field image whose peak width at half minimum is between $\xi_g/3$ and $\xi_g/5$, where ξ_g is the extinction distance, often many hundreds of angstroms in size (HIRSCH et al., 1965).

For many years these wide images from strain-related defects were considered to be a standard result of transmission electron microscopy, although the consequences of wide images are many. In fact, wide images produce an effective resolution limitation on microstructural observations which is much more severe than the operating restrictions created by the optical resolution of a modern-day transmission electron microscope. The microscopist's ability to resolve dense arrays of dislocations, second phases in the vicinity of dislocations, closely spaced Shockley partial dislocations, etc. is directly related to the image width of the defects being observed. If two dislocation images significantly overlap, then, even though the separation of the dislocations may be ~ 150 Å, they will not be resolved as two individual dislocations. If wide defect images can be replaced with narrower images, an improvement in "practical" resolution will follow.

This chapter deals with the high resolution technique of weak-beam electron microscopy whereby narrow (∿ 15 Å) image widths are produced from strain-related defects. The technique was pioneered by COCKAYNE and coworkers (COCKAYNE, 1972; COCKAYNE, RAY, and WHELAN, 1969) in the late 1960's and early 1970's.

20.2 THEORETICAL BACKGROUND

A significant insight into the strengths of the weak-beam technique can be gleaned from a relatively simple version of the dynamical theory of electron diffraction, namely, the Howie-Whelan equations without absorption. This is the approach chosen in the present case. A more fully developed theoretical approach is available (COCKAYNE, 1972).

Strong-Beam Images

The Howie-Whelan equations for a perfect crystal not including absorption can be written in the form:

$$\frac{d\phi_o}{dz} = \frac{\pi i}{\xi_g} \phi_g \tag{20.1}$$

$$\frac{d\phi_g}{dz} = \frac{\pi i}{\xi_g} \phi_o + 2\pi i s \phi_g \tag{20.1}$$

where ϕ_0 and ϕ_g are the transmitted and diffracted amplitudes for the two beams under consideration, ξ_g the extinction distance for reflection g, i is the imaginary number, z is a positional vector in the beam direction, and s is the deviation parameter defined from the Ewald sphere construction (to be discussed in a later section) which measures the degree to which the Bragg condition is satisfied.

The dependence of the resultant transmitted and diffracted intensities, $I_0 (= \phi_0\phi_0^*)$ and $I (= \phi_g\phi_g^*)$, on the magnitude of the deviation parameter are made clear in Fig. 20.1. Figure 20.1(a) shows the behavior of I_0 and I_g as a function of postion, z, below the crystal surface (z=0) when the crystal is oriented at the Bragg angle (s=0). A strong dynamical interaction between the transmitted and diffracted beams is evident. For instance, at positions z = (2n + 1) $\xi_g/4$ (minima in I_0 or I_g occur at depth intervals $\Delta z = \xi_g$), $I_0 = I_g$ emphasizing the dynamical nature of the scattering. The result from Fig. 20.1(a) is that electron intensity can be readily redistributed between transmitted and diffracted beams when the crystal is oriented for strong Bragg diffraction.

The depth dependence of I_0 and I_g for the situation where the crystal is far from the Bragg condition ($|s| \gg 0$) is considerably different. Figure 20.1(b) shows the intensity variation with depth in this

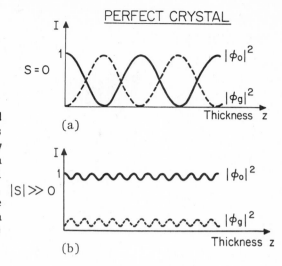

PERFECT CRYSTAL

(a)

(b)

20.1.   Transmitted ($|\phi_o|^2$) and diffracted ($|\phi_g|^2$) intensities as a function of position, z, below the entrance surface of a thin foil.   The   two-beam   dynamical theory, not including absorption, has been used to calculate these results.   In   (a)   the   deviation parameter,   s = 0   where as in (b) $|s| \gg 0$.

case.   It is clear that the transmitted intensity modulates about unity and the diffracted intensity modulates about zero.   The strong dynamical coupling seen when s=0 is absent and, in fact, this is a kinematical scattering condition.

When an imperfection is introduced into the crystal, the Howie-Whelan equations are modified due to the presence of the defect.   The character of the defect is defined by its displacement field, $\bar{R}$.   The form of Eqn. 20.1 is now modified:

$$\frac{d\phi_o}{dz} = \frac{\pi i}{\xi_g} \phi_g$$

$$\frac{d\phi_g}{dz} = \frac{\pi i}{\xi_g} \phi_o + 2\pi i(s + \bar{g} \cdot \frac{d\bar{R}}{dz}) \phi_g$$

(20.2)

When presented in this form it can be seen that the defect enters into the equations through the term $\bar{g} \cdot d\bar{R}/dz$ which is added to the geometrically defined deviation parameter, s.   The term $d\bar{R}/dz$ is a plane-bending term, and the dot product with $\bar{g}$ describes how the plane bending is affecting the diffraction condition.   That is to say that in the vicinity of the defect the atomic planes can be locally reoriented (bent) into the Bragg condition.   In this case, the term $\bar{g} \cdot d\bar{R}/dz$ is equal in magnitude but opposite in sign to the deviation parameter, s.   Therefore, s' = s + $\bar{g} \cdot d\bar{R}/dz$ = 0, and by analogy Eqn. 20.1, strong dynamical coupling of I_0 and I_g is expected.   Note that s and $\bar{g} \cdot d\bar{R}/dz$ can have like signs in which case the planes in the vicinity of the defect are bent farther from the Bragg condition.

Weak-Beam Images

Using the information presented above, consider the change in diffracted intensity when a crystal containing a dislocation, and oriented
away from the Bragg condition, is observed. Such a geometry is shown
schematically in Fig. 20.2(a). Here it can be seen that the "perfect"
part of the thin foil is far from the Bragg condition.

In Fig. 20.2(b) is shown the change in the diffracted intensity at
each position, z, below the foil surface for a specific column labeled
(in Fig. 20.2(a)) A, B, C, D. Note that from position A to B, the diffracted intensity varies slightly from zero. This is a consequence of
the large deviation from the Bragg condition and follows directly from
Fig. 20.1(b). However, from position B to C, the planes have been bent,
due to the presence of the dislocation. In fact, the plane bending is
such that, locally, the exact Bragg condition is satisfied, that is, s
and $\bar{g} \cdot dR/dz$ are equal in magnitude but opposite in sign so that the
deviation parameter $s' = s + \bar{g} \cdot d\bar{R}/dz = 0$. Refering to Fig. 20.2(b),
since a dynamical diffracting condition now exists for the region between points B and C, electrons are transferred from the transmitted
beam to the diffracted beam, similar to Fig. 20.1(a), and the diffracted
intensity increases. From points C to D, a kinematical condition again
exists and the intensity gained from B to C, if it is not large, is retained.

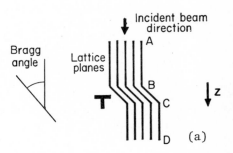

Fig. 20.2. (a) A schematic edge
dislocation where the perfect part
of the crystal is oriented far from
the Bragg angle. (b) The diffracted
intensity as a function of depth, z,
below the entrance surface of the
thin crystal.

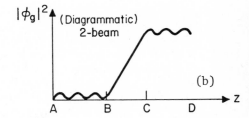

The result of this small exercise is that if a dark-field experiment is performed where the crystal is far from the Bragg condition (the
diffracted beam is weak) a small segment of the displacement field of a
defect for which $s + \bar{\bar{g}} \cdot dR/dz = 0$ can give rise to a significant intensity above background. The width of this image peak and its position

can be better defined with respect to Fig. 20.3. In Fig. 20.3 a dislocation is being observed. In the bottom part of this figure various plots of $\bar{g} \cdot d\bar{R}/dz$ as a function of z are presented at varying distances from the dislocation core. Superimposed upon the $\bar{g} \cdot d\bar{R}/dz$ plots is the value of -s. On one side of the dislocation it can be seen that -s and $\bar{g} \cdot d\bar{R}/dz$ can never be equal (s' ≠ 0). However, on the other side of the dislocation, and specifically for the column pp', -s = $\bar{g} \cdot d\bar{R}/dz$, therefore s' = 0 and a significant perturbation in the weak-beam occurs at that position.

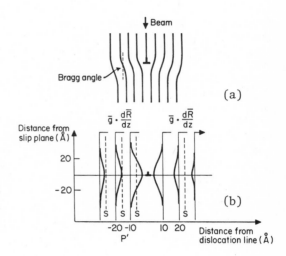

Fig. 20.3. (a) A schematic edge dislocation. (b) Plots of $\bar{g} \cdot d\bar{R}/dz$ versus z as a function of distance away from the edge dislocation. Note the relation of -s to $\bar{g} \cdot d\bar{R}/dz$ for each plot.

The weak-beam technique will produce narrow (∿ 15 Å), intense (peak/background > 100), positionally well defined image peaks from strain-related defects. A fundamental explanation for this phenomenon has been presented above. A more rigorous treatment is available and the serious practitioner is referred to that treatment (COCKAYNE, 1972).

20.3 THE PRACTICE OF WEAK-BEAM MICROSCOPY

Instrumental Needs

In order to use the weak-beam technique the microscopist must have available a good condition transmission (or scanning transmission) electron microscope equipped with a beam deflection system, a two-axis tilting stage, and preferably equipped with thin film objective apertures and a bright electron source such as a pointed W filament or LaB_6 filament. The last item is by no means a necessity but, if available, can be very useful in the application of this technique.

A two-axis tilting stage is desirable because this allows the sample to be oriented so that a specific reflection can be used to produce an image. Since this is also a typical aim of the microscopist working in bright-field, most materials science instruments are equipped

with such a stage. It is important that the weakly diffracted beam be
aligned on the optic axis of the objective lens to minimize spherical
aberration and aid in astigmatism corrections. To accomplish this, a
beam deflection system is required. Again, most modern-day instruments
are equipped with this device. Thin film objective apertures are desirable because they minimize short-term astigmatism associated with aperture contamination. Finally, the brighter the electron source available, the shorter the exposure time to produce an acceptable, contrasty
photographic negative.

The items listed above are named in an effort to insure that the
instrument is placed in a configuration capable of using the improved
resolution available by this technique. As might be expected, many of
the experimental difficulties associated with putting the weak-beam technique into practice are a direct result of the small absolute intensities in weak-beam images even though the relative intensity (peak to
background) is very large. This condition puts requirements on the
mechanical stability of microscope, specimen stage, and specimen which
are much more stringent than the requirements for bright-field microscopy because of the long exposure times required. The usual intensities available in weak-beam microscopy are such that an image cannot
be observed on the screen of the microscope and   require exposure
times of ∿ 30 seconds or longer to properly record. All efforts to improve instrument mechanical stability, or reduce exposure times, such as
faster plates, brighter electron sources, etc., are potentially worthwhile in attempting to put weak-beam microscopy into practice.

Establishing a Weak-Beam Condition

It has been determined (see COCKAYNE, 1972) that a beam is weak
enough to maintain all of the advantages of the technique if $s \geq 2$ x
10^{-2} $\overset{\circ}{A}{}^{-1}$. Below will be described techniques for establishing a weak-
beam condition, but the reader is cautioned that not all reflections
used will yield a sufficiently large s.

i The Ewald Sphere Construction for Strong-Beam Microscopy

It might be most educational to refer to three "spaces" when considering diffraction conditions. These "spaces" are the real space in
which the object resides, Ewald "space," and the screen of the microscope observed while the instrument is in the selected area diffraction
mode. Figure 20.4(a) shows these three spaces, schematically, for the
case of a good two-beam diffraction geometry where the transmitted beam
is being used to form a bright-field image. The uppermost segment of
Fig. 20.4(a) shows the thin crystal oriented near the exact Bragg condition. Various diffracted rays are produced, but only g is strongly diffracted.

The middle section of Fig. 20.4(a) shows the Ewald sphere construction for this geometry. As it must, the origin of the reciprocal
lattice, o, is on the sphere of reflection, as is g in this case.

Reflection $\bar{g}$ being on the sphere of reflection, is consistent with the
Bragg condition being satisfied. In the lower portion of Fig. 20.4(a)
is a schematic of the diffraction pattern observed for this geometry.
The transmitted beam and the reflection $\bar{g}$ are intense and the Kikuchi
band (shown here by vertical lines at the edge of the Kikuchi band) intersects both transmitted and strongly diffracted beams. An aperture is
placed around the transmitted beam, which lies on the optic axis, and a
bright-field image would be formed. Figure 20.4(b) shows an actual diffraction pattern for such a geometry. Note the two intense spots (o and
$\bar{g}$) and the position of the Kikuchi band.

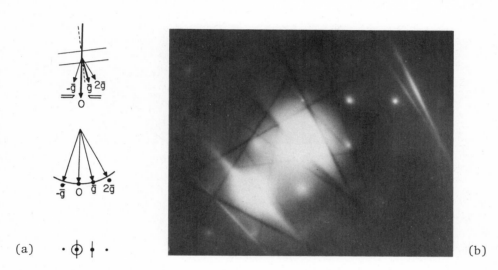

(a)   (b)

Fig. 20.4. (a) A crystal oriented at the Bragg angle for reflection $\bar{g}$
displayed in real space (upper), "Ewald" space (middle), and in a schematic diffraction pattern (lower). An actual diffraction pattern for
such a case is shown in (b) where $\bar{g}$ = (220) for an Al crystal.

ii The Ewald Sphere Construction for Weak-Beam Microscopy

Starting from the two-beam geometry above, consider tilting the incident illumination so that reflection $\bar{g}$ now lies on the optic axis.
This is shown in the upper part of Fig. 20.5(a). The Ewald sphere construction for this geometry is shown in the middle section of Fig.
20.5(a). By the definition of the Ewald sphere, when the incident beam
direction was changed, so, too, must the position of the Ewald sphere
change. As shown, the Ewald sphere position has changed such that $3\bar{g}$
falls near the Ewald sphere. Note that the reciprocal lattice has not
moved since the crystal has not been moved, only the Ewald sphere.
Note, too, that in moving $\bar{g}$ onto the optic axis it has, by necessity,
been moved off of the Ewald sphere, i.e., it is now weak. The diffraction pattern schematic now shows all beams weak except the transmitted
beam. The Kikuchi band has not moved since the crystal has not been
tilted. Now the objective aperture is placed around $\bar{g}$, which lies on

the optic axis, to form a weak-beam image. An actual diffraction pattern for this geometry is shown in Fig. 20.5(b).

(a)   (b)

Fig. 20.5. (a) The same crystal as in Fig. 20.4(a), but here the incident illumination has been tilted to put reflection g on the optic axis of the objective lens. Again, real space (upper), "Ewald" space (middle), and a schematic diffraction pattern (lower) are shown. In (b) an actual diffraction pattern is shown for this geometry. Note that s3g > 0 as it should be for best results.

This technique, known as $\bar{g}(3\bar{g})$ meaning that one starts with $\bar{g}$ satisfying the Bragg condition and after beam tilting has $3\bar{g}$ satisfying the Bragg condition, has one distinct advantage. Since Fig. 20.3(a) is the proper condition for bright-field microscopy <u>and is also one starting point for a weak-beam geometry</u>, simply tilting the illumination, starting with a two-beam condition, to bring the initially strongly diffracted beam on optic axis, via beam deflection, will produce the weak-beam geometry. Therefore, by switching between the "no-beam deflection" and "beam deflection" conditions the operator switches between bright-field and weak-beam microscopy. <u>No specimen tilt is required</u>.

Other geometries are possible. For instance, the operator can start in a two-beam case, bring the weakly diffracted beam on axis via beam deflection and then tilt the crystal to make that beam weak. This is called the $\bar{g}(-\bar{g})$ approach and was the approach used in the early weak-beam work (COCKAYNE, RAY and WHELAN, 1969). The obvious disadvantage is that specimen <u>and</u> beam tilting is required.

iii Determining the Deviation Parameter, s

As already stated, a deviation parameter of $s \geq 2 \times 10^{-2} \text{ Å}^{-1}$ is needed to use the weak-beam technique optimally. The deviation parameter can be found from the equation

$$S = \frac{(1-n)g^2}{2k} \qquad (20.3)$$

where n is the order of the reflection intersecting the Ewald sphere (n need not be integer), g is the magnitude of the first order reflection in the systematic row and $k = |1/\lambda|$ where λ is the electron wavelength. Note that s can be made larger by either increasing n [this can be done by having $\underline{s} > o$ in the two-beam case and then going to the $\underline{g}(3\underline{g})$ condition] or g [that is, s will increase as one goes from $\underline{g}$ = (111), to (200), to (220) while using the $\overline{g}(3\overline{g})$ condition].

20.4 APPLICATIONS OF WEAK-BEAM MICROSCOPY

All of the examples that follow exhibit the advantages of weak-beam microscopy. The major advantage, as previously stated, is that a narrow (∿ 15 Å), intense image peak can be produced from a strain-related defect. An additional advantage is that the position of this image peak can be accurately determined by finding where -s = $\underline{g} \cdot d\underline{R}/dz$. The reader should realize that the usual conditions on image formation hold. For instance, if a dislocation is observed, $\underline{g} \cdot \underline{b}$ or $\underline{g} \cdot \underline{b} \times \underline{u}$ must be non-zero in order that an image be produced.

Separation of Partial Dislocations

When two partial dislocations are spaced ∿ 150 Å apart or less, a single image is usually observed in bright-field. If two Shockley partials exhibit their equilibrium separation, and these partials can be observed, an accurate value for stacking fault energy can be determined. The earliest weak-beam work centered on just such observations. Care must be taken to avoid partials that are near either surface of the thin foil and a "core peak" has been theoretically predicted (HUMPHREYS et al., 1978) which can cause complications (that are not unsurmountable) for partial spacings below ∿ 25 Å.

Using the weak-beam technique, Shockley partials have been observed, and stacking fault energies determined for Cu-Al (COCKAYNE, RAY and WHELAN, 1969), Cu (COCKAYNE et al., 1971; STOBBS and SWORN, 1971), Ag (COCKAYNE et al., 1971), Au (JENKINS, 1972), Si (RAY and COCKAYNE, 1971), Ge (RAY and COCKAYNE, 1973; HÄUSERMANN and SCHAUMBERG, 1973; GOMEZ et al., 1975; PACKEISER and HAASEN, 1977); and CdTe (HALL and VANDER SANDE, 1978). An example of an occasionally constricted pair of partial dislocations in CdTe is shown in Fig. 20.6

In addition, observations of superlattice dislocation dissociations have been made in Fe-Al (CRAWFORD and RAY, 1977), NiAl (CAMPANY et al., 1973), and Cu_2MnAl (GREEN et al., 1977).

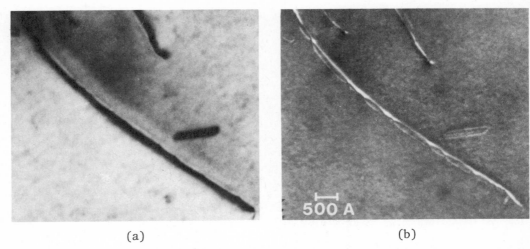

(a)   (b)

Fig. 20.6. A bright-field (a)/weak-beam (b) micrograph pair showing
Shockley partial dislocations in CdTe. The diffraction condition for
the weak-beam micrograph is $\bar{g}(3\bar{g})$. The partial separation is $\sim$ 100 Å.
Occasional constriction of the partial dislocations can be seen.

Dense Defect Arrays: Dislocation Dipoles, Dislocation Tangles, Dislocation Cell Walls

     Subsequent to deforming a crystalline material, dislocations and
stacking faults can be arrayed in dipole form, dislocation tangles or
bundles, dislocation cell walls, etc. Again, since the weak-beam technique provides a narrow image for each dislocation viewed, even closely
spaced dislocations can be resolved as individual dislocations.

     The application of the weak-beam technique to faulted dipoles
(CARTER, 1977) and dense dislocation networks (NORDLANDER and THÖLEN,
1973) has been reported, although the true promise of the technique
applied to dense dislocation networks, after heavy deformation, cyclic
deformation, or recovery, has not yet been fully explored.

     Examples of two areas containing dislocation tangles in deformed Al
are shown in Figs. 20.7 and 20.8 (ALLEN, 1979). In each figure a comparison between bright-field and weak-beam can be made.

Precipitation on Dislocation Lines: Second Phase Particle Interfaces

     Recent work (ALLEN and VANDER SANDE, 1978) reports the results of
experiments in which weak-beam microscopy was used to examine the growth
kinetics of fine second-phase particles which had formed heterogeneously
along dislocation lines in an Al-Zn-Mg alloy. In the early stages of
growth in this system a spherical G.P. zone is produced which is Mg and
Zn rich (the equilibrium precipitate is $MgZn_2$). A weak-beam/bright-
field micrograph pair of a sample aged for 1/2 hour at 150°C is shown in
Fig. 20.9. The arrows in the weak-beam image indicate various locations

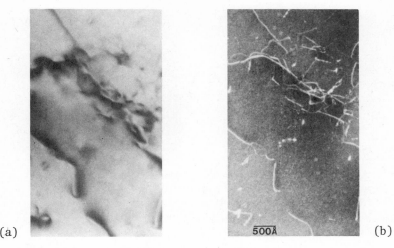

(a)   (b)

Fig. 20.7. A bright-field (a)/weak-beam (b) micrograph pair showing a dislocation tangle in deformed Al. Here $\bar{g}$ = (220) for each and the weak-beam micrograph was taken using the $\bar{g}(3\bar{g})$ condition.

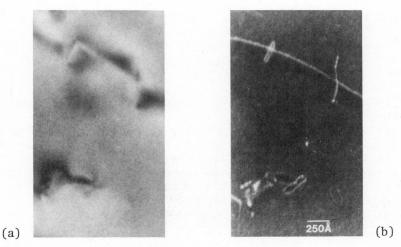

(a)   (b)

Fig. 20.8. A bright-field (a)/weak-beam (b) micrograph pair showing a dislocation tangle in deformed Al. For each of these micrographs g = (220). The condition $\bar{g}(3\bar{g})$ was used to produce the weak-beam micrograph.

where "gaps" have appeared in the otherwise continuous dislocation image. A comparison of the bright-field and weak-beam micrographs in Fig. 20.9 clearly shows that the gaps in the weak-beam image mark those locations along the dislocation line, where precipitates are visible in bright-field. The observation of a gap in the dislocation line image implies that the local strain field is no longer capable of responding to the weak-beam condition. From this it may be concluded that the precipitate associated with the gap has interacted with the dislocation in

 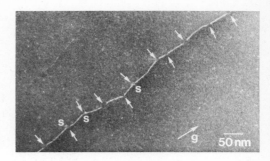

Fig. 20.9. A bright-field (a)/weak-beam (b) micrograph pair taken of an
Al-Zn-Mg sample aged for 1/2 hour at 150°C. For the bright-field micrograph, g = (311). For the weak-beam micrograph, g̅ = (220). Precipitate
gaps are marked with arrows and discussed in the text.

a manner which locally relaxes the lattice strain below the level needed
to produce weak-beam image intensity. In addition, for the large deviation parameters employed in weak-beam it is easily shown that structure
factor contrast induces only an insignificantly small perturbation in an
already low background matrix intensity explaining the absence of particle contrast.

Using the weak-beam technique, gap size observations as small as 15
Å are claimed. A weak-beam/bright-field pair demonstrating such an observation is shown in Fig. 20.10.

The later stages of precipitation on dislocations have also been
investigated (ALLEN and VANDER SANDE, 1979) where the precipitate
develops a lath-like morphology and the dislocation is forced to follow
the crystallography of the lath growth. An example of a lath on a dislocation is shown in Fig. 20.11 where it is easily seen that the dislocation has been reoriented in the vicinity of the lath.

 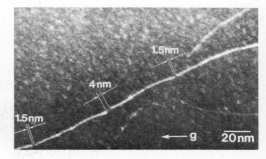

Fig. 20.10. Similar to Fig. 20.9 [bright-field is (a), weak-beam is
(b)] but this sample was aged for 3 minutes at 100°C. For both micrographs, g̅ = (220). Gaps as small as 15 Å are observed.

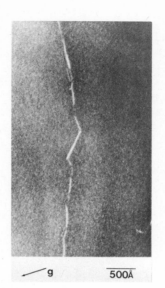

Fig. 20.11. An η(MgZn$_2$) lath in an Al-Zn-Mg alloy heterogeneously nucleated on a dislocation. The dislocation is being forced to conform to the growth crystallography of the lath. Note the G.P. zone "gaps" along other segments of the dislocation line.

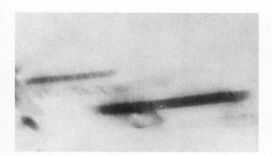 

Fig. 20.12. A bright-field (a)/weak-beam (b) micrograph pair showing large, homogeneously nucleated η(MgZn$_2$) laths in an Al-Zn-Mg alloy. Note the dislocation loops around the laths and the glissile dislocation which has nearly wrapped itself around the perimeter of the right-most lath.

Finally, these authors have also shown the strength of the weak-beam technique when observing dislocations involved in the development of the particle/matrix interface during precipitate growth. An example of the interface of η(MgZn$_2$) laths in Al-Zn-Mg is shown in Fig. 20.12. Note that, in comparison, the weak-beam micrograph in Fig. 20.12 exhibits much more detailed information about the dislocation structure at the interface than does the bright-field micrograph. In particular, at the lath in the lower right-hand corner of the weak-beam micrograph, a glissile dislocation has nearly completely become involved with the precipitate interface. Only a small segment of that dislocation can be seen to reside in the matrix.

Comments

The examples presented above are neither expected to provide a complete review of applications to date nor establish a limit to the potential areas where this powerful technique can be fruitfully used. Rather, the reader should now recognize that any strain-related defect can be observed by the weak-beam technique and higher "practical" resolution realized thereby.

ACKNOWLEDGMENTS

The author would like to thank Ernest L. Hall and Robert M. Allen for discussions and for providing previously unpublished results. The interaction of the author with Drs. D. J. H. Cockayne and I. L. F. Ray was important in his formulating the material presented here.

REFERENCES

Allen, R. M., 1979, unpublished research.

Allen, R. M. and Vander Sande, J. B., 1978, Met. Trans., 9A, 1251.

Allen, R. M. and Vander Sande, J. B., 1979, to be published.

Campany, R. G., Loretto, M. H. and Smallman, R. E., 1973, J. Microsc., 98, 174.

Carter, C. B., 1977, Phil. Mag., 36, 147.

Cockayne, D. J. H., Ray, I. L. F. and Whelan, M. J., 1969, Phil. Mag., 20, 1265.

Cockayne, D. J. H., 1972, Z. Naturf. A, 27, 452.

Cockayne, D. J. H., Jenkins, M. L. and Ray, I. L. F., 1971, Phil. Mag., 24, 1383.

Crawford, R. C. and Ray, I. L. F., 1977, Phil. Mag., 35, 549.

Green, M. L., Chin, G. Y. and Vander Sande, J. B., 1977, Met. Trans., 8A, 353.

Gomez, A., Cockayne, D. J. H., Hirsch, P. B. and Vitek, V., 1975, Phil. Mag., 31, 105.

Hall, E. L. and Vander Sande, J. B., 1978, Phil. Mag., 37, 137.

Hausserman, F. and Schaumberg, H., 1973, Phil. Mag. 27, 745.

Hirsch, P. B., Howie, A., Nicholson, R. B., Pashley, D. W. and Whelan, M. J., 1965, Electron Microscopy of Thin Crystals, Butterworths, London.

Humphreys, C. J., Drummond, R. A., Hart-Davis, A. and Butler, E. P.,
    1977, Phil. Mag., 35, 1543.

Jenkins, M. L., 1972, Phil. Mag., 26, 747.

Karnthaler, H. P. and Wintner, E., 1975, Acta Met., 23, 1501.

Nordlander, I., and Thölen, A., 1973, J. Microsc., 98, 221.

Packeiser, G. and Haasen, P., 1977, Phil. Mag., 35, 821.

Ray, I. L. F. and Cockayne, D. J. H., 1971, Proc. Roy. Soc., A325, 534;
    1973, J. Microscopy, 98, 170.

Stobbs, W. M. and Sworn, C. H., 1971, Phil. Mag., 24, 1365.

CHAPTER 21

THE ANALYSIS OF DEFECTS USING COMPUTER SIMULATED IMAGES

*PETER HUMBLE**

DEPARTMENT OF MATERIALS SCIENCE and ENGINEERING

UNIVERSITY OF FLORIDA, GAINESVILLE, FLORIDA 32611 USA

21.1 INTRODUCTION

Most of the other contributions to this workshop consider the analytical use of the electron microscope in a chemical or structural/crystallographic sense. In this section, we consider the analysis of extended defects which are nearly always present in crystalline solids. These defects are those which are visible in conventional transmission electron microscopy as a result of their extended nature or their long range elastic displacement fields, e.g. dislocations, stacking faults, dislocation loops, coherent and semicoherent precipitates, extrinsic grain boundary dislocations, etc. It is important to note at the outset that the images of such defects are not dependent only on the form of their displacement field, but also on factors such as their geometry or orientation in the crystal. For example, the image of a small dislocation loop depends, among other things, upon the shape and size of the loop, the orientation of the Burgers vector relative to the plane of the loop, and the orientation of the loop with respect to the electron beam. Even for a straight dislocation in an elastically anisotropic crystal, the image depends not only on the Burgers vector and its orientation with respect to the dislocation line, but also on the direction in the crystal of the dislocation line itself. Thus, a total analysis of a defect involves the determination of these other mainly geometric and crystallographic factors in addition to the displacement field.

The images which defects produce in transmission electron microscopy arise from the complex and nonlinear interaction of the electrons with the distorted crystal. In some cases (usually "null" cases where, despite the presence of the defect, the electrons are diffracted as if from a perfect crystal), it is possible to predict what the image of a

---

*On leave from CSIRO, Division of Chemical Physics, P.O. Box 160, Clayton, Victoria 3168, Australia.

given defect will look like, but in general this is not possible. Conversely, it follows that in general one cannot intuitively identify a defect from its images; some computational technique must be employed to achieve this.

It has been shown in principle (HEAD, 1969a, b) that under certain idealized conditions it is possible to deduce the displacement field of some types of defects from a knowledge of the intensities in their images. Unfortunately, these idealized conditions are not approached in practice, so that this direct analysis and identification technique is not yet possible. Thus, in order to identify defects from their TEM images, one must resort to modelling the defect, together with the diffracting conditions in a computer, calculating the resulting contrast, and comparing this with the contrast actually observed. This is essentially a trial and error method and as such could involve considerable time and effort. It is the purpose of this contribution to outline several aspects of the theory, computer programming, and experimental procedure which help to minimize this.

Depending upon the degree of sophistication and the detail required in the analysis, there are several simplifications which may be made to aid the process of identifying defects from their TEM images. The following have been found suitable for a wide range of defects in different materials.

We choose to work within the framework of a two-beam theory using a column approximation because this is the easiest and cheapest to compute, it is a good approximation for many materials, and it is easy to set up experimentally. We choose to work with bright field images because they contain the same information as their dark field counterparts and the equivalent quality dark field micrographs, with the incident electron beam deflected so that the subsequent diffracted beam coincides with the optical axis of the microscope, are much more tedious to obtain experimentally. Lastly, we choose to display the results of the calculations in the form of simulated electron micrographs because this is the usual way the experimental images are recorded and therefore comparison is made easier.

21.2 THEORY AND COMPUTATIONAL CONSIDERATIONS

The theory of electron diffraction commonly used in this type of work is a two-beam dynamical theory using a column approximation which was formulated by HOWIE and WHELAN in 1961. This assumes that the crystal is divided up into independent, noninteracting columns parallel to the electron beam direction. The crystal is oriented such that it is close to the Bragg condition for only one set of planes so that in each column just two electron wave amplitudes T and S are generated, in the incident and diffracted directions respectively. These interact with each other dynamically through the agency of the crystal lattice potential as they pass through the specimen. In the presence of a defect, the atoms in each column are displaced from their perfect crystal positions and each column has a different set of displacements compared with neighboring columns. The theory results in a pair of coupled, linear,

first order differential equations describing the rate of change of T
and S with respect to a variable z down the column. The equations include parameters N, representing the normal absorption of electrons;
A, representing the anomalous absorption (or preferred transmission)
of electrons; w, a measure of the deviation of the crystal orientation
from the exact Bragg condition; and β', a function describing the rate
of change of the distortions down a column. The equations are

$$\frac{dT}{dZ} = -N T + (i - A)S$$

$$\frac{dS}{dZ} = (i-A)T + (-N + 2iw + 2\pi i\beta')s \qquad (21.1)$$

where $Z = z\pi/\xi_g$. ξ_g is the extinction distance appropriate to the diffracting vector $\underline{g}$, and β' = d($\underline{g} \cdot \underline{R}$)/dZ.

$\underline{R}$ is the vector displacement field due to the defect. The electron
wave amplitudes T and S are complex, and the bright field or dark field
intensity at the bottom of a column is obtained by integrating the equations down the column and then multiplying the final values of T or S
respectively by their complex conjugates. Because of the form of β',
the equations cannot be integrated analytically and one has to resort to
numerical methods. The computational difficulties surrounding Eqn. 21.1
are concerned mainly with a suitable choice of integration routine and
also with the form which the displacement field $\underline{R}$ and its associated geometry should take in order to model the defect adequately. These considerations are somewhat interrelated as will become evident in what
follows.

The integration method we have used (HEAD, HUMBLE, CLAREBROUGH,
MORTON, and FORWOOD, 1973) is the Runge-Kutta fourth order process incorporating a modification by Merson (LANCE, 1960). This routine automatically adjusts the size of integration step taken to meet a stipu-
lated error criterion. In our programs, this has been chosen so that
for computations involving specimens about five extinction distances
thick, and for a value of w of about 0.5, the final intensities are accurate to ∿2%. Our work has been mostly with straight or planar defects, and as we shall see later, by making one further assumption concerning β', it is possible to greatly reduce the amount of computing
time involved in numerical integration for such defects. However, for
other defects, different approaches may be made in order to try to reduce the computing time.

One approach is to notice that in Eqn. 21.1 one may combine the
terms involving w and β':

$$2iw + 2\pi i\beta' = 2i(w + \pi\beta')$$

$$\cong 2iw_{eff}$$

for a small interval ΔZ. That is, for sufficiently small intervals ΔZ
down the column, β' may be considered constant and then its effect is
merely to alter locally the deviation from the Bragg condition from w to

w_{eff}. Thus the column may be thought of as divided into elements of perfect crystal each with a different w_{eff}. The "integration" of the electron wave amplitudes can then be carried out using repeated applications
of the scattering matrix (HIRSCH, HOWIE, NICHOLSON, PASHLEY, and WHELAN,
1965, p.223). This has been used by THÖLĒN, (1970a, b) for dislocations
and dislocation networks, and by COOPER, (1977) for small dislocation
loops.

Another variant has been used by WILKENS and RÜHLE (1972). They
use a Bloch wave formulation, and, by assuming β' is small and slowly
varying, are able to evaluate the wave amplitudes analytically. This
approximation seemed to be adequate for small defects such as the small
dislocation loops which they were considering.

As indicated earlier, the choice of integration method can be related to the form of β' and for linear and planar defects there is an
assumption which may be made concerning β' which reduces the integration
time by more than an order of magnitude. HEAD (1967) realized that in
integrating down columns in a plane in the crystal parallel to a
straight dislocation, the displacements in neighboring columns were
largely the same. This is illustrated schematically in Fig. 21.1 which
shows a section through a parallel sided specimen; D'S' is the projection of a dislocation line running between the bottom and top surface of
the specimen.* If one assumes that D'S' is a segment of a straight dislocation in an infinite solid (i.e. that the specimen has been "cut out"
of an infinite solid and the stresses on the new surfaces have not been
allowed to relax to their free surface values), then the displacements
in elements along any line parallel to D'S' (e.g. ds) are exactly the
same. It follows that the displacement field may be reduced from three
dimensions to two by projection along the direction ds onto some convenient plane. A convenient plane as far as the integrations are concerned
is one containing the electron beam direction and we have called this
the generalized cross-section.

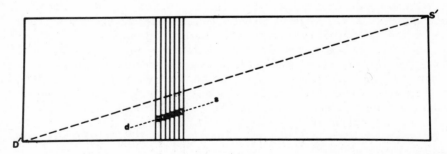

Fig. 21.1. A side elevation of an untilted foil containing a dislocation D'S' not in the plane of the paper. If the effects of the free surfaces are neglected, the displacements in neighboring columns along
lines such as ds (parallel to D'S') are identical.

---

*The notation of HEAD et al. (1973) has been used through this section.

A generalized cross-section for a single dislocation in an untilted foil is shown in Fig. 21.2. The dislocation is represented by the circle in the middle of the diagram; the dislocation is coming out of the

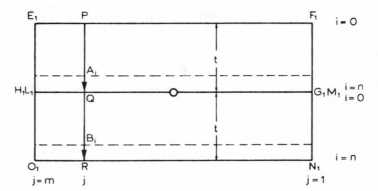

Fig. 21.2. Generalized crosssection for a single dislocation in an untilted parallel sided specimen.

paper but is not normal to it. The region $E_1F_1N_1O_1$ contains all the displacements found in the volume of crystal in which the dislocation just runs from the bottom of the specimen to the top. Note that the height of the cross-section is 2t, where t is the thickness of the specimen. The width of the cross-section is chosen so that the image may be suitably displayed.

By considering pairs of horizontal lines in the generalized cross-section, it is possible to select displacements associated with the dislocation at a particular level in the specimen. For example, the displacements between E_1F_1 and H_1G_1 are relevant to the case when the dislocation is just at the bottom of the foil, while the displacements between L_1M_1 and O_1N_1 refer to the case where the dislocation is just emerging from the top of the foil. The pair of dashed lines in Fig. 21.2 represents a typical intermediate case. Integration of Eqn. 21.1 down columns such as that marked j gives intensities along a row in the simulated micrograph parallel to the projection of the dislocation line.

The organization of the integration depends on the fact that Eqn. 21.1 are linear and so a general solution may be obtained by taking a linear combination of two particular solutions. Two independent integrations of Eqn. 21.1 are performed down each column, such as j, to obtain the two particular solutions. To ensure independence, one integration is started with T = 1, S = 0 and the other with T = 0, S = 1. The column is divided into two equal parts PQ and QR each of which is further divided into n equal steps. The end of each of the n steps in the first part of the column, PQ, will represent points on the entrance surface of the real crystal, so that in this region the results of the two integrations have to be combined to give the boundary conditions pertaining to the top surface of the real specimen, i.e. T = 1, S = 0. Thus, at a representative position, say A , if the amplitudes for the first integration are currently $T_{Ai}^{(1)}$ and $S_{Ai}^{(1)}$ and for the second integration are $T_{Ai}^{(2)}$ and $S_{Ai}^{(2)}$, then

$$a_i T_{A_i}^{(1)} + b_i T_{A_i}^{(2)} = 1$$

and

$$a_i S_{A_i}^{(1)} + b_i S_{A_i}^{(2)} = 0$$

where a_i and b_i are constants which may be determined from the above relationships. Note that since T and S are complex, a and b are also
complex. The constants a_i and b_i are evaluated and stored at each of
the n steps in the integrations from P to Q.

The integrations are continued in the second half, QR, of the
column, the process again being interrupted at the end of each of the n
steps, this time to calculate the final amplitudes corresponding successively to the positions t above each. Thus, at position B_i, t below
A_i, the final amplitudes T^F and S^F are calculated using the constants a_i
and b_i for position A_i:

$$T_i^F = a_i T_{B_i}^{(1)} + b_i T_{B_i}^{(2)}$$

and

$$S_i^F = a_i S_{B_i}^{(1)} + b_i S_{B_i}^{(2)}$$

Squaring the modulus of T^F or S^F gives the bright field or dark field intensities respectively for a point i along a row parallel to the dislocation. By letting i run from 1 to n we obtain an entire row of intensity
values. In our programs, n is usually 64, which, with linear interpolation, gives 129 intensity points along each row. We usually compute 60
rows, i.e. j = 1 to 60. These values were chosen so that we could produce a micrograph conveniently on one page of line printer paper.

We chose to display the micrographs this way for two reasons.
Firstly, a line printer is a fairly standard output device in most computing systems, so that the programs are fairly easy to transfer from
one system to another. Secondly, hard copy micrographs are preferred
(to, say, a TV display) and the line printer provides these with the
minimum of operator action. Hard copy simulated micrographs are preferred because intercomparison may then be made between several images
and between them and the experimental image. This is an important part
of the rejection/selection process leading to a best match.

When using the formulation of Howie and Whelan represented by Eqn.
21.1 in conjunction with the generalized cross-section to produce images
of stacking faults, the integrations cannot be allowed to continue
through the fault since β' would become infinite. Instead, the integrations are stopped at the fault, the amplitudes T and S adjusted in

phase, and the integrations restarted with the new values of T and S. The new values are given by

$$T_{new} = T$$

$$S_{new} = S\exp(2\pi i\underline{g}\cdot\underline{R}) \tag{21.2}$$

where $\underline{R}$ is the translation vector of the part of the crystal which the electrons are about to enter with respect to the part of the crystal they are just leaving.

A generalized cross-section can only be defined for linear and planar defects or composite defects consisting of parallel configurations of these (including planar grain boundaries). However, with its use, it is possible to produce a simulated electron micrograph of 129 x 60 = 7740 intensity values with just 60 x 2 integrations down a length of column equal to twice the thickness of the specimen, an equivalent saving of 7500 numerical integrations through the thickness of the specimen. This accounts for the speed, and therefore cheapness, of theoretical micrographs of this type of defect. Currently, on a CDC 7600 computer, each micrograph takes less than one second to compute.

For defects which are not linear or planar, every point in the simulated micrograph has to be calculated by separate integrations of Eqns. 21.1 through the thickness of the specimen. Although, of course, some points may be obtained by interpolation, micrographs of other defects generally require an order of magnitude more integrations than the 240 which are sufficient for linear and planar defects. Despite the large increase in computing time, several studies involving simulated micrographs of other defects have been published. For example, BULLOUGH, MAHER, and PERRIN (1971), MAHER, PERRIN, and BULLOUGH (1971), HÄUSSERMANN, RÜHLE, and WILKENS (1972), YOFFE (1972), OHR (1976), and COOPER (1977) have all calculated the images of small dislocation loops and SASS, MURA, and COHEN (1967), DEGISCHER (1972), and MELANDER (1976) have considered small precipitates. In all of these cases, much effort was spent on keeping the computing time to a minimum by making suitable choices of algorithms, integration routines, by adopting scattering matrix or pseudo Bloch wave approaches, or by using more sophisticated interpolation procedures, etc. Nevertheless, only relatively few simulated micrographs have appeared in the literature and these have mostly been calculated in attempts to establish trends in image behavior rather than identification of specific defects. Fig. 21.3(a) is an illustration of the comparison between experimental and simulated images of small dislocation loops taken from the work of Bullough et al. and Fig. 21.3(b) is a similar illustration for small precipitate particles published by Degischer. Apart from the increased computing time concerned with integrating Eqns. 21.1 for defects such as small precipitates and dislocation loops, these defects present an additional problem in that, in general, their geometry cannot be determined separately from their strain field as is usually the case for defects for which a generalized cross-section can be defined. Thus, in any analysis concerning small defects it is necessary to include, and vary over some range, parameters specifying the shape, orientation, position in the specimen, etc., as well as the parameters defining the vector displacement field.

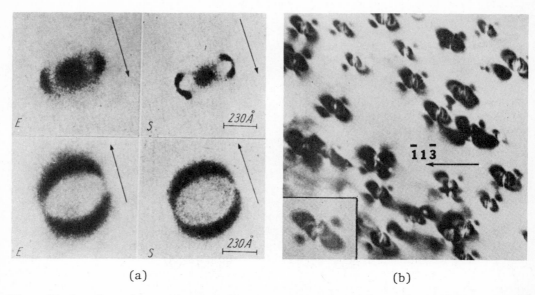

(a)   (b)

Fig. 21.3. Examples of the comparison of experimental and simulated images for defects for which a generalized cross-section cannot be defined.
(a) Fig. 7 of BULLOUGH et al. (1971) comparing experimental, E, and simulated, S, images of dislocation looks for g = 310 (upper pair) and g =
3̄10 (lower pair) (b) Figure 3 of Degische (1972) comparing experimental
micrograph of coherent cobalt precipitates in copper with the simulated
image (inset) computed using anisotropic elasticity assuming the particles to be spherical.

The displacement fields of defects which have been used in
image simulation programs so far have all been derived on the basis
of linear continuum elasticity theory. In a few cases (e.g. HEAD et al.,
1973; SASS et al., 1967; DEGISCHER, 1972; MELANDER, 1976), anisotropic elasticity has been used, but mostly the displacement
fields have been derived on the basis of defects in an infinite isotropic medium. Anisotropic elasticity is a more realistic approximation on which to base image calculations concerning crystalline
solids and although it usually increases the complexity of the programming, it does not increase significantly the time taken to compute a micrograph. It also requires the specification of the geometry of the defect with respect to the crystal and elastic constants axes. However, if the geometry can be determined independently of the strain field then it can be entered into the data as a
known value and does not become an extra variable parameter in i-
dentifying the defect. In some cases, such as the identification of
straight dislocations in β-brass, the use of anisotropic elasticity
theory is absolutely necessary in simulating the images (HEAD,
LORETTO, and HUMBLE, 1967a, b). In other cases it is necessary in
order to predict the defect geometry correctly, e.g. see the example
concerning dislocation dipoles in nickel in section 4.

21.3 EXPERIMENTAL METHOD AND THE COLLECTION OF INFORMATION

An inspection of Eqn. 21.1 will show most of the parameters which have to be known or determined in order to run an image simulation program. Consider first the parameters which may be involved in the term $\beta' = d(\underline{g} \cdot \underline{R})/dZ$.

The expression involving the displacement field of a defect, either β' or $\underline{R}$, is usually written into the program in a general form and in such a way that only a few key parameters such as the Burgers vector and the dislocation line direction need to be specified as data for each micrograph. More complex defects may of course require the specification of other parameters such as the spacing of dislocations, the plane of a defect, or its shape and orientation.

However, in all cases, irrespective of the geometry of a defect, its displacement field always occurs in the two-beam theory as the scalar product with the diffracting vector $\underline{g}$. In other words, in the two-beam approximation, a single micrograph samples only the component of $\underline{R}$ parallel to $\underline{g}$. Thus, in order to fully determine $\underline{R}$ and thus identify the defect, it is necessary to sample as completely as possible the full three dimensional nature of $\underline{R}$. It will be apparent that the minimum condition for determining $\underline{R}$ is three micrographs taken with noncoplanar diffracting vectors. In practice, it is advisable to take micrographs with as many different diffracting vectors as widely spaced as possible. This inevitably involves tilting the specimen in the electron microscope. Tilting to different known orientations is also necessary in order to determine the geometry of the defect by viewing it in different projections. Thus the experimental technique involved in the identification of defects consists of tilting the specimen to different known orientations, setting up two-beam diffraction conditions with different diffracting vectors, and photographing the image (to record a component of the displacement field and a projection of the geometry) and the diffraction pattern (to record the diffraction and orientation information). A complete description of this technique with worked examples is given in Chapter 3 of HEAD et al..(1973).

Parameters which are not usually determined from case to case but are needed for the computation of simulated micrographs include the normal and anomalous absorption coefficients, N and A , the elastic constants of the crystal, c_{ij}, and the extinction distance, ξ_g. Values for the last three quantities may be found in various reference works, for example, HEAD et al., (1973), HIRSCH et al. (1965), HUNTINGTON, (1958). For two-beam micrographs and specimens of constant thickness, the intensities may be normalized to background intensity (the intensity transmitted by an undistorted crystal set in the same diffracting condition) and the normal absorption coefficient divides out. However, in order to evaluate Eqn. 21.1 a value of N must be specified: for speed of computation, in our programs N is put equal to A .

21.4 EXAMPLES OF THE USE OF SIMULATED IMAGES IN THE ANALYSIS OF DEFECTS

There are two main ways in which computed images may be used as an aid to analyze and identify defects. The first consists of computing micrographs for various "representative" cases; that is, one compiles a catalog. This can be very useful, particularly in looking for a specific type of contrast behavior or a specific type of defect experimentally. However, it has two disadvantages. The main one is that even a modest number of values for the different variables (different beam directions, diffracting vectors, deviations from the Bragg condition, thicknesses of specimen, not to mention the crystallography, geometry, orientation, and displacement field parameters of the defect itself) results in an inordinately large amount of computing. The other disadvantage is common to all numerical computing methods as opposed to analytical ones. It is only possible to produce micrographs for specific cases and it can only be assumed that the trend indicated by neighboring cases is smooth. If it is not, then interpolation leads to the wrong deduction about how the image varies with a particular parameter and hence may cause a wrong identification. The danger increases, of course, if, to alleviate the amount of computing, the representative cases in the catalog are widely spaced.

We favor the use of computed images to analyze experimental observations on specific defects. This method leads to the examination of relatively few particular cases, but these may be analyzed in detail. There is more certainty in the interpretation, because the end point of the process is a set of best matching computed images for a definite set of parameters and no interpolation is required. The following three examples should give an idea of the scope of the technique when used this way.

Figure 21.4 shows an area of a thinned nickel specimen which contains several dislocation dipoles. Dipoles are the simplest form

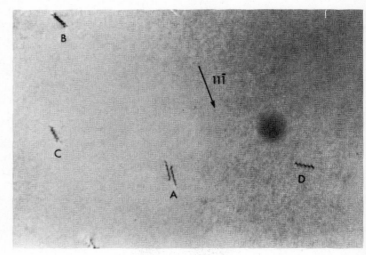

Fig. 21.4. Electron micrograph of a thinned specimen of nickel illustrating parallel dislocation dipoles at A,B and D.

of dislocation array and consist of an interacting pair of dislocations of opposite sign (Burgers vector) on parallel slip planes. They form a stable configuration of low energy and thus contribute to the resistance to plastic flow of the crystal.

The individual dislocations in dipole A are sufficiently separated that they could be identified separately (FORWOOD and HUMBLE, 1970). They have Burgers vectors of $\pm 1/2[10\bar{1}]$ and since their line direction is close to $[\bar{1}\bar{1}2]$ they are 30° from screw orientation. Isotropic elasticity theory was used to predict their stable angle of interaction, ϕ (see Fig. 21.5) but when the images corresponding to this were computed,

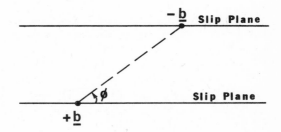

Fig. 21.5. Schematic diagram of a dislocation dipole illustrating the measurement of the stable angle ϕ of the configuration.

bad mismatches were obtained. Figure 21.6 shows one of these. The equilibrium calculation was re-evaluated using anisotropic elasticity theory. It can be seen from Fig. 21.7 that the difference in equilibrium angle ϕ obtained from the two theories is most marked for dislocations close to screw orientaton up to about 40° from screw.* Dipole A lies in this region, and when images for the appropriate equilibrium angle, ϕ = 60° (rather than ϕ = 90° obtained on isotropic elasticity) was used, good agreement between theoretical and experimental images was obtained. Figure 21.8(b) shows the set of computed micrographs which

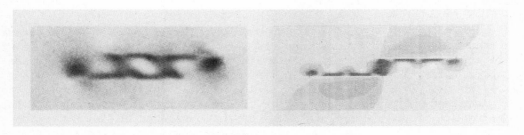

Fig. 21.6. Comparison of an experimental image of dipole A (Fig. 21.4) with a theoretical image computed using isotropic elasticity to determine the stable configuration.

---

*This is quite a surprising result since nickel is not usually considered to be very elastically anisotropic. Its anisotropy factor $A = 2C_{44}/(C_{11}-C_{12}) = 2.52$; for isotropy, $A = 1.0$.

best matches five selected experimental micrographs, Fig. 21.8(a). From
the parameters used to compute these images dipole A may be identified
as lying along the [1̄1̄2] direction having Burgers vectors ±1/2[101̄].

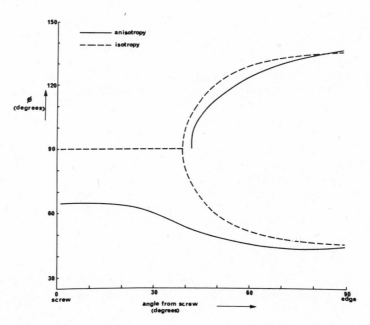

Fig. 21.7. Curves giving the equilibrium angle of ϕ for ± 1/2 <110> dislocations forming dipoles in nickel for isotropic (γ = 0.32) and anisotropic elasticity.

The dislocations are therefore glissile on (111) slip planes; the planes
are separated by about 420 Å* and the dislocations are aligned at an
angle ϕ to the slip planes of 60°.

    The value of ϕ = 60° was obtained from theory, and it is therefore
interesting to consider whether the dislocations are lying exactly at
this equilibrium value. It is possible to test this by computing theoretical images for a range of values of ϕ. This has been done in Fig.
21.9 for the $\underline{g}$ = 2̄02 image of Fig. 21.8; the range of ϕ goes from 55° to
65° in steps of 2 1/2°. It may be seen that the best match occurs for
the middle of the range and the images corresponding to the extreme
values ϕ = 55° and ϕ = 65° are not good matches to the experimental image. That is, the value of ϕ may be determined as ϕ = 60±2 1/2°. Thus,
assuming the theory to be accurate, the dislocations in dipole A may be
up to 2 1/2° away from equilibrium without any marked effect on this image. Assuming that dipole A is in fact this far from equilibrium, it is
then possible to calculate the restoring forces on the dislocations:
this would be balanced by, and is a measure of, the lattice frictional

_____

*Using values for the extinction distances ξ_{111} = 236 Å, ξ_{200} = 275 Å,
ξ_{220} = 409 Å, and ξ_{311} = 499 Å.

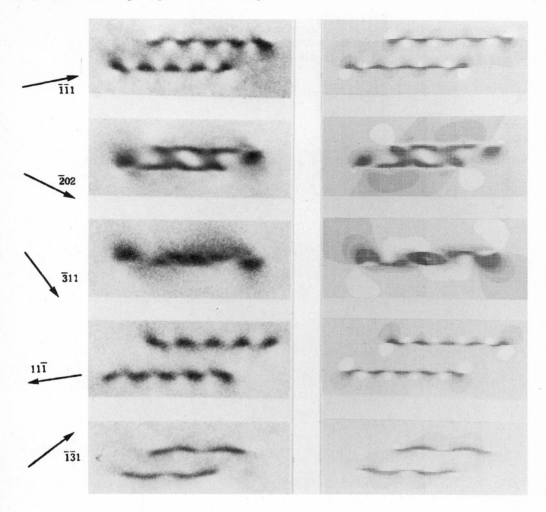

Fig. 21.8. Comparison of experimental (a) and computed (b) images of dipole A (Fig. 21.4) for five different diffraction conditions. For full details, see FORWOOD and HUMBLE (1970).

force on these dislocations. In this case it was found to be <0.5 dyne cm^{-1} or about 3 x 10^{-5} of the shear modulus.

The dipoles B and D in Fig. 21.4 were analyzed and identified in a similar manner. Dipole D was particularly interesting. The dislocations lie along [121] and initial matching showed that their Burgers vectors were ±1/2[10$\bar{1}$]. The orientation angle ϕ was about 44°, but the separation of the slip planes was extremely small, approximately 20 Å. The images computed for these values, Fig. 21.10(b), did not match the experimental images, Fig. 21.10(a), very well, but they were the best which could be obtained. When due account of the splitting of each of the dislocations into partial dislocations with stacking fault in between (even though the stacking fault energy of nickel is quite large) a much better match was

obtained,* particularly for the bottom image. This is shown in Fig.
21.10(c). Dipole D was identified as having a line direction of [1$\bar{2}$1],
being composed of partial dislocations with Burgers vectors ±(1/6[1$\bar{1}$2] +
1/6[211]) separated by intrinsic stacking fault about 25 Å wide, orientation angle φ of 46.5° and a distance between slip planes of about 22
Å.

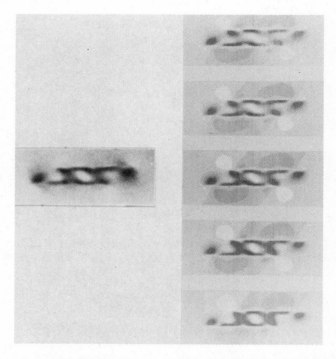

Fig. 21.9. An experimental image of dipole A (Fig. 21.4) compared with
theoretical images computed for φ = 55o (top) to φ = 65° (bottom) in
steps of 2-½°.

These examples are fairly typical of the detailed information which
may be obtained by analyzing the TEM images of defects with the help of
computer simulated images.

This is perhaps a good place to illustrate the usefulness of presenting the computed information in the form of a half-tone simulated
micrograph rather than in the form of a set of intensity profiles across
the defect. This latter was the usual form of presentation before 1967.
Figure 21.11(a) shows one of the experimental images of dipole A shown
in Fig. 21.8 together with the matching theoretical micrograph and a set
of profiles compiled from exactly the same intensity values contained in

---

*Separation into partials probably also occurs for dipoles A and B, but
the splitting will be smaller since they are not edge dislocations, and
will also be less important in these cases because the slip planes are
much further apart.

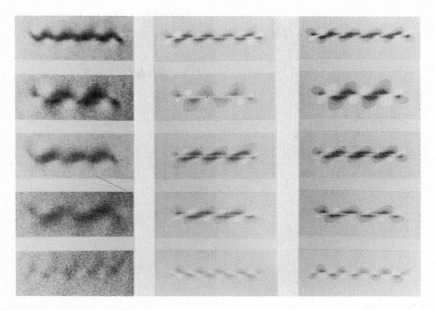

Fig. 21.10. Comparison of five experimental images of dipole D of Figure 21.4 (a) with the corresponding simulated images computed assuming the dislocations are unsplit (b) and dissociated into partial dislocations (c). For full details see FORWOOD and HUMBLE (1970).

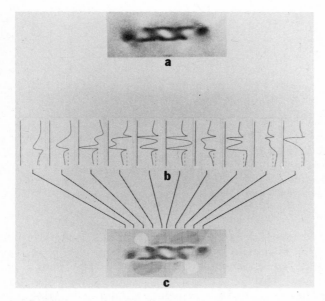

Fig. 21.11. Comparison of an experimental micrograph (a) with intensity calculations using the two-beam theory when the same information is present in the form of intensity profiles (b) or as a simulated micrograph (c).

the theoretical micrograph. The rows of the micrograph corresponding to each of the profiles are indicated by the lines between Fig. 21.11(b) and (c). There can be no doubt that comparison between theory and experiment is more easily made using the theoretical micrograph.

The second example of defect analysis using simulated images concerns overlapping stacking faults. In f.c.c. metals and alloys having low stacking fault energy, plastic deformation often involves the slip of widely separated partial dislocation pairs on closely spaced parallel slip planes. When these are viewed in a specimen thinned for TEM, the overlapping regions of stacking fault give rise to fringes of various intensity and symmetry. Two examples are shown at ABC and DE in Fig. 21.12. The problem is to try to identify what sequence of stacking faults gives rise to the observed contrast.

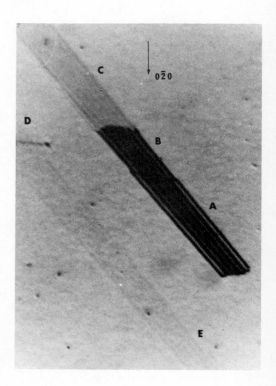

Fig. 21.12. Electron micrograph of a lightly deformed specimen of Cu + 8 at % Si alloy showing faint asymmetrical images at C and fainter symetrical images from D to E.

It was noted earlier in Eqn. 21.2 that if the displacement vector associated with a stacking fault is $\underline{R}$, then the phase angle α in the electron wave amplitudes across the plane of the fault is given by $\alpha = 2\pi g \cdot \underline{R}$. WHELAN and HIRSCH (1957a, b) considered the case of closely spaced overlapping faults and suggested that the phase angle could be written

$$\alpha = 2\pi \underline{g} \cdot \sum_{i=1}^{n} \underline{R}_i \qquad (21.3)$$

where $\sum_{i=1}^{n} R_i$ is the (vector) sum of the displacement vectors of the n individual overlapping faults. For f.c.c. crystals, the fault vectors are shear vectors of the type 1/6<112> and it follows that α goes through a sequence of three repeating values. When the number of overlapping intrinsic faults is 3n-2, the total shear displacement is equivalent to that of one intrinsic stacking fault; when it is 3n-1, the total shear is equivalent to one extrinsic stacking fault; and when it is 3n, the total shear displacement is a proper translation vector of the lattice, thus $g \cdot \sum_{i=1}^{n} R_i$ is integral, α is an integral multiple of 2π, and no contrast results.

On this basis, the interpretation of the upper set of fringes in Fig. 21.12 is that in region A there is a single intrinsic stacking fault, in region B there are two intrinsic faults (equivalent to one extrinsic fault) and in region C there are three overlapping faults. However, it is clear that there is some contrast in region C consisting of a faint set of asymmetric fringes. The reason for the discrepancy between the observations in Fig. 21.12 and the prediction of Eqn. 21.3 lies in the fact that in Eqn. 21.3 it has been assumed that the amplitudes and the phases of the electron waves remain constant in between successive stacking faults, so that one may replace successive phase changes of the type described in Eqn. 21.2 by a single one, Eqn. 21.3. While this is a reasonable approximation if the contrast is strong (e.g. region B of Fig. 21.12) it is not good where the predicted contrast is zero, e.g. regions C and DE. In these cases, images must be evaluated on the basis that the electron waves are altered in phase at each fault in turn with the appropriate propagation being carried out in the usual way between faults.

Figure 21.13 shows theoretical micrographs computed in this way for three overlapping stacking faults and the specific diffracting conditions and crystallography corresponding to Fig. 21.12. The micrographs in Fig. 21.13 are computed for various combinations of spacing x and y (in units of the slip plane spacing, d) of the faults starting at one fault on each of three successive planes, i.e. x = d, y = d, and progressing upwards. It is apparent from Fig. 21.13 that three overlapping faults do produce some contrast although it is quite weak (the grey scale used to print these micrographs has visibility limits of 2% below background and 5% above). From the micrographs it may be seen, and it may be shown analytically, that when x = y a symmetrical set of fringes results; when x ≠ y the fringes are asymmetric. The contrast of the fringes increases as x or y increases. Comparison of the symmetry properties, width of fringes and intensity of contrast (HUMBLE, 1968) showed that the very weak symmetrical fringes DE arise from three intrinsic stacking faults on neighboring planes (i.e. a microtwin) and the fringes at C are due to faults in which the upper pair are on neighboring planes, but the lower one is five planes away. Thus, in this case, the analysis using computer simulation is mostly concerned with determining the geometry of the defect, the nature of the defect being more or less known.

The last example which will be given here of the identification of defects using computer simulated images concerns the analysis of extrin-

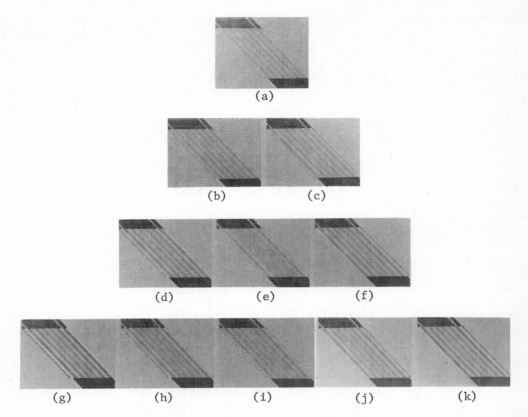

Fig. 21.13. Theoretical electron micrographs for three overlapping stacking faults computed for the diffracting conditions of Fig. 21.12. The spacings x and y between the three faults are given below in units of the slip plane spacing:

	(a)	(b)	(c)	(d)	(e)	(f)	(g)	(h)	(i)	(j)	(k)
x	1	1	2	1	2	3	1	2	3	4	5
y	1	2	1	3	2	1	5	4	3	2	1

sic grain boundary dislocations left at the boundary between two crystals of different orientation after plastic deformation. The most definite condition in which to make observations on a bicrystal is when both crystals are set up for two-beam diffraction (HUMBLE and FORWOOD, 1975), but of course not in general with the same diffracting vector. This condition, although restrictive, is quite general enough to provide sufficient information (i.e. enough different diffracting vectors) for identification to be possible. In order to test this technique on a boundary whose crystallography, and that of its associated extrinsic dislocations, was relatively well-defined, we made observations on the dislocations which occurred at coherent twin boundaries in copper after light deformation (FORWOOD and HUMBLE, 1975). The copper specimens had been previously annealed at 800°C for 1 hour so that the coherent twin

boundaries were initially dislocation free. Figure 21.14 shows two
simultaneous two-beam micrographs of a coherent twin boundary in copper
after light deformation. Several linear defects which appear to be extrinsic GBD's may be seen along the boundary and some of the more isolated examples are enumerated 1-9 in Fig. 21.14(a). There are several
possibilities for the way that these GBD's arise. The most likely concern the slip of 1/2<110> dislocations in each grain and in the boundary, the slip of 1/6<112> dislocations in the boundary (either as a residue of the transfer of a normal slip dislocation from one grain to the
other, or a straightforward twinning dislocation) and pairwise interactions of these. The 26 lowest energy possibilities for the Burgers vectors of GBD's formed in this way are: six ±1/2<110> in crystal 1, six
±1/2<110> in the plane of the boundary, six ±1/2<110> in crystal 2 (indexed as ±1/6<411> relative to crystal 1), six ±1/6<112> in the plane of
the boundary and two ±1/3<111> normal to the boundary. All these possibilities were considered in analyzing the GBD's in Fig. 21.14.

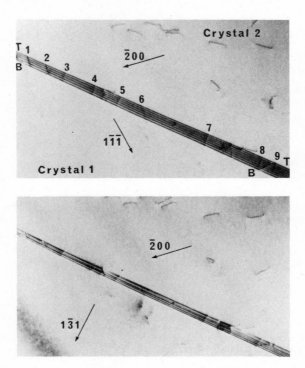

Fig. 21.14. Same area of a coherent twin boundary in lightly deformed
copper imaged under two different simultaneous two-beam conditions.

    Figure 21.15 shows six experimental images of GBD 7 in Fig.
21.14(a) together with two corresponding sets of computed images, one
for b = 1/6[1̄2̄1̄] and the other for b = 1/6[121]. The set of images for
b = 1̄/6[1̄2̄1̄] matches the set of experimental images. Since GBD 7 lies
close to the [121] direction in the boundary it is therefore a left-
handed screw twinning partial dislocation. It is glissile in the boundary and since it can be identified as left-hand rather than right-hand

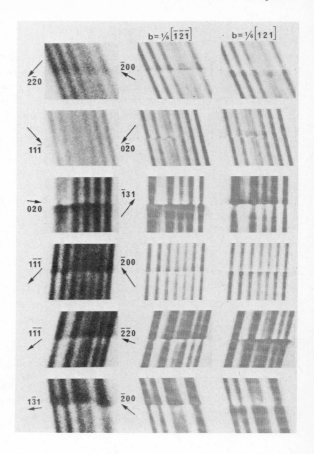

Fig. 21.15. Six experimental images of grain boundary dislocation 7 (Fig. 21.14), left column, with the matching set of micrographs for b = 1/6 [12̄1̄], centre column. The right hand column is computed for b = 1/6 [121].

(b = 1/6[1̄2̄1̄] rather than b = 1/6[121]) it is possible to define in which sense the boundary moves when GBD 7 slips. If it were to move towards GBD 6, the area of boundary traversed would be displaced from crystal 1 towards crystal 2 by one {111} interplanar spacing, specifically, 1/3[1̄1̄1̄].

GBD's 3, 4, 5, 7, and 8 are all twinning partials of the left-handed screw type with b = 1/6[1̄2̄1̄]. GBD 6 was also identified as being a twinning partial, but in this case b = 1/6[11̄2̄]. Since it lies close to the [011] direction in the boundary, it is 30° from left-handed screw orientation.

GBD's 1 and 2 also lie close to the [011] direction, but they were identified as having the Frank partial Burgers vector of 1/3[1̄1̄1̄]. This is demonstrated by the matching set of experimental and computed images for GBD 2 in Fig. 21.16. Being an edge dislocation with its Burgers vector normal to the boundary, this partial is sessile.

Thus, in this example, information is gained mainly about the types of extrinsic GBD's which are generated at a coherent twin boundary during the early stages of plastic deformation, together with some geometric information such as whether they are potentially sessile or

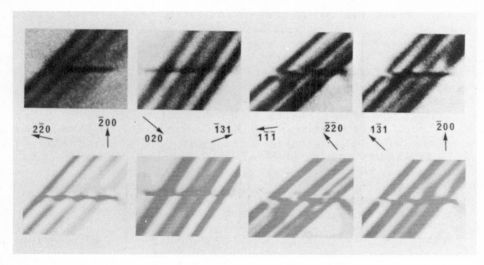

Fig. 21.16. Four experimental images of grain bonding dislocation 2 of Fig. 21.14 together with the matching set of computed images which identify it as a Frank dislocation, $\underline{b}$ = 1/3 [$\bar{1}$1$\bar{1}$].

glissile and in what way the structure of the boundary is altered by their presence.

21.5 THE CONTEXT OF THIS TECHNIQUE IN AEM

From the foregoing examples, it can be seen that the major use of computer simulated images lies in obtaining information from experimental observations on actual defects in crystals. Used in this way, the technique fits well into the context of Analytical Electron Microscopy. Note also that it has the desirable feature that the information obtained is at a fairly basic level so that the subsequent use to which the information is put (e.g. the estimation of lattice frictional stress or a value for the effective stacking fault energy of nickel in the dipole case; perhaps the mechanism and ease of formation of mechanical twins in the overlapping fault example; or speculation on the modes of transfer or blocking of plastic strain between two crystals in the last case) is largely independent of the means of gathering the information.

The other technique in this workshop which deals exclusively with defects is the weak beam technique. It also uses TEM and has the advantage over a two-beam bright field technique in that it is possible to obtain better resolution of the geometry of defects. Like the bright field method of observation, the images are not generally intuitively interpretable and at some stage they have to be subjected to calculation. However, since the weak beam method is a multibeam technique, the appropriate calculations are much more time consuming than two-beam ones. Sometimes the calculations have been approximated using the simpler kinematic theory. As a technique for identifying the strain fields of defects, it has no advantages over the two-beam simulation method and, indeed, as it has been practiced to date, has the disadvantage that it has commonly been based on isotropic elasticity and $\underline{g \cdot b}$ rules. This, of course, is not a necessary (nor a desirable) feature of the theory of the weak beam method. Weak beam observations may be used as an adjunct to computer simulated images either to confirm that the assumptions used there are reasonable (that the dislocations are in fact straight, for instance) or to obtain greater resolution in the geometry (the relative separations of the four dislocations in dipole D, for example). Conversely the image matching technique may be used as an adjunct to the weak beam technique to identify the strain fields of the defects. In summary, it is probable that no single technique can provide all the information to analyze all types of defects in all situations and it will be necessary to choose the technique best suited to the problem. In this context mention should be made, at least briefly, that information concerning defects may be obtained from diffraction patterns. This is particularly so for planar objects such as surfaces (e.g. CHERNS, 1974), grain boundaries (e.g. SPYRIDELIS, DELAVIGNETTE and AMELINCK, 1967; TAN, SASS and BALLUFFI, 1975; FORWOOD and CLAREBROUGH, 1977) or stacking fault arrays (e.g. CLAREBROUGH and FORWOOD, 1976). This method of obtaining information about defects is likely to be used increasingly in future, especially in conjunction with microdiffraction techniques.

REFERENCES

Bullough, R., Maher, D. M. and Perrin, R. C., 1971, Phys. Stat. Sol. (b), 43, 689.

Cherns, D., 1974, Phil. Mag., 30, 549.

Clarebrough, L. M. and Forwood, C. T., 1976, Phys. Stat. Sol. (a), 33, 355.

Cooper, W. D., 1977, Ph.D. Dissertation, University of Florida.

Degischer, H. P., 1972, Phil. Mag., 26, 1137.

Forwood, C. T. and Humble, P., 1970, Aust. J. Phys., 23, 697.

Forwood, C. T. and Humble, P., 1975, Phil. Mag., 31, 1025.

Forwood, C. T. and Clarebrough, L. M., 1977, Phil. Mag., 36, 1131.

Häusserman, F., Rühle, M. and Wilkens, M., 1972, Phys. Stat. Sol. (b), 50, 445.

Head, A. K., 1967, Aust. J. Phys., 20, 557.

Head, A. K., 1969a, Aust. J. Phys., 22, 43.

Head, A. K., 1969b, Aust. J. Phys., 22, 345.

Head, A. K., Loretto, M. H. and Humble, P., 1967a, Phys. Stat. Sol., 20, 505.

Head, A. K., Loretto, M. H. and Humble, P., 1967b, Phys. Stat. Sol., 20, 521.

Howie, A. and Whelan, M. J., 1961, Proc. Roy. Soc., A263, 217.

Humble, P., 1968, Phys. Stat. Sol., 30, 183.

Humble, P. and Forwood, C. T., 1975, Phil. Mag., 31, 1011.

Huntington, H. B., 1958, Solid State Phys., 7, 213.

Lance, G. N., 1960, Numerical Methods for High Speed Computers (Iliffe) (London).

Maher, D. M., Perrin, R. C. and Bullough, R., 1971, Phys. Stat. Sol. (b), 43, 707.

Melander, A., 1976, Phys. Stat. Sol. (a), 33, 255.

Ohr, S. M., 1976, Phys. Stat. Sol. (a), 38, 553.

Sass, S. L., Mura, T. and Cohen, J. B., 1967, Phil. Mag., 16, 679.

Spyridelis, J., Delavignette, P. and Amelinckx, S., 1967, Mater. Res. Bull., 2, 615.

Tan, T. Y., Sass, S. L. and Balluffi, R. W., 1975, Phil. Mag., 31, 575.

Thölén, A. R., 1970a, Phil. Mag., 22, 175.

Thölén, A. R., 1970b, Phys. Stat. Sol. (a), 2, 537.

Whelan, M. J. and Hirsch, P. B., 1957a, Phil. Mag., 2, 1121.

Whelan, M. J. and Hirsch, P. B., 1957b, Phil. Mag., 2, 1303.

Wilkens, M. and Rühle, M., 1972, Phys. Stat. Sol., 49, 749.

Yoffe, E. H., 1972, Phil. Mag., 25, 935.

MAJOR REFERENCES

Head, A. K., P. Humble, L. M. Clarebrough, A. J. Morton and C. T. Forwood, "Computed electron micrographs and defect identification," North Holland, Amsterdam, 1973.

Covers virtually all the topics in this article, but in much greater detail. Good chapter on the techniques involved in gathering the experimental data. Gives listings of the programs

for computing micrographs. "Electron microscopy of thin crystals,"
P. B. Hirsch, A. Howie, R. B. Nicholson, D. W. Pashley and M. J.
Whelan, Butterworths, London, 1965. Classic reference for almost
all aspects of transmission electron microscopy. Theory of electron diffraction is discussed from several points of view (kinematic, dynamic, Bloch wave, scattering matrix, two-beam, multibeam,
etc.) usually with reference to the observation of defects.

CHAPTER 22

THE STRATEGY OF ANALYSIS

RON ANDERSON and J.N. RAMSEY

INTERNATIONAL BUSINESS MACHINES CORPORATION

HOPEWELL JUNCTION, NEW YORK 12533

22.1 INTRODUCTION

This chapter addresses itself to the task of optimizing a research-er's approach to a materials Analytical Electron Microscopy (AEM) prob-lem. We will briefly discuss the AEM and related non-AEM techniques a-vailable and what each can tell us. The questions of where to start, the optimum strategy for problem solving, routes to avoid and useful shortcuts will be discussed. After we discuss strategies for analysis, there will be several examples of successful and unsuccessful applica-tions of AEM. These will show the synergism of complementary tech-niques.

22.2 WHERE TO BEGIN

The question of where one starts ultimately reduces to the question of how much specimen material is available to the researcher.

If you have an ample supply of uniform, homogeneous specimen mater-ial, you can afford the luxury of conducting your investigation in par-allel. By this we mean that you can divide your specimen and conduct time-consuming and/or destructive ancillary specimen analyses, such as x-ray diffraction or ion microprobe studies, at the same time that AEM studies are underway.

This is an optimum strategy because there is always the chance that some esoteric destructive analysis, that you wouldn't dare perform with a limited specimen volume, will yield an important clue or suggest a fruitful avenue of approach.

The alternative to an experiment conducted in parallel, not surpri-singly, is one conducted in series. Here we must make the distinction between having enough specimen material available that some ancillary techniques are possible and so little specimen volume that a very limited number of experiments can be conducted. For example, if all a researcher has is one square centimeter of a thin film deposited on a benign substrate, AEM studies would have to be postponed until after the completion of x-ray diffraction experiments, as a square centimeter of thin film is the minimum required for conventional x-ray diffraction techniques. Likewise, 1cm2 is probably a minimum amount of thin film to have before beginning a destructive series of AEM heating experiments.

Particulate or powdered samples are easier to divide and a well controlled AEM experiment requires remarkably few particles. However, several ancillary powder techniques consume considerable amount of specimen. It is therefore well advised to consider all possible alternative technique requirements before committing a limited amount of specimen to an AEM specimen preparation procedure. This is especially so when the ancillary technique cannot return the specimen unaffected by its analysis.

For the last case, that of an extremely limited specimen volume all large beam techniques and most non-AEM spectroscopies are impossible. With full understanding of the possibilities and careful planning, a series of experiments, with AEM last, is still possible. Here, specimen selection and control is the key. The methods of specimen selection and handling pioneered at the McCrone Institute [McCRONE,(1973)], in conjunction with McCrone's small sample light optical analysis techniques, are invaluable to the AEM scientist.

To answer the question "Where to begin?", in light of the above, first ask "How much specimen do I have?" and then plan accordingly. Rushing to commit limited specimen resources immediately to the AEM is tempting, but in many cases, a poor strategy for analysis.

22.3 SPECIMEN TYPE STRATEGY

With few expections, there are three basic specimen types encountered in material's AEM analysis: particles, thin films, and bulk samples. A specimen type strategy is concerned with preparing these specimens for analysis in the AEM, making full utilization of all AEM features. It is not writers' intent to go into TEM specimen preparation procedures, as these are readily accessible in the literature. There are, however, some minor modifications to existing specimen preparation techniques that may be helpful to the AEM researcher. These modifications are made necessary by the extreme sample thinness required for EELS analysis.

Turning first to particulate specimens. Conventional specimen preparation usually involves a modified extraction replication procedure: the particles are dispersed on a benign substrate, picked up in collodion or some other medium, carbon coated, the collodion dissolved, leaving the particles embedded in a carbon film. This type of specimen is fine for morphological, electron diffraction, and EDS examination,

but no good for EELS analysis. In EELS analysis, we are always dealing
with false carbon signals arising from specimen contamination and samples that are too thick [JOY,(Ch. 7); and MAHER, (Ch. 9)]. Particles
imbedded in a carbon film cannot be analyzed for carbon. Also, resulting EELS spectra suffer from poorer signal to noise ratio, due to increased background from the trailing off of the carbon plasmon and edge
signals. The simple solution to the problem is to prepare particle samples by sprinkling the powder onto holey carbon films of the type used
for checking objective lens astigmatism. The object is to find an isolated particle stuck in a hole, so that a thin edge can be subjected to
EELS analysis without intervening substrate. AEM EELS accessories have
sufficient spatial resolution that they are minimally influenced by the
nearby carbon film. Check the vacuum spectrum on a nearby, similiar
sized, hole for carbon just in case. Particle agglomeration is not a
problem, as a single isolated particle in a hole can usually be found.
You can perform conventional particulate dispersive specimen preparation
for particle size statistics. One disadvantage is that not all AEM
owners may be willing to have loose particles falling off holey films in
their instruments.

For the case of thin films, conventional TEM techniques work well
in the AEM except for the no-substrate, extreme thinness requirements
for EELS. Useable EELS thickness is less than 100nm, with less than 50nm
preferred for most cases. This may be a problem when a film is to be
floated off a substrate to be picked up on a grid. In the authors laboratory, we rarely use the floatation technique, turning instead to the
use of oxidized silicon wafers as substrates. A very small 200-500nm
hole can be opened under the oxide film by jet etching [KEAST and
WILSON, (1966)]. The oxide removed by chemical etching or ion-milling,
leaving the specimen film unsupported over a 100-200nm hole in the
oxide. Highly stressed films that break and roll up on the oxide when
it is perforated can be dealt with by leaving a very thin oxide residual
layer, or by opening up the smallest possible hole in the oxide.

Bulk samples can be very easy or very hard to prepare for AEM analysis. Again, it's the EELS requirements that cause problems. If you
are using a specimen preparation technique that yields very thin representative films, there is no problem. If, however, one must analyze
small precipitates or crystallites in a matrix, special techniques, or
very precise application of standard techniques must be employed.

A recent example [ANDERSON and KUMAR,(1977)] of bulk AEM analysis
is seen in Fig. 22.1, which shows a lithia-alumina-silica ceramic grain
containing small precipitates. An EDS spectra taken at a precipitate
site is seen in Fig.22.2, it shows Al and Si plus a Ti system peak. The
EELS spectra, Fig.22.3, taken at the same precipitate shows Ca, Li, Al,
Si and O edges superimposed on a large carbon plasma edge. Knowledge of
the light element make-up of the precipitates made identification of the
accompanying selected area electron diffraction pattern far simpler and
more certain. The purpose of repeating the report of these findings in
this paper is to illustrate the difficulties of dealing with EELS spectra of precipitates in a matrix. Note in Fig. 22.3 that the signal to
noise ratio of the absorption edges is at the vanishing point due to the

large background from the plasma peak and this is the "best" spectra to come out of a month-long effort. When lithia, alumina and silica powders were ground to 30-50nm particles and dispersed on a holey film the resulting EELS spectras were very satisfactory-establishing that the Fig. 22.1 - 22.3 analysis was specimen-preparation limited.

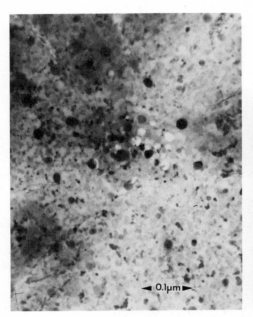

Fig.22.1. Lithia, Alumina, Silica ceramic material showing numerous precipitates.

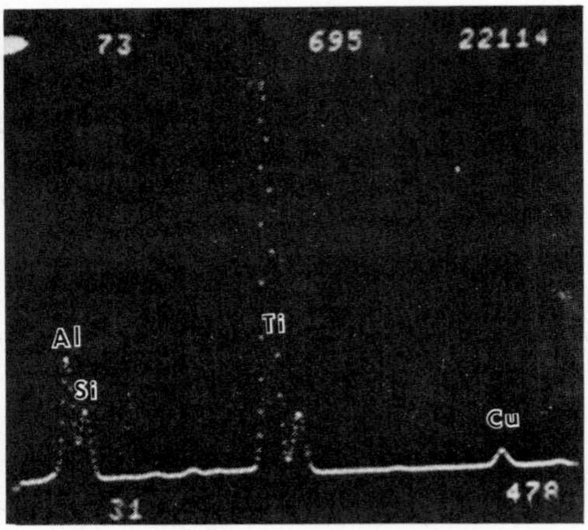

Fig.22.2 EDS spectra from Figure 22.1 small precipitate.

A still more stringent specimen case is encountered when one wishes to examine a combination of specimen types: a thin film on a bulk substrate for example. Epitaxial relationships can always be found using conventionally prepared samples and looking normal to the plane of the film. However, the interrelation of film and substrate morphology can best be examined using the vertically oriented transmission samples prepared by the method of SHENG and MARCUS at Bell Labs (1976). An example is shown in Fig. 22.4.

Having reviewed specimen "type" strategy let us now turn to a specimen "state" strategy. Specimen "state" is related to the degree of crystallinity exhibited by the specimen. The AEM approach to a single crystal sample is completely different from the approach taken with an amorphous one. One cannot overemphasize the desirability of corroberative x-ray diffraction analysis of your samples when possible. Single crystal zonal orientation studies using Kikuchi lines are fast and easy when Kikuchi lines are present. But a Laue photograph of the sample adds confidence in the results and gives information on the extent of crystallinity unobtainable in the AEM.

ture of amorphous materials and aluminum chlorides and oxychlorides as determined by electron probe microanalysis, TEM and SAD on extraction replicas [CAMERON et al.,(1978)]. Recently, we have had atmospheric corrosion problem due to chlorine when process tooling set-up an electromotive cell with the metallurgy on the silicon wafer. Figure 22.9 shows the same region of corrosion product which was analyzed by Wavelength Dispersive Analysis (WDS) in an electron microprobe, by EDS in the AEM and by Scanning Auger Microscopy [JOSHI,et al.,(1975)]. X-ray analysis showed only aluminum and oxygen, with trace chlorine in only some of the particles. The excellent energy resolution possible today in scanning Auger systems is shown in Fig. 22.10, [SMITH,et al.,(1979)] where chemical shift information of the Al_{KLL} transition can be seen between aluminum as metal and aluminum as an oxide orhydroxide: previously this distinction was possible only by X-ray Photoelectron Spectroscopy, formerly called ESCA [RIGGS and PARKER,(1975)]. The reaction product was amorphous, even to electron wavelengths, so diffraction was not possible to further deliniate the structure. It appears that in this case of corrosion, the inadvertent EMF set-up in the process fixturing overwhelmed the usual, well known [BROWN and HUNTER,et al.,(1941)] EMF system, which produces grain boundary corrosion in Al-Cu alloys. Now the chlorine was not in the process, per se, (as extensive electron microprobe analyses showed the tooling and wafers to be clean) but was from the laboratory air. The chlorine played a catalytic, but vital, role in the surface/ pitting corrosion but was not enough to transform the corrosion product. Such cases of Al-Cu corrosion as these two, are processing problems and not reliability problems because it is customary in industry to protect the device with a passivation layer.

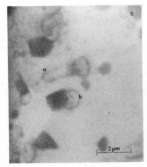
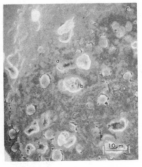

Fig. 22.9. TEM (A), STEM (B), and SEM (C) micrographs from the same area. The corrosion particle (b) is seen to originate at an Al Cu precipitate.

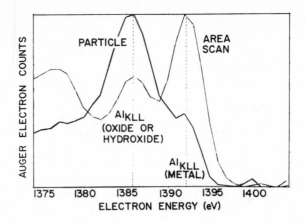

Fig. 22.10. Auger spectra of metal area and particle, showing differences in bonding of the aluminum.

Al-Cr Films and Al-Hf Films

Chromium, as well as other transition metal films, have been used successfully for adhesion layers and diffusion barriers in semiconductor processing. However, there have been apparent anamolies with different processing parameters, and it became desirable to determine the kinetics of formation of the various layers at the interface between chromium and aluminum thin films [HOWARD,et al.,(1976)]. The TEM portion of this study was complicated by the nearly identical diffraction patterns of $CrAl_7$ and Cr_2Al_{11}. Therefore, another analytical technique had to be used to sort out phase identification.

Nuclear Backscatter (also called Rutherford Backscatter) techniques, using 90° incident 2.8 MeV He ions [CHU,et al.,(1973)], in which the energy lost in large angle backscattering is a measure of the atomic number of the material, its concentration and its depth beneath the surface. Profiling thin film Al-Cr couples and Al-Cr-Al sandwiches after various heat treatments provided diffusion data and the atomic ratios of the various phases formed. This allowed interpretation that the $CrAl_7$ phase (aluminum rich) occurs only when excess aluminum exists, and converts to Cr_2Al_{11} when the aluminum is depleted.

In a related study of Al-Hf thin film couples [LEVER,et al.,(1977)] to explain the enhanced electromigration in this metal system, it required both nuclear backscattering and Auger electron spectroscopy to support the TEM analyses. The needle-like precipitates of $HfAl_3$ formed at the Al-Hf interface (with large excess aluminum) do not form a continuous layer to actas a diffusion barrier.

Organic Residue on Fired Thick Film Conductors

An experimental electroplated fired thick film conductor was showing poor adhesion of the plating in some areas which had been masked off in certain steps prior to being plated. There was the question whether an organic layer remained from the RTV material being used as a masking gasket at 50°C. AEM showed carbon and oxygen, but because there was no contrast or diffraction from the "amorphous" film, the results couldn't be separated from system contamination. Therefore, other complementary techniques were employed. SEM was used to infer the presence of a thin layer, while Electron Spectroscopy for Chemical Analysis (ESCA) [RIGGS and PARKER,(1976)]; and Plasma Chromatography/Mass Spectroscopy (PC/MS) [CARR,(1977)] were used to analyze it. Figure 22.11 shows the same area of the fired paste before and after the gasket operation. The left side shows small platelets of insulator material charging up (the specimen was left uncoated for SEM, of course) whereas on the right, the same particles are not charging after the gasket operation. This implies that the area is coated with a thin layer of material, sufficently conductive to allow any charge to bleed off (the layer must be very thin, because there is no indication of thickness in the micrograph). ESCA showed only materials of the conductive land on the left, and only silicone on the right: as the escape depth of the Si 2p electrons is at least 3.5 to 5 nm, we can only say that the organic film thickness is greater than this value. Plasma Chromatography is a relatively new, large area, very sensitive (to femtograms) but difficult to quantify surface analysis technique. It is based on the evaporation of material from the surface into a gas stream which is then ionized and separated into its molecular components in a plasma column (hence the term Plasma Chromatography). Fingerprints of the molecular components can be obtained and compared with known materials, or further analysis of the molecular components can be done by Mass Spectroscopy, which allows one to infer the starting molecules from the fragments as in the usual MS of organics. PC/MS showed not only silicone, but phthalates as well (di-octyl-phthalate is a known adhesion demoter). Subsequently, cleaning procedures were developed to clean off the residues, and their efficiencies were verified by ESCA and PC/MS.

Fig. 22.11. SEM of same area of fired paste conductor before gasket application (left) and after (right).

Premature Collector-Base Breakdown

As almost everyone today knows, a transistor usually consists of three layers in a silicon crystal in which the alloying materials and their levels of concentration are different. The two junctions between the three layers are electrically rectifying and must sustain certain voltage levels in the "back-bias" condition to make the device operable. On one of our development device lines this voltage should have been 18-22 volts, but was actually breaking down in the range of 5 to 9 volts. To locate where in the lateral area of the device that the breakdown occurs, it is necessary to use one of several techniques, e.g. micro plasma [KRESSEL,(1967)], flying light spot scanning [KASPRZAK,(1975)] or Electron Beam Induced Current (EBIC) [SHICK,(1975)]. EBIC is done in an SEM, essentially by measuring the transistor current induced by the electrons in the beam: the current fluctuations are used to modulate the grid of the scanning CRT in the same manner as fluctuations of secondary electrons are used to develop a video picture. Thus, defects can be located and mapped for analysis by some other technique, usually TEM. In this particular case, the analyses were made in the AEM, using TEM, SEM and EDS after conventional backside thinning techniques of the regions of interest [BOHG,MIRBACH and SCHNEIDER (1978)]. Figure 22.12 is a TEM micrograph showing a dislocation, produced by the embedded large particle, running to the upper surface, where a small pit has formed during an etching step in device processing. The particle, while rather

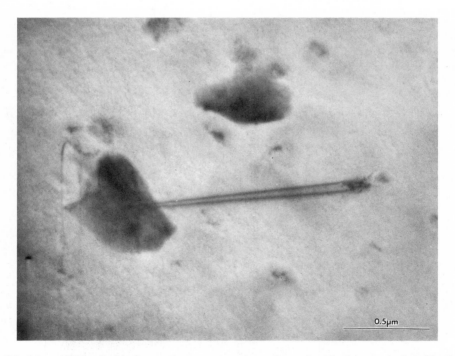

0.5µm

Fig. 22.12. TEM of dislocation emanating from buried particle (left) to surface (right) where pit has formed.